Lecture Notes in Physics

Springer
Berlin
Heidelberg
New York
Barcelona
Budapest
Hong Kong
London
Milan
Paris
Santa Clara
Singapore
Tokyo

The Editorial Policy for Proceedings

The series Lecture Notes in Physics reports new developments in physical research and teaching – quickly, informally, and at a high level. The proceedings to be considered for publication in this series should be limited to only a few areas of research, and these should be closely related to each other. The contributions should be of a high standard and should avoid lengthy redraftings of papers already published or about to be published elsewhere. As a whole, the proceedings should aim for a balanced presentation of the theme of the conference including a description of the techniques used and enough motivation for a broad readership. It should not be assumed that the published proceedings must reflect the conference in its entirety. (A listing or abstracts of papers presented at the meeting but not included in the proceedings could be added as an appendix.)

When applying for publication in the series Lecture Notes in Physics the volume's editor(s) should submit sufficient material to enable the series editors and their referees to make a fairly accurate evaluation (e.g. a complete list of speakers and titles of papers to be presented and abstracts). If, based on this information, the proceedings are (tentatively) accepted, the volume's editor(s), whose name(s) will appear on the title pages, should select the papers suitable for publication and have them refereed (as for a journal) when appropriate. As a rule discussions will not be accepted. The series editors and Springer-Verlag will normally not interfere with the detailed editing except in fairly obvious cases or on technical matters.

Final acceptance is expressed by the series editor in charge, in consultation with Springer-Verlag only after receiving the complete manuscript. It might help to send a copy of the authors' manuscripts in advance to the editor in charge to discuss possible revisions with him. As a general rule, the series editor will confirm his tentative acceptance if the final manuscript corresponds to the original concept discussed, if the quality of the contribution meets the requirements of the series, and if the final size of the manuscript does not greatly exceed the number of pages originally agreed upon. The manuscript should be forwarded to Springer-Verlag shortly after the meeting. In cases of extreme delay (more than six months after the conference) the series editors will check once more the timeliness of the papers. Therefore, the volume's editor(s) should establish strict deadlines, or collect the articles during the conference and have them revised on the spot. If a delay is unavoidable, one should encourage the authors to update their contributions if appropriate. The editors of proceedings are strongly advised to inform contributors about these points at an early stage.

The final manuscript should contain a table of contents and an informative introduction accessible also to readers not particularly familiar with the topic of the conference. The contributions should be in English. The volume's editor(s) should check the contributions for the correct use of language. At Springer-Verlag only the prefaces will be checked by a copy-editor for language and style. Grave linguistic or technical shortcomings may lead to the rejection of contributions by the series editors. A conference report should not exceed a total of 500 pages. Keeping the size within this bound should be achieved by a stricter selection of articles and not by imposing an upper limit to the length of the individual papers. Editors receive jointly 30 complimentary copies of their book. They are entitled to purchase further copies of their book at a reduced rate. As a rule no reprints of individual contributions can be supplied. No royalty is paid on Lecture Notes in Physics volumes. Commitment to publish is made by letter of interest rather than by signing a formal contract. Springer-Verlag secures the copyright for each volume.

The Production Process

The books are hardbound, and the publisher will select quality paper appropriate to the needs of the author(s). Publication time is about ten weeks. More than twenty years of experience guarantee authors the best possible service. To reach the goal of rapid publication at a low price the technique of photographic reproduction from a camera-ready manuscript was chosen. This process shifts the main responsibility for the technical quality considerably from the publisher to the authors. We therefore urge all authors and editors of proceedings to observe very carefully the essentials for the preparation of camera-ready manuscripts, which we will supply on request. This applies especially to the quality of figures and halftones submitted for publication. In addition, it might be useful to look at some of the volumes already published. As a special service, we offer free of charge LATEX and TEX macro packages to format the text according to Springer-Verlag's quality requirements. We strongly recommend that you make use of this offer, since the result will be a book of considerably improved technical quality. To avoid mistakes and time-consuming correspondence during the production period the conference editors should request special instructions from the publisher well before the beginning of the conference. Manuscripts not meeting the technical standard of the series will have to be returned for improvement.

For further information please contact Springer-Verlag, Physics Editorial Department II, Tiergartenstrasse 17, D-69121 Heidelberg, Germany

Miguel Rubí
Conrado Pérez-Vicente (Eds.)

Complex Behaviour of Glassy Systems

Proceedings of the XIV Sitges Conference
Sitges, Barcelona, Spain, 10–14 June 1996

 Springer

Editors

Miguel Rubí
Conrado Pérez-Vicente
Dept. Física Fonamental
University of Barcelona
Diagonal 647
E-08028 Barcelona, Spain

Cataloging-in-Publication Data applied for.

Die Deutsche Bibliothek - CIP-Einheitsaufnahme

Complex behaviour of glassy systems : proceedings of the XIV
Sitges Conference, Sitges, Barcelona, Spain, 10 - 14 June 1996 / José
Miguel Rubí ; Conrado Pérez-Vicente (ed.). - Berlin ; Heidelberg ;
New York ; Barcelona ; Budapest ; Hong Kong ; London ; Milan ;
Paris ; Tokyo : Springer, 1997
   (Lecture notes in physics ; Vol. 492)
   ISBN 3-540-63069-4

ISSN 0075-8450
ISBN 3-540-63069-4 Springer-Verlag Berlin Heidelberg New York

Typesetting: Camera-ready by the authors
Cover design: *design & production* GmbH, Heidelberg
SPIN: 10550798   55/3144-543210 - Printed on acid-free paper

Preface

There has been pronounced historical interest in the study of glassy systems as well as a considerable amount of experimental data and industrial applications collected during the last years. Nowadays, these systems constitute one of the most interesting fields of condensed matter physics. Although its understanding is of high potential interest from theoretical and applied points of view and much progress has been achieved during recent years, the essentials of the underlying physical mechanisms are still not understood. The present theoretical approaches to glassy systems are based on similar techniques. This fact allows for a common language and the existence of unified background. In all of these systems, frustration, i.e., competition between interactions of different signs, play a crucial role. This generates the appearance of strong complex dynamical behaviour under variation of some external parameters such as temperature or magnetic field. Systems exhibiting glassy behaviour include for example: structural glasses, spin glasses, vortex glasses, Josephson junction arrays, and protein folding. Based on the interest in establishing a forum to unify the present ideas about the subject and to debate new perspectives, we organized the XIV Sitges Conference. We wish to express our gratitude to all the speakers and participants who certainly contributed to create a high scientific level and a very pleasant atmosphere.

The conference was sponsored by CEE (Euroconference) and by institutions who generously provided financial support: DGCYT of the Spanish Government, CIRIT of the Generalitat of Catalunya, University of Barcelona, CSIC, and "La Caixa". The city of Sitges allowed us, as usual, to use the Palau Maricel as the lecture hall.

Finally, we are also very grateful to all those who collaborated in the organization, Drs. F. Ritort, A. Pérez, I. Pagonabarraga, C. Miguel, R. Pastor, J.M. Vilar, and A. Corral as well as to Profs. J. Bermejo, J. Brey, J. Marro, and M. San Miguel.

Barcelona, April 1997 _The Editors_

Contents

Entropy, Fragility, "Landscapes", and the Glass Transition

C.A. Angell

Department Of Chemistry, Arizona State University, Tempe, AZ 85287-1604

1 Introduction

The focus of this meeting is on a subject which I have had the privilege of watching grow over some three and a half decades from a passing interest of a few physicists, to the preoccupation of many. The glass transition, and the peculiar slow dynamics and ergodicity breaking that the term now implies, seems to be of very broad occurrence in nature. In my opinion its best known example, the liquid-to-glass transition (now called the "structural" glass case (as opposed to "spin" glass, "vortex" glass, etc.) is also its most challenging case, mainly because of the existence of the entropy paradox which does not arise in most other cases. It is the phenomenology of this case or, rather, of a key part of it, on which I will concentrate in this lecture.

An account of the phenomenology of glass formation from cooling liquids and solutions requires consideration of two major problem areas which can be summarized by the two questions:
1. Why do certain metastable liquids fail to generate crystals, the thermodynamically stable state, during cooling at reasonable rates?.
2. Why do liquids, in which crystals fail to form during cooling, quite abruptly become non-diffusive, hence britle, at temperatures of roughly 2/3 of the expected (or observed) melting points?

Here I will only address the second of these questions. Some detailed attention to the first is given in the proceedings of two 1996 summer shcool proceedings [1], to which those interested in this aspect of the problem are referred. Some deep questions about crystal lattice stabilities and liquid mixing relation are involved.

1.1 What is a glass?

Nowadays, many different sorts of "glasses" are under discussion. The general concept of "glass" relates to systems which have some degree of freedom that (a) fluctuates at a rate which depends strongly on temperature or pressure and that (b) becomes so slow at low T or high P that the fluctuations become frozen. At this, point, properties determined by the slow degree of freedom change value more or less abruptly, giving the "glass transition". For instance, in spin glasses, it is the magnetic susceptibility which decreases suddenly as the fluctuations in magnetization freeze in, while with glassforming liquids, it is the heat capacity, compressibility, expansivity, and dielectric susceptibility. Thus the most general definition we can give for a "glass' is a follows:

> "A glass is a condensed state of matter which has become non-ergodic by virtue of the continuous slow-down of one or more of its degrees of freedom."

Satisfying this definition, there are spin glasses and orientational glasses (dipole, quadrupole, octapole) and vortex glasses, as well as the classical glasses which themselves have now become known as "structural" glasses. Even ordinary crystals can be considered "glasses" by this definition, since, when defect concentrations become frozen-in during cooling (because defect population change require migration of defects to or from the crystal surface -which can be slow), there is a quite sudden change in properties, e.g. the temperature dependence of electrical conductivity in the case of ionic crystals. While this is a simple case as far as the freezing in of the degree of freedom (the defect population) is concerned, it is in fact atypical because the "freezing-in temperature" depends on the crystal size. In the glasses which are getting the most attention today, the time scale for establishing equilibrium is intrinsic to the substance, provided the samples are large enough that surface layer effects are negligible: the "defects" or "structural states" are generated internally.

"Structural" glasses are distinguished from most others by the large change in heat capacity (thermal susceptibility) which accompanies the freezing in of a particular structural state, or defect population. Consequently, the Kauzmann paradox to which so much attention has been given in structural glass studies (as detailed below) is not discussed in other glass physics circles. The only cases of structural glasses which lack the heat capacity jump are those at the "strong" extreme of the overall strong/fragile behaviour pattern (see below) and these have been considered the least interesting (at least until recently when "polymorphism" in strong liquids was recognized [2], [3]. The importance of the heat capacity jump in the phenomenology of "structural" glasses, and the importance of the equilibrated state being kinetically stable enough to study and characterize, are both included in the definition of a "structural" glass preferred by this author:

> "A (structural) glass is an amorphous solid which is capable of passing continuously into the viscous liquid state, usually, but not necessarily, accompanied by and abrupt increase in heat capacity."

Note that this definition relegates most of the "metallic glass" materials to the grey world of "amorphous solids" because, although formed from a liquid (by ultra fast quenching), they crystallize before ever achieving the supercooled liquid state (exceptions are now becoming known). On the other hand, the definition, admits many substance produced initially by routes which never involve a liquid state [3].

The various possible routes to the glassy state are summarized in Fig. 1. Note that the non-liquid routes all involve some more or less drastic departure from the initial state, and this complicates systematic analysis of the process. Thus for the purposes of this lecture, we will restrict attention to structural glasses formed from liquids and furthermore, from liquids in which no changes of composition occur during vitrification.

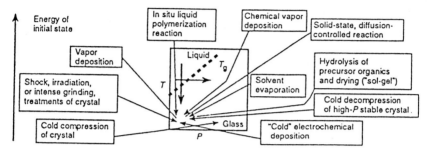

Fig. 1. Various routes to the glassy state, roughly indicating the energies of the initial states relative to the final glassy states. The route of crystal compression below the glass transition temperature (T_g) may yield glasses that are thermodynamically distinct from those obtained by the other route but that may transform to them via non-equilibrium first-order transitions. [Reproduced with permission from Ref. 2 (copyright American Association for the Advancement of Science).]

2   Ergodicity-breaking and the glass transition

The freezing-in of a structural state during cooling of the liquid, means that the state of internal equilibrium possessed by the initial liquid is lost. To modernize the classical description, "vitrification", this process has recently [4] been called "ergodicity breaking." This is because states at temperatures above the glass transition satisfy the ergodic hypothesis of statistical mechanics [5] (i.e. systems in equilibrium in the course of fluctuations revisit, or may revisit, the same state within the observation period), hence are called "ergodic states". Glasses are "non-ergodic".

The definitions given above are both consistent with the latter statement. However, is should be recognized that common usage among physicists is obscuring this distinction. There is a strong tendency to use the term "glassy" for any aspect of a system capable of generating a glass on sufficient cooling. Thus one sees frequent reference to (a) "glassy dynamics" -by which is usually meant the dynamics of viscous *liquids* approaching the glass transition temperature from above, and (b) "strong or fragile glasses" in reference to phenomenology of supercooled liquids above T_g. While such a blurring of the distinction between liquids and glasses is regrettable, it may be inevitable.

The crossover from liquid to glassy (ergodic to non-ergodic) behaviour occurs over a range of temperature called the "glass transformation range". This is natural for a kinetically controlled phenomenon, and it is probably inappropriate that the tern glass "transition" is used to describe the ergodicity-breaking phenomenon. The term is, however, firmly entrenched, and there is little question of changing it. Because of the range of temperature involved and also because of its dependence on cooling rate, there is a problem in attaching a characteristic "glass transition temperature" to the phenomenon for any particular material. This is compounded when, as is commonly the case, the "glass temperature" T_g is determined during heating because there is then a further (poorly appreciated) dependence on annealing history as well as on heating rate. Nevertheless, when appropriately measured, T_g is very reproducible and has become recognized as an important material parameter.

The glass transition temperature can be defined arbitrarily as the temperature at which the viscosity reaches a certain high value, traditionally 10^{13} poise, near which changes in heat capacity are observed during heating or cooling. This is unambiguous because the viscosity is an equilibrium property. It can also be defined more meaningfully, but now ambiguously, as the temperature at which a break in the heat capacity is observed during heating or cooling. This temperature is ambiguous because, if observed during cooling it depends on the cooling rate, and if observed during heating it depends,. as noted above, on the entire thermal history since the time ergodicity was first broken. While it may seem more logical to use the cooling definition, the lack of reliable temperature calibration methods [6] has lead to the common use of the heating definition. This is illustrated in Fig. 2 for the case of isopropyl benzene [7].

Because of the very different time scales used in the measurement of heat capacity in adiabatic measurement and differential scanning calorimetry (DSC) measurements, it is common to see the glass transition temperatures measured by groups using these alternative measures differing by several degrees (higher for the higher effective heating rate DSC measurements). The most commonly cited values are obtained from DSC or differential thermal analysis (DTA) measurements conducted during heating at 10K/min (often after an ill- defined cooling process such as quenching the sample in liquid N_2) and correspond closely to the temperature at which the structural relaxation time reaches 100 s.[8].

It is the jump in heat capacity, and the closely related deviations from Arrhenius temperature dependencies of the relaxation times (hence of the viscosities and diffusivities), which are the most provocative aspects of the glass transition problem. The challenge is to describe precisely the nature of the new degrees of freedom that are being excited above T_g, explain why does it take so long (100 s of seconds at T_g) to excite them, and understand why that time scale changes so rapidly with temperature around T_g? An important question to answer is, what can we learn about the nature of the liquid state in general from the observations we make near the glass transition temperature? These are the questions which we will address in the remainder of this paper. Before the next section, it should be pointed put that, while most of the non- vibrational heat capacity is lost at the glass transition, a part (which is important to developing micro-heterogeneous models of the supercooled liquid state) remains, and is only lost at much lower temperatures, see Fig. 2. This is a weak component associated with a fast process (or processes) which is less cooperative than the principal relaxation and has different dynamic character. These are called secondary relaxations ($\beta-$, $\gamma-$relaxations) to distinguish them from the primary ($\alpha-$) relaxation which carries most of the thermodynamic strength. They have been long known in polymer science where they are usually associated with side chain motion, but their general occurrence in simple glasses was not expected until Goldstein [9] predicted them as a necessary consequence of his "landscape" picture of glassforming systems. Their calorimetric characterization is a recent development [10], [11], [12]. In many cases, they appear to be the continuation of the high temperature relaxation process before any cooperative processes set in [12], but this matter can only be dealt with properly after the subject has been further developed.

3   Dividing up the liquid state

While it is common to think of the liquid state as the state which exists between the melting point and the boiling point, this is conditioned by our familiarity with the two first order processes which are involved in the melting and boiling boundaries. For molecular liquids, the latter boundary completely disappears if we consider pressure conditions just a little above normal relative to the range of pressures now available for experimentation. The lower boundary is clearly only a consequence of fast crystallization kinetics for the majority of common liquids, and for the many liquids of interest to this conference, it can easily by bypassed. In this broader context, the liquid state must be viewed as that state, which exists between GAS, in which a description in terms of isolated binary collisions is adequate, and "IDEAL GLASS" in which the molecular packing density is so great that only a statistically insignificant number of alternative packings of equal of lower energy is possible. In the latter case, the non-vibrational entropy would be no greater than in a perfect crystal, and the substance near absolute zero would obey the Nernst Heat Theorem.

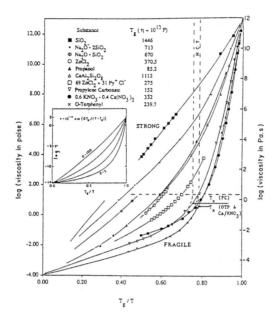

Fig. 2. Heat capacity of isopropyl benzene in liquid, glass and crystal states, determined by adiabatic calorimetry. T_g is defined as the temperature at which the jump in heat capacity occurs. Note how the excess heat capacity of gass over crystal below T_g diminishes with decreasing temperature and effectively disappears below 50 K. Around 80K, a further time dependence of the calorimetric response reveals a secondary relaxation process of time scale $\sim 10^{-3}$ sec coinciding with the extrapolation of the dielectric $\beta-$ relaxation to this time scale. (after Kishimoto et al. Ref. 7, permission of Japanese Chemical Society).

Within this broad density range, it seems that the liquid state can be divided into two major dynamically distinct regimes, [1] the *normal* (freely diffusing) regime, and [2] the *viscous* ("landscape- dominated") regime, with which we are primarily concerned in this article. The evidence for a crossover between two such regimes [13], [14] is best seen by mens of two scaling schemes for transport of relaxation time data (the Rössler viscosity scaling and Fujimori-Oguni relaxation time scaling) which are discussed below. Two subdivisions are possibly useful. At high temperatures, a subdivision of the "normal" regime into Arrhenius and non- Arrhenius regimes is possible. At lower temperatures, a more artifical subdivision is possible into *accessible viscous liquids* (for $T > T_g$) and *inaccessible liquids* (for $T_g > T > T_K$ where T_K is the Kauzmann temperature, explained below). The term "inaccessible" is to be regarded liberally because the regime below T_g is that in which the practically important processes called "annealing" and "aging" take place as the initially out-of-equilibrium system ("glass") slowly rearranges its molecules into configurations appropriate to the lower temperature equilibrated liquid state (which it may or may not reach). In the inaccessible liquid temperature domain, and below, the substance is usually referred to as a "glass".

To appreciate these limits and divisions better, it is desirable to review the origin of the concept of a configurational ground state or "ideal glass' to be reached (at T_K) during infinitely slow cooling below the normal T_g.

3.1   The Kauzmann paradox

In Fig. 3, we reproduce Kauzmann's original presentation of the entropy problem in supercooling liquids[15]. Kauzmann used the difference in entropy between the crystal and liquid at the melting point, δS_f, as a scaling parameter to permit a comparative display of the manner in which the difference in entropy between the liquid and crystal states, for six different substances, varies during their supercooling. All cases show positive slopes, reflecting the fact that liquids have higher heat capacities than the corresponding crystals. What is interesting is the magnitude of the slope and the related extrapolation to zero excess entropy.

In the case of boron trioxide, now much spoken of as a "strong" liquid[13], [16], [17], [18], the slope is such that the excess entropy of the liquid over crystal is only tending to disappear in the vicinity of $0K$ - which raises no concern at all. On the other hand, to various degrees the other liquids in the figure show provocative behaviour. In the case of lactic acid, which we would now call the most "fragile" of the six, the excess heat capacity of the liquid over crystal is causing the excess entropy to decrease so quickly that a simple extension of its behaviour to lower temperatures would lead that excess to vanish at a temperature which is only $\sim 2/3$ of the fusion temperature. As far as can be told from the data, all that prevents this at-first-sight-mind-boggling *thermodynamic* inversion from occurring, is the occurrence of a *kinetic* phenomenon, the glass transition (i.e. the trapping) at the temperature T_g —hence the term "paradox".

Since we must always imagine that a cooling process can be carried out more and more slowly, Fig. 3 suggests the possible [19] existence of an equilibrium transition to the configurationally non-degenerate glassy state referred to above. Such a state would represent the configurational ground state for a non-crystalline system, hence merits the term "ideal glass".

Two quite profound theoretical problems are presented by the data of Fig. 3. The first[20], [21] is the problem of constructing an equilibrium theory for the liquid state which contains an explanation of how, on infinite time scales, the system evolves so as to undergo a rather abrupt, if not singular, change in heat capacity at some temperature between the glass transition temperature and absolute zero. Part of this problem involves the interpretation of fragility of liquids and the coupling of vibrational to configurational degrees of freedom. The second [20], 22] is the problem of constructing a theory which

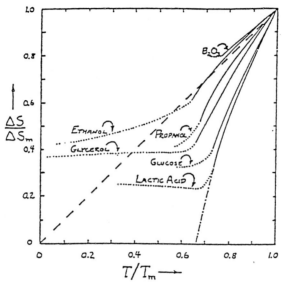

Fig. 3. Kauzmann's presentation of the entropy crisis which bears his name [15]. The figure shows the rate at which the difference in entropy between liquid and crystal, normalized at the fusion point, disappears as T is lowered towards absolute zero. For B_2O_3, now known as a "strong" liquid, the liquid would always be of higher entropy than the crystal, even if the glass transition did not intervene, at high T/T_m, to change the heat capacity. At the other extreme, lactic acid loses its excess entropy so rapidly on cooling that if T_g did not interven to arrest the loss, the liquid would arive at the same entropy as the crystal at 2/3 of the melting point. [This is the temperature usually associated with the temperature of the glass transition itself (the 2/3 rule which this set of data only weakly supports).]. Lactic acid is an example of a "fragile" liquid. Other examples of these plots for fragile liquids are given in [13].

explains in a satisfying manner the reason why, in every case known case, the *kinetic* characteristics of the liquid (which can to first approximation be represented by its diffusivity) evolve with temperature in such a way as always to generate equilibration times of the order of experimental time scales *before* the thermodynamic crisis arrives. It is first necessary to consider the general aspects of liquid behaviour as glass transition is approached from above.

3.2 High frequency crossovers

There have been suggestions over the years that some change of transport or relaxation mechanism occurs at high temperatures in the vicinity ~ 1 Pa.s viscosity. It was argued in a 1969 paper by Goldstein [9] that, as molecules in the liquid pack more closely with decreasing temperature, there should come a point where free diffusion, characteristic of simple liquids and dense gases, can no longer occur because the molecules begin to "jam up", and energy fluctuations, whose probability is a Boltzmann function of temperature, are needed to free them, i.e. the molecular mobility becomes activated. This change in mechanism has been much discussed recently [13], [14][23], [25], and evidence for its reality is accumulating.

The most direct evidence for some change in mechanism is probably the splitting off, in most fragile liquids at least, of a weak (called secondary or β-) dielectric relaxation from the main α-relaxation, the β relaxation often appearing as a continuation of the high temperature approximately Arrhenius process. This is seen most clearly in the Fujimori- Oguni scaling scheme discussed below. The idea of a high temperature crossover to activated transport has been strongly reinforced through more recent evaluations of mode coupling theories (MCT) of the glass transition which have provided an interesting and detailed prediction of the transition from free diffusion to a dynamically jammed state identified with the glass. Because the jammed state is not found but many other predicted features of the initial slowdown are, a natural conclusion has been hat the jamming is avoided by the intervention of activated processes which are not provided for in the idealized theory. Finally, at a temperature close to this T_c, a breakdown in the Stokes.-Einstein relation between diffusion and viscosity [23] and reorientation time and viscosity [24] has been reported. This crossover phenomenon and its characteristic temperature T_x has been the basis of a second of the scaling schemes we now discuss.

3.3 Scaling, relations, fragility and cooperativity

Much effort has been made to find appropriate ways of highlighting the general or universal features of glassforming systems. We now describe here three alternative schemes for collecting together in a single diagram the behaviour of a wide variety of systems in a way which stimulates thinking about the nature and features of these systems.

The first is the T_g-scaled Arrhenius plot introduced by Uhlmann [26] and Angell [27] in the '70s and popularized under the title of "strong and fragile" liquids in the '80s by Angell [16]. An example of this data presentation is provided in Fig. 4. "Strong" liquids are those with network structure. Interestingly enough, the recently developed bulk glassforming metal systems based on Zr-rich Zr-Cu alloys[28] prove to be strong liquids and may contain a quasi covalent 4-bonding Zr network. Fragile liquids, on the other hand, tend to be more highly coordinated ionic liquids, or aromatic hydrocarbons.

The second, a recent variant of this scaling, is that due to Rössler and coworkers [25] in which all data are collapsed onto a single curve by introducing the crossover temperature referred to above, into the scaling. For strong liquids, the crossover temperature T_c identified by Rössler and Novikov lies far above T_g, for fragile liquids lies close to T_g, and in each case is close to the T_c derived from fits to MCT[13], [29], [30], [31]. By plotting the data of Fig. 4 against the new temperature variable $(T_g - T)/T \ [T_c/(T_c - T_g)]$, which scales

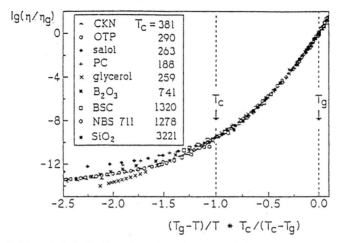

Fig. 4. T_g-scaled Arrhenius plots for viscosities of different glassforming liquids by Rössler and coworkers showing spread of data between strong and fragile extremes. Predictions of mode coupling theory power law are shown by dashed line for two cases. Inset shows the pattern of behaviour obtained by varying the D parameter in the modified Vogel-Fulcher equation. (From ref. 13 by permission)

out the differences in reduced glass-to- crossover temperature $(T_c - T_g)/T_g$, the difference between strong and fragile liquids is removed and a universal behaviour is obtained, as shown in Fig. 5. The fragility should thus be represented by the quantity $T_g/(T_c - T_g)$ (though I prefer the use of T_x over T_c, since T_x symbolizes "crossover" [13]).

A third and very interesting scaling, which introduces new information into the total picture, is that proposed by Fujimori and Oguni [12], based on their calorimetric identification of the $\beta-$glass transition (a calorimetric manifestation of the Johari-Goldstein β-relaxation) at temperatures far below that of the $\alpha-$glass transition. These authors then used T_g, β as the temperature-scaling parameter in their Arrhenius plot of relaxation time data, and obtained Fig. 6.

The value of Fig. 6 is the emphasis it gives to the concept of the α process as a process which grows out of a background "sea" of simpler activated processes as a result of the increasing cooperativity forced on the systems by its increasingly dense packing. The liquids which are most fragile by the Fig. 4 scaling are seen in Fig. 6 to be those in which the α- process first splits off from the background -hence which are the most cooperative. The possible connection between this phenomenology and the microheterogeneity being detected near the glass transition temperature will be discussed later in sections 4.1 and 5.3. The overall scenario is strongly reminiscent of that exposed in spin model analogs by Butler and Harrowell[32] and recently de-

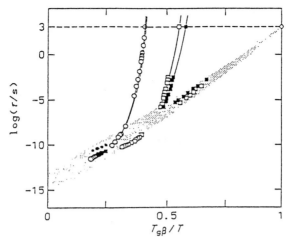

Fig. 5. Scaling of viscosity data for a variety of liquids showing how fragile and strong liquids can be collapsed onto a master curve by using a second scaling temperature at T_c, which coincides closely with the mode coupling theory of critical temperature. $CKN = Ca_{0.4}K_{0.6}(NO_3)_2$; OTP= 0-terphenyl; PC=propylene carbonate; BSC= boron silicate crown glass. Crossover temperature T_c is listed in the insert. (From Roessler and Sokolov, Ref. [25], by permission)

veloped into an interpretation of strong and fragile liquid patterns by Perrera and Harrowell[33].

3.4   The landscape paradigm

A useful way of thinking about the complex behaviour discussed in the previous section is in terms of the potential energy hypersurface representative of the liquid[9], [34]. This is a 3N+1 dimensional map of the systems potential energy as a function of its particle coordinates. In a glass, the system clearly resides within a local minimum (or "basin") on this hypersurface and executes collective vibrational modes of motion in accord with basin shape. Near but below T_g, i.e. during annealing, the systems can explore nearby lower energy minima, while lowering its entropy. At very high temperatures, it either "sees" no energy wells at all or relaxes by mechanisms which avoid them, and in consequence, vibrational modes of excitation pass over to independent binary collisions via a regime described very well by mode coupling theory [29], [30], [35]. This situation, in which entropy and relaxation time are connected to the topology of the energy "landscape," is depicted in Fig. 7[36]. It shows how diverging relaxation times and vanishing excess entropy (i.e. the entropy in excess of the vibrational entropy which is essentially that of the crystal) are linked to the shortage of configurational states at energies near RT/mole. For every potential inf interaction, there is, according to the "landscape paradigm", an energy of order RT_K below which there are no

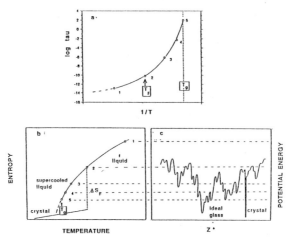

Fig. 6. Scaling of relaxation time data for four molecular liquids (left to right): propylene carbonate, PMS (a disiloxane), o-terphenyl, and isotropylbenzene), using the T_g (β) temperature as scaling parameter (defined as the temperature where the enthalpy relaxation time of the β-relaxation becomes 10^3s, see also Fig. 2. (For data refs., see ref. 12) From Ref. 12, by permission).

states other than vibrational states and those involved in the β relaxation, so any property linked to the availability of configurational states will show anomalies as T_K is approached. Whether the liquid is strong or fragile will depend on how densely packed these states are. The interval $T_x - T_g$ in the Rössler scaling of Fig. 5, for instance, is a crude measure of the temperature interval from the bottom to the top of the Fig. 7(c) landscape (or a least to the point at which free diffusion becomes more efficient than basin hopping). It is for theory to tell us what features in the intermolecular potential determine how dense this packing of states, and indeed what the total number of states, will be (estimates $\sim e^N$ [36], $e^{1.2N}$ [37]).

3.5 The "height" of the landscape, and the fragility

It is possible to evaluate the "height" of the landscape semiquantitatively by the following simple argument, and the results are quite informative. The question to ask is:

Over what temperature range above the Kauzmann temperature must the systems be driven before all the $\sim e^N$ microstates are accessible and the entire configurational entropy S_c

$$S_c = klnW_c = klne^N = Nk_B, \qquad \text{or } R \text{ per mole} \qquad (1)$$

is fully excited?

13

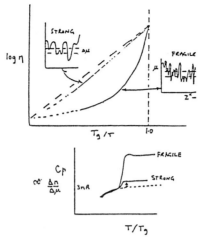

Fig. 7. Summary of phenomenology of glassformers showing diverging relaxation times related, by points 1, 2, 3... on the plot, to vanishing excess entropy and to the postulated relation to the energy minima on the potential energy hypersurface. The temperature T_K corresponds with the energy of the lowest minimum in the amorphous phase megabasin. Many vertical spikes, corresponding to configurations in which particle core coordinates overlap, are excluded from this diagram for clarity. Note the domain, at high temperatures, which is free of any features of the landscape. (From Ref. 36 by permission).

The answer is roughly accessible via a calculation of the upper limit temperature T_u in the expression (in which the quantities are all molar quantities)

$$S_c = R = \int_{T_K}^{T_u} \Delta C_p(T) dlnT \tag{2}$$

For the simplest case in which ΔC_p is taken as constant, then

$$R = \Delta C_p ln[T_u/T_K] \tag{3}$$

and

$$T_u = T_K e^{(R/\Delta C_p)} \tag{4}$$

which we argue below is a minimum value. For a monatomic glass we can approximate ΔC_p by the value 18 J/K per mole of atoms found for the simplest known molecular glassformer, the four atom-molecule S_2Cl_2[38] [which has the classical Dulong-Petit (3R per g-atom) value in the glassy state]. This ΔC_p value for S_2Cl_2 is close to that found for LJ argon by Clarke, 16.6 J/K-mole [39], which in turn is confirmed by the more detailed studies on LJ mixtures by Vollmayr et al. [40]. Using this value in Eq. 4, we obtain $T_u = 1.59T_K$. This compares with the typical fragile liquid crossover temperature, T_x in Fig 5, of about $1.3T_g$ [see Fig. 2 in Ref. 13] hence about $1.56T_K$

(since $T_K = 0.8T_g$)[3], [13]. The correspondence can hardly be coincidental and is certainly consistent with the idea that T_x is the temperature at which the dynamics ceases to be landscape-dominated. Other correspondences are noted below.

If we choose instead the better approximation for molecular liquids $\Delta C_p = K/T$ (discussed in the next section), then

$$R = \Delta C_{p(atT_K)}(T_u - T_K)/T_u \tag{5}$$

and

$$T_u = T_K/[1 - R/\Delta C_{p(T_K)}] \tag{6}$$

ΔC_p at T_K (taken as $0.8T_g$), evaluated from the hyperbolic expression and the above ΔC_p value of 18 J/molK at T_g, is 22.5 J/K per mole of heavy atoms. With this value, we obtain landscape "saturation" at $T_u = 1.59T_K$ which is unchanged from the first estimate (due to entropy compensation above and below T_g), hence can be described as a robust result. It should weakly underestimate the actual limit because a small part of ΔC_p, perhaps 10%, should be due to differences in vibration frequencies between ground state and the shallower (and less harmonic) excited states. 1.7 T_K would probably be a better "top of landscape (T.O.L.)" estimate for fragile liquids.

The T.O.L. temperature 1.7 T_K is far below the value of the boiling point for a typical fragile liquid [41] ($\sim 3T_g$ or $3.75T_K$). Thus it would appear that the representation of the fragile liquid given in Fig. 7, with a landscape limit at $\sim 1.7T_K$ and a large region at high energy free of any landscape features, is a reasonable one.

This "robust" estimate of the height of the landscape falls rather close to other temperaturs of significance. For instance, an early collection [13] of the mode coupling theory T_c values for fragile liquids (obtained from viscosities) showed them falling between 1.2 and 1.3 T_g, which means between 1.44 and 1.56 T_K, just a little below the T_u estimate. The associated correspondence with the Rössler upper scaling (or crossover) temperatures of Fig. 5 has already been mentioned. Finally, the $\alpha - \beta$ bifurcation temperatures of Fig. 6 range from 1.2 to 1.3 T_g hence are also close to T_u though also a little below it. The latter coincidence links the generation of landscape features in configuration space to the generation of clusters of molecules in physical space. That the various dynamic crossovers fall a little below T_u seems reasonable since the crossover involves only a change in *dominant* influence rather than arrival at some terminal condition. In fact, the "top of the landscape" should be only fuzzily defined unless the energy topology is closer to a "pitted plateau" than to a "landscape".

In less fragile liquids, the crossover from landscape-dominated relaxation to free diffusion should occur at higher T/T_K. For a liquid like $ZnCl_2$, intermediate in Fig. 3 ΔC_p is only 6.3 J/K per mole (of heavy atoms) and its either approximately constant or increasing with increasing temperature [38], so T_u by Eq. (4) would fall at 3.76 T_K at the maximum. T_X for this liquids falls at 1.41 T_g, or 2.14 T_K, since T_g/T_K in this case is 1.52 [36]. Provided the number of configuron states per mole of heavy atoms is approximately constant, it would therefore seem that fragility must be determined by the density of configurons per unit of energy. In fragile liquids, then, all $\sim e^N$ states fall in a narrow band of energy hence become excited in a short range of T above T_K.

At the limit of high fragility, there would be a first order transition from liquid to glass, as pointed out before[42]. While such a liquid to glass transition has yet to be observed, first order transitions from one isotropic liquid state to another have been observed[43], [44] or implied[45] and these are indeed associated with dramatic changes of particle mobility. For instance, in the case of liquid silicon, the term "liquid-to-amorphous transition" has been used, and in the case of water (in which the first order character of the transition at ambient pressure is in dispute) the pretransition behaviour is associated with such rapid changes in relaxation time that a power law rather than the V-F law (eq. (2)) must be used to describe them[46]. This aspect of the phenomenology will not be discussed here; the interested reader is referred to articles on polyamorphism published elsewhere [47].

4 Relaxation and entropy

The above line of thought connecting relaxation behavior to entropy excitation was made more quantitative by the development of the "entropy theory" of Adam and Gibbs22], and the related entropic relaxation argument by Mohanty et al.[48]. We now discuss their usefulness in more detail.

4.1 Relaxation in the ergodic state

The Adam-Gibbs theory22], which was based on a modification of conventional transition state theory to accommodate the notion that, in viscous liquids, the rearrangements over energy barriers must be cooperative, led to an expression for the relaxation time which contains the excess (configurations) entropy, S_c of the Kauzmann paradox, in the exponent denominator.

$$\tau = \tau_0 \exp\left(\frac{C'\Delta\mu}{TS_c}\right) \tag{7}$$

where $\Delta\mu$ is the conventional free energy barrier to rearrangements, and C' is a constant. The familiar departure from Arrhenius behaviour comes from the temperature dependence of S_c which itself depends on the value of the configurational heat capacity, i.e., the ΔC_p manifested at the glass transition. Thus the thermodynamic fragility helps determine the fragility measured by reference to relaxation time temperature dependence. The latter is also influenced by the value of $\Delta\mu$ in Eq. (7).

In their original treatment, Adam and Gibbs made the simplest assumption for the excess heat capacity, which determines the configurational entropy, viz. that it is a constant with temperature. This yields

$$S_c = \Delta C_p \ln T/T_K \tag{8}$$

Substitution into Eq. (7) then yielded the Vogel-Fulcher equation as an approximation, valid near T_K. However $\Delta C_p=$ constant does not describe many molecular systems. More accurate[49], [50] is $\Delta C_p = K/T$, from which

$$\Delta S(T) = \Delta C_p(at T_K)(T - T_K)/T_K \tag{9}$$

which leads to the Vogel-Fulcher equation as an identity

$$\tau = \tau_0 \exp\left[DT_0/T - T_0)\right] = \tau_0 \exp -(F\varepsilon) \tag{10}$$

where $F(= D^{-1})$ es a fragility parameter, $0 < F < 1$, and $\varepsilon = (T/T : -1)$.

There has always been dispute concerning the validity of Eq. (11), and this has been revived with vigor recently in the light of a temperature derivative analysis by Stickel et al.[51], [52], [53]. Stickel et al. show that, particularly for fragile liquids, there is a region at relatively high temperatures where the $V - F$ equation fits quite well but yields a T_0 that is considerably higher than T_K and often is also larger than T_g (a result which is unphysical). This analysis emphasizes the high temperature data above the crossover temperature of Fig. 6 and above the bifurcation temperature of Fig. 7. An analysis focusing on the last five to six decades in relaxation time (hence in the landscape-dominated domain- which covers a small range of ordinate values on the Stickel plot) yields a lower value of T_0, a value which usually agrees rather well with the Kauzmann temperature, while also usually yielding a physical (phonon-like) pre-exponential τ_0. We have documented this correspondence elsewhere [36] for a large number of different glassformers, finding $T_K/T_0 = 1.02\pm, 0.03$. For intermediate liquids such as glycerol, even the Stickel analysis yields a T_0 in good accord with T_K[51], probably because the landscape remains influential to much higher T/T_g (section 4.1).

The table in Ref. 36 shows that T_K/T_0 values close to unity are obtained for glassformers with T_g varying between 50 and 1,000 K, utilizing the low temperature T_0 when available and insisting on a physical τ_0 value (sin 10^{-14}s). Since the liquids represented range from molecular through covalent (Se, As_2Se_3) to complex ionic oxides, we judge the case for the Adam-Gibbs approach to relaxation in glassforming liquids to be quite a strong one.

The fragility of the different liquids, is expressed in two ways in the cited table [36]: the first is by the parameter $D(D = F^{-1})$ in Eq. (11)[54], and the second by the slope of the Arrhenius plot for each liquid at its T_g normalized by T_g (i.e. dlog $\tau/dT_g/T$) - which is designated m [55], [56], [57]. A third definition which, like F, has the advantage of varying between 0 and 1, is T_K/T_g which is usually the same as T_0/T_g, especially if, as noted above, data fits are constrained by fixing τ_0 at the physical value of 10^{-14}s. For polymers in which T_K is usually not available, an equivalent fragility index F' is obtainable from the common Williams Landel Ferry (WLF) equation parameter C_2 via the relation

$$F' = 1 - C_2/T_g = T_0/T_g \tag{11}$$

This is obtained from the well-known equivalence of the WLF and VTF equations. For this relation to be a reliable index of fragility, C_1 must be fixed at 16 for relaxation data which, as explained elsewhere[58], is the equivalent of fixing $\tau_0 = 10^{-14}$, the quasi-lattice vibration period.

4.2 Primary and secondary relaxations and the relaxation function

The above relations provide a "broad brush" picture of the relaxation in which only two time scales are considered; the first is the short time scale of the vibrational motions which are responsible for the background glass heat capacity and which provide the time scale for attempts to escape from the potential minima, while the second and longer time is that for structural relaxation itself, i.e. for escape from the minima. Such a scenario would lead to a two-step relaxation function similar to that anticipated by mode coupling theory[29], [31] (and also generally invoked in phenomenological descriptions of viscous relaxation[59], [60]. Its form is illustrated by the curve (b) in Fig. 8 (which is seen in many accounts fo computer studies of relaxation in glass-forming liquids) and also by curve (g).

However, these two curves contain no account of the bigger picture of relaxation provdied in Fig. 6. At least for some fragile liquid cases, the bifurcation into α- and β-processes occurs, as noted already, at the same temperature as the upper characteristic temperature identified by the scaling of Fig. 5. For instance, o-terphenyl has $T_{\alpha-\beta} = 290K$ (at $\log f \sim 7$)[61] vs $T_c = 290K$[62] and $T_x = 290K$ (Fig. 5). The relaxation curve for such liquids must therefore develop new features at temperatures where the system contacts the landscape and molecular clustering becomes pronounced. Since the β-relaxation should be strongest at temperatures where most of the molecules are *not* in clusters, we suppose the β relaxation should initially be dominant (as seen in neutron scattering[63]) and that the complete relaxation functions in this temperature domain should look like curves (e-h) in Fig. 8. The β relaxation wil get progressively weaker as fewer and fewer molecules remain in intercluster positions and, by T_g, its contribution to the total strength may be difficult to detect (curves i and j). Certainly its presence was only marginally detectable at the T_g, β of the study of Fujimori and Oguni[12]. Thus the second step in the curves near $T_{\alpha-\beta}$ should vary with temperature as shown by curves (e-h) in Fig. 8. Such functions differ from those of Götze and Sjögren for the case of MCT with hopping processes included [64]

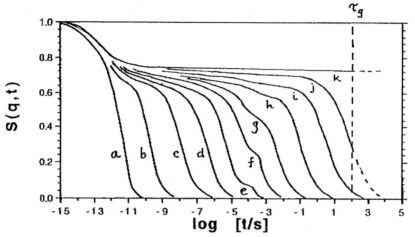

Fig. 8. Total structural relaxation function (experimentally obtained as the "intermediate scattering function" $S(q, t)$ for different regimes (a) far above the landscape-dominated domain, (b,c,d) approaching the landscape- dominated regime, heralded by the splitting off of the Johari-Goldstein β_{JG}-relaxation (the β_{JG} relaxation is still a strong component of the total relaxation), (h.i) deep in the landscape-dominated domain where the β_{JG} relaxation has weakened (fraction of non-clustered molecules has diminished), (i.j) near T_g where β_{JG} relaxation has vanishing relaxation strength on the scale of the figure.

and have yet to be recorded experimentally. They will be difficult to observe in computer simulation because they tend to fall in a time region which is currently inaccessible. Design of systems with particularly high fragility may relieve this problem, according to Fig. 6. A cooperative cluster model which, at one extreme, is capable of giving a first order liquid-to-glass transition, will be described elsewhere[65].

4.3 Relaxation in the non-ergodic state near T_g, and the nature of annealing

Near T_g, relaxation is complex because the quantity S_c of Eq. (7) is time-dependent. Thus a measurement performed at a constant temperature will yield a relaxation time for recovery of the equilibrium state which is never a linear function of the displacement because S_c itself relaxes according to an Adam-Gibbs equation. When S_c finally reaches its equilibrium value, the relaxation time, of course, becomes time-independent. Behaviour in this region has been treated in detail in the review by Hodge[66].

5 Concluding remark

The understanding of visous liquids as they progress towards the glassy state is evolving in interesting ways. In this evluation, the development of configuration space thermodynamic treatments, and real space structural and dynamic analyses, should be complimentary.

Acknowledgments

This work has been supported by the National Science Foundation under Solid State Chemistry Grant Nos. DMR 9108028-002 and 9614531. In the preparation of the manuscript, the author has benefited from some provocative discussions with Peter Harrowell of Sidney University during a sabbatical leave support by NSF under Grant No. INT.9602884.

References

[1] Angell, C. A., in *Amorphous Insulators and Semiconductors*,eds. M.F. Thorpe and M.I. Mitkova, NATO-ASI Series, Plenum Press (1997) pp. 1-20; and Angell, C. A., Proc. 1996 Enrico Fermi Summer School in Physics, Italian Physical Society, in press.

[2] Angell, C. A., Poole P.H.m and Shao J., *Nuovo Cimento***16D** (1994) 993.

[3] Angell, C. A., *Science* **267** (1995) 1924.

[4] Palmer, R. J. and Stein, D. C., in *Relaxations in Complex Systems*, edited by K. Ngai and G.B. Wright (National technical Information Service, U.S. Department of Commerce, Springfield, VA 22161) 1985, p. 253.

[5] McQuarrie, D. A., *Statistical Mechanics* (Harper & Row, New York) 1973, P. 554. The *"ergodic hypothesis...*states that for a stationary random process, a large number of observations made on a single systems at N arbitrary instants of time have the same statistical properties as observing N arbitrarily chosen systems at the same time from an ensemble of similar systems."

[6] Lu, Q., Velikov, V. and Angell, C.A. (to be published).

[7] Kishimoto, K., Suga, H. and Seki, S., *Bull. Chem. Soc. Japan* **46** (1973) 3020.

[8] Moynihan, C. T., Macedo, P. B., Montrose, C. J., Gupta, P. K., DeBolt, M. A., Dill, J. F., Dom, B. E., Drake, P. W., Easteal, A. J., Elterman, P. B., Moeller, R. P., Sasabe, H. A. and Wilder, J. A., *Anna. NY, Acad. Sci.,* **279**, (1976) 15.

[9] Goldstein, M., *Chem. Phys.,***51** (1969) 3728.

[10] Suga, H. and Seki, S., *Non-Cryst. Solids,***16** (1974) 171.

[11] Fujimori, H. and Oguni, M., *J. Chem. Thermodynamics,***26** (1994) 367.

[12] Fujimori, H. and Oguni, M., *Solid State Commun.* **94** (1995) 157.

[13] Angell, C. A.,*J. Phys. Chem. Sol.***49** (1988) 863.

[14] Ediger, M. D., Angell, C. A. and Nagel, S. R.,*J. Phys. Chem.***100** (1996) 13200.

[15] Kauzmann, W., *Chem. Rev.***43** (1948) 219.

[16] Angell, C. A.,*J. Non-Cryst. Sol.***131-133** (1991) 13.

[17] Sidebottom, D. and Torell, L.,*Phys. Rev. Lett.***71** (1993) 2260.

[18] Sokolov, A. P., Kislink, A., Soltwisch, M. and Quitmann, D., *Physic. Rev. Lett.***69** (1992) 1540.

[19] While this is a frequently made observation, and one espoused over a long period by the present author [e.g. ref. 13 and C.A. Angell, *J. Chem. Ed.***47**, 583 (1970)], it was not the view of Kauzmann himself. Kauzmann ref. 15 preferred the possibility [based on the theoretical (and experimentally supported) idea that the size of the critical nucleus for transformation to the crystalline state decreases with decreasing temperature] that at low enough temperature the barrier would effectively disappear. In this case, the crystal state would be slowly established. We [C.A. Angell, D.R. MacFarlane, and M. Oguni, *Ann. N.Y. Acad. Sci.* **484** (1986) 241.] argued against this by showing that for certain cases it seemed that the relaxation time for evolution towards the equilibrated amorphous state would always be shorter than the transient nucleation time, but recently evidence has been presented that nucleation can occur by activity of the β- relaxation [T. Hikima, M. Hanaya and M. Oguni, *Bull. Chem. Soc. Japan 69* (1996) 1863] which occurs on a much shorter time scale than that of the α-relaxation and which furthermore has an Arrhenius temperature dependence (see Fig. 7). In this case Kauzmann's argument gains a new dimension of validity, and rather than an ideal glass, a cryptocrystalline state consisting of films of near-critical nuclei surrounding maximum density amorphous clusters might be the slow coiling limit for liquids.

[20] Gibbs, J. H., in *Modern Aspects of the Vitreous State*, edited by J. D. McKenzie (Butterworths, London) 1960, ch. 7.

[21] Gibbs, J. H. and Dimarzio, E. A., *J. Chem. Phys.***28** (1958) 373.

[22] Adam, G. and Gibbs, J. H.,*J. Chem. Phys.***43** (1965) 139.

[23] Fujara, F., Geil, B., Sillescu, H. and Fleischer, G., *Z. Phys. B88* (1992) 195.

[24] Rössler, E., *Phys. Rev. Lett.***65** (1990) 1595.

[25] Rössler, E. and Sokolov, *Chem. Geol.***128** (1996) 143 ; and Novikov, V. N., Rössler, E. Malinovsky, V. K. and Surovtev, N. V., *Europhys. Lett.***35** (1996) 289.

[26] Laughlin, W. T. and Uhlmann, D. R., *J. Phys. Chem.***76** (1972) 2317.

[27] Angell, C. A. and Tucker, J. C., in *Chemistry of Process Metallurgy*, Richardson Conference (Imperial College of Science, London, 1973), Eds., J. H. E. Jeffes and R. J. Tait, *Inst. Mining Metallurgy Publ.,* (1974) 207.

[28] Busch, R., Scheider, S., Peker, A. and Johnson, W:L., *Appl. Phys. Lett.***67** (1995) 1544.

[29] Götze, W. in *Liquids, Freezing, and the Glass Transition*, Eds. Hansen, J.P. and Levesue, D., NATO-ASI, North Holand (Amsterdam) (Les Houches 1989) 287-503.

[30] Du, W.M., Li, G., Cummins, H.Z., Fuchs, M., Toulouse, J. and Knauss, L.A., *Phys. Rev. E49* (1994) 2192; Borjesson, L. and Howells, W.S., *Non-Cryst. Solids***131-133** (1991) 53.

[31] Petry, W., Bartsch, E., Fujara, F., Kiebel, M. Sillescu, H., Farrago, B.Z., *Phys. B83* (1991) 175.

[32] Butler, S. and Harrowell, P., *Chem. Phys.***95** (1991) 4466.

[33] Perrera, D. and Harrowell, P.,*Phys. Rev. E54* (1996) 1652.

[34] Stillinger, F. S. and Weber, T., *Science***225** 983 (1984); Stillinger, F. S.,*Science***267** (1995) 1935.

[35] Kob, W. and Andersen, H.C., *Phys. Rev. E***51** (1995) 4626; Kob, W. this volume.

[36] Angell, C. A., APS Symposium Procedings, *J. Res. NIST* (in press).

[37] Speedy, R.J., Debenedetti, P.G., *Mol. Phys.***88** (1996) 1293.

[38] Angell, C.A. and Tuker, J.C. (to be published); Angell, C.A., Williams, E., Rao, K.J. and Tucker, J.C.,*J. Phys. Chem.***81** (1977) 238.

[39] Clarke, J.H.R., *Trans. Far. Soc.***2** 76 (1976) 1667.

[40] Vollmayr, K., Kob, W. and Binder, K., *Phys. Rev. B***54** (1996) 15808.

[41] Alba, C., Busse, L. E. and Angell, C. A.,*J. Chem. Phys.***92** (1990) 617-624.

[42] Angell, C. A., Boehm, L., Oguni, M. and Smith, D. L.,*J. Mol. Liquids***56** (1993) 275.

[43] Thompson, M. O., Galvin, G. J., Mayer, J. W., Peercy, P. S. Poate, J. M., Jacobson, D. C., Cullis, A. G. and Chew, N. G., *Phys. Rev. Lett.***52** (1984) 2360.

[44] Aasland, S. and McMillan, P. F., *Nature***369** (1994) 633.

[45] Poole, P. H., Sciortino, F., Essmann, U. and Stanley, H. E., *Nature* London **360** (1992) 324.

[46] Angell, C. A. *Ann. Rev. Phys. Chem.***34** (1983) 593.

[47] Poole, P.H., Grande Tor, Angell, C.A. and McMillan, P.F., *Science*, in press (Jan. 1997); Angell, C.A., Poole, P.H. and Shao, J., *Nuovo Cimento*16D (1994) 993.

[48] Mohanty, U., Oppenheim, I. and Taubes, C. H., *Science***266** (1994) 425.

[49] Privalko, Y., *J. Phys. Chem.***84** (1980) 3307.

[50] Alba, C., Busse, L. E. and Angell, C. A., *J. Chem. Phys.***92** (1990) 617.

[51] Stickel, F., Fischer, E. W. and Schönhals, A., *Phys. Rev. Lett.***73** (1991) 2936.

[52] Stickel, F. and Fischer, E. W., *Physica A***201** (1993) 263.

[53] Stickel, F., Fischer, E. W. and Richert, R., *J. Chem. Phys.***104** (1996) 2043.

[54] Angell, C. A., Alba, C., Arzimanoglou, A., Böhmer, R., Fan, J., Lu, Q., Sánchez, E., Senapati, H. and Tatsumisago, M., *Am. Inst. Phys. Conference Proceedings***256** (1992) 3.

[55] Böhmer, R. and Angell, C. A., *Phys. Rev. B.***45** (1992) 10091.

[56] Böhmer, R., Ngai, K. L., Angell, C. A. and Plazek, D. J., *J. Chem. Phys.***99** (1993) 4201.

[57] Plazek, D.J. and Ngai, K.L., *Macromolecules***24** (1991) 1222.

[58] Angell, C. A., *Polymer* (submitted).

[59] Moynihan, C.T., Macebo, P.B., Montrose, C.J., Gupta, P.K., DeBolt, M.A., Dill, J.F., Dom, B.E., Drake, P.W., Easteal, A.J., Elterman, P.B., Moeller, R.P., Sasabe, H.A. and Wilder, J.A., *Anna. NY, Acad. Sci.***2799** (1976) 15.

[60] Wong, J. and Angell, C.A., *Glass: Structure by Spectroscopy* Marcel Dekker, New Yokr, New York (1976).

[61] Johari, G.P. and Goldstein, M., *J. Chem. Phys.***53** (1970) 2372; **55** (1971) 4245.

[62] Petry, W. et al. *J. Phys. B. Condensed Matter***83** (1991) 175.

[63] Frick, B. and Richter, D.,*Science***267** (1995) 1939 (see Fig. 9).

[64] Götze, W. and Sjögren, L., *J. Phys. C***21** (1988) 3407.

[65] Angell, C.A. (to be published). A preliminary account will appear in an overview paper by the author in Supercooled Liquids: Advances and Novel Applications, Ed. J. Fourkas, U. Mohanty, K. Nelson and D. Kivelson, ACS Symposium Series; ACS Books, Washignton, D.C. 1997 (in press).

[66] Hodge, I. M., *J. Non-Cryst. Sol.***131-133** (1991) 435; and **169** (1994) 211.

Computer Simulation of Models for the Structural Glass Transition

K. Binder, J. Baschnagel, W. Kob, K. Okun, W. Paul, K. Vollmayr, and M. Wolfgardt

Institut für Physik, Johannes Gutenberg-Universität, D-55099 Mainz, Staudinger Weg 7, Germany

Abstract. In order to test theoretical concepts on the glass transition, we investigate several models of glassy materials by means of Monte Carlo (MC) and Molecular Dynamics (MD) computer simulations. It is shown that also simplified models exhibit a glass transition which is in qualitative agreement with experiment and that thus such models are useful to study this phenomenon. However, the glass transition temperature as well as the structural properties of the frozen-in glassy phase depend strongly on the cooling history, and the extrapolation to the limit of infinitely slow cooling velocity is nontrivial, which makes the identification of the (possible) underlying equilibrium transition very difficult. In addition we demonstrate that microscopic properties are much stronger cooling rate dependent than macroscopic properties like the enthalpy or the density.

These points are exemplified with results for three types of models: The first one is a model for silica, a prototype of a strong glass former, the second is a Lennard-Jones model, which is a fragile glass former and the third is the bond-fluctuation model of polymer melts. For this third model we also review evidence for a growing correlation length at low temperatures resulting from finite size and surface effects. Furthermore we compute the configurational entropy of this lattice model as a function of temperature, which in turn allows us to perform a critical test of the Gibbs-di Marzio entropy theory. It is shown that the vanishing of the entropy in the latter theory gives a reasonable estimate of the glass transition region, but that the actual entropy stays positive down to zero temperature.

1 Introduction: An Overview of Theoretical Concepts

Understanding the nature of the glassy state of materials and the character of the glass transition is one of the biggest scientific challenges of our time [1-7]. The structural information that results from elastic scattering of X-rays or neutrons shows only small differences between the supercooled fluid and the glass, although dynamical properties are so different. It is also very remarkable that the dynamics of chemically very different systems, such as amorphous polymers, semiconductors, silica, metallic glasses etc., show qualitatively a quite similar behavior.

Despite these similarities, one does, however, not yet know whether there exists a static (structural) quantity that would distinguish fluid and glass just as an "order parameter" distinguishes different phases in conventional phase transitions. Also it is not known under which conditions a material can exist only in one type of glassy state or in several of them [8], distinguished, e.g., by classifications such as "strong vs. fragile glass formers" [9]. Furthermore the physical basis for ubiquitous phenomenological descriptions of glassy relaxation, such as the Vogel-Fulcher-"law" [10] describing the increase of the relaxation time upon cooling, the Kohlrausch-Williams-Watts "law" [11] for the relaxation at long times, or scaling ideas like the "time temperature superposition principle" [2] is still obscure.

Analytical theories are incomplete and controversial. In spin glasses [12] and orientational glasses [13], a correlation length $\xi_G(T)$ which describes fluctuations of a glass order parameter, can be identified, and the growth of the relaxation time τ is linked to the increase of $\xi_G(T)$ via a "dynamic exponent" z [14],

$$\tau \propto [\xi_G(T)]^z \tag{1}$$

Hence in these systems [12], [13] the standard picture of critical slowing down valid for ordinary second order phase transitions [14] is responsible also for glassy relaxation (because z is very large). However, it is not clear how much these systems have in common with ordinary structural glasses since in the former the disorder is quenched, i.e. it is present even above the freezing transition [12], [13], whereas in supercooled fluids no such frozen-in degrees of freedom have been identified for temperatures above the glass transition temperature. Also, evidence for the existence of a growing static correlation length for the structural glass transition is essentially lacking [15], [16], although there have been claims to the contrary [17], [18]. Also models have been constructed which lack any static correlations but characteristic length scales appear nevertheless as a result of constraints ruling the moves of the particles [19] (this "dynamic length" hence measures the size of a cooperatively rearranging region [20]). However, these models are usually rather abstract and it is not clear to what extent they capture the essential features of real materials.

This paper will be mostly concerned with the test of two theories: The first is the entropy theory of Gibbs *et al.* [21-24] which suggests a thermodynamic transition at a temperature $T_0 < T_g$, where the extrapolated excess configurational entropy of the fluid seems to vanish (Fig.1). Note that the experimental glass transition temperature T_g is usually defined phenomenologically by requiring that the viscosity $\eta(T_g) = 10^{13}$ Poise [1-3]: since for $T < T_g$ relaxation times quickly become astronomically large, the vicinity of T_0 is inaccessible. But the vanishing of the extrapolated entropy difference ΔS in this theory is taken as an explanation of the "Kauzmann paradox" [25], and it has also been linked to the Vogel-Fulcher law [10] for the viscosity,

$$\ln \eta \propto E_{VF}/(T - T_{\mathrm{VF}}), \tag{2}$$

since often the equality $T_0 = T_{\mathrm{VF}}$ seems compatible with the experimental data [1], [26]. However, although some resulting predictions seem to agree with experiment [26], the approach invokes approximations of questionable validity [27-29].

The second theory we consider is the only microscopic theory of glassy relaxation available today, the so-called mode coupling theory (MCT) [4], [5]. Its idealized version postulates a dynamical transition at a critical temperature $T_c > T_g$, where η shows a power law divergence, cf. Fig. 1,

$$\tau \propto \eta \propto (T - T_c)^{-\gamma}. \tag{3}$$

This singular behavior results from the solution of an approximate nonlinear kinetic equation for the dynamic structure factors $S(q, \omega)$ [4], [5], but unlike Eq.(1) it is believed that there is no underlying static correlation length divergence causing Eq.(3). At the same time, a glass "order parameter" (the "nonergodicity parameter" f) appears discontinuously. In the more refined version ("extended MCT" [30]) these singularities are rounded off, there is only a smooth crossover (T_c being the center of the crossover region), from a power law (Eq.(3)) at T around T_c to simple Arrhenius behavior at temperatures distinctly lower than T_c,

$$\ln \tau \propto E/T. \tag{4}$$

If Eq.(4) remains the asymptotic behavior down to $T \to 0$, there would be no well defined glass transition at all: rather the system falls out of equilibrium at some $T_g(t_{obs})$ where the time-scale of observation (t_{obs}) becomes of the same order as the intrinsic relaxation time: if one could increase t_{obs} substantially, the metastable fluid would still be in equilibrium at lower and lower temperatures.

Despite its microscopic origin, it is still a matter of debate whether MCT actually gives a correct description of the dynamics of supercooled liquids. In particular, the description of what happens near T_g remains an open problem.

In this paper we discuss what computer simulations can contribute to elucidate these problems. Three models are presented which differ in the extent to which they take into account chemical detail: the most realistic model deals with amorphous silica [31], based on the potential proposed by van Beest, Kramer and van Santen [32] ("BKS potential"). Computationally considerably more efficient is a model of a binary Lennard-Jones (LJ) mixture [33], [34]. For both types of models we investigated how the glass transition temperature T_g as well as the properties of the resulting glass depend on the cooling rate with which the system was cooled. This cooling was done by coupling the systems to a heat bath whose temperature $T_b(t)$ is decreased linearly with time from an initial high temperature T_i, where the system is a liquid, to zero temperature, where the system is a glass, $T_b(t) = T_i - \gamma t$, γ being the cooling rate [31,35-37]. We demonstrate that for both models the location of the glass transition temperature $T_g(\gamma)$ and the structural correlations and physical properties of the resulting glass depend significantly on the cooling rate [31,35-37]. It has to be remembered, however, that in such

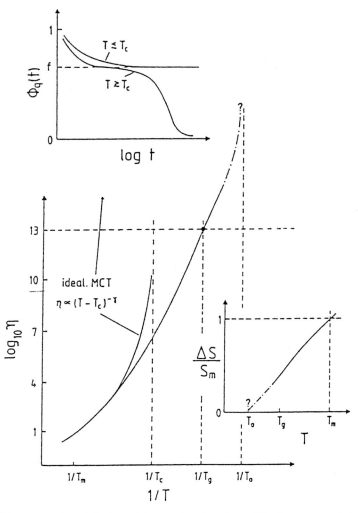

Fig. 1: Schematic plot of the logarithm of the viscosity $\eta(T)$ of a supercooled glass-forming fluid versus inverse temperature. From the melting temperature T_m to the glass transition temperature T_g the viscosity rises by many decades. The upper inset illustrates qualitatively the main prediction of the idealized MCT [4] for the relaxation function $\phi_q(t)$ of density fluctuations at wave-numbers q corresponding to the peak of the static structure factor $S(q)$. While for $T \geq T_c$ $\phi_q(t)$ decays to zero with time t in two steps, for $T \leq T_c$ it decays only towards the nonergodicity parameter $f_q > 0$. The lower inset shows the excess configurational entropy of the fluid ΔS (relative to the crystal), normalized by the entropy at melting S_m, versus temperature. T_0 is found by linear extrapolation of ΔS vs. T (or by a fit of $\eta(T)$ to the Vogel Fulcher law, Eq.(2), assuming $T_{VF} = T_0$). Both extrapolations are questionable, as emphasized by the question marks in the figure, since below T_g there is a substantial interval where no data are available ($T_g - T_0 \approx 30 - 50K$ typically), and thus T_0 depends on the temperature region used to make the fit.

MD simulations cooling rates γ exceed experimental ones by many orders of magnitude, and only for the LJ model can one equilibrate the system at temperatures close to the critical temperature of the MCT [33], [34], certainly one cannot obtain well-equilibrated data at $T < T_c$.

Thus it may be worthwhile to sacrifice even more detail and study lattice models, which can be equilibrated even at low temperatures by using unphysical moves (configurational changes that do not attempt to mimic the physical dynamics) [38]. An example of this approach has been worked out for the bond fluctuation model of glassy polymers [39-49]. The so-called "slithering snake" algorithm [50] can be used to produce equilibrated states of glassy polymers at low temperatures [38]. These states can subsequently be used as initial configurations for runs where one uses the "random hopping" algorithm [39-44,46-49], which is a reasonable coarse-grained description of the Brownian motion of polymers in dense melts [50], [51].

2 Modelling the glass transition of silica (SiO$_2$)

Silica is an excellent glass former, it is a technologically important material [1-3] and amorphous SiO$_2$ is a prototype of a glass of the "continuous random network" type [1]. Thus there has been enormous interest to model this material, and we cannot review all these efforts here (see [31] for more references). Our recent study [31] uses the BKS potential [32], which is known to reproduce the physical properties of the various crystalline phases of silica very well [52], and has the huge computational advantage that no three-body forces are required to model the directional covalent bonds. Instead one uses only pair potentials of the form

$$\phi(r_{ij}) = \frac{q_i\,q_j\,e^2}{r_{ij}} + A_{ij}\exp(-B_{ij}\,r_{ij}) - C_{ij}/r_{ij}^6, \tag{5}$$

where r_{ij} is the distance between atoms i, j (=Si,O), e is the charge of an electron and the values of the parameters q_i, A_{ij}, B_{ij} and C_{ij} are chosen as given in Ref.[32]. Potentials of the form of Eq.(5) are still computationally very demanding, since the long range Coulomb interactions require the implementation of Ewald summation techniques [31], [53]. Constant pressure ($p_{\text{ext}} = 0$) MD runs with 1002 particles (334 Si atoms, 668 O atoms) were carried out, using the velocity Verlet algorithm [53] with a step size of 1.6 fs. For the equilibration, the system was coupled to a stochastic heat bath with the Andersen algorithm [54]. The initial temperature was $T_i = 7000$ K, and for the cooling procedure we used cooling rates from $1.14 \cdot 10^{15}$ K/s to $4.44 \cdot 10^{12}$ K/s. In order to improve the statistical quality of our results we averaged for all cooling rates over at least 10 independent runs (note that for the smallest cooling rate one run took already about two weeks of CPU time on a IBM RS 6000/370, thus investigating cooling rates that are significantly smaller than the ones considered here is currently beyond the computational possibilities).

As expected from corresponding experimental evidence [1], [2], [3], [55], the properties of the glass produced in this manner depend sensitively on the cooling rate. In Fig.2 we show the cooling rate dependence of the enthalpy and we see that this quantity clearly depends on γ, although the dependence is slower than logarithmic. A similar result holds for the density [31]. This dependence is consistent with two theoretical formulae suggested in the context of spin glasses [56], and stated in the figure, but our data certainly cannot distinguish between the two of them. What is probably more important is the fact that also structural characteristics relating to the local order, such as the distribution function $P(n)$ of rings formed from successive covalent Si-O bonds, or the distribution function of the angle θ between two Si-O bonds at the same Si atom, etc., depend on the cooling rate very sensitively [31].

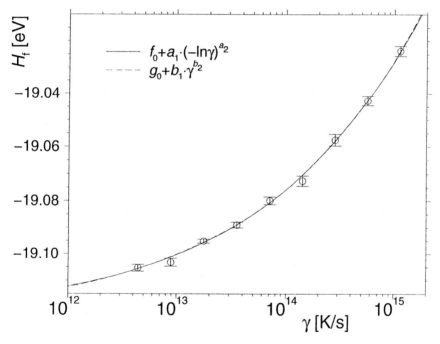

Fig. 2: Enthalpy after the quench plotted vs. the cooling rate, for the model of amorphous silica. The solid and dashed curves are fits with the functional forms given in the figure which were proposed in Ref. [56]. From Vollmayr *et al.* [31].

If a subtle glass correlation length $\xi_G(T)$ would develop in metastable equilibrium of supercooled SiO_2 near its glass transition, the rapid quenching necessary in MD work (even our slowest cooling rate is still many orders of magnitude larger than experimental cooling rates) almost certainly will suppress it. In this context we also note that the glass transition temperature $T_g(\gamma)$, which we extract phenomenologically from the intersections of suitable functions fitted to the liquid branch and the glass branch of the enthalpy [31], depends very strongly on γ (Fig.3), and the observed values of $T_g(\gamma)$ in the accessible range of γ differ from the actual glass transition temperature of SiO_2 reported in experiments ($T_g = 1446$ K [57]) by about a factor of two! This discrepancy, in our opinion, can only in part be attributed to the fact that the potential model might not be realistic enough but instead might stem from the fact that the range of cooling rates accessible is extremely different from the one used in real experiments. Furthermore finite size effects could affect the value of T_g [59]. Thus we conclude that at least for certain properties extracting useful information from computer simulation can be very difficult. This contrasts the sometimes expressed view [58] that all what

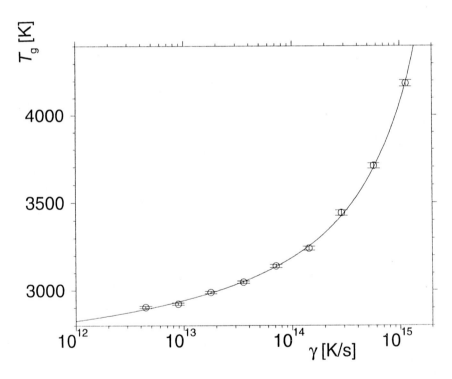

Fig. 3: Glass transition temperature of simulated silica plotted vs. cooling rate γ. The solid line is a fit with the form $T_g(\gamma) = T_{VF} - B/\ln(\gamma A)$, which is obtained if one assumes a Vogel-Fulcher form for the relaxation time, $\tau(T) = A \exp[B/(T - T_{VF})]$, and assumes that $T_g(\gamma)$ follows from $\tau(T_g) = \gamma^{-1}$ [58]. Note that $T_{VF} = 2525K$ would result here, i.e. more than 1000 K higher than the experimental T_g. From Vollmayr et al. [31].

rapid quenching does is to shift $T_g(\gamma)$ to higher temperatures and we also note the warning example of spin glasses [12] where nothing reliable is learned about the spin glass transition in rapid quenching computer experiments.

3 Study of the glass transition in a binary Lennard-Jones mixture

While the above model for SiO_2 belongs to a "strong glass" in the Angell classification [9], the Lennard-Jones binary mixture is a model of a "fragile glass" and thus its study is interesting in its own right. Moreover, due to the absence of long range forces, the simulation code is about a factor of 30 faster than for the above model, and thus a somewhat wider range of cooling rates is accessible.

The system is a binary (A,B) mixture of point particles of equal mass m, which interact with Lennard-Jones potentials

$$V_{\alpha\beta}(r) = 4\epsilon_{\alpha\beta}[(\sigma_{\alpha\beta}/r)^{12} - (\sigma_{\alpha\beta}/r)^6], \qquad (6)$$

with parameters $\sigma_{AA} = 1$, $\epsilon_{AA} = 1$ (defining units of lengths and energy), $\sigma_{AB} = 0.8$, $\epsilon_{AB} = 1.5$, $\sigma_{BB} = 0.88$ and $\epsilon_{BB} = 0.5$. These potentials are truncated and shifted at $r = 2.5\sigma_{\alpha\beta}$. This potential is chosen such to avoid crystallization [33], [34]. The particle numbers were chosen $N_A = 800$, $N_B = 200$ (in a few cases larger systems were also studied, but the data reported here did not seem to be affected by finite size effects [35]). The unit of time is $(m\sigma_{AA}^2/48\epsilon_{AA})^{1/2}$ here. The simulation technique is the same as described above; as initial temperature we use $T_i = 2$ (in units of ϵ_{AA}, note $k_B \equiv 1$).

Now, the general conclusions are fairly similar to those obtained for the model of SiO_2: again we find a pronounced dependence of many physical properties in the glassy state (density, enthalpy, structural properties like the probability that a particle has a coordination number z of a specified type of nearest neighbors — distinguishing AA pairs or BB pairs, etc.) on the cooling rate [35-37]. It is also remarkable that the cooling rate dependence of *microscopic* properties, such as the mentioned coordination number, is significantly larger than the one of *macroscopic* properties, such as the density or the enthalpy. There is also a pronounced dependence of T_g on the cooling rate γ, and in the accessible range of γ there even may be a slight systematic dependence on the type of quantity, that is fitted (Fig.4).

Thus the location of the glass transition temperature in the limit of vanishing cooling rate $T_g(\gamma \to 0)$ clearly still is rather uncertain. The remarkable success that idealized MCT has in describing data for this model is described elsewhere [33], [34].

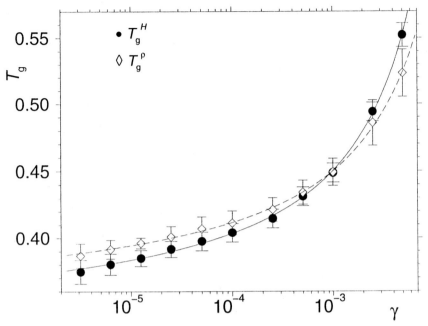

Fig. 4: Glass transition $T_g(\gamma)$ plotted vs. γ, as determined from the enthalpy (circles) and the density (diamonds), for the Lennard-Jones mixture. Curves are fits of the same type as included in Fig. 3, yielding $T_{VF} = 0.334$ and 0.348 for the enthalpy and the density, respectively. From Vollmayr *et al.* [37].

4   Lattice models for the glass transition of polymer melts

Polymer melts are often excellent glass formers, since their tendency to crystallize is sometimes extremely weak. Hence they are ideal materials to study the question of the glass transition in supercooled melts experimentally [26]. On the other hand, polymer coils are large slowly relaxing objects, already at temperatures far above the glass transition, which makes their simulation very difficult [50], [51]. Hence conclusive results can only be expected from simulations using very simplified models [50]. We do this by a "coarse-graining" of the polymer by combining several chemical bonds along the backbone of the (linear) macromolecule into one effective bond [49-51]. Depending on the conformation of these chemical groups, this effective bond may be longer or shorter. Therefore the "bond fluctuation model" [60] allows the length b of these effective bonds to fluctuate, $2 \leq b \leq \sqrt{10}$ if we now use the lattice spacing of a simple cubic lattice as the unit of length (Fig.5). If one would take the coarse-graining step from a chemically realistic model to such a schematic lattice model literally, the real potentials (potentials for the length of covalent bonds, angles between them, etc.) would be lumped into effective potentials for the lengths of effective bonds, for the angles between them, etc. Such a mapping procedure is possible (see Ref. [61]) but for the present context of no interest. Since the glass transition of most flexible polymers is rather similar, irrespective of the diversity in the size and shape of their chemical monomers and the corresponding potentials, we feel that a universal simplified model as that of Fig.5 is a reasonable idealization.

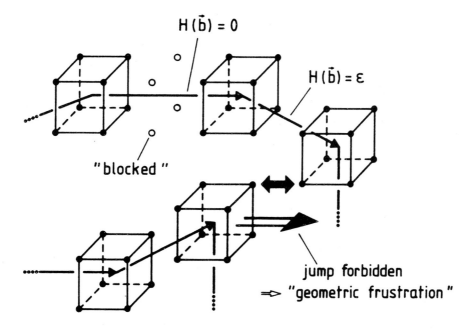

Fig.5 : Sketch of a possible configuration of effective monomers belonging to different chains in the melt in order to illustrate the effect of the energy function $\mathcal{H}(\mathbf{b})$ and the concept of "geometric frustration". Each effective monomer blocks the 8 sites of an elementary cube of the lattice from further occupation, and neighboring effective monomers along the chain are connected by a bond vector $\mathbf{b}$ joining their centers of mass. All $\mathbf{b}$ have the energy $\mathcal{H}(\mathbf{b}) = \epsilon$ ($= 1$ in our units) except $\mathbf{b}_0 = (\pm 3, 0, 0)$ {or permutations thereof} for which $\mathcal{H}(\mathbf{b}_0) = 0$. However, this vector $\mathbf{b}_0$ "blocks" 4 sites (marked by white circles) from further occupation, since two monomers may not overlap. E.g., due to this excluded volume interaction the jump indicated by the large arrow is forbidden. As more and more bonds take their ground state $\mathbf{b}_0$, more and more free volume is wasted (blocked sites), and for some bonds in a dense system it becomes impossible to reach their groundstate [40].

The dynamics of these systems is modelled by a Monte Carlo method in which we select an effective monomer at random, and attempt to move it by a lattice unit in a randomly selected direction. This move is possible only if excluded volume restrictions are obeyed, and if the bond vectors after the move still belong to the allowed set. Then the transition probability $W = \exp(-\delta\mathcal{H}/T)$ is evaluated, $\delta\mathcal{H}$ being the energy change due to the move. Only if W exceeds a random number z (uniformly distributed between zero and unity) is the attempted move actually performed. As is well-known, this Metropolis algorithm [50] generates (for times $t \to \infty$) a distribution of states with the correct Boltzmann weight. The motivation for the chosen type of moves, of course, is that in the real chains configurational changes come about by random hops of small groups of chemical bonds over barriers in the torsional potential [50]. Hence the random hops of the effective monomers are the counterpart of these moves on a coarse-grained mesoscopic level. The advantage of this crudely simplified model (in comparison to more realistic models) is that it can be simulated very efficiently (two millions of attempted hops per seconds on a single CRAY-YMP processor). Moreover, one can equilibrate the system using algorithms with an unphysical dynamics, e.g. the "slithering snake" propagation of chains [38], [50]. In this algorithm, one attempts a move in which a monomer is removed from one end of the chain and randomly added at the opposite end of the chain, choosing from the set of allowed bond vectors at random. This method allows to obtain the entropy of our model even at such low temperatures [45] where the relaxation time and diffusion constant, determined with the conventional MC algorithm, can no longer be measured reliably.

Despite the crude character of this lattice model and the simplified nature of the "random hopping"-algorithm, e.g., oscillatory small amplitude motions have been completely lost, it still shares with nature, and the more realistic models treated above, that a dramatic slowing down occurs as one approaches low temperatures. Furthermore also this system shows a pronounced dependence of physical properties on the cooling rate [40]. In this work, one starts with "athermal melts" ($T \to \infty$, only excluded volume and bond length constraints are still effective), which are easily equilibrated for short chains. We use chain lengths $3 \le N \le 16$ [42], and in most cases data for $N = 10$ will be presented [40,41,43-45,47-49]. Then the inverse temperature is increased with time as $1/T(t) = \Gamma_Q t/T_f$, with $T_f = 0.05$ being the final temperature of the glass. The quench rates are in the range $4 \cdot 10^{-7} \le \Gamma_Q \le 4 \cdot 10^{-5}$. Again it is found that physical quantities, such as the enthalpy, the average length of effective bonds, the gyration radius of the chain, etc., systematically depend on this cooling rate, and also the glass transition temperature $T_g(\Gamma_Q)$ decreases with decreasing Γ_Q slower than logarithmically, see Fig.6. The resulting Vogel-Fulcher temperature $T_{VF} = 0.17$, however, is significantly too high since, as we shall demonstrate below, when T_{VF} is determined from the temperature dependence of various relaxation times, which were measured with states that were equilibrated with the slithering snake algorithm, one obtains $0.12 \lesssim T_{VF} \lesssim 0.13$, and fits to MCT yield [43], [44] $T_c \approx 0.15$.

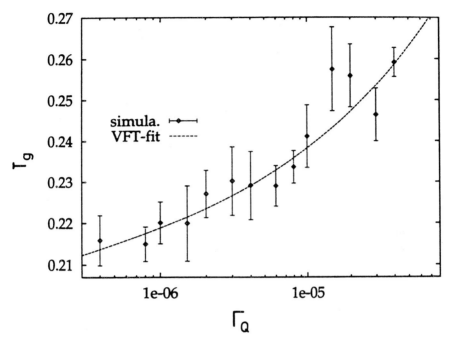

Fig. 6: Glass transition temperature $T_g(\Gamma_Q)$ of the bond fluctuation model plotted vs. the cooling rate Γ_Q, as obtained from extrapolated data for the "internal temperature" (extracted from the probability of excited bonds, cf. [40]). The curves is a fit with the function $T_g(\Gamma_Q) = T_{VF} - B/\ln(\Gamma_Q A)$, cf. Fig.3, with $T_{VF} = 0.17$. For $30 \times 30 \times 30$ lattices averages were taken over 160 replicas (i.e., the statistics was based on 288.000 effective monomers). From Baschnagel *et al.* [40].

We now briefly summarize the main results of the studied model [40-49]. internal energy decreases monotonically with decreasing temperature, but does not reach its ground state value $E_b(T = 0) = 0$: the smaller the quench rate Γ_Q is, the lower is the plateau value that is reached for $T < T_{VF}$. The bonds which are frozen in excited states represent the "frustration" present in the system. Analogous results are found for the mean-square length $\langle b^2 \rangle$ of effective bonds and the mean square radius of gyration $\langle R_g^2 \rangle$. Generally, one finds that the larger the considered scales is the higher is the temperature at which the corresponding quantity falls out of equilibrium [40]. This has an interesting consequence when one considers quantities like the characteristic ratio $C_N = 6\langle R_g^2 \rangle / \langle b^2 \rangle$ which measures the effective stiffness of the chain (Fig.7): For fast cooling C_N decreases, with decreasing temperature, because $\langle R_g^2 \rangle$ is frozen in at such a high temperature, that on cooling $\langle R_g^2 \rangle$ does not increase at all, while there is some increase of the effective bond length, since the latter is on a smaller length scale. At intermediate cooling rate there is just enough increase of $\langle R_g^2 \rangle$ to apparently cancel the expansion of $\langle b^2 \rangle$, so C_N stays constant, while only for small cooling rate do long range correlations in the structure build up sufficiently strongly upon cooling, that one sees an increase of C_N with decreasing temperature.

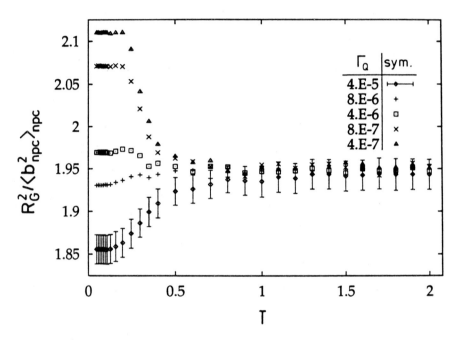

Fig. 7: Characteristic ratio $C_N/6 = \langle R_g^2 \rangle / \langle b^2 \rangle$ plotted vs. temperature, for the bond fluctuation model with $N = 10$ and at a volume fraction $\phi = 0.533$ of occupied sites. Five different quench rates are included, as indicated in the figure. From Baschnagel *et al.* [40].

Although at the chosen volume fraction of occupied lattice sites ($\phi = 0.533$) the system could in principle reach an ordered crystal ground state (with $E_b(T = 0) = 0$), there is no detectable tendency towards crystallization. The collective structure factor (Fig.8) depends only weakly on temperature, and is qualitatively compatible with experiment. For small wavenumber q, $S(q)$ is very small, reflecting the fact that polymer melts are nearly incompressible. Then it rises to the "amorphous halo" at $q = 3$. Since lengths are measured in units of the lattice spacing, and physically a lattice spacing corresponds to about 2Å [61], the peak occurs at about 1.5 Å^{-1}, as observed in experiments [62]. Of course, the structure at $q \approx 4-6$ in Fig.8 can no longer be taken seriously, one sees one lattice spacing at $q = 2\pi$, so there lattice artifacts must come in, but it is gratifying that one still can reproduce the amorphous halo and thus roughly the packing of the amorphous structure.

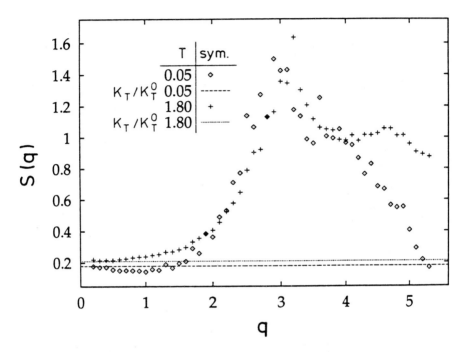

Fig. 8: Collective structure factor $S(q)$ plotted vs. q for the reduced temperatures $T = 0.05$ (diamonds) and $T = 1.8$ (crosses) for the cooling rate $\Gamma_Q = 4 \cdot 10^{-6}$. The dashed and dotted horizontal lines represent estimates of the reduced compressibility obtained from an analysis of density fluctuations in large subsystems of the total simulated system. From Baschnagel and Binder [41].

Although for every lattice model of polymers the presence of lattice artifacts clearly is a problem that deserves careful consideration, it was found that the bond fluctuation model is a surprisingly good approximation to continuum behavior [40-51]. This is also obvious from a study of the single chain structure factor, which is described very well by the Debye function [41], as it is expected for Gaussian coils. Thus although short chains of only $N = 10$ effective monomers are studied, the asymptotic properties of very long chains ($N \rightarrow \infty$) are already nicely reproduced. Even the linear variation of $T_g(N)$ with $1/N$ predicted by the theory of Gibbs and di Marzio [22], [23] and seen in experiments (e.g. [63]) is confirmed [42].

To analyze dynamic properties near the glass transition, many properties have been considered: relaxation times $\tau(q)$ extracted from the dynamic single-chain structure factor, from the end-to-end vector correlation function τ_{ete}, from the Rouse modes of the chain τ_p for $p = 1, 2, 3$ {Rouse modes are defined [64] in terms of the positions $\mathbf{r}_n(t)$ of the n'th monomer in a chain as $\mathbf{X}_p(t) = \sum_{n=1}^{N} \mathbf{r}_n(t) \cos[p\pi(n - 1/2)/N]$}, etc. [65], see Figs.9,10. For these data and their mutual consistency it turned out to be crucial of not using data (as described above in Figs. 6-8) resulting from a slow cooling procedure, but rather to equilibrate each temperature thoroughly with long runs using the "slithering snake" algorithm. These well-equilibrated states are used then as initial configurations for generating a time series of configurations with the "random hopping"-algorithm, that mimics the real dynamics, as emphasized above. The data for the relaxation times thus obtained (Fig.10) can be fitted well only with the Vogel-Fulcher law, Eq.(2), whereas the Arrhenius law, Eq.(4), fits the data not well at all, and also the Bässler law [66] {$\ln \tau \propto 1/T^2$} provides a fit of inferior quality than the Vogel-Fulcher law.

While originally [40] it was erroneously concluded that $T_{VF} = 0.17 \pm 0.01$, because data on the diffusion constant was used that was equilibrated not as well and at high temperatures, the reanalysis of the dynamics by Okun $et\ al.$ [65] has yielded a considerably lower estimate for T_{VF}, $0.12 \leq T_{VF} \leq 0.13$. This experience shows again that it is very difficult to draw quantitatively reliable conclusions from simulations of the structural glass transition. However, these ambiguities of accurately estimating T_{VF} are also typical of experiments [66].

5   What do we learn about the theoretical concepts?

It is of interest to analyze the time dependence of the normalized intermediate incoherent scattering function

$$\phi_q^s(t) = \frac{1}{n} \sum_i \langle \exp[i\mathbf{q} \cdot (\mathbf{r}_i(t) - \mathbf{r}_i(0))] \rangle_T, \tag{7}$$

where the sum is over all n monomers in the system, since this is the basic quantity of MCT [4], [5], cf. Fig.1. Monte Carlo data [44] for $q = 2.92$, where $S(q)$ has its peak, were analyzed with the extended version of the MCT [30]. It turned out that the idealized theory seemed not to be able to rationalize all these data in a consistent way [43]. Thus the conclusion from this analysis was that one must use the extended version of MCT, where the critical singularity at T_c is rounded off (tentatively this rounding is associated with the "hopping processes" which are needed in order to allow the particles to escape from the cages formed by their surrounding particles, but the precise connection of the fitted parameters to this picture remains to be given.).

The other theory that we wanted to test (Fig.1), the concept of a vanishing configurational entropy [21-24] is found to be not very satisfactory either.

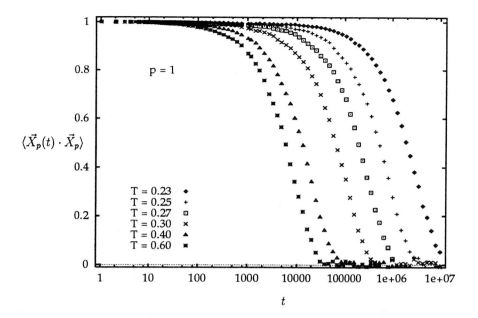

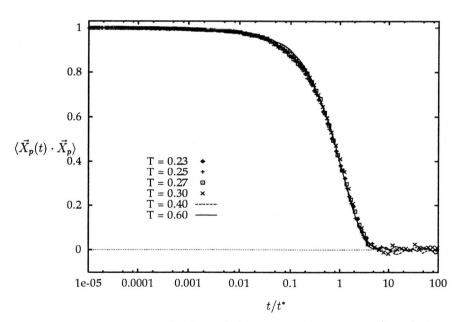

Fig. 9: Correlation function $\langle \mathbf{X}_p(t \cdot \mathbf{X}_p(0) \rangle$ for the first Rouse mode ($p = 1$) plotted vs. time for a variety of temperatures (upper part) and rescaled according to the time-temperature superposition principle (lower part). A relaxation time τ_p is defined from the condition that this correlation function has decayed to a value of 0.4. From Okun *et al.* [65].

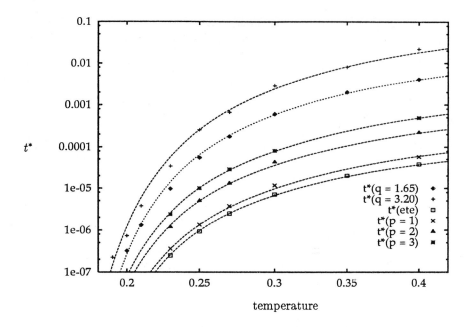

Fig. 10: Plot of inverse relaxation times $\tau^{-1}(q = 1.65)$, $\tau^{-1}(q = 3.20)$, τ_{ete}, and τ_p (for $p = 1,2$, and 3) plotted vs. temperature. Curves indicate Vogel-Fulcher fits, where in all cases T_{VF} falls in the interval $0.12 \leq T_{VF} \leq 0.13$. From Okun *et al.* [65].

By special Monte Carlo techniques [45] one can obtain the configurational entropy and its temperature dependence for our lattice model (Fig.11). It is also possible to obtain all the parameters that enter the Gibbs – di Marzio theory [21-24], namely the fraction of bonds that are not in the ground state, the effective coordination number and the effective vacancy concentration, from the simulation itself, and thus one can test the theory without any adjustable parameter at all (Fig.11). It is seen that the theories of Gibbs and di Marzio [22], [23] and Flory [67] yield a temperature dependence that is quite similar to the simulation — but the entropy of the athermal melt (at $T = \infty$) clearly is too low. Since the decrease of the entropy with decreasing temperature is predicted roughly correctly, the theoretical entropy curves hit the abscissa (zero entropy) already at a nonzero temperature T_0. Obviously, this "entropy catastrophe" at T_0 is an artifact due to the inaccurate estimation of the entropy of the athermal melt, and thus has no real bearing on the glass transition. Milchev's correction [28] removes the "entropy catastrophe" but is not perfect either since now the entropy is overestimated. While Fig.11 shows that indeed the strong decrease of the entropy occurs when one approaches the glass transition region, a linear extrapolation from T_g to zero entropy is wrong, since the curve bends over and levels off. Thus, the Kauzmann paradox is only due to an inappropriate extrapolation and is therefore, at least for our model, not a real problem.

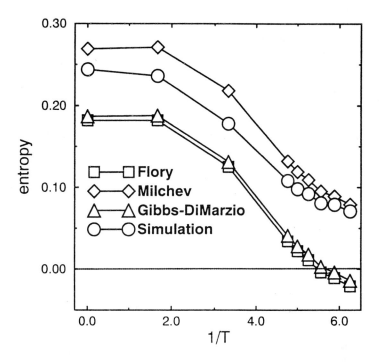

Fig.11 : Entropy per lattice site of the bond fluctuation model at $\phi = 0.533$, $N = 10$, plotted vs. inverse temperature. Theoretical predictions based on the theories of Gibbs and di Marzio [22], [23], Flory [67] and Milchev [28] are included. From Wolfgardt *et al.* [45].

These findings, however, not necessarily rule out the possibility that there is an underlying static transition that explains the properties of the glass transition: There is indirect evidence from our simulations that this model exhibits a static correlation length that increases strongly upon cooling [46], [47]. This length can be extracted indirectly from finite size effects on the relaxation time [46] or from simulations of our model in a thin film geometry confined by hard walls [47]. These hard walls produce density oscillations over a range ξ that is much smaller than the gyration radius at high temperatures (Fig.12) but strongly increases when one approaches the glass transition. At

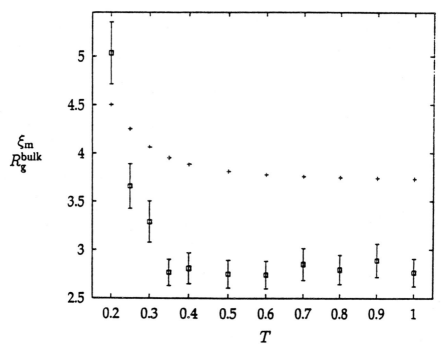

Fig.12 : Temperature dependence of various characteristic lengths associated with wall effects on our polymer model: decay length of the monomer density profile (squares), of the end monomer density profile ($\times$), the decay length of the profile of monomers belonging to the same chain (triangles), and the gyration radius in the bulk (+). All quantities are given in units of the bulk bond length b^{bulk}. From Baschnagel and Binder [47].

$T = 0.2$ this length already is the largest length in the system (Fig.12). One finds that it is this length that controls the orientation of the chains relative to the wall [47], rather than the gyration radius as for athermal melts.

Unfortunately, we do not have reliable data in the regime near T_{VF} on ξ, and hence it is unclear whether ξ really diverges. Also it is still an unsolved problem to identify the proper (necessarily high-order) correlation function in the bulk, from which this length could be directly extracted, without recourse to wall effects.

Thus, the theoretical concepts tested have some success in accounting for at least certain aspects of these simulations. However, none of the theories describes coherently all the phenomena seen in the simulations. Encouragingly, the models do reproduce many features of the experiment qualitatively. But from a computational point of view the accurate location of the glass transition from simulations of a supercooled fluid in metastable equilibrium is very difficult. While it is easy to freeze-in some glass-like structures in rapid quenching simulations, these structures do depend significantly on the cooling rates, as it is also observed in real experiments. Thus if one wants to investigate such cooling rate dependencies, computer simulations can be very useful, since they allow to access the whole microscopic information. However the fundamental question regarding the existence and the properties of an "ideal" glass transition are much harder to study and it remains to be seen whether computer simulations are able to make a valuable contribution in this area as well.

Acknowledgements:

This work was supported by the Deutsche Forschungsgemeinschaft (DFG), SFB 262. K.V. was supported by SCHOTT Glaswerke via the SCHOTT Fonds. J.B. was supported by the Bundesministerium für Forschung und Technologie (BMFT, No. 03M4076A3).

References

[1] R. Zallen *The Physics of Amorphous Solids*, (Wiley, New York, 1983).

[2] J. Jäckle, Rep. Progr. Phys. **40**, 171 (1986).

[3] J. Zarzycki (ed.) *Materials Science and Technology, Vol. 9*, (VCH Publ., Weinheim, 1991).

[4] W. Götze, in *Liquids, Freezing and the Glass Transition*, edited by J. P. Hansen, D. Levesque, and J. Zinn-Justin, (North-Holland, Amsterdam, 1990).

[5] W. Götze and L. Sjögren, Rep. Prog. Phys. **55**, 241 (1992).

[6] A. J. Dianoux, W. Petry and D. Richter (eds.) *Dynamics of Disordered Materials II*, (North-Holland, Amsterdam, 1993).

[7] K. L. Ngai (ed.) *Proc. 2nd International Discussion Meeting on Relaxations in Complex Systems*, J. Non-Cryst. Solids **172-174** (1994).

[8] C. A. Angell, in this volume.

[9] C. A. Angell, in: K. L. Ngai and G. B. Wright (eds.) *Relaxation in Complex Systems*, (US Dept. Commerce, Springfield, 1985).

[10] H. Vogel, Phys. Z. **22**, 642 (1921); G. S. Fulcher, J. Amer. Ceram. Soc. **8**, 339 (1925).

[11] R. Kohlrausch, Ann. Phys. (Leipzig) **12**, 393 (1847); G. Williams and D. C. Watts, Trans. Faraday Soc. **66**, 80 (1980).

[12] K. Binder and A. P. Young, Rev. Mod. Phys. **58**, 801 (1986).

[13] K. Binder and J. D. Reger, Adv. Phys. **41**, 547 (1992).

[14] P. C. Hohenberg and B. I. Halperin, Rev. Mod. Phys. **49**, 435 (1977).

[15] C. Dasgupta, A. V. Indrani, S. Ramaswamy and M. K. Phani, Europhys. Lett. **15**, 307 (1991).

[16] R. M. Ernst, S. R. Nagel and G. S. Grest, Phys. Rev. **B43**, 8070 (1991).

[17] E. Donth, J. Non-Cryst. Solids **53**, 325 (1982).

[18] E. W. Fischer, E. Donth and W. Steffen, Phys. Rev. Lett. **68**, 2344 (1992).

[19] J. Jäckle, J. Phys.: Condens. Matter **8**, 2733 (1996).

[20] G. Adam and J. H. Gibbs, J. Chem. Phys. **43**, 139 (1965).

[21] J. H. Gibbs, J. Chem. Phys. **25**, 185 (1956).

[22] J. H. Gibbs and E. A. Di Marzio, J. Chem. Phys. **28**, 373 (1958).

[23] E. A. Di Marzio and J. H. Gibbs, J. Chem. Phys. **28**, 807 (1958).

[24] E. A. Di Marzio, J. H. Gibbs, P. D. Fleming III, and I. C. Sanchez, Macromolecules **9**, 763 (1976).

[25] W. Kauzmann, Chem. Rev. **43**, 219 (1948).

[26] G. B. McKenna, in: C. Booth and C. Price (eds.) *Comprehensive Polymer Science, Vol. 2*, (Pergamon Press, Oxford, 1990).

[27] P. D. Gujrati, J. Phys.**A 13**, L437 (1980); P. D. Gujrati and M. Goldstein, J. Chem. Phys. **74**, 2596 (1981).

[28] A. I. Milchev, C. R. Acad. Bulg. Sci. **36**, 1413 (1983).

[29] H. P. Wittmann, J. Chem. Phys. **95**, 8449 (1991).

[30] S. P. Das and G. F. Mazenko Phys. Rev **A 34**, 2265 (1986); W. Götze and L. Sjögren, Z. Phys. **B 65**, 415 (1987).

[31] K. Vollmayr, W. Kob and K. Binder, Mainz University preprint KOMA-96-10; ; K. Vollmayr and W. Kob, Ber. Bunsenges. Phys. Chem. (1996, in press).

[32] B. W. van Beest, G. J. Kramer and R. A. van Santen, Phys. Rev. Lett. **64** 1955, (1990).

[33] W. Kob and H. C. Andersen, Phys. Rev. Lett. **73**, 1376 (1994); Phys. Rev. **E 54**, 4626 (1995); *ibid.* **52**, 4134 (1995).

[34] W. Kob and M. Nauroth, in this volume; M. Nauroth and W. Kob, Mainz University preprint KOMA-96-20.

[35] K. Vollmayr, W. Kob and K. Binder, p. 117 in *Computer Simulation Studies in Condensed Matter Physics VIII*, ed. D. P. Landau, K. K. Mon, and H. B. Schüttler, (Springer, Berlin, 1995).

[36] K. Vollmayr, W. Kob and K. Binder, Europhys. Lett. **32**, 715 (1995).

[37] K. Vollmayr, W. Kob and K. Binder, J. Chem. Phys. (1996, in press).

[38] M. Wolfgardt, J. Baschnagel and K. Binder, J. Phys. **II** (Paris), **5**, 1835 (1995).

[39] H.-P. Wittmann, K. Kremer and K. Binder, J. Chem. Phys. **96**, 6291 (1992).

[40] J. Baschnagel, K. Binder and H.-P. Wittmann, J. Phys.: Condens. Matter **5**, 1597 (1993).

[41] J. Baschnagel and K. Binder, Physica **A 204**, 47 (1994).

[42] B. Lobe, J. Baschnagel and K. Binder, Macromolecules **27**, 3654 (1994).

[43] J. Baschnagel, Phys. Rev. **B 49**, 135 (1994).

[44] J. Baschnagel and M. Fuchs, J. Phys.: Condens. Matter **7**, 6761 (1995).

[45] M. Wolfgardt, J. Baschnagel and K. Binder, J. Chem. Phys. **103**, 7166 (1995); M. Wolfgardt, J. Baschnagel, W. Paul and K. Binder, Phys. Rev. **E** (1996, in press).

[46] P. Ray and K. Binder, Europhys. Lett. **27**, 53 (1994).

[47] J. Baschnagel and K. Binder, Macromolecules **28**, 6808 (1995).

[48] J. Baschnagel and K. Binder, J. Phys. **II** (Paris), (1996, in press).

[49] For a brief review, see K. Binder, Ber. Bunsenges. Phys. Chem. (1996, in press).

[50] K. Binder (ed.) *Monte Carlo and Molecular Dynamics Simulations in Polymer Science*, (Oxford University Press, New York, 1995).

[51] K. Binder, Makromol. Chem., Macromol. Symp. **50**, 1 (1991).

[52] J. S. Tse and D. D. Klug, Phys. Rev. Lett. **67**, 3559 (1991); J. S. Tse, D. D. Klug and D. C. Allan, Phys. Rev. **B 51**, 16392 (1995), and references therein.

[53] M. P. Allen and D. J. Tildesley, *Computer Simulation of Liquids*, (Oxford University Press, Oxford, 1987).

[54] H. C. Andersen, J. Chem. Phys. **72**, 2384 (1980).

[55] R. Brüning and K. Samwer, Phys. Rev. **B 46**, 11318 (1992).

[56] G. S. Grest, C. M. Soukoulis and K. Levin, Phys. Rev. Lett. **56**, 1148 (1986); D. A. Huse and D. S. Fisher, Phys. Rev. Lett. **57**, 2203 (1986).

[57] C. A. Angell, J. Chem. Phys. Solids **49**, 863 (1988).

[58] C. A. Angell, J. H. R. Clarke and L. V. Woodcock, Adv. Chem. Phys. **48**, 397 (1981).

[59] J. Horbach, W. Kob, K. Binder and C. A. Angell, Mainz University preprint; W. Kob, p. 1 in *Annual Reviews of Computational Physics, Vol. III*, (ed.) D. Stauffer, (World Scientific, Singapore, 1995).

[60] I. Carmesin and K. Kremer, Macromolecules **21**, 2819 (1988); H.-P. Deutsch and K. Binder, J. Chem. Phys. **94**, 2294 (1991).

[61] W. Paul, K. Binder, K. Kremer, and D. W. Heermann, Macromolecules **24**, 6531 (1991); W. Paul and N. Pistoor, Macromolecules **27**, 1249 (1994); V. Tries, W. Paul, and K. Binder, J. Chem. Phys. (1996, in press).

[62] D. Richter, B. Frick and B. Farago, Phys. Rev. Lett. **61**, 2465 (1988); B. Frick, B. Farago and D. Richter, Phys. Rev. Lett. **64**, 2921 (1990).

[63] J. M. G. Corrie and P. M. Toporowski, Eur. Polym. J. **4**, 621 (1968).

[64] M. Doi and S. F. Edwards, *The Theory of Polymer Dynamics*, (Oxford University Press, Oxford, 1986).

[65] K. Okun, J. Baschnagel, M. Wolfgardt and K. Binder, Mainz University preprint.

[66] F. Stickel, E. W. Fischer and R. Richert, J. Chem. Phys. **102**, 1 (1995).

[67] P. J. Flory, Proc. Roy. Soc. London, Series A, **234**, 60 (1956).

Microscopic Dynamics in Glasses in Relation to That Shown by Other Complex Systems

F J Bermejo[1], H E Fischer[2], M A Ramos[3], A de Andrés[4], J Dawidowski[1] and R Fayos[1]

[1] Instituto de Estructura de la Materia, Consejo Superior de Investigaciones Científicas, Serrano 123, Madrid E-28006, Spain
[2] Institut Laue Langevin, BP 156, F-38042 Grenoble, France
[3] Departamento de Física de la Materia Condensada, Facultad de Ciencias, Universidad Autónoma de Madrid, E-28049 Cantoblanco, Spain
[4] Instituto de Ciencia de Materiales, Consejo Superior de Investigaciones Científicas, Campus de Cantoblanco, E-28049 Cantoblanco, Spain

Abstract. The relevance of studies on orientational and/or rotationally disordered crystals is discussed within the context of glass-forming and other systems exhibiting the basic aspects of glassy behaviour.

1 Introduction

The quest for a simple microscopic model which retains the basic ingredients giving rise to most of the phenomenological manifestations of "glassy behaviour" (e.g. the nearly universal low-temperature thermal properties) has recently been the focus of an increased research activity. As a result, a number of physical systems have been explored quite often without explicit recourse to disorder (i.e. whether this be of stuctural leading to an amorphous state or dynamic origin). These encompass a variety of conceptually simple systems which go from a finite-temperature Frenkel-Kontorova chain (Shumway and Sethna 1991), to mean-field descriptions even devoid of multiple free-energy barriers (Ritort 1995). In the search for physical realizations for such idealizations of "glassy dynamics" (i.e. relaxation of two-level systems), several proposals have appeared where the sought behaviour is even present in periodic (eg. ordered) objects such as long-range arrays of Josephson junctions (Chandra 1996).

However, care should be exercised when identifying some of the dynamical properties commonly regarded as "fingerprints" of glassy dynamics with the presence of an underlying "glassy state" (defined as a state which has been formed by either cooling from the liquid through a process not showing discontinuities in the fisrt order thermodynamic properties or any other route involving the loss of long-range spatial regularities via solid-state transformations), as recent experimental results on the acoustic properties of polycrystals within the mK and μK ranges (Esquinazi et al. 1996) vividly exemplifies.

In fact, the striking similarity between the low-temperature acoustic properties of simple monoatomic polycrystalline metals in their normal state (Ag, Cu, Pt, Al) superconductors (Al, Ta, Nb, NbTi) and glasses seems to point out the existence, in all these systems, of a significant amount of low-lying excitations widely distributed as far as their energies and relaxation times are concerned. Even more, such similarity between the elastic behaviour of glasses and polycrystals have been also found to apply not only to crystals well known to show glass-like anomalies such are orientationally or compositionally disordered crystalline solids but also to quasicrystals (White and Pohl 1996).

The presence within crystals subjected to large internal strains of tunneling states very much akin to those portrayed by the standard two-level systems (TLS) model which dominate the low-temperature properties of these materials has also been recently demonstrated by experiment (Watson 1996) or simulational means (Wolf et al. 1995). As a consequence, it has been evidenced that the broad energy distributions which give rise to glass-like properties at low temperatures can be controlled at will by the introduction of random strain fields on crystalline lattices.

Such "glass-like anomalies" may thus seem to be far more universal than previously expected and the very fact of their ubiquity poses some questions about our present understanding of the "glassy state". It is also worth noticing that even within the realm of time-dependent properties, some models have recently appeared (Van de Walle 1996) able to account for one of the most common features in the dynamics of disordered systems as it is the presence of stretched exponential relaxations. Such relaxation patterns were postulated as one of the "universal" characteristics of glassy dynamics, and in the simplest case were thought to arise from relaxation of non-interacting units taking place with a distribution of timescales. However, the fact that a simple model for dispersive decay not invoking the assumption of an underlying statistical distribution of relaxation times as that of (Van de Walle 1996) accounts for the stretched-exponential behaviour, clearly shows that the presence of such distributions which are consequence of a complex energy landscape, is not a fundamental ingredient.

A class of condensed matter systems which has been shown to contain all the characteristics attributable to glass dynamics, and which has constituted the focus a good number of studies over the recent couple of decades regards partially ordered crystals where only the translational order is preserved while the relative orientations of the particles forming the crystal show either rotational or orientational disorder (Lynden-Bell and Michel 1994). Because of the presence of an underlying lattice, such crystalline systems offer the possibility of a fully analytical treatment for a number of static and dynamical properties also regarded as characteristic of the glassy state such are the softening of the elastic constants due to orientational disorder, the frequency dependence of the sound velocity or the well known anomalies in the dielectric behaviour evidenced by most glasses. In addition, such systems are also known to exhibit a complicated phase diagram (Loidl abd Böhmer 1994) and in a variety of respects can rightly be considered as realistic model systems to study the glass transition in far more complex systems. In fact, many of those crystals (Loidl abd Böhmer 1994) show a remarkable polymorphism, very much alike that found in some canonical glass-formers (eg. vitreous silica), a fact which appears to be linked to the ability of such liquids to form a glass (Jäckle 1986).

In the search for a system where the effects brought forward by different kinds of disorder on its dynamic response could be isolated, we have carried out an extensive study on the different condensed phases of ethanol, well known to form a structural glass as well as some intermediate-order phases as revealed by heat capacity measurements (Haida et al. 1972). There, the formation of such an intermediate phase was achieved by means of controlling the cooling from the liquid phase at rates just quick enough to prevent crystallization into the stable, orientationally ordered, monoclinic crystal (SCP) phase ($\simeq 2$ K min^{-1}). In addition, further cooling of this metastable phase leads to another crystal phase still above the crystalline groundstate (SCP) and, strikingly enough, such a transition occurs at a temperature which is basically indistinguishable from that corresponding to the liquid-to-glass transition (T_g), and shows a jump in the heat capacity somewhat above eighty per cent of that characteristic of the canonical glass-transition (Haida et al. 1972). The remarkable similarities of those calorimetric transitions lead those authors to characterize such a solid-solid transition as an additional glass-transition, a fact latter found in a larger number of materials (Haida and Seki 1980). ¿From calorimetric as well as evidences from neutron scattering (see below), the phase formed by annealing the supercooled liquid (SCL) was found to correspond to a rotator-phase crystal (RPC) which upon further cooling freezes into an orientationally disordered state (ODC).

The relevant calorimetric transitions for one of such materials (a fully deuterated sample employed for neutron diffraction) are depicted in Figure 1.

From there several points seem worth remarking. First concerns the fact that both, the liquid-glass transition temperature T_g and the RPC$\longrightarrow$ODC freezing transition one T_{odc} basically take place within the same range of temperatures. The latter transformation also accounts for about eighty eight per cent of the C_p jump found for that of liquid to glass. Finally, it seems also pertinent to remark the onset of a noticeable increase in the calorimetric entropy of the SPC solid, as judged by the riseup in $C_P(T)$ at temperatures also close to those of the alluded transformations but still some 65 K below melting.

Our aim is then twofold. First, to relate the observations regarding static and dynamic correlations in the intermediate-order phases to those of the canonical (i.e. amorphous) glass obtained from rapid quench of the liquid,

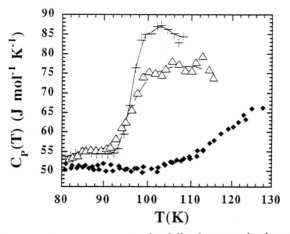

Fig. 1. Heat capacity measurements for fully deuterated ethanol around the temperatures of supercooled-liquid to glass (crosses) and rotator to orientationally disordered crystal (RPC-ODC) (triangles) transformations. The solid lozenges depict the values for the same thermodynamic function for the stable crystalline solid (SPC). Solid and dashed lines are spline approximations to the data and are drawn as guides to the eye.

and second, to explore in further detail the relevance of the solid-solid transition as a convenient ground to be explored towards the understanding of the liquid-glass transition. Notice that because such a material seems to be the only one where the different "glassy" phases can be prepared at ease (eg. others showing similar phase diagrams require special handling of the material to prepare the canonical glass phase), the relationship between features appearing in the glass and disordered crystal forms should be considered as a direct one, thus implying that what is observed in the disordered crystals should explain, at least in part, what is characteristic of the canonical glass.

2   Static Structure

The nature and spatial extent of the static (concerning the positions of the molecular centres-of-mass as well as their orientations) correlations in all the five, topologically disordered (liquid and glass), fully ordered and intermediate-order phases found by calorimetric means was investigated first by means of radiation scattering techniques. From preliminary X-ray measurements (Srinivasan et al. 1996) the crystal structure of the intermediate-order phases was shown to correspond to a body-centered-cubic (b.c.c.) with two particles per unit cell and a macroscopic density not far from that characteristic of the supercooled liquid (SCL). As regards the kind of orientational correlations appearing in such phases, Raman scattering measurements evidenced remarkable similarities between the quantities

$$I(E)E/[n(E) + 1] \propto Z_R(E) \tag{1}$$

for the disordered crystal and glass. These are basically the Raman intensity normalized by the Bose factor and thus, $Z_R(E)$ can be taken as an approximation to the true $Z(E)$ generalized frequency distribution, which will be exact if the light-to-vibrations coupling followed that of an elastic continuum (i.e. $C(E) \propto E^2$). The fact that the presence of the underlying b.c.c. lattice does not have a strong effect on the spectral frequency distribution of the crystal gives a direct indication of the existence of a large amount of orientational disorder, since the crystal selection rules would have restricted significantly the number of Raman-active modes otherwise.

To provide additional details enabling a quantitative characterization of the orientational and/or positional correlations a number of neutron diffraction measurements were carried out. The most significant results of such a study are shown schematically in Figure 2, which shows the reduced intensity functions for all the five condensed phases.

As can easily be inferred from visual inspection of the Figure 2, three different types of atomic ordering can be identified. The stable crystal (SCP) phase shows a large number of Bragg reflections superimposed to a long-period oscillation generated by the polycrystalline average of short-ranged contacts. In contrast, the glass and liquid diffraction patterns evidence the absence of long-range periodicity by the absence of any strong reflection, being the intense, first sharp diffraction peak (FSDP), the only remnant of order beyond nearest neighbours. On the other hand, the rotator phase (RPC) and orientationally disordered (ODC) crystals show the presence of five Bragg peaks which index as resulting from b.c.c. structures with lattice constants of 5.32 Å and 5.28 Å respectively. The most remarkable fact about all this regards the close proximity between the location Q_p of the main peaks in the disordered crystal phases and that for the glass (or even the liquid) as it is shown in the insert, whereas those arising from the SCP are located sideways. Because of the presence of an underlying b.c.c. lattice as well as the macroscopic isotropy condition, the position Q_p in the RPC and ODC crystals should define a characteristic distance $R = 2\pi/Q_p$ as given by a Scherrer approximation, whereas the width of such peaks Δ_Q, once corrected by finite-resolution effects can be taken as a coherence length $R_c = 2\pi/\Delta_Q$. In fact, the first intense peak in the disordered crystals patterns is located at wavenumbers having a correlate in real space which is nothing else than the distance between [110] planes. As regards R_c, a lower bound of some 36 nm could be guessed from the peak widths. In contrast, the diffraction pattern for the structurally disordered phases would, if the identification still applies, have "coherence lengths" of some 20 Å (glass) and 15 Å (liquid), whereas the "characteristic distances" still are close to about 3.7 Å that is the distance between planes in the ODC and RPC.

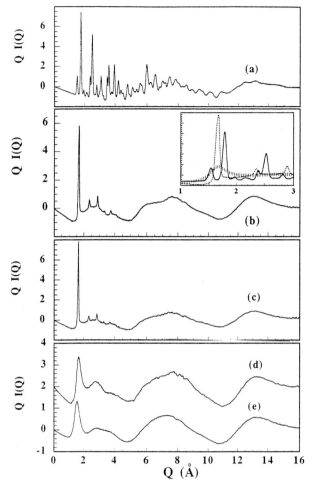

Fig. 2. Static, Q I(Q), reduced intensity functions for: a) the stable (monoclinic) crystal phase, b) the orientationally disordered crystal, c) the rotator (plastic) crystal, d) the structural glass and e) the normal liquid. The inset displays a comparison between the location of the first intense peak of the stable crystal (solid), that of the orientationally disordered solid (dashes) and that of the glass (vertical bars).

A model-free assessment of estimations based on the interference patterns can be readily made from calculation of the $g_{tot}(r)$ radial distribution functions which are related to the quantities discussed previously by

$$g_{tot}(r) = 1 + \frac{1}{2\pi^2 \rho \, r} \int Q \, I(Q) \sin(Qr) dQ \qquad (2)$$

where the subscript indicates that the average is taken over all the different nuclei and ρ is the macroscopic number density. The results are summarized in Figure 3.

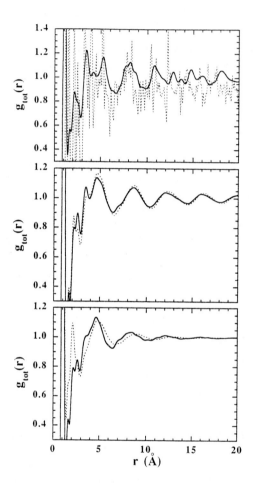

Fig. 3. Radial distributions for the SCP (top), disordered crystals (middle) and structurally disordered phases (bottom). The solid traces represent the experimental functions for the monoclinic crystal, ODC and structural glass respectively. The strong peaks below some 2.5 Å arise from the internal particle form-factors and are not shown in detail. The dashed lines depict : the distribution corresponding to a zero-temperature crystal as calculated from a harmonic, lattice-dynamical model (Criado et al. 1994), the RPC and the normal liquid.

The structure shown in the top frame of Figure 3 compares the experimental distribution function for the low-temperature (5K) SCP, with what could be expected for a harmonic (T=0K) solid. As seen by comparison between experiment and calculation anharmonic (thermal expansion) effects become noticeable even at such low temperatures, as judged by the progressive dephasing of oscillations in the experimental curve in relation to those resulting from calculation. Because of the complicated crystal Pc structure a large number of Fourier components appear in the distribution function. In contrast, the disordered crystals follow a far more regular behaviour for distances of about 4.5 Å onwards, showing an oscillation with an average period of some 3.74 Å (ODC) and 3.78 Å (RPC), which persists until distances larger than those covered by our experimental window. It is then evident that such long-period oscillations mainly arise from the intense (110) peaks in the RPC and ODC phases. If one now turns to the topologically disordered phases, it becomes clear that the period and phase of the oscillations above 4 Å are rather close to those of the ODC and RPC crystals, the small difference in phase being easily accountable by a small reduction in the macroscopic density. On the other hand, the most dramatic difference between the structural and orientationally disordered phases regards the persistence of those long-period oscillations, which is in the former case limited to distances below $\approx$ 15 Å, a bound not far from that guessed from the peak widths.

The origin of the oscillations in the radial distributions of the disordered crystals is easily ascribed, at least on semiquantitative grounds as arising from scattering from a b.c.c. lattice where the particles (molecules) located at each lattice node show a large degree of randomness. As a matter of fact, the results from a calculation employing a large array of b.c.c. unit cells containing at each node a rigid molecule with arbitrary orientation are compared with the experimental two-point static correlation function

$$D(r) = -4\pi \rho \, r[g_{tot}(r) - 1] \tag{3}$$

in Figure 4. As seen there, apart from the differences located below 3 Å due to the crude model (rigid) employed to represent the molecular geometry, the main difference between experiment and calculation concerns the peak at some 3.6 Å, which cannot be accounted for employing a model of complete decorrelation of orientations. Some orientational correlation between nearest neighbours needs then to be assumed and, in fact, looking at how the intensity of such a feature nearly withers away in passing from the SCP to the normal liquid seems to indicate a way of quantifying the extent of orientational order.

The results discussed above thus provide a direct and model-free verification of the meaning of parameters characterizing the FSDP of structurally disordered matter by means of relating the intermediate-range-order (IRO) (Gaskell and Wallis 1996, Salmon 1994), that is that involving distances beyond nearest-neighbours, to that characteristic of condensed phases showing a stability next to that of the glass or supercooled liquid. In fact, what our results are evidencing is the fact that a metastable phase such as a supercooled of glassy liquid a few degrees above the calorimetric glass transition,

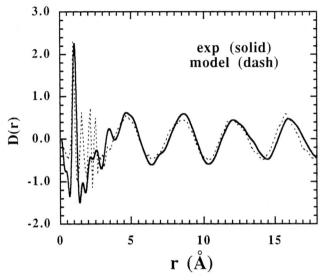

Fig. 4. Comparison between experimental data for the ODC phase, with results from a calculation involving rigid molecules located within an array of 512 unit cells

will adopt the structure of a phase of not too different macroscopic density, provided that a balance between competing interactions (eg. in our case highly directional electric multipole interactions versus short-ranged van der Waals contacts) exists which enables the stabilization of a frustrated structure such is that of the RPC solid. Formation of such crystals then proceeds by atomic rearrangements enabled by thermally activated hopping processes, which need to surmount barriers whose height was estimated to be of about 0.4 eV.

The fact that such a solid does not undergo, as the temperature is lowered, a transition driven by the electrostatic (hydrogen-bond) interactions into an orientationally ordered state (the SCP) as known for a large number of rotator-phase (plastic) crystals (Loidl abd Böhmer 1994) but to a rotationally frozen ODC phase can surely be a consequence of the presence within such a material of particles adopting different internal geometries (conformations), thus resembling the case of molecular alloys with random bonds.

Rather than being a result pinned down to a specific kind of samples, the present example seems to point towards finding a microscopic origin for some physical objects discussed in the recent literature such are the frustration-limited-clusters and other soft-structures (Melcuk et al. 1995) postulated to exist in glass-forming liquids (Kivelson et al. 1994, Fischer 1993), the origin of which has been pursued on mesoscopic scales (Kawasaki 1995). In the light of the present results such objects could be thought of as the result of ordering processes within the normal or SCL liquid phases which would end up forming a disordered crystal if the relative strengths of the relevant interactions could be tunned in to the required level, being their rather short lifetimes a mere consequence of the failure to stabilize a disordered cystal phase.

3 Dynamics

Once the relevance of studying the disordered crystal phases as models for structural glasses is established, the dynamic behaviour of those phases was investigated. As shown in Figure 5 where the low-frequency reduced Raman intensities for the ODC, structural glass and SCP are plotted, the main differences between the structural glass and ODC spectra regards, apart from the presence in the former as well as in the SCP of a strong Rayleigh component arising from macroscopic inhomogeneities, the appearance of a broad feature, usually referred as the "Boson peak" covering a frequency range up to 20 meV, and shows a maximum at about 6 meV in the glass and about 8 meV in the ODC.

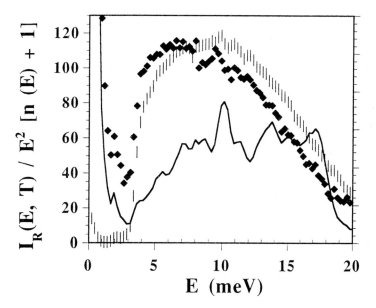

Fig. 5. Low-frequency reduced Raman intensities for the monoclinic (SCP) crystal (solid line), structural glass (filled lozenges) and ODC (vertical bars). The measurements correspond to a temperature of 14 K.

Since such frequency distributions are expected to be related with the low temperature heat capacity, several measurements of such thermodynamic function were carried out (Ramos et al. 1996), the most representative results shown in Figure 6.

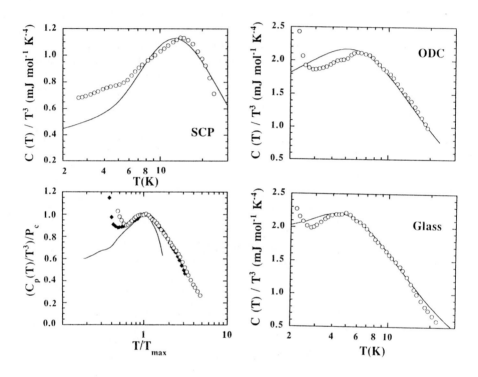

Fig. 6. Low-temperature heat capacity curves drawn as $C_p(T)/T^3$ for the stable (SCP) crystal, disordered (ODC) and glassy solids as measured by adiabatic calorimetry (symbols). The solid lines represent values for the same functions calculated from the generalized vibrational frequency distributions as measured by inelastic neutron scattering (see below). The lower-left frame shows the scaling of the heat capacity for the three solids scaled by the position of the maxima T_{max} and its height P_c (Liu and von Löhneysen 1996). Filled symbols represent data for the ODC ,open ones for the structural glass and the solid line for the stable crystal.

The most remarkable fact about the $C_p(T)/T^3$ curves displayed in Figure 6 regards the close proximity of data regarding the glassy and ODC solids. As seen there, the height of maxima (labelled as P_c in the sequel) of the SCP crystal, is about half of that of the glass and about 0.6 times that of the disordered crystal, whereas the corresponding temperatures are of ≈ 15

K (SCP), ≈ 6.5 K (ODC) and ≈ 4.9 K (glass). The shape of the curves for the glass and ODC solids conform to that shown by structurally disordered matter or even some other Complex Systems such as crystals supporting a charge (CDW) (or a spin (SDW)) density wave (see for instance Biljakovic et al. 1986), where the maximum is usually attributed to a phason contribution and the lower temperature riseup, as in glassy solids, shows a linear temperature dependence. The latter commonly attributed in glasses to tunnelling between two-level-systems (TLS) seems to arise in CDW crystals from phonon activated excitations between metastable CDW energy levels. What seems here important in the light of recent empirical results (Liu and von Löhneysen 1996) which show that regardless of whether the material is in crystal or amorphous form the two parameters characterizing the maxima in $C_p(T)/T^3$ scale as $P_c \propto T_{max}^{-1.6}$ is the fact that a scaling rather close to that $P_c \propto T_{max}^{-1.52}$ is found from data measured so far (for three hydrogenated and three fully deuterated solids), and that the correlations for the hydrogenated or deuterated samples taken independently seem to follow an exact power-law dependence. The finding of such correlations can also be pursued into the frequency domain since, as described previously by us (Bermejo et al. 1994(a)) a related power-law dependence was found for a restricted number of materials, between the frequency where the "boson peak" shows its maxima and that of the first intense inelastic peak in the corresponding crystalline solids, which in turn, scales with the "fragility" index of the materials in glassy form. Since such a scaling only depends upon material constants (i.e. density and stiffness) it is then not surprising to see how the curves for the glass and ODC solids are basically superimposable from $T/T_{max} = 0.6$ onwards. Also and in full analogy with what was found by comparison of several forms of silica with varying density with some crystalline polymorphs (Liu et al. 1995), once the scaling of $(C_p(T)/T^3)/T_{max}$ vs. T/T_{max} is done, the most relevant differences between the stable crystals and glass forms regard the absence of the low-temperature (TLS) contribution as well as the narrower distribution for the crystal.

Finally, the close similitude between the dynamic behaviour of glassy and ODC solids is vividly exemplified by data shown in Figure 7 which constitute a summary of the inelastic neutron scattering (INS) studies on such materials.

As seen in the upper frame of Figure 7, while a large difference between the INS spectra of glass and ODC versus the stable crystal is readily seen for frequencies below about 5 meV, only a rather small one differentiates the spectra of glassy and ODC phases, and that is located at frequencies about 2 meV. A clearer assessment of the latter can be seen from the reconstructed $Z(E)$ generalized frequency distributions shown in the lower frame of Figure 7. From there, it is seen that in partial accordance with the Raman data shown in Figure 5, the low-frequency peak for the ODC solid is located somewhat above that of the structural glass and that the latter shows a small but significant excess of intensity below 3 meV.

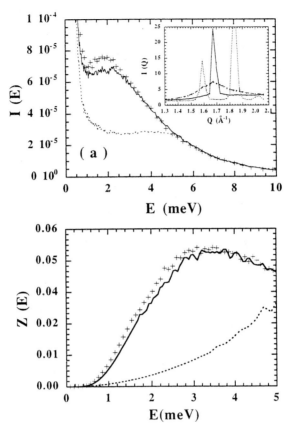

Fig. 7. The top frame shows a comparison between the angle-averaged spectral distributions from a sample of fully deuterated ethanol in fully amorphous (crosses) ODC (solid) and SCP (dots) forms. The choosing of a deuterated sampled enabled to attest the ordering state of the solids as shown in the inset which displays the FSDP for the glass (dashes with vertical bars) ODC (solid) and SCP (dots) , see above for details. The lower frame shows the low-frequency portions of the $Z(E)$ generalized frequency distributions for glass (crosses) ODC (solid) and SCP (dots) for a fully hydrogenated (i.e. incoherent neutron scatterer) sample. Notice that these $Z(E)$ are the ones used to calculate within the harmonic approximation the $C_p(T)/T^3$ curves shown in Figure 6.

What seems also worth pointing is the remarkable appearance of glassy behaviour in the rotator crystal phase, as judged by the large number of "excess modes" clearly seen in Figure 8 by comparison with the frequency distribution of the stable crystal measured at T=105K, that is above the calorimetric T_g and close to the onset of the steep increase in the SPC crystal $C_p(T)$. The finding of glassiness within the rotator phase crystal provides a clear physical realization of some simple models like the hard-needle one

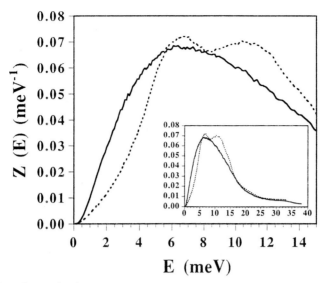

Fig. 8. Experimental vibrational frequency distributions for the SPC (dots) and rotator (RPC) phases at 105 K. The inset shows the comparison between both densities of states across the accesible range of energy-transfers.

(Renner et al. 1995), that is a model of infinitely thin, non-interacting needles with their centers-of-mass fixed to the sites of a cubic lattice. Such a simple, analytically tractable system has also been shown to exhibit a glass-transition of a purely dynamic origin, which occurs when the ratio l of needle length to lattice constant approaches a critical value, a transition which is then dominated by the number of colliding neighbours. The found transition reminds that found in mixed $Ar_{1-x}(N_2)_x$ crystals where the orientational degrees of freedom freeze gradually upon increasing x, the percentual composition of dumbells (Müser and Nielaba 1995). Decreasing the temperature has, in the system here analyzed, the same role than increasing l or x in the two systems just commented on, and therefore it seems to constitute a nearly ideal benchmark against which predictions regarding glass-transitions could be scrutinized.

In summary, from structural as well as thermal and dynamic data it should become clear that *positional* randomness does not constitute a *conditio sine qua non* for the manifestation of glassy behaviour to a extent comparable with that observed for the fully disordered system, at least as far as microscopic scales are concerned. Rather, the present work pinpoints to competing interactions as well as to incommensurability, this taken in terms of the very disparate spatial and point-group symmetries, as the relevant ingredients giving rise to glassy dynamics.

4 Discussion and conclusions

The results herein presented may be of help to compare the predictions of kinetic-theories of the mode-coupling family (MCT) (Götze 1991) to systems other than the usual glass-forming liquids, where because of the highly complicated molecular or ionic structures, such tests have been carried in a way where many relevant degrees of freedom are unaccounted (Criado et al. 1994). As shown by the seminal paper of (Aksenov et al. 1987), the MCT approach could well be applied to structural solid-solid transitions often mediated by the rise within the critical domain of a strong quasi-elastic signal. While the spirit of such an approach has been incorporated within some phenomenological treatments such are those pertaining to the "soft-mode" descriptions of the dynamics of glasses (Karpov et al. 1987), it should become clear that a study on a real material showing a freezing transition such the RPC $\longrightarrow$ ODC one should serve as a benchmark to perform such a test on the predictive capabilities of several MCT scenarios under full control of the relevant parameters. In fact, as shown in studies of model systems such as the ϕ^4 one oftentimes used for the description of structural phase transitions, a formal analogy can be found between MCT models which predict the existence of a dynamical transition at some point of the phase space and those derived from the use of ϕ^4 models (Duering et al. 1996) whose application to a rotational freezing transition can be made following well specified routes.

A particularly appealing endeavour would be to explore the possibility of finding a relevant order parameter(s) able to describe the RPC $\longrightarrow$ ODC transition in terms of a standard (Landau-type) approach following steps akin to those employed for the study of order-disorder transitions in other orientationally-disordered crystals (Lynden-Bell and Michel 1994). In such cases, the Landau free-energy can be written in terms of a set of order parameters usually taken as the mean values of $\bar{u}(\mathbf{Q})$ phonon displacement coordinates for the molecular centres-of-mass as well as $\bar{\Theta}(\mathbf{Q})$ rotator functions which specify the instantaneous orientation of one particle with respect to the lattice site. The interest of such an exercise would then be the possibility of describing a glass-like transition of a real material using well established tools of condensed matter sciences. As a matter of fact, once the order parameters have been set, all the relevant microscopic information (susceptibilities directly comparable with those measured by neutron scattering) should be available from the canonical partition function calculable from a microscopic Hamiltonian such as (Lynden-Bell and Michel 1994)

$$H = V + T + \sum_{\mathbf{Q}} \bar{u}(\mathbf{Q})^{\dagger} \cdot G_u(\mathbf{Q}) + \sum_{\mathbf{Q}} \tilde{\Theta}(\mathbf{Q})^{\dagger} \cdot G_{\Theta}(\mathbf{Q}) \qquad (4)$$

which is expressed in terms of the potential and kinetic energies of the crystal, the G_u, G_{Θ} virtual fields which interact with the phonon and rotator order parameters and the tilde stands for collective phonon and rotator functions.

From there, following steps delineated by Lynden-Bell and Michel (Lynden-Bell and Michel 1994) the (macroscopic) Landau free eenergy is expressed in compact form in terms of the Helmholtz free energy A as

$$F = A - \sum_Q \left[G_u(\mathbf{Q})\bar{u}(\mathbf{Q})G_\Theta(\mathbf{Q})\bar{\Theta}(\mathbf{Q}) \right] \tag{5}$$

that is, a bilinear form of the order parameters which constitute the starting point for the calculation of a number of magnitudes relevant to glassy dynamics (eg. dielectric behaviour, softening of elastic constants etc.)

The results here discussed along with some preliminary measurements of the low-temperature thermal conductivity of ODC and SCP crystals (Ramos et al. 1996) where the heat conduction in the disordered crystal is shown to show also all the features exhibited by structural glasses, pinpoint the importance of localized excitations ("localons" in the therminology of (Krumhansl 1993)) whether these be molecular librations, internal-mode excitaions or some hybrid between these two, as the relevant modes limiting heat flow which otherwise would be transported ballistically. An example of the dominant role of localized, low-frequency modes as the main mechanism for scattering of heat carrying phonons in a structural glass has already been given (Bermejo et al. 1994(b)), as well as that of the coupling between internal and external modes *within a ordered crystal* as the main one generating a substantial thermal contraction at low temperatures (i.e. large negative Grüneisen parameter) (Ramos et al. 1995), a phenomenon often considered as one of the fingerprints of glassy dynamics. The importance of a truly microscopic consideration of low-frequency, localized modes is also of interest in the light of some recent proposals which go beyond the standard TLS phenomenology (Burin adn Kagan 1996) in order to account for the quantitative universality of some of the properties of amorphous solids such is the thermal conductivity outside of the "plateau" region (Yu and Leggett 1989), in terms of a model of collective excitations in an ensemble of defects with an internal degree of freedom.

As a final remark, it seems here pertinent to recall the analogy here established between long ranged ordered systems as are the RPC and ODC solids and those analyzed by (Chandra 1996), both exhibiting the most cherished fingerprints characterizing the glassy state but pointing towards the possibility of describing the transition in terms of a jump in some order parameter which in the case of the Josephson array turns out to be identical to the Edwards-Anderson one whereas in the crystals here discussed would certainly have to be related to some statistical average of the $\bar{\Theta}(\mathbf{Q})$ rotator functions.

5 Acknowledgements

The authors wish to express their gratitude for the substantial help given by A. Criado, (Univ.Seville), A. de Andrés (ICMM-CSIC), Prof. S. Vieira

(UAM, Madrid), H. Schober (ILL), J. Zúñiga (UPV, Bilbao), M.A. González (ILL), R.Fernández-Perea and M.L. Senent (IEM-CSIC), C.K. Loong and D.L. Price (ANL). Work performed in part under D.G.I.C.Y.T (Spain) grant PB92-0114-C03-01.

References

Aksenov V.L., Bobeth M., Plakida N.M., Schreiber J. (1987): J.Phys. **C20** 375

Bermejo F.J., Criado A., Martinez J.L. (1994): Phys.Lett. **A195** 236

Bermejo F.J., Garcia-Hernandez M., Martinez J.L., Enciso E., Criado A. (1994): Phys.Rev.**B49** 8689

Biljakovik K., Lasjaunias J.C., Zougmore F., Monceau P., Levy F., Bernard L., Currat R. (1986): Phys.Rev.Lett. **57** 1907

Burin A.L., Kagan Y. (1996): Phys.Lett. **A215** 191

Chandra P. These proceedings and Chandra P., Feigelman M.V., Ioffe L.B. (1996): Phys.Rev.Lett. **76** 4805 and references therein.

Criado A., Bermejo F.J., de Andrés A. Martinez J.L. (1994): Molec.Phys. **82** 787, and Bermejo F.J. , Criado A., de Andrés A., Enciso E., Schober H. (1996): Phys.Rev. **B53** 5259. For a critique of some of the tests of MCT predictions without fully accounting for internal and rotational modes see.

Duering E.R., Schilling R., Wittman H.P. (1996): Zeit.Phys. **B100** 409

Esquinazi P., KönigŔ.,Pobell F. Physica **B 219-220**, 247, and references therein.

Fayos R., Bermejo F.J., Dawidowski J., Fischer H.E., González M.A. , Criado A. (1996): Phys.Rev. (submitted)

Fischer E.W. (1993): Physica **A201** 183 and references therein.

Gaskell P.H., Wallis D.J. (1996): Phys.Rev.Lett. **76** 66

Götze W. (1991): In : Hansen J.P. el at.in *Liquids, freezing and the glass transition*, North Holland, Amsterdam

Haida O., Suga H. , Seki S. (1972): J.Chem.Thermodyn. **9**, 1133

Haida O., Seki S. (1980): Farad.Dissc. **69** 221

Jäckle J. (1986): Rep.Prog.Phys. **49** 171

Karpov V.G., Klinger M.I., Ignatiev F.N. (1987): Sov.Phys.JETP **57**, 439

Kawasaki K. (1995): Physica **A217** 124

Kivelson S., Zhao X., Kivelson D., Fischer T.M., Knobler C.M. (1994): J.chem.Phys. **101** 2391

Krumhansl J.A. (1993): in *Phonon Scattering in Condensed Matter VII*, Meissner M., Pohl R.O. (Eds.) Springer Series in Solid State Sciences 112, Springer 1993, pag.3.

Liu X., von Löhneysen H., Weiss G., Arndt J. (1995): Zeit.Phys.**B99** 49

Liu X., von Lohneysen H. (1996): Europhys.Lett. **33** 617

Loidl A, Böhmer R (1994): in *Disorder Effects on Relaxational Processes*, Richert R. and Blumen A (Eds.), Springer, Berlin, pag. 659

Lynden-Bell R.M., Michel K.H. (1994): Rev.Mod.Phys.**66** 721

Melcuk A.I., Ramos R.A., Gould H., Klein W., Mountain R.D. (1995): Phys.Rev.Lett. **75** 2522

Müser M.H., Nielaba P (1995): Phys.Rev.**B52** 7201. See also Xie J ., Knorr K. (1994): Phys.Rev.**B50** 12977

Ramos M.A., Zou Q.W, Vieira Bermejo F.J. (1996): *Proc. XXI International Conf. on Low.Temp.Physics*, Prague, 1996. See also *Proc. Workshop on Non-Equilibrium phenomena in Supercooled liquids, Glasses and Amorphous Materials*, World Scientific (in press)

Ramos M.A., Vieira S., Bermejo F.J., Martinez J.L. (1995): Molec.Phys. **85** 1037

Renner C., Löwen H., Barrat J.L. (1995): Phys.Rev.**E52** 5091

Ritort F. (1995): Phys.Rev.Lett. **75** 1190. See also Kim B.J., Jeon G.S.,Choi M.Y. (1996): Phys.Rev.Lett. **76** 4648

Salmon P.S. (1994): Proc.R.Soc.Lond. **A445** 351

Shumway S.L.,Sethna J.P. (1991): Phys.Rev.Lett. **67**, 995

Srinivasan A., Bermejo F.J., de Andrés A., Dawidowski J., Zúñiga J., Criado A. (1996): Phys.Rev. **B53** 8172

Van de Walle C.G. (1996): Phys.Rev. **B53**, 11292

Watson S.K. (1996): Phys.Rev.Lett. **75** 1965

White B.E., Pohl R.O. (1996): Zeit.Phys.**B100**, 401

Wolf D., Wang J., Phillpot S.R., Gleiter H. (1995): Phys.Lett. **A205**, 274

Yu C.C.. Leggett A.J. (1989): Comm.Cond.Matter Phys. **14** 231. Also Leggett A.J. (1991): Physica **B169** 322

Microscopic Dynamics of $A_1 C_{60}$ Compounds

H. Schober[1], B. Renker[2], R. Heid[2], F. Gompf[2] and A. Tölle[1]

[1] Institut Laue-Langevin, F-38042 Grenoble, France
[2] INFP, Forschungszentrum Karlsruhe, D-76021 Karlsruhe, Germany

Abstract. We present the dynamics of $A_1 C_{60}$ compounds as investigated by in-elastic neutron scattering using $Rb_1 C_{60}$ as example. Special emphasis is given to the influence of disorder on the dynamical response in the various phases. In the high-temperature fcc phases we find hardly any differences in the dynamics between $Rb_1 C_{60}$ compounds and pristine C_{60}, i.e. we observe the typical signatures of a plastic crystal phase. In particular, it turns out that at equal temperatures the rotational diffusion constants for pristine C_{60} and $Rb_1 C_{60}$ ($D_r = 2.4 \ 10^{10} \ s^{-1}$ at 400 K) do not differ within the experimental errors. Upon cooling we witness strong changes in the inelastic spectra indicative of intra-molecular bond formation. The build-up of intensities in the gap region separating internal and external vibrations in pure C_{60} is the most striking of these changes. The details of the spectra in the low-temperature phases are compatible with the formation of linear polymers (upon slow cooling from the fcc phase) and dimers (upon fast cooling), respectively. Using lattice dynamical model calculations reliable information about the restoring forces is obtained. We find that the inter-cage bonds are weaker in the dimer than in the polymer. The same holds for the cage deformations. The conversion of the metastable dimer phase into the polymer phase is followed in real time and turns out rather complex. The transition between the polymer and fcc phases is accompanied by inelastic precursor effects which show a strong hysteresis. These precursors can be identified with fast reorientational movements of monomers. There is strong evidence that the polymer phase is heterogeneous at elevated temperature, containing monomer regions, the amount of which is time and temperature dependent and can be determined quantitatively from the inelastic spectra.

1 Introduction

$A_1 C_{60}$ (A=K,Rb,Cs) compounds exist in a great variety of phases which differ appreciably in their physical properties. This relative richness of the phase diagram, if compared with other alkali-metal-doped C_{60} crystals, has its origin in the solid state chemical reactions leading to covalent bonding between the C_{60} cages. Several experimental techniques have been combined to characterize those phases. X-ray diffraction (Stephens et al. 1994, Pekker et al. 1994, Chauvet et al. 1994, Zhu et al. 1995, Oszlány et al. 1995, Faigel et al. 1995) has been used successfully to determine the structural parameters. It turns out that at sufficiently high temperatures ($\approx$ 400K for $Rb_1 C_{60}$) all $A_1 C_{60}$ compounds form fcc crystals with the alkali-ions occupying the octahedral interstitial sites. When these materials are slowly cooled the fcc structures deform into orthorhombic structures via compression along one of the cubic < 110 > directions. The resulting short center-to-center separations of the cages along the orthorhombic **a** axis of about 9.1 Å (compared to approximately 10 Å in most other fullerides) strongly suggest that the C_{60} molecules combine to form linear polymeric chains. These polymer phases are thermodynamically stable for Rb_1C_{60} and Cs_1C_{60}. They also have the outstanding property to be chemically stable when exposed to air (in contrast to other alkali-fullerides). By rapidly quenching A_1C_{60} crystals from the high temperature fcc phase polymer formation can be suppressed. The diffraction patterns of this new phase show superstructure peaks indicative of $(C_{60})_2^{-2}$ dimer formation (Zhu et al. 1995, Oszlány et al. 1995). The center-to-center separation of 9.3 Å in the dimer is appreciably larger than in the polymer. The dimer phase is thermodynamically metastable and transforms irreversibly to the polymer phase with a strongly temperature dependent transformation rate. In K_1C_{60} the situation is more complex. The fcc phase becomes thermodynamically unstable at about 440 K. The result is an intermediate phase featuring K_3C_{60} regions in a continuous C_{60} lattice.

The formation of inter-molecular bonds must lead to appreciable distortions of the cages in the vicinity of the bonds[1]. A confirmation of the polymer and dimer formation can, therefore, be provided by optical spectroscopy (Winter and Kuzmany 1995, Martin et al. 1995). Raman and infrared measurements show indeed strong variations in the C_{60} intra-molecular excitation spectrum signaling the break-down of icosahedral symmetry caused by inter-C_{60} covalent bond formation.

Quantum chemical calculations (Adams et al. 1994) favor a [2+2] cycloaddition mechanism for the polymer bond formation (see figure 1). Parallel double (6-6) bonds on adjacent C_{60} molecules break up to form 4-membered rings.

Although the polymer- and dimerization in these materials is by now well established many open questions remain concerning the formation processes themselves. It should be underlined that the traditional route to polymerization goes through solution phases. In the case of fullerenes this traditional route has so far been granted only limited success. Polymerization in a solid phase is rather unusual and introduces a high degree of translational order into the system. E.g. all the polymeric $(C_{60})_n$ chains and $(C_{60})_2$ dumbbells are collinear in a well ordered crystal structure. A lot less is known about the degree of orientational order of the chains or dumbbells. This is particularly true close to the phase transitions (polymer $\leftrightarrow$ fcc and dimer $\rightarrow$ polymer). As we will show a main source of disorder in these systems is the presence of a finite amount of monomers in the polymer phase. While in equilibrium the monomer content is only a function of temperature it strongly depends on time during the formation processes itself.

---

[1] As the icosahedral symmetry of the C_{60} molecule is incompatible with Bravais-lattice symmetries distortions may already be caused by the crystal fields. It is, however, known that those are sufficiently small to be neglected here.

Inelastic neutron scattering (INS) proves a very powerful tool to follow the phase transitions in detail. The dynamical range of a time-of-flight spectrometer covering the complete vibration spectrum is in this context of particular advantage.

The paper is structured as follows. We will start by giving a short outline of the experimental technique (INS) used. The first part of the discussion will be concerned with the dynamics of the different phases of A_1C_{60} compounds using Rb_1C_{60} as example. The results obtained there then enable us to comment in detail the changes observed during the various phase transitions. The notion of disorder will play a key role in this second part.

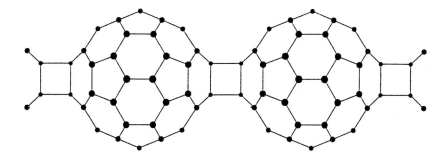

Fig. 1. Schematic presentation of a linear $(C_{60})_n$ polymeric chain and a $(C_{60})_2$ dimer (dumbbell) according to Adams et al. (1994). The monomer cages are bound by a [2+2] cycloaddition mechanism involving parallel 6-6 double bonds on adjacent monomers. In the orthorhombic structure the chains are aligned along the short **a** axis obtained by a compression along one of the cubic $< 110 >$ directions.

2   Experiment

We produced homogeneous A_1C_{60} samples by annealing stoichiometric quantities of C_{60} and A_6C_{60} powders at T $\approx$ 600 K for several days. The samples were investigated by INS using two spectrometers: the cold neutron instrument IN6 at the high-flux reactor of the Institute Laue-Langevin and the thermal neutron instrument DN6 at the Siloé reactor of the CENG (Grenoble

(France)). Both instruments posses multiple detectors and use the time-of-flight method to determine the energy transfers. Together they cover a broad range in energy and wave vector. This is essential for the investigation of the dynamics in fullerenes (for more details about the spectrometers see Renker et al. 1993a, Renker et al. 1996).

In order to avoid misunderstandings concerning the presentation of the data certain details concerning the data analysis should be mentioned. The discussion of the dynamics will be based either on the generalized susceptibility $\chi''[\omega]$ or the generalized density-of-states $G(\omega)$. The term *generalized* indicates that in both cases the contributions of the various ions are weighted by their respective scattering powers σ/M. The susceptibility (or linear dynamical response) $\chi''[\omega]$ is obtained from the scattering law $S(Q, \omega)$ by integration[2] over Q and division through the temperature factor (Lovesey 1984).

The vibrational density-of-states (VDOS) is a particularly useful quantity because of its direct physical meaning. For a monatomic system the VDOS is directly related to the one phonon part of the incoherent scattering function (Lovesey 1984). In the case of A_1C_{60} compounds we are, however, dealing with coherent scatterers. Therefore, in order to extract the VDOS from the INS spectra we are obliged to invoke the incoherent approximation, i.e. we assume:

$$\int_{Q_{min}}^{Q_{max}} \left(\frac{d^2\sigma}{d(\hbar\omega)d\Omega(Q)} \right)_{coh} dQ \approx \int_{Q_{min}}^{Q_{max}} \left(\frac{d^2\sigma}{d(\hbar\omega)d\Omega(Q)} \right)_{inc} dQ \quad (1)$$

for a sufficiently large Q-sampling. The integral on the right side refers to a hypothetical sample made up of incoherent scatterers. The validity of the incoherent approximation in the present case was established by several tests (see Schober et al. 1996). Multi-phonon contributions to $S(Q, \omega)$ are determined self-consistently.

To a very good approximation the quantity $\omega^{-1}\chi''[\omega]$ is proportional to $\omega^2 G(\omega)$ as Debye-Waller and multi-phonon corrections partially compensate each other (Wuttke et al. 1993). The main advantage of $\omega^{-1}\chi''[\omega]$ is its direct connection with the actual experimental data

$$\omega^{-1}\chi''[\omega] \propto S(\omega)/T \quad \text{for} \quad \hbar\omega << k_b T. \quad (2)$$

The relative errors in the data are thus preserved.

---

[2] Please note that due to the integration $\chi''[\omega]$ as defined here is not completely instrument independent. This poses no problem for the following discussion as $\chi''[\omega]$ will serve only to illustrate relative changes in the dynamics as a function of time or temperature and will never be used in absolute terms.

3   Dynamics of the Different Phases

3.1   Fcc Phase

As the symmetry $Fm\overline{3}m$ found for the high-temperature phase of A_1C_{60} compounds contains the four-fold axis not present in the point group of the C_{60} cages, this phase must be orientationally disordered. As in pristine C_{60} this disorder is of dynamical origin (Renker et al. 1993a, Neumann et al. 1991), i.e. the monomer cages perform fast reorientational motions. In the INS spectra these motions lead to strong quasi-elastic signals as shown in figure 2. The line-shape (sum over Lorentzians) and intensity of the quasi-elastic signal as a function of the wave vector transfer Q can be described very satisfactorily in the framework of a rotational diffusion model. In this classical model the rotations of adjacent molecules are assumed to be uncorrelated and statistically distributed, i.e. they can be described by a single function $p(\Omega, \Omega', t)$, which contains the information of finding the molecule in the orientation Ω at time t if at time 0 it was at Ω'. For the isotropic case this probability function is required to satisfies the differential equation:

$$D_{\text{rot}} \nabla_\Omega^2 p(\Omega, \Omega', t) = \frac{\partial}{\partial t} p(\Omega, \Omega', t). \tag{3}$$

The only free parameter of the model is the rotational diffusion constant D_{rot} which for Rb_1C_{60} at 400 K comes out as $D_{\text{rot}} = 2.4 \ 10^{10} \text{s}^{-1}$. This implies very fast reorientations, faster than the ones found for C_{60} in solution. No differences can be detected between the rotational diffusion speeds in Rb_1C_{60} and pristine C_{60} at equal temperatures. This is somewhat surprising as the inter-cage distances in the doped samples are appreciably reduced due to the Coulomb attractions. The distances found in A_1C_{60} at about 400 K (Zhu et al. 1995) are close to the ones in pristine C_{60} near the order-disorder transition (David et al. 1992). This result suggests that the rotational potential felt by the monomers is hardly affected by changes in the inter-cage distances or the presence of alkali-ions at the octahedral sites. It should be noted that it is impossible to unambiguously exclude any correlation between the motions of adjacent molecules on the sole ground of powder measurements (Pintschovious et al. 1995). However, if present correlations must be of very short-range nature. The fast reorientation of the C_{60} molecules in the absence of short-range orientational order is certainly very favorable for the formation of polymers. It brings together C_{60} molecules in all possible relative configurations, among which those featuring parallel double bonds then may be cycloadded (Pusztai et al. 1995).

In figure 3 we show the generalized density-of-states of Rb_1C_{60} in the fcc phase[3]. As expected for a monomer system the low-frequency vibrations for

---

[3] Please note that the excitations corresponding to the rotational degrees of freedom of the monomer are underestimated. The concept of a density-of-states is inadequate for the representation of diffusive motion.

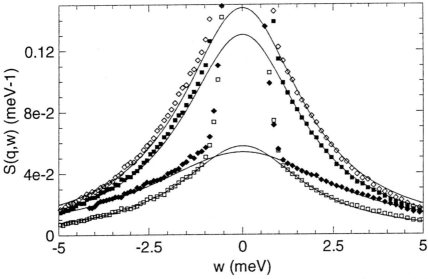

Fig. 2. Quasi-elastic intensities observed for Rb_1C_{60} at 400 K as a function of Q. The experiment was done using DN6 at an incident energy of 17.4 meV. The full lines are Lorentzian functions fitted to the data. The analysis of the quasi-elastic widths and intensities in terms of a rotational diffusion model gives a rotational diffusion constant of $D_{rot} = 2.4 \cdot 10^{10} s^{-1}$ to be compared with $D_{rot} = 1.8 \cdot 10^{10} s^{-1}$ for pristine C_{60} at 300 K.

which the cages can be considered as rigid are well separated from the internal vibrations. The gap reflects the two energy scales set by the weak intermolecular Van der Waals and Coulomb forces on one side and the covalent intra-molecular interactions on the other side.

3.2 Polymer Phase

When monomers add to form polymers the low-frequency spectrum changes completely. This applies, in particular, to the gap region separating external from internal vibrations. As can be seen from figure 3 the A_1C_{60} compounds are no exception to this general rule. A broad distribution of intensities is observed between 10 meV and 23 meV in the polymer phase while there are no intensities in this region in the fcc phase. These extra intensities translate the fact that due to the inter-cage bonding the inter- and intra-molecular energy scales come closer together. Although many features of the GDOS can already be understood in the framework of a linear chain quantitative estimates of the restoring forces need adequate lattice dynamical calculations. The dispersion curves of the polymer phase calculated with a rather simple model are shown in figure 4. In this model the cages are stabilized by a force field developed by Jishi et. al (1992). Two types of inter-cage interactions

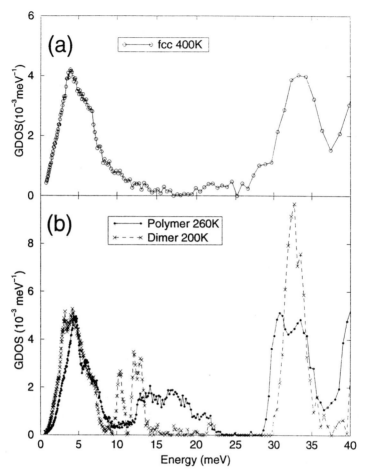

Fig. 3. Generalized density of states of Rb_1C_{60} as obtained with the spectrometer IN6 in the energy region $\hbar\omega < 40$ meV. Upper panel (a): high-temperature fcc phase. The rigid-molecule vibrations ($\hbar\omega < 8$ meV) are well separated from the cage deforming internal excitations by a gap (only the first internal modes corresponding to the $H_g(1)$-modes of icosahedral C_{60} are shown). Lower panel (b); low-temperature dimer and polymer phases. The presence of intensities in the gap region is a direct proof of covalent inter-cage bonds.

are present: Van der Waals forces, which can to a large extent be transferred from other doped fullerides, plus additional short-range interactions among the members of the four-fold ring forming the polymer bonds. The parameters after adjustment to the GDOS are given in tables 1 and 2. For more details about the models see (Schober et al. 1996).

The interactions between the cages are still sufficiently weak to identify the 18 dispersion branches in the region below 25 meV with external vi-

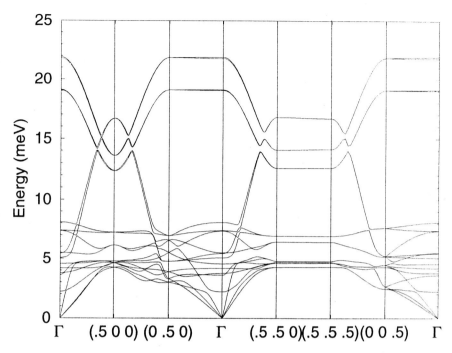

Fig. 4. Dispersion relation for the polymer phase of Rb_1C_{60} as obtained with a lattice dynamical model after adjustment of the parameters to the experimental density of states. The lattice geometry has been taken from Stephens et. al (1994) with two RbC_{60} units in the primitive cell. This leads to the 18 external dispersion curves shown. The polymer chains are aligned parallel to the orthorhombic **a** axis. Note the steep slop of the longitudinal branch along the chain direction. The highest external mode branches correspond to librational movements about axes perpendicular to the chain direction while the twisting modes of the chain itself are predicted by the model at surprisingly low frequencies.

brations of the cages. These branches are highly dispersive along the chain directions. The highest external branches correspond to librations of the cages about axes perpendicular to the chain direction while the chain twisting modes are predicted by the model at surprisingly low frequencies in the whole Brillouin zone leading to the peak in the density of states at 4 meV.

3.3 Dimer Phase

As can be seen in figure 3 the low-frequency part of the density-of-states in the dimer phase is characterized by three sharp peaks at 10.4 meV, 12.2 meV

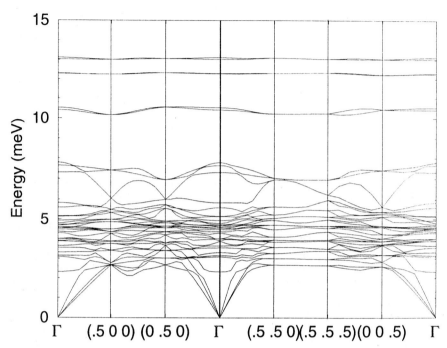

Fig. 5. Dispersion relation for the dimer phase of Rb_1C_{60}. The lattice geometry has been taken from Oszlány et. al (1995) with four RbC_{60} units in the primitive cell. This leads to the 36 external dispersion curves shown. In contrast to the polymer phase the optical branches are flat leading to sharp peaks in the density of states. The highest branches correspond to librations about axes perpendicular to the dumbbell long axis, in analogy to the polymer case. The dumbbell stretching mode is found at 10 meV. The acoustic spectrum is significantly softer than in the polymer as expected for a molecular system ($(C_{60})_2^{-2}$) bound by Van der Waals forces.

and 13.0 meV. According to the lattice dynamical calculations[4] (see figure 5) the 10.4 meV peak corresponds to a stretching mode of the dimer, while the two higher peaks have their origin in librations about axes perpendicular to the dimer bond. The restoring forces in the dimer are twice as weak as in the polymer (see table 2). This is by no means surprising given the appreciable differences of the center-to-center separations. Also very suggestive this finding does not allow us to unambiguously identify the bonds in the dimer as single bonds. When comparing the dimer with the polymer bonds it is important to note that we are dealing with anions. The symmetry requirements encountered by the donated electrons are completely different for $(C_{60})_2^{-2}$ and $(C_{60})_n^{-n}$. While in the polymer there is a one to one correspondence be-

---

[4] Lacking structural information on the geometry of a possible single bonded dimer the calculation was carried out for [2+2] cycloadded dimers.

Table 1. Van der Waals parameters for the different type of bonds. We use a Lennard-Jones parametrization of the form $V(r) = D[(\sigma/r)^{12} - 2(\sigma/r)^6]$.

Bond	σ (Å)	D (meV)
C-C	3.82	2.85
Rb-C	4.42	1.3
Rb-Rb	5.11	1.95

Table 2. Force constants for the polymer bonds. The angle force constants are defined as $K = 1/(r_1 r_2)\partial^2\Phi/\partial\alpha^2$, i.e. as the second derivative of the lattice potential Φ with respect to the angle α, devided by the lengths r_1 and r_2 of both bonds attached to the angle. They therefore possess the same units as the longitudinal Born von Karman force constant F_{inter}. K_1 refers to the 90 degree angles inside the 4-fold inter-cage rings and K_2 to the angles between intra- and inter-cage bonds.

Parameter	Force constant (10^3 dyn/cm)	
	Polymer	Dimer
F_{inter}	200	100
K_1	40	23
K_2	30	20

tween the number of donated electrons and the number of inter-cage bonds there are two donated electrons per inter-cage bond in the dimer. Differences in the inter-cage bond strengths are, therefore, to be expected. Recent semi-empirical calculations by Kürti and Németh (1996) seem to indicate, that the charged dimers are bound by single bonds while neutral dimers feature four-fold rings, analogous to the polymers.

The weaker inter-cage bonds in the dimer mean that we are dealing with a system characterized by three nicely separated energy scales: weak Van der Waals forces between single dimers, intermediately strong forces between cages forming dimers and strong covalent forces between carbon atoms forming cages.

3.4 Internal Modes

So far we have not discussed the internal modes. In principle optical spectroscopy out-performs INS in resolution at elevated frequencies. However, as INS data are not sensitive to optical selection rules and cover the complete Brillouin zone they supply complementary information. In addition, the fact that the evolution of the internal and external mode spectrum can be observed simultaneously under exactly the same experimental conditions is an invaluable advantage for the study of the phase transitions.

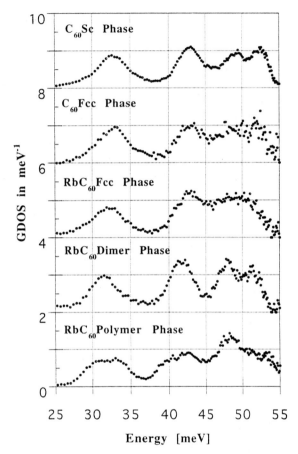

Fig. 6. Generalized density-of-states in the lower internal mode region for $Rb_1 C_{60}$ in the fcc (rotator), polymer and dimer phases and comparison with pristine C_{60} in the fcc and sc phases. As can be seen both charge transfer, albeit rather weakly, and cage deformation affect the internal modes. Compared to the fcc spectrum (which is rotationally broadened) the polymer spectrum changes appreciably more than the dimer spectrum. This may be taken as an indication for stronger cage deformations in the polymer.

In figure 6 we show the internal density of states in the region 25 meV < 55 meV for pristine C_{60} in the fcc and sc phases as well as for $Rb_1 C_{60}$ in the fcc, polymer and dimer phase. The data have been obtained in downscattering with the instrument DN6 using an incident energy of 70 meV. As can be seen by comparing pristine C_{60} with Rb_1C_{60} (both in the fcc orientationally disordered phases) the internal modes react, albeit rather weakly, to the charge transfer. This electron-phonon coupling was already observed for metallic Rb_3C_{60} where it is a lot stronger. On the other hand the internal modes clearly reflect the deformation of the cages. This is e.g. demonstrated by the splitting of the $H_g(1)$ mode at 33 meV in the polymer phase visible in figures 3 and 6. This splitting is absent in the fcc[5] and dimer phases. More generally it can be stated that, apart from some general softening, the dimer spectrum of Rb_1C_{60} strongly resembles the spectrum of pristine C_{60} in the orientationally ordered simple cubic phase, while an appreciable redistribution of intensities is observed for the polymer phase. This can be taken as a hint for less cage deformation in the middle part where the two monomers are connected, as predicted for charged dimers bound through single bonds (Kürti and Németh 1996). Care must, however, be taken not to over-interpret the internal mode changes. The various phase transitions in A_1C_{60} systems always involve a combination of both charge redistribution and deformation. Experimentally it is not possible to separate these two possible causes. Ab initio calculations of the vibration spectrum for all the charged systems would, therefore, be highly desirable.

4 Phase Transitions

Although the presence of polymer chains in A_1C_{60} compounds is by now firmly established little is known about the actual formation process. As we have outlined above the fingerprints left by the polymer phase in the INS spectra are highly specific. It is, therefore, possible to follow the formation process in real time under the condition that it is sufficiently slow to allow for adequate data statistics. We will demonstrate that this condition is met for A_1C_{60} systems.

4.1 Dimer $\longrightarrow$ Polymer

As already mentioned the dimer phase is meta-stable and converts irreversibly into the polymer phase. The conversion rate of this process is highly temperature dependent (Martin et al. 1994, Gránásy et al. 1996): about 18 hours at 260 K and 2 hours at 270 K. In figure 7 we show the time evolution of the generalized susceptibility $\omega^{-1}\chi''[\omega]$ of the quenched sample at 260 K. Each of the 6 curves corresponds to 3 hours of measuring time. We have chosen to decompose the spectrum into three parts which represent the different energy scales in the system.

The middle part[6] covering 8 meV $< \hbar\omega <$ 25 meV can be used as a direct measure for the quantity of cages bound either in dimers or polymers. As we

---

[5] The fcc signal is broadened due to the rotational diffusion of the cages.

[6] The energies are negative indicating that the data were taken in the energy loss mode.

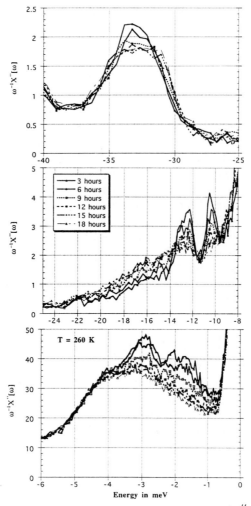

Fig. 7. Time evolution of the generalized susceptibility $\omega^{-1}\chi''[\omega]$ for the quenched Rb_1C_{60} sample at 260 K. Each curve corresponds to 3 hours of measuring time. Changes typical for the dimer $\longrightarrow$ polymer transformation are present on all three energy scales. Units in this and all other graphs showing $\omega^{-1}\chi''[\omega]$ are arbitrary.

have approached 260 K very slowly the amount of dimers as measured by the intensities of the 10 meV to 14 meV dimer peaks had already slightly decreased at the beginning of the 260 K measurement. As time goes on these peaks continue to become weaker signaling the gradual disappearance of the dimers. Remnants of the peaks can be observed even after 18 hours. The loss of the typical dimer intensities is accompanied by a build-up of intensities in the polymer region (12 meV $< \hbar\omega <$ 25 meV).

The evolution of the low-frequency spectrum ($\hbar\omega$ ¡ 8 meV) takes place in two steps. During the first 9 hours the structures at 2 meV and 3 meV typical for the dimer phase disappear leaving a broad hump which qualitatively already strongly resembles the final polymer spectrum. This hump then relaxes: fast during the next three hours and rather slowly afterwards.

In the case of the first internal modes ($H_g(1)$) the main variations happen during the first 9 hours[7] and thus correlate with the changes in the low frequency region.

If the transformation proceeded via the growth of polymer domains within the dimer phase then the spectra obtained for the intermediate stages could be expressed, to a very good approximation, as just a linear combination of the spectra obtained at the initial and final stages. This condition is violated for the runs taken after the first 6 hours in what concerns the low-frequency region of the spectra (see lower panel of figure 7). Therefore, at those stages the system has to be in a microscopically disordered state far from thermal equilibrium or a third phase must come into play. Microscopic disorder is most properly described by a statistical chain length distribution $p(n)$ which evolves gradually with time. E.g., more than 50 percent of the cages have become part of polymers after 9 hours. This seems to constitute a threshold beyond which we completely loose the dimer signatures in the low-frequency spectrum ($\hbar\omega < 8$ meV).

As we are dealing with very slow evolutions the disordered intermediate systems can be frozen into glassy states by quenching the samples to temperatures below 250 K.

The existence of transient cubic monomer phases during the transformation process is claimed by several authors (see Gránásy et al. 1996). This question is of great importance for the understanding of the polymer formation. On one hand freely rotating monomers should speed up the polymer formation by producing correct alignments of bonds on adjacent cages with a high probability. On the other hand the cubic symmetry of the monomer phase bears the possibility of large geometrical frustrations which may hinder a complete polymerization. As the pseudo-orthorhombic dimer phase already features a preferred direction along which to nucleate the growth of polymer chains geometrical frustrations plays no role in the absence of a monomer phase. In the inelastic spectra at 260 K we observe, at no time, strongly enhanced signals below 2 meV which would indicate fast reorientations of monomer cages. Therefore, if the polymer formation process includes the break-up of dimers into monomers, these monomers are either not rapidly reorienting or, if rapidly reorienting, instantaneously[8] rebound into polymers. In the case of dimers bound by single bonds such a break-up into monomers

---

[7] Due to the poorer resolution in the case of this experiment as compared to the previous one the nice splitting of the $H_g(1)$ modes shown in figure 3 for the polymer phase reduces to a broadening.

[8] On the time scale of the complete transformation process.

becomes a precondition for polymerization due to the fact that the singly
bound dimers cannot enter another bonding.

4.2   FCC $\longleftrightarrow$ Polymer

The fcc $\longleftrightarrow$ polymer transition in A_1C_{60} shows a strong hysteresis. It is,
therefore, essential to follow the evolution of the dynamics both upon heating
of the polymer phase and cooling from the fcc phase. The situation is rendered
even more complicated by the fact that, as we will show, the polymerization
process is highly time dependent and complete thermodynamic equilibrium
may be reached only after several hours.

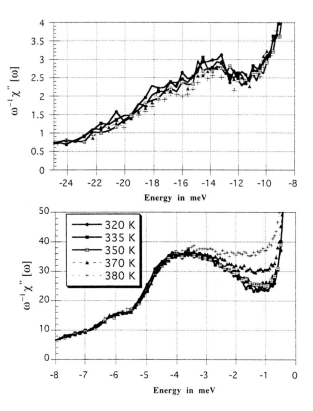

Fig. 8. Evolution of the generalized susceptibility $\omega^{-1}\chi''[\omega]$ for Rb_1C_{60} upon heating. The sample was fully equilibrated at the beginning of the measurements. Each
run took 2 hours. In the complete temperature range investigated no kinetic effects
could be detected on the time scale of the experiment. As can be seen the susceptibilities do not change up to 350 K. The system behaves harmonically. Starting
at about 350 K an appreciable build-up of inelastic intensities is observed in the
region below 4 meV (lower panel).

We have performed several series of measurements. Figure 8 shows the generalized susceptibility $\omega^{-1}\chi''[\omega]$ for consecutive two hour runs starting from the fully equilibrated sample. Below 350 K we observe no changes in $\omega^{-1}\chi''[\omega]$, i.e. the system behaves harmonically. Above 350 K intensities build-up stronger than expected for a harmonic system in the region below 4 meV.

An important point for the interpretation of the dynamics is the question of thermal equilibrium. As we do not witness any evolution of the spectra as a function of time we consider that the sample is equilibrated during the heating part of the cycle for the temperatures shown in figure 8. This is definitely not the case when cooling from the fcc phase. Figure 9 summarizes the results of a cooling series starting at 430 K. 45 minute runs separated by 15 minute cooling intervals have been taken in the order 430 K, 400 K, 360 K, 370 K, 350 K, 330 K and 300 K. Above 400 K the generalized susceptibilities $\omega^{-1}\chi''[\omega]$ are typical for the rotator phase: strong quasi-elastic scattering at low frequencies and vanishing gap intensities. The gap intensities typical for the polymers build-up rather abruptly upon cooling and later evolve slowly. The same holds for the low-frequency part, where we witness a rather abrupt loss of quasi-elastic intensity below 400 K. The susceptibilities relax afterwards visibly over several hours and down to 330 K. Thermal equilibrium is certainly not achieved for the 360 K measurement as clearly evidenced by the inversion of the 370 K and 360 K curves. The 360 K run preceded the 370 K run thus leaving the system less time to evolve.

In figure 10 we show the variation of the susceptibility relative to the low-temperature function. The results have been scaled to a common area in the dominantly quasi-elastic region from -4 meV to 2 meV. Please note that due to the scaling the relative errors for the lower temperature data are relatively high. As can be seen, all curves, inclusive the one obtained for the 430 K rotator phase, coincide within the experimental uncertainties. This observation holds both in the low-frequency quasi-elastic (lower panel) region and the gap region (upper panel). The asymmetry of the quasi-elastic signal arises from the Q-integration, which covers inequivalent regions on the down- and up-scattering side. This result constitutes clear evidence for the presence of monomer clusters if one takes into consideration the following facts:

1. The contributions to the susceptibilities from ideal harmonic vibrations are independent of temperature. They, therefore, do not change as long as we stay with a well-defined phase.
2. The line-shape of the quasi-elastic signal is basically determined by the rotation speed and does change very little with temperature (see section 3.1). It can be assumed constant in the T-range of interest here.
3. The intensities of the contributions to the susceptibilities arising from rotational diffusive motion scale with T^{-1} for the temperatures and widths considered here. For the small spread ΔT of about 50 K compared with a T of about 350 K the variations are small.

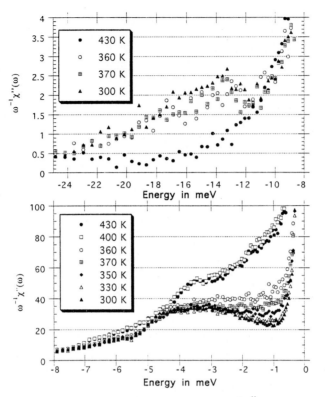

Fig. 9. Evolution of the generalized susceptibility $\omega^{-1}\chi''[\omega]$ for Rb_1C_{60} upon cooling. One hour runs have been performed in the order indicated in the legend. Above 400 K the sample is in the rotator phase as demonstrated by the quasi-elastic scattering at low frequencies (lower panel). Polymerization as measured by the gap intensities (upper panel) sets in rather abruptly. However, the system takes several hours to achieve equilibrium as seen in the low-frequency part. This is most clearly evidenced by the inversion of the 360 K and 370 K curves. The 360 K run was done prior to the 370 K run leaving the system more time for relaxation.

Therefore, the susceptibility of a system consisting of monomer clusters embedded in a polymer matrix should be a superposition of the low-temperature polymer and high-temperature rotator results. And that must hold for all temperatures. This is exactly what we observe.

It is impossible to predict the size of the monomer regions from the inelastic data. However, the observation that the differences in figure 10 scale over the complete low-frequency spectrum including the collective translational excitations is incompatible with clusters containing only very few monomers. From the scaling factors of the quasi-elastic intensities, it is possible to determine the percentage of monomers present in the system as a function of time and temperature. The result is summarized in table 3 after correction

Table 3. Amount of monomers in the polymer phase. For the cooling sequence we indicate the mean time elapsed between the 430 K measurement and the data acquisition. The 300 K (cooling) and 320 K (heating) spectra are taken as the respective reference points, i.e. zero percent monomers. The experimental errors are about 20 %.

Temperature [K]	Time elapsed [min]	Fraction of monomers
360 (cooling)	15	0.29
370 (cooling)	60	0.15
350 (cooling)	120	0.07
330 (cooling)	180	0.02
350 (heating)	—	0.02
370 (heating)	—	0.09
385 (heating)	—	0.21

for the T^{-1} dependence discussed above. In summary we find that the polymer phase of Rb_1C_{60} becomes unstable at about 350 K making way for a heterogeneous phase consisting of monomer clusters embedded in a polymer matrix. The monomer content changes with temperature up to 390 K. This heterogeneous phase shows no signs of instability over several hours and the amount of monomers may reach 20 percent. Upon cooling the system takes a long time to reach equilibrium. More experiments are necessary to determine the T-dependence of the conversion rates. The polymer content slowly increases at the expense of the monomer clusters. E.g., after 30 minutes at 360 K close to 30 percent of the cages are still not bound to polymers. The onset of excess intensities upon heating correlates exactly with the observation of endothermic signals in the differential scanning calorimetry measurements by Gránásy et al. (1996). The DSC results are on the other hand confirmed by the Monte-Carlo-type calculations of Pusztai et al. (1995) which based on known values of activation energies predict the presence of monomers.

It is interesting to note that the changes in the dynamics observed here resemble in many aspects the ones found in glass-forming polymers (Zorn et al. 1995). As in our case the changes have their origin in the heterogeneity one may ask the same question for the glass-formers.

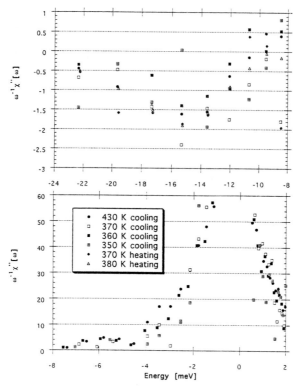

Fig. 10. Difference spectra of the generalized susceptibility $\omega^{-1}\chi''[\omega]$ for Rb_1C_{60} as obtained by subtracting the lowest temperature results and after scaling to a common area in the range from -4 meV $< \hbar\omega <$ 2 meV. As can be seen the spectra coincide within the experimental uncertainties both in the quasi-elastic region (lower-panel) and in the gap region (upper panel). This scaling property extends to the collective translational excitations, e.g. around -6 meV, and is a clear sign of the presence of extended monomer clusters within the polymer phase.

5 Conclusion

We have shown that by studying the dynamics of A_1C_{60} materials it is possible to obtain detailed information about the chemical and physical mechanism acting in these system. It turns out, that disorder plays a key role in all the intermediary stages of the transition processes. The possibility of freezing those intermediary stages at will, and thus to tune the degree of disorder in a well-defined way in the future opens up the field for more general investigations. The high symmetry of the molecule combined with the fact that all processes take place in a solid crystalline state convey, to our opinion, a model character for studying ill-ordered materials upon those systems.

References

Adams, G.B., Page, J.B., Sankey, O.F., O'Keefe, M., Phys. Rev. B **50**, 17471 (1994)

Chauvet, O., Oszlànyi, G., Forro, L., Stephens, P.W., Tegze, M., Faigel, G., Jànossy, A., Phys. Rev. Lett. **72**, 2721 (1994)

David, W.,I.,F., Ibberson, R.M., Dennis, T.J.S., Hare, J.P., Prassides, K., Europhys. Lett. **18**, 219 (1992)

Faigel, G., Bortel, G., Tegze, M., Gránásy, L., Pekker, S., Oszlányi, G., Chauvet, O., Baumgartner, G., Forró, L., Stephens, P.W., Mihály, G., Jánossy, A., Phys. Rev. B **52**, 3199 (1995)

Gránásy, L., Kemény, T., Oszlány, G., Bortel, G., Faigel, G., Tegze, M., Pekker, S., Forró, L., Jánossy, A., Solid State Commun. **97**, 573 (1996)

Lovesey, S.W., *Theory of Neutron Scattering from Condensed Matter*, Vol. 1, Oxford Science Publishers, Oxford p. 121, 1984.

Martin, M.C., Koller, D., Du, X., Stephens, P.W., Mihály, L., Phys. Rev. B **49**, 10818 (1994)

Martin, M.C., Koller, D., Rosenberg, A., Kendizora, C., Mihaly, L., Phys. Rev. B **51**, 3210 (1995)

Neumann, D.A., Copley, J.R.D., Cappelleti, R.L., Kamitakahara, W.A., Lindstrom, R.M., Creegan, K.M., Cox, D.M., Romanow, W.J., Coustel, N., McCauley, J.P., Maliszewskyj, N.C., Fischer, J.E., Smith, A.B., Phys. Rev. Lett. **67**, 3808 (1991)

Kürti, J., Németh, K., Chem. Phys. Lett. **256**, 119 (1996)

Oszlányi, G., Bortel, G., Faigel, G., Tegze, M., Gránásy, L., Pekker, S., Stephens, P.W., Bendele, G., Dinnebier, R., Mihály, G., Jánossy, A., Chauvet, O., Forró, L., Phys. Rev. B **51**, 12228 (1995)

Pekker, S., Forró, L., Mihály, L., Jánossy, A., Solid State Commun. **90**, 349 (1994)

Pintschovious, L., Chaplot, S.L., Roth, G., Heger,G., Phys. Rev. Lett. **75**, 2843 (1995).

Pusztai, T., Faigel, G., Gránásy, L., Tegze, M., Pekker, S., Europhys. Lett. **32**, 721 (1995)

Renker, B., Gompf, F., Heid, R., Adelmann, P., Heiming, A., Reichardt, W., Roth, G., Schober, H., Rietschel, H., Z. Phys. B **90**, 325 (1993).

Renker, B., Gompf, F., Schober, H., Adelmann, P., Bornemann, H.J., Heid, R., Z. Phys. B **92**, 451 (1993).

Renker, B., Schober, H., Gompf, F., Heid, R., Ressouche, E., Phys. Rev. B **53**, 14701 (1996)

Schober, H., Renker, Heid, R., Gompf, F., Tölle, A., submitted to Phys. Rev. B

Stephens, P.W., Bortel, G., Faigel, G., Tegze, M., , Jánossy, A., Pekker, S., Oszlányi, G., Forró, L., Nature **370**, 636 (1994)

Winter J., and Kuzmany, H., Phys. Rev. B **52**, 7115 (1995)

Wuttke, J., Kiebel, M., Bartsch, E., Fujara, F., Petry, W., Sillescu, H., Z. Phys. B **91**, 357 (1993).

Zhu, Q., Cox, D.E., Fischer, J.E., Phys. Rev. B **51**, 3966 (1995)

Zorn, R. Arbe, A. Colmenero, J. Frick, B. Richter, D. Buchenau, U. Phys. Rev. B **52**, 781 (1995)

Dynamics of a Supercooled Lennard–Jones System: Qualitative and Quantitative Tests of Mode-Coupling Theory

Walter Kob and Markus Nauroth

Institut für Physik, Johannes Gutenberg-Universität, D-55099 Mainz, Germany

Abstract. Using a molecular dynamics computer simulation we investigate the dynamics of a supercooled binary Lennard-Jones mixture. At low temperatures this dynamics can be described very well by the ideal version of mode-coupling theory. In particular we find that at low temperatures the diffusion constants show a power-law behavior, that the intermediate scattering functions obey the time temperature superposition principle, and that the various relaxation times show a power-law behavior. By solving the wave-vector dependent mode-coupling equations we demonstrate that the prediction of the theory for the wave-vector dependence of the nonergodicity parameters and the r-dependence of the critical amplitudes are in good agreement with the one determined from the computer simulation, which shows that the theory is also able to make quantitative correct predictions.

1 Introduction

In the last decade there has been an impressive effort to understand the dynamics of strongly supercooled liquids. Because the times scales involved span about 10-12 decades in time the technical difficulties one faces when studying these systems experimentally are quite formidable. For the same reason also the theoretical description of these systems is very difficult and quite a few approaches have been put forward. One of the most successful theories is the so-called mode-coupling theory (MCT), proposed in 1984 and subsequently worked out mainly by Götze, Sjögren and coworkers [1]. A compendium of articles in which many of the test of MCT are collected can by found in Ref. [2]. Here we report some of the results of a molecular dynamics computer simulation of a binary Lennard-Jones liquid in which we investigated whether the low-temperature dynamics of this system is described correctly by MCT. In order to find out whether the agreement between the results of the simulation and the predictions of the theory are correct only in a *qualitative* or even in a *quantitative* way, we also solved the wave-vector dependent mode-coupling equations and compared their solutions with the results of the simulation.

2 Model and Numerical Details

The system we are considering is a binary mixture of Lennard-Jones particles (type A and type B) both types having the same mass m. The interaction of two particles of type α and β $(\alpha, \beta \in \{A, B\})$ is given by $V_{\alpha\beta}(r) = \epsilon_{\alpha\beta}\{(\sigma_{\alpha\beta}/r)^{12} - (\sigma_{\alpha\beta}/r)^6\}$. The parameters $\epsilon_{\alpha\beta}$ and $\sigma_{\alpha\beta}$ are given by $\epsilon_{AA} = 1.0$, $\sigma_{AA} = 1.0$, $\epsilon_{AB} = 1.5$, $\sigma_{AB} = 0.8$, $\epsilon_{BB} = 0.5$, and $\sigma_{BB} = 0.88$. In the following we will use σ_{AA} as the unit of length, ϵ_{AA} as the unit of energy (setting $k_B = 1$) and $(m\sigma_{AA}^2/48\epsilon_{AA})^{1/2}$ as the unit of time. The size of the simulation box was 9.4 and the number of A and B particles was 800 and 200, respectively. The equations of motion were integrated with the velocity form of the Verlet algorithm using a step size of 0.01 and 0.02 at high and low temperatures, respectively. More details concerning the simulation can be found in Ref. [3].

¿From a separate computer simulation we determined the temperature dependence of the partial structure factors [4]. Using these as input we solved the wave-vector dependent mode-coupling equations for the system [5] by considering 300 wave-vectors in the interval $[0, 40]$. Outside this interval the partial structure factors can be considered as constant and therefore do not contribute to the memory kernel. More details regarding these calculations can be found in Ref. [4].

3 Results

One of the main prediction of the idealized version of MCT is that there exists a critical temperature T_c at which the dynamics undergoes a transition from an ergodic behavior for $T > T_c$ to a nonergodic behavior for temperatures $T < T_c$. This means that for $T < T_c$ time correlation functions do not decay to zero any longer even at long times, and that transport quantities, such as the diffusion constant D, vanish. In particular MCT predicts that in the vicinity of T_c the temperature dependence of the diffusion constant is given by a power-law, i.e.

$$D \propto (T - T_c)^\gamma \tag{1}$$

where the exponent $\gamma > 1$ depends on the system, i.e. is not universal.

By measuring the mean squared displacement of a tagged particle, we have determined the diffusion constant for the A as well as for the B particles and tested whether at low temperature its temperature dependence is of the form of Eq. (1), were we used T_c as a fit parameter. In Fig. 1 we show the diffusion constant for the A and the B particles versus $T - T_c$ in a double logarithmic plot, with $T_c = 0.435$. From this figure we recognize that at low temperatures the diffusion constant shows indeed the power-law behavior predicted by MCT. Furthermore we notice that the critical temperature T_c is the same for the A and for the B particles, also this in accordance with MCT. The critical exponent γ is not quite the same for the two species, as it should be according to MCT, but the difference is only on the order of 15%, which is not a big discrepancy.

The dynamics of the system can be investigated well by means of the intermediate scattering function. In Fig. 2 we show the time dependence of

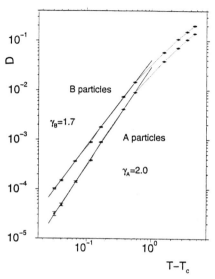

Fig. 1. Diffusion constant for the A and B particles versus $T - T_c$, with $T_c = 0.435$.

the incoherent part of the intermediate scattering function for all temperatures investigated. The value of q corresponds to the first peak in the static structure factor of the AA correlation function. ¿From this figure we see that at high temperatures the correlators decay essentially exponentially after the microscopic regime. For intermediate temperatures a shoulder becomes noticeable at intermediate times and at low temperatures this shoulder has developed into a plateau. Thus we find that the correlation function shows, at low temperatures, a two step relaxation behavior, in accordance with MCT.

We define the α-relaxation time $\tau(T)$ of a time correlation function as the time it takes the correlation function to decay to e^{-1} of its initial value. MCT predicts that at low temperatures a plot of a time correlation function versus the rescaled time $t/\tau(T)$ will yield a master curve in the α-relaxation regime, i.e. in the time regime for which the correlation functions start to decay below the above mentioned plateau. That this prediction of the theory is correct is shown in Fig. 3 where we plot the same correlation functions as in Fig. 2 versus the rescaled time $t/\tau(T)$.

MCT predicts that in the vicinity of the plateau the master curve is described well by the so-called β-correlator, a functional form provided by the theory and which shape depends on the so-called exponent parameter λ. We find that in the *late* β-relaxation regime, which is the time scale where the correlation function has fallen below the plateau, the master curve is indeed well described by the β-correlator (dashed curve in Fig. 3). However, in the *early* β-relaxation regime this is clearly not the case, indicating that for this system this prediction of the theory is not correct, likely because correction

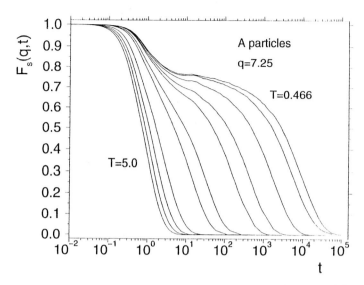

Fig. 2. Incoherent intermediate scattering function for the A particles versus time t for all temperatures investigated.

terms to the asymptotic scaling formula have not been taken into account.

For long times the theory predicts that the master curve is approximated well by the so-called Kohlrausch-Williams-Watts function $A \exp(-(t/\tau)^\beta)$. A fit to our data with this functional form showed that also this prediction of the theory is correct (dotted line in Fig. 3).

The theory also predicts that close to T_c the α-relaxation times $\tau(T)$ should show a power-law dependence on temperature with an exponent γ that is the same as the critical exponent for the diffusion constant (Eq. 1). We have computed this relaxation time for various time correlation functions, considering the different coherent and incoherent intermediate scattering functions for several values of q, and in Fig. 4 we show that at low temperatures they show indeed the predicted power-law behavior. The critical exponent is independent of the correlator, also this in agreement with MCT, and its value is around 2.6. This value is significantly different from the one found for the diffusion constant, see Fig. 1, in contradiction with the prediction of the theory that they should be the same. MCT also provides a connection between the exponent parameter λ, which determines the shape of the β-correlator (see Fig. 3), and the critical exponent γ. If this connection is used for the value of $\lambda = 0.78$, see Fig. 3, the theory predicts γ to be 2.7, in very good agreement with the value found in the simulation.

We have also tested many other predictions of the theory and found that MCT is able to give a surprisingly good description of the low-temperature dynamics of this system [3], [6]. In order to check whether this agreement is only of a *qualitative* nature or whether MCT is also able to make quantitative

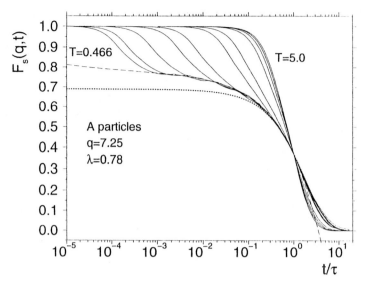

Fig. 3. Incoherent intermediate scattering function for the A particles versus rescaled time $t/\tau(T)$ for all temperatures investigated. Dashed curve: fit with the β-correlator proposed by MCT. Dotted curve: Fit with KWW law.

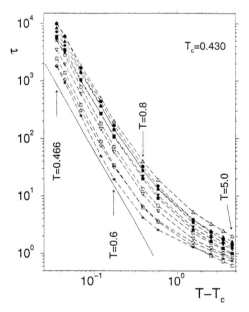

Fig. 4. Temperature dependence of various α-relaxation times. The straight solid line is a power-law with exponent 2.6.

correct predictions, we have solved the wave-vector dependent mode-coupling equations in the long time limit [4]. The only input in these calculations was the temperature dependence of the static structure factors which was determined from a computer simulation.

In Fig. 5 we show the q-dependence of the nonergodicity parameter for the coherent as well as the incoherent intermediate scattering function as measured in the simulation (dashed lines) as well as computed via MCT (solid lines). From this figure we see that the agreement between the simulation and the theory is surprisingly good for wave-vectors in the vicinity of the first peak and first minimum in the structure factor (note that there is no fit parameter involved!). For larger values of q the agreement is not as good, but still fair. The reason for the deteriorating agreement between MCT and the simulation is likely the fact that in the derivation of the mode-coupling equations a factorization ansatz has been made which is good for large distances, i.e.

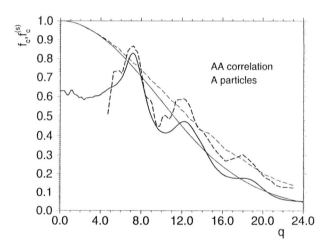

Fig. 5. Nonergodicity parameter for the incoherent intermediate scattering function (thin lines) and for the coherent intermediate scattering function (bold lines) as measured in the simulation (dashed lines) and as computed from MCT (solid lines).

small and intermediate values of q, but is probably unreliable for distances that are much smaller than the diameter of a particle, i.e. very large values of q. However, we have also shown [4] that leaving out the contributions to the mode-coupling equations that stem from large values of q altogether, leads to results that are significantly inferior than the present ones. This shows that, although not quite appropriate, these terms are important to make a quantitatively correct prediction.

¿From the solution of the mode-coupling equations it is also possible to compute the critical amplitudes $H_{ij}(r)$ $(i, j \in \{A, B\})$ [5]. These functions describe the relaxation behavior of the system in the β-relaxation regime. Often one reports not $H_{ij}(r)$, but $H_{ij}(r)/H_{ij}(r')$, where r' can be chosen arbitrarily, since this quotient is easier to measure than $H_{ij}(r)$ [3]. In Fig. 6 we show the quotient for the case of the AA correlation function (with $r' = 1.095$). ¿From the simulation we get a lower and upper bound for this function (thin solid lines) and the prediction of MCT is given by the bold dashed line. We see that for values of r that are not too small, the agreement between simulation and theory is very good and also small features, like the small dip at $r \approx 1.9$ is reproduced correctly. For small values of r the agreement between theory and simulation is not very good. The reason for this is likely the same that gave rise to similar discrepancies in the nonergodicity parameter at large wave-vectors, i.e. small distances.

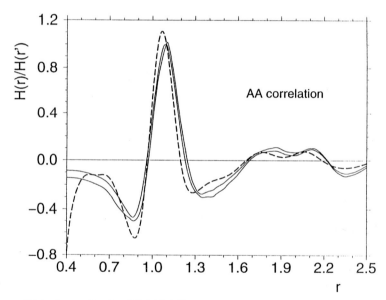

Fig. 6. Critical amplitude $H(r)/H(r')$ for the coherent intermediate scattering function as predicted by MCT (bold dashed line) and the upper and lower bound for this function (thin lines) as determined from the simulation.

4   Summary

We have investigated the dynamics of a binary Lennard-Jones liquid in order to test whether MCT gives a correct description of this dynamics at low temperatures. We find that the temperature dependence of the diffusion constant and of the α-relaxation times of various correlators are in accordance with the predictions of the theory. At low temperatures also the time dependence of the incoherent intermediate scattering function is in agreement with the theory, as long as one restricts oneself to the α-relaxation regime. For earlier times we find discrepancies between MCT and the simulations, that are likely to be due to corrections to the asymptotic scaling laws. Although not presented here, we have also investigated the time dependence of various other time correlation functions, such as the coherent intermediate scattering function, and found that they behave qualitatively similar to the one presented here [3], [6]. Thus the conclusions regarding the agreement between the theory and the time correlation function discussed here hold also for the other time correlation functions.

In order to test whether the predictions of MCT are correct also from a *quantitative* point of view, we solved the wave-vector dependent mode-coupling equations in the long time limit. Comparing the so obtained solutions with the results of the computer simulation we find that the theory is able to predict correctly the q-dependence of the nonergodicity parameter, if the value of q is not too large. Furthermore we have shown that MCT makes a surprisingly accurate prediction of the r-dependence of the critical amplitudes, if r is not too small.

Thus we can conclude that for the binary Lennard-Jones system considered here MCT gives a very good qualitative description and a good quantitative description of the dynamics at low temperatures.

References

[1] U. Bengtzelius, W. Götze, and A. Sjölander, J. Phys. C **17**, 5915 (1984); E. Leutheusser, Phys. Rev. A **29**, 2765 (1984); W. Götze, p. 287 in *Liquids, Freezing and the Glass Transition* Eds.: J. P. Hansen, D. Levesque and J. Zinn-Justin, Les Houches. Session LI, 1989, (North-Holland, Amsterdam, 1991); W. Götze and L. Sjögren, Rep. Prog. Phys. **55**, 241 (1992).

[2] Theme Issue on Relaxation Kinetics in Supercooled Liquids-Mode Coupling Theory and its Experimental Tests; Ed. S. Yip. Volume **24**, No. 6-8 (1995) of *Transport Theory and Statistical Physics*.

[3] W. Kob and H. C. Andersen, Phys. Rev. E **51**, 4626 (1995).

[4] M. Nauroth and W. Kob, Mainz University preprint KOMA-96-20

[5] J.-L. Barrat and A. Latz, J. Phys.: Condens. Matter **2**, 4289 (1990); M. Fuchs, PhD Thesis, TU Munich, 1993.

[6] W. Kob and H. C. Andersen, Phys. Rev. Lett. **73**, 1376 (1994); Phys. Rev. E **52**, 4134 (1995); Nuovo Cimento D **16**, 1291 (1994).

An Ideal Glass Transition
in Supercooled Water?

F. Sciortino[1], S.H. Chen[2], P. Gallo[2] and P. Tartaglia[1]

[1] Dipartimento di Fisica and Istituto Nazionale per la Fisica della Materia, Universitá di Roma *La Sapienza*, P.le Aldo Moro 2, I-00185, Roma, Italy
[2] Department of Nuclear Engineering, Massachusetts Institute of Technology, Cambridge, MA 02139

Abstract. Analyzing recent molecular dynamics simulations in deeply supercooled liquid states, we have found that the single particle dynamics in water can be interpreted in terms of Mode Coupling Theory, in its so-called ideal formulation. In this paper we review such evidence and discuss the relevance of this finding for the debated thermodynamic behavior of supercooled water. The experimental apparent power-law behavior of the transport coefficients in water, diverging or going to zero at the so-called Angell temperature could indeed be interpreted as a kinetic, as distinct from thermodynamic, phenomena. This finding removes the need of a thermodynamic singularity for the explanation of the anomalies of liquid water. We also comment on the development of a significant harmonic dynamics on cooling the liquid, which could indicate a transition from a fragile to a strong behavior in liquid water.

In this paper we review some of the results obtained in a series of very long simulations of supercooled liquid SPC/E water. We aim to test predictions of Mode Coupling Theory (MCT)[1], [2], [3] for the correlation functions of single particle dynamics in water with corresponding quantities calculated from Molecular Dynamics (MD) simulations. In doing so, we try to assess to what extent the MCT, which has been proposed to describe simple liquids is applicable also to the description of the single particle dynamics of supercooled water, a hydrogen bonded liquid with a strong non-isotropic interactions among molecules. For a detailed study of the ability of MCT to describe the dynamics of simple liquids see for example [4], [5].

This work addresses also the thermodynamics behavior of liquid water on deep supercooling. This topic has been the subject of a long-standing scientific debate in the last thirty years[6], [7], [8]. It has been found that there are anomalous increases of thermodynamic quantities and apparent divergences of dynamic quantities on approaching a singular but experimentally unreachable temperature T_s of about $227K$ at ambient pressure[9], [10]. This discovery has stimulated an enormous amount of experimental, theoretical and computational work in an attempt to clarify the origin of the singularity. A recent detailed review of all these studies can be found in Ref.[8]. The possibility of interpreting T_s as the critical temperature postulated by

MCT[11], [12], [13], and thus as a result of a fully kinetic effect, is also reviewed in this article.

We also present new results on the harmonic dynamics which develops in supercooled liquid water at very low temperature. Such behavior suggests that close to the MCT ideal glass transition, when stable thetrahedral coordinated cages are formed, water undergoes a cross-over from a fragile liquid toward a strong harmonic liquid behavior.

The basic quantity in the study of self-dynamics in liquids is the van-Hove self correlation function $G_s(r,t)$. $G_(r,t)d\mathbf{r}$ is the probability of finding a molecule in $d\mathbf{r}$ at time t, knowing the same molecule was at the origin at time 0. The space-time Fourier Transform of $G_s(r,t)$ can be measured via quasi elastic incoherent scattering experiments.

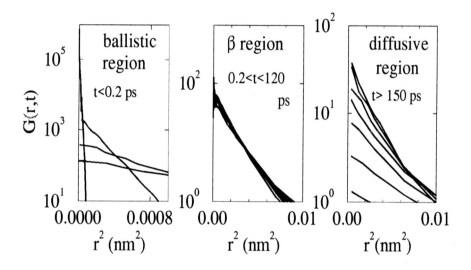

Fig. 1. Self part of the Van Hove distribution function for the three time regions at the lowest studied temperature. (a) $t < 0.2\ ps$; (b) β-relaxation region; $0.2 < t < 120\ ps$ (c) α-relaxation region $t > 150\ ps$.

Fig.1 shows $G_s(r,t)$ vs r^2 evaluated from the simulations at the lowest simulated temperature. We note three different broad behaviors. The very early times (the microscopic regime), where $G_s(r,t)$ is approximatively gaussian; the intermediate times, where $G_s(r,t)$ is almost constant in time (the β-regime); the long times, where $G_s(r,t)$ resumes changing in time (the diffusive regime). During the microscopic regime, the molecule explores its cage and the variance of the gaussian increases with time. For very early times, the variance increases with t^2, independently of the type of motion the molecule is performing (i.e. oscillations or free-particle motion). In the β regime, the molecule is almost trapped in the cage created by its neighbors. The molecule keeps moving in the cage but no new space is explored. In this regime $G_s(r, t)$ does not change significantly with time. In the diffusive regime, the probability of leaving the cage becomes significant and $G_s(r, t)$ starts changing with time appreciably again. These three different behaviors characterize the dynamics of the water molecule and show up in all evaluated dynamical quantities.

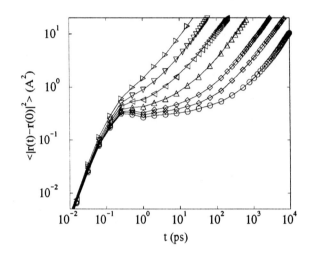

Fig. 2. MSD as a function of time for $\bigcirc T = 206.3\ K$, $\square T = 209.3\ K$, $\diamondsuit T = 213.6\ K$, $\triangle T = 224.0\ K$, $\triangleleft T = 238.2K$, $\triangledown T = 258.5K$, $\triangleright T = 284.5K$). The curves show the cage effect, starting at 0.25 ps, followed by the eventual diffusion of the molecule. Arrows indicate the time at which the non gaussian parameter $\alpha_2(t)$ (see text) is maximum.

We show next (Fig. 2) the time dependence of the second moment of $G_s(r, t)$, which coincides with the mean squared displacement $\langle r^2(t) \rangle = \langle |\mathbf{r}(t) - \mathbf{r}(0)|^2 \rangle$. Different curves correspond to different temperatures. All curves are characterized by an initial $\sim t^2$ law (ballistic motion) and by a long time linear law Dt, where D defines the diffusion coefficient of the molecule. At low enough temperatures, these two regimes are separated by a region in time characterized by a t-independent value of $\langle r^2(t) \rangle$. In analogy with the interpretation of $G_s(r, t)$, we interpret the behavior of $\langle r^2(t) \rangle$ in terms of progressively increasing lifetime of the cages confining the molecule on cooling. When the cage lifetime (e.g. the average time it takes a molecule to leave the cage), becomes longer than the microscopic time, the normal-liquid crossover from ballistic motion to diffusive motion breaks into two crossover: from ballistic motion to constant and later on from constant to diffusive motion.

It is interesting to note at this point that $\langle r^2(t)\rangle$ at the crossover between the β and diffusive regime shows a sub-diffusive behavior. In other words, as shown in Fig. 3, $\langle r^2(t)\rangle$ is not simply given by the simple $constant + Dt$ law.

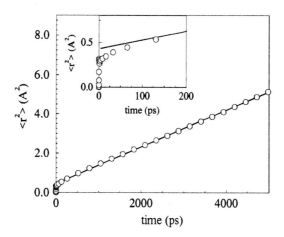

Fig. 3. MSD as a function of time for $T = 206.3\ K$ in linear scale. Full line is the fit from which the diffusion coefficient is calculated. The inset is an enlargement of the short times, to highlight the initial sub-diffusive behavior

One possible physical interpretation of this sub-diffusive behavior at the crossing between the β and the diffusive regime is offered by the ideal MCT. This theory[2] predicts that any correlation function coupled to density, beyond the microscopic regime, is controlled by the behavior of the density fluctuations (the slow modes of the system) and thus it has an universal behavior, independent on the physical quantity under scrutiny. According to MCT, all correlation functions $\phi(t)$ decay to a plateau value f_{EA}(the so-called Edwards-Anderson non-ergodicity factor), at intermediate times. The approach and departure from such plateau value are controlled by a self-similar dynamics, i.e. the correlation function decays as $\phi(t) - f_{EA} \sim t^{-a}$ on approaching the plateau and $\phi(t) - f_{EA} \sim -t^b$ on leaving it. a and b are two related positive exponents.

The sub-linear behavior of $\langle r^2(t)\rangle$ is consistent with the prediction of MCT. Being $\langle r^2(t)\rangle$ the second moment of the $G(r,t)$ correlator, it can be expanded according to the same laws close to the plateau. In this respect, MCT predicts a subdiffusive increase of $\langle r^2(t)\rangle$ in the late β-regime. The exponent of the subdiffusive behavior is T independent, sufficiently close to the glass transition. Moreover, this exponent is related to the exponent controlling the power-law T dependence of the diffusion coefficient.

Stochastic models for the glass transition have also been proposed to interpret the subdiffusive increase of $\langle r^2(t)\rangle$. In the so-called trapping model[14], molecules are assumed to vibrate in cages and occasionally jump out from their cage. The distribution of jumping times is assumed to be a power-law distribution extending from a fixed microscopic time up to infinity. The temperature dependence in this model enters only in the power-law exponent. A kinetic glass transition is predicted at the temperature where the first moment of the waiting time distribution diverges. Temperature at which higher moment of the distribution diverge are also given a physical interpretation[15]. In the trapping model, above the glass transition $\langle r^2(t)\rangle$ also crosses from a subdiffusive to a diffusive behavior but, differently from MCT, the exponent of the subdiffusive behavior is T dependent.

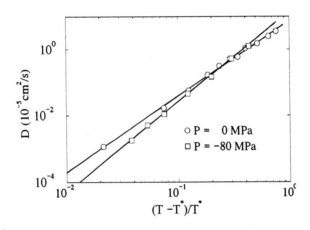

Fig. 4. Temperature dependence of the diffusion coefficient D for two isobars. □ are from [12], ○ from [16]. Full lines are power-law fit respectively given by $D = 13.93\ (T/198.7 - 1)^{2.73}$ and $D = 7.39\ (T/186.3 - 1)^{2.29}$ where D is in cm^2/s and T is in K.

From our MD simulations we also find that the diffusion coefficient, evaluated form the long time slope of $\langle r^2(t)\rangle$ has a power-law dependence on T. Fig. 4 shows the T-dependence of D from our work[12] and from independent calculations from [16]. For both isobars, the temperature dependence of D is well described by a power-law, as in real experiments[17]. T^* and γ are pressure dependent. The differences in γ between the two simulated isobars are consistent with the experimentally observed sensitivity of γ on pressure (see Ref.[17]) and with MCT. Indeed, according to MCT, γ depends on the specific point of the glass transition line $T_c(P)$ approached. Moreover, according to MCT, γ is fully determined by the knowledge of the b exponent controlling the self-similar dynamics in the late β-regime. For $b = 0.5$, the value we measured for our simulations[13], MCT predicts a value of γ of 2.74, in extremely good agreement with our findings.

As we said above, experimental data at different positive pressures have been fitted according to $D \sim |T - T_s(P)|^{\gamma(P)}$. The $T_s(P)$ values have been found to run parallel to the line of density maxima (TMD), shifted by about $50K$. $\gamma(P)$ values increase on increasing pressure from 1.8 to 2.2. In SPC/E water the TMD line is shifted compared to real water. This notwithstanding, the difference between T^* and the SCP/E TMD is about $50\ K$ for both pressures, as in real water. The value of γ we calculated are higher that the experimental one. In this respect, it would be very valuable to perform runs at higher pressures to compare with the experimental trends.

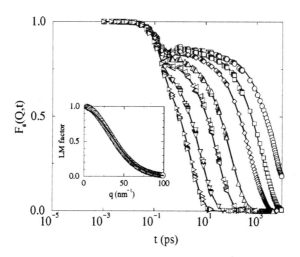

Fig. 5. $F_s(Q_{max}, t)$ vs time (symbols as in Fig.2). Solid lines are calculated according to a initial gaussian decay plus a stretched exponential decay (for further details see [13]). The inset shows the Q dependence of the plateau value at $T = 206.3$. Full line in the inset is a fit with a gaussian function (see text).

We now move to the incoherent intermediate scattering function $F_s(Q, t)$, the spatial Fourier transform of the van Hove self correlation function $G_s(r, t)$. Fig. 5 shows the T dependence of $F_s(Q_{max}, t)$. Note that for all T investigated, $F_s(Q_{max}, t)$ decays to zero in the long time limit. This confirms that the simulations were long enough to guarantee the complete decay to zero of the test particle thermal fluctuations, i.e. that all simulations where in the liquid state and in equilibrium. At low T, $F_s(Q_{max}, t)$ also shows the presence of the same three different regimes, the initial one characterized by a fast decay, followed by a plateau region, and by a final decay to zero. The approach to zero can well be represented by a stretched exponential decay, with a stretching exponent which tends to one as Q goes to zero and on increasing T. A complete discussion on the T and Q dependence of the stretching exponent can be found in Ref. [13]. The departure from the plateau value, in agreement with MCT, is better fitted with a power-law than with a stretched exponential function[13].

The value of the plateau and its Q dependence is, in Fourier space, an indication of the shape and size of the confining cage. The Q dependence of the plateau value at the lowest T (the so-called Lamb-Mössbauer factor) is shown in the inset, together with a solid line fit to $exp(-Q^2 a^2/6)$. We find $a^2 = 0.25 Å^2$, which is a measure of the size of the confining cage. The quality of the fit for large Q values supports the view that the shape of the cage in which molecules are constrained is significantly harmonic.

This observation suggests to study in more detail the hypothesis that the short time dynamics in water is well described by harmonic oscillations within the cages. Such undamped harmonic motion would survive for times comparable to the cage lifetime, i.e. it would persist more and more on cooling the system. One way of performing this kind of study is to evaluate the so-called Instantaneous Normal Mode (INM) spectrum[18]. The INM spectrum is calculated from equilibrated configurations at finite T and requires the diagonalization of the Hessian matrix[19]. This procedure allows to estimate the curvature of the potential energy along 6N-6 independent directions[20]. Because the T is finite, the system is not sitting in a local minima of the phase space and thus the curvature along a few directions can be negative, implying that along these directions the potential is anharmonic. The negative curvature directions may lead to changes in local minima, and in this case these directions are related to structural changes in the system, as well as to confined anharmonic oscillations.

Fig.6-a shows the INM spectrum. Negative frequencies are associated with directions with negative curvature. The spectrum can be decomposed in two regions, above and below 400 cm^{-1}, which separate the mainly translational modes from the mainly libration modes. In the translational region, which we are most interested in, we find the (strongly harmonic) peak related to modes of the Oxygen-Oxygen-Oxygen bending type and an intense shoulder, associated to modes related to Oxygen-Oxygen stretching[19]. The dynamical behavior of the fictitious harmonic system obtained by exciting with $k_B T$ all modes with positive frequency is shown in part (b) and (c). Part (b) compare the time-dependence of $\langle r^2(t) \rangle$. We note, in both curves, the presence of an overshoot around 1ps. This overshoot is due to the superposition of

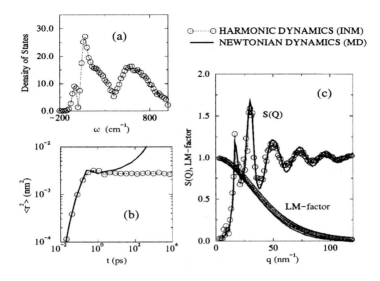

Fig. 6. (a) Instantaneous Normal Mode spectrum for the $T = 206.3K$ system. (b) ○ $\langle r^2(t) \rangle$ for the harmonic system, full line $\langle r^2(t) \rangle$ from the MD simulation. (c) Structure factor and Lamb-Mössbauer factor for the harmonic system (○) and from the MD simulation (full line).

many independent modes in the determination of the molecular displacement. Indeed, each mode contributes a term proportional to $k_B T(1 - cos(\omega t))/\omega^2$ to $\langle r^2(t) \rangle$, i.e. an oscillating function of time. All these independent undamped contributions, weighted by the density of states, add coherently for short times and incoherently for large time, giving rise to the initial t^2 behavior, to the overshoot and to the finite $\langle r^2(t) \rangle$ value at large times. This sequence is not peculiar of the INM distribution shown in Fig.6-a but is also observed, for example, in the Debye model on the simple cubic lattice[21].

The difference in the position of the peak between MD and INM $\langle r^2(t) \rangle$ is probably related to having neglected the confined anharmonic oscillations (which show up in the negative frequency region) and to finite size effects. Indeed, the low frequency modes (which have the largest amplitudes) are cut by the finite size of the simulated box. The amplitude and the position of the peaks may thus change on going to larger system sizes, although the feature seems to be a truly genuine effect produced by the harmonic motion.

The presence of an overshoot in $\langle r^2 \rangle$ has been tentatively associated[22] also with the so-called boson-peak[23]. It would be very valuable to perform a comparative analysis of the presence of an overshoot in $\langle r^2 \rangle$ with the shape of the confining potential and with the harmonic character of the dynamics as evaluated via INM. It would be interesting to compare such quantities for strong and fragile liquids[24], to shed light on the inter-relations between the presence of a boson-peak, the shape of the potential confining molecules in their cage and their fragility. In the case of water, the development at low temperature of a significant harmonic dynamics may well be connected to the transition between a fragile and a strong behavior on supercooling water, as proposed by C.A. Angell [25].

In summary, we have presented evidence that SPC/E water undergoes a kinetic glass transition 50 degrees below the TMD. The molecular dynamics is well accounted for by the idealized MCT of supercooled liquids, suggesting an interpretation of the so-called Angell temperature as the critical temperature of MCT[12], [13]. Such hypothesis was first proposed in Ref. [11] on the basis of an analysis of experimental Raman spectra. In this regard, the apparent power-law increase of transport coefficient in liquid water on supercooling is traced to the formation of cages and to the associated slow dynamics resulting from the presence of long living cages. In other words the divergence of transport coefficients does not need to rely on a thermodynamical instability, either connected to the re-entrance of the gas-liquid spinodal or to the presence of a critical point at high pressure and low temperature. The SPC/E behavior described in this work thus supports the recently presented thermodynamic analysis of liquids with a negatively sloped TMD in the absence of singularities[26].

The scenario described above bears a strong resemblance to the results of MD simulation for the mixed Lennard-Jones spheres carried out recently to test the MCT of kinetic glass transition. In this respect, the prediction of the idealized MCT seems to be robust, and able to describe fragile liquids[24], for molecules interacting via spherical as well as highly directional potentials. It is indeed surprising that a simple Lennard-Jones system, in which molecules are confined in cages with a large coordination number behaves, close to its glass transition, similarly to a tethrahedrally coordinated system, like a SPC/E water, in which the cages are formed more by the deep hydrogen-bonding potential than by the excluded volume constraint.

Acknowledgment: FS and PT would like to thank the organizers for the very stimulating conference. FS wishes to thank C.A. Angell and W. Kob for comments and suggestions on the work presented here.

References

[1] E. Leutheusser, Phys. Rev. A, **29**, 2765 (1984), U. Bengtzelius, W. Götze and A. Sjölander, J. Phys. C **17** 5915 (1984).

[2] W. Götze and L. Sjögren, *Rep. Prog. Phys.* **55**, 241 (1992).

[3] W. Götze and A. Sjögren , *Transport Theory and Statistical Physics* **24**, 801 (1995).

[4] W. Kob and H. C. Andersen, *Phys. Rev. E* **51**, 4626 (1995) and *Phys. Rev. E* **52B**, 4134 (1995).

[5] W. Kob, *Ann. Rev. Comp. Physics*, Vol III, D. Stauffer Editor, World Scientific 1995.

[6] C.A. Angell, *Ann. Rev. Phys. Chem.* **34**, 593 (1983).

[7] C.A. Angell, in *Water: A Comprehensive Treatise*, Ed. F. Franks (Plenum, New York, 1981), Ch. 1.

[8] P. G. Debenedetti, *Metastable Liquids* (Princeton University Press, 1996), in press.

[9] R.J. Speedy, *J. Chem. Phys.* **86**, 982 (1982).

[10] R.J. Speedy and C.A. Angell, *J. Chem. Phys.* **65**, 851 (1976).

[11] A. P. Sokolov, J. Hurst and D. Quitmann, *Phys. Rev. B* **51**, 12865 (1995).

[12] P. Gallo, F. Sciortino, P. Tartaglia, S. H. Chen, *Phys. Rev. Letts.* **76** 2730 (1996).

[13] F. Sciortino, P. Gallo, P. Tartaglia, S. H. Chen, *Phys. Rev. E* **xx**, xxxx (1996).

[14] T. Odagaki and Y. Hiwatari, *Phys. Rev. A* **43**, 1103 (1991).

[15] T. Odagaki, *Phys. Rev. Lett* **75**, 3701 (1995).

[16] L. A. Baez and P. Clancy, *J. Chem. Phys.* **101**, 9837 (1994).

[17] E. W. Lang and H. D. Lüdemann, *Angew. Chem. Int. Ed. Engl.* **21**, 315 (1982).

[18] B. Madan, T. Keyes and G. Seeley, *J. Chem. Phys.* **92** 7565, (1990). B. Madan, T. Keyes and G. Seeley, *ibidem* **94** 6762, (1991)

[19] F. Sciortino and S. Sastry, *J. Chem. Phys.* **100**, 3881 (1994).

[20] The SPC/E model is a rigid model with constrains on the oxygen-hydrogen bond and hydrogen-oxygen-hydrogen angle. Thus, each molecule contributes only 6 degree of freedom.

[21] G. H. Vineyard, *Phys. Rev. A* **110**, 999 (1958).

[22] C. A. Angell *Science* **267** 1924 (1995).

[23] W. Götze and L. Sjögren, special issue of *Chem. Phys* on *Rate processes with kinetic parameters distributed in time and space* Y.A. Berlin, J.R. Miller and A. Plonka Editors, in press (1996)

[24] C. A. Angell in *Relaxations in Complex Systems*, edited by K. Ngai and G. B. Wright. (National Technical Information Service, U.S. Dept. of Commerce: Springfield, VA, 1985) p. 1; C. A. Angell, J. Non-Cryst. Solids **13**, 131.

[25] C. A. Angell, J. Phys. Chem. **97**, 6339 (1993).

[26] S. Sastry, P. G. Debenedetti, F. Sciortino and H.E. Stanley, *Phys. Rev. E* , **53** 6144 (1996).

Glass Transition in the Hard Sphere System

Chandan Dasgupta[1] and Oriol T. Valls[2]

[1] Department of Physics, Indian Institute of Science, Bangalore 560012, India and Jawaharlal Nehru Center for Advanced Scientific Research, Bangalore 560064, India

[2] School of Physics and Astronomy and Minnesota Supercomputer Institute, University of Minnesota, Minneapolis, Minnesota 55455

Abstract. The glass transition in a hard sphere system is studied numerically, using a model free energy functional that exhibits glassy local minima at sufficiently high densities. The numerical methods used in our work include free-energy minimization, direct integration of Langevin equations and Monte Carlo simulation. At relatively low densities, the system is found to fluctuate near the uniform liquid minimum of the free energy and to exhibit mode-coupling behavior. At densities higher than a *first crossover density*, the dynamics is governed by thermally activated transitions between glassy free-energy minima. The typical time scale for such transitions grows very rapidly as a *second crossover density* is approached from below. Interpretation of existing molecular dynamics data in the light of our results is discussed.

1 Introduction

Existing theories of the glass transition [Angell (1988)] in supercooled liquids may be broadly classified into two categories. The first category consists of theories which describe the glass transition as a purely dynamic phenomenon. Mode coupling (MC) theories of the glass transition [Götze (1991)] are the most prominent ones in this class. In MC theories, the slowing down of the dynamics near the glass transition is attributed to a nonlinear feedback mechanism arising from correlations of density fluctuations in the liquid. MC theories provide a detailed and qualitatively correct description of the dynamic behavior observed in experiments and numerical simulations over a temperature range that covers the first few decades of the growth of the characteristic relaxation time of so-called "fragile" [Angell (1988)] liquids in the supercooled regime. The original version of MC theories predicts a power-law divergence of the relaxation time at an *ideal glass transition temperature* T_c. Experimentally, however, this divergence is not found, and the predictions of conventional MC theories do not provide a correct description of the actual behavior at temperatures close to or lower than the T_c extracted from power-law fits to the data at higher temperatures. It is generally believed that the breakdown of conventional MC theories at temperatures near T_c arises because "activated processes" become important at such temperatures. However, the nature of such "activated processes" has not been elucidated so far.

In the second class of theories [Wolynes (1988)] of the glass transition, the starting point is the assumption that the free energy of the liquid, expressed as a functional of the time-averaged local density, develops a large number of "glassy" local minima as the temperature is decreased below the equilibrium crystallization temperature. When the system is rapidly cooled from a high-temperature liquid state, it gets trapped in one of these glassy local minima from which it eventually relaxes. The relaxation is slow because it involves thermally activated transitions over free energy barriers. In this picture, therefore, the slow dynamics near the glass transition is attributed to activated transitions among metastable glassy minima of the free energy. This description is similar to that developed in recent years for a number of quenched random systems such as spin glasses [Binder and Young (1986)]. This analogy suggests that the behavior observed near the glass transition may be the precursor of a true thermodynamic phase transition which would take place at a temperature lower than the conventional glass transition temperature T_g (defined [Angell (1988)] as the temperature at which the viscosity reaches a value of 10^{13} P) if one could maintain thermodynamic equilibrium all the way down to this temperature. The possibility of a thermodynamic glass transition is also suggested by experimental observations [Angell (1988)] such as a Vogel-Fulcher growth of the viscosity (the Vogel-Fulcher form predicts a divergence of the viscosity at a temperature $T_0 < T_g$) and the apparent vanishing of the extrapolated entropy difference between the supercooled liquid and the crystalline solid at a temperature close to T_0.

Recently, we have carried out a number of numerical studies [Dasgupta (1992), Dasgupta and Ramaswamy (1992), Lust *et al* (1993), Lust *et al* (1994), Dasgupta and Valls (1994), Valls and Dasgupta (1995), Dasgupta and Valls (1996)] of the glass transition in a hard sphere system which suggest that elements of these two apparently dissimilar descriptions of the glass transition should be combined for the development of a full understanding of the observed phenomena. In this paper, we review the results obtained in these studies. Our work is based on a free energy functional of the form proposed by Ramakrishnan and Yussouff (RY) [Ramakrishnan and Yussouff (1979)]. This free energy functional is known [Dasgupta (1992), Dasgupta and Ramaswamy (1992)] to possess, at sufficiently high densities, many "glassy" local minima (i.e. local minima with inhomogeneous but aperiodic distribution of the time-averaged local density) in addition to the homogeneous liquid minimum and the crystalline minimum with a periodic distribution of the local density. Our work provides a fairly complete understanding of the role played by these different minima of the free energy in the glassy dynamics of the system and explains a number of features observed in Molecular Dynamics (MD) simulations of the same system.

The rest of this paper is organized as follows. After a brief review (section 2) of the results of existing MD simulations (which provide the "experimental" data with which we compare the results of our numerical calculations), we summarize the results of (i) an investigation of the glassy local minima of the RY free energy functional for the hard sphere system (section 3); (ii) a numerical study of a set of Langevin equations which describe the nonlinear fluctuating hydrodynamics of the hard sphere liquid near the glass transition (section 4); and (iii) a Monte Carlo study of the time scales associated with transitions between glassy local minima of the RY free energy functional (section 5). The main conclusions derived from these studies are summarized in section 6.

2   Summary of Molecular Dynamics Results

The control parameter for a hard sphere system is the dimensionless density $n^* \equiv \rho_0 \sigma^3$, where ρ_0 is the average number density and σ is the hard-sphere diameter; increasing (decreasing) n^* has the same effect as decreasing (increasing) the temperature of systems for which the temperature is the relevant control parameter. MD simulations [Woodcock (1981), Woodcock and Angell (1981)] show that the hard sphere liquid undergoes an equilibrium crystallization transition at $n_f^* \simeq 0.943$. The liquid can be locally equilibrated in the "supercooled" state over simulational time scales if the density is lower than a characteristic density $n_1^* \simeq 1.08$. The diffusion constant D in the supercooled regime ($n_f^* < n^* < n_1^*$) decreases rapidly with increasing n^*. The dependence of D^{-1} on n^* is consistent [Woodcock and Angell (1981)] with a Vogel-Fulcher law, $D^{-1} \propto \exp[a/(v - v_0)]$, where $v \equiv 1/n^*$, a is a constant and $n_0^* \equiv 1/v_0 \simeq 1.21$, and also with a power law, $D^{-1} \propto (n_c^* - n^*)^{-\gamma}$ with $n_c^* \simeq 1.15$ and $\gamma \simeq 2.5$. For $n^* > n_1^*$, the system spontaneously freezes into a near-crystalline state within the time scale of MD simulations. If, on the other hand, the system is rapidly quenched from the liquid state at a density lower than n_1^* to a density above a second characteristic density $n_2^* \simeq 1.2$, then it ends up in a glassy state with no long-range translational order. The structure in the glassy states obtained in this way is similar to that in random close packing of hard spheres. As discussed below, our work provides an understanding of the physical origin of the existence of these two characteristic densities n_1^* and n_2^*.

3   Glassy Local Minima of the Free Energy

The starting point of our calculation is the RY free energy functional which has the form

$$F[\rho] = F_l(\rho_0) + k_B T \left[\int d\mathbf{r}[\rho(\mathbf{r}) \ln(\rho(\mathbf{r})/\rho_0) - \delta\rho(\mathbf{r})] \right.$$
$$\left. - \frac{1}{2} \int d\mathbf{r} \int d\mathbf{r}' C(|\mathbf{r} - \mathbf{r}'|)\delta\rho(\mathbf{r})\delta\rho(\mathbf{r}') \right], \tag{1}$$

where $\delta\rho(\mathbf{r}) \equiv \rho(\mathbf{r}) - \rho_0$ is the deviation of the time-averaged number density at the point $\mathbf{r}$ from its average value ρ_0, k_B is the Boltzmann constant, F_l the grand canonical free energy of a uniform liquid of density ρ_0, T is the temperature, and $C(|\mathbf{r}-\mathbf{r}'|)$ is the direct pair correlation function [Hansen and MacDonald (1986)] of the uniform liquid at density ρ_0. We use the Percus-Yevick form [Hansen and MacDonald (1986)] for the direct correlation function of a hard sphere liquid. In our numerical study, the density functional of Eq.(1) is discretized by dividing the system into cubic cells, each of volume a_0^3, and defining "coarse-grained" variables $\{x_i\}$ where x_i represents the integral of the density over the ith cell. In terms of these variables, the free energy has the following approximate form:

$$F(\{x_i\}) \simeq F_l + k_B T \left[\sum_i [x_i \ln(x_i/x_0) - (x_i - x_0)] \right.$$

$$\left. - \frac{1}{2} \sum_{i,j} C_{ij}(x_i - x_0)(x_j - x_0) \right], \tag{2}$$

where $x_0 \equiv \rho_0 a_0^3$ and the "interaction matrix" C_{ij} is defined as

$$C_{ij} = a_0^{-6} \int_{v_i} d\mathbf{r} \int_{v_j} d\mathbf{r}' C(|\mathbf{r} - \mathbf{r}'|), \tag{3}$$

the two integrals being over the volumes v_i and v_j of the cells i and j, respectively. The minima of this free energy are located by applying a numerical iterative procedure [Dasgupta (1992)] which converges only to local minima. We consider systems with both periodic and free boundary conditions, values of a_0 in the range $0.1\sigma - 0.3\sigma$, and values of the system size L in the range $3\sigma - 6\sigma$. While the same qualitative behavior is found in all cases, the quantitative results show a weak dependence on the choice of parameter values and boundary conditions.

We find a crystalline minimum with fcc structure at sufficiently high densities if the values of a_0 and L are commensurate with the lattice structure. The free energy of this minimum becomes lower than that of the uniform liquid as the density is increased above n_f^*. The values of n_f^* obtained in our calculations lie in the range $0.79 - 0.83$, depending on the values of a_0 and L used. The difference between these values of n_f^* and the result of MD simulations is a consequence of the fact that the discretized form of $C(r)$ used in our calculation overestimates the short-range correlations in the liquid. We also find a very large number of glassy local minima at densities higher than n_f^*. In a glassy minimum, the density is sharply peaked in a small number of cells, but the locations of these cells do not show any crystalline order. Free energies of these glassy minima cross that of the uniform liquid at densities which are slightly higher that n_f^*, indicating a weakly first-order liquid-to-glass transition in the mean-field sense. Typical data for the free energies of

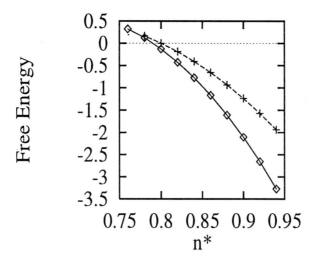

Fig. 1. The dimensionless free energy per particle, measured relative to that of the uniform liquid, is shown as a function of the dimensionless density n^* for the crystalline minimum (diamonds, solid line) and a typical glassy minimum (crosses, dashed line) obtained for $a_0 = 0.25\sigma$ and $L = 3\sigma$.

the crystalline solid and a typical glassy minimum, obtained for a system with $a_0 = 0.25\sigma$ and $L = 3\sigma$, are shown in Fig.1. The structure of the glassy minima is found to be close to that of glassy states obtained by fast compression in MD simulations [Woodcock (1981)]. Measurements of several quantities which characterize the local bond-orientational symmetry of the glassy minima yield [Dasgupta (1992)] results expected for random close packing of spheres. We find that glassy minima with lower free energies generally have a higher degree of short-range order. This is illustrated in Fig.2 where we have shown the normalized two-point correlation function of the frozen-in density at two glassy minima obtained for $a_0 = \sigma/4.6$ and $L = 15a_0$.

4 Langevin Dynamics Near the Uniform Liquid Minimum

In this section, we describe the results of a numerical study of a set of Langevin equations in which information about the static structure of the liquid is incorporated through the RY free energy functional. The main objective of this study was to understand the role of the different free-energy minima in the dynamic behavior of the system near the glass transition. The Langevin equations for a system with equilibrium properties described by the RY free energy can be derived [Lust *et al* (1993)] following standard methods [Das and Mazenko (1986), Farrell and Valls (1989)]. One obtains Langevin

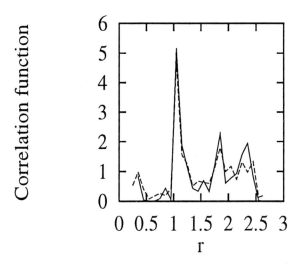

Fig. 2. The normalized two-point correlation function of the frozen-in local density for two glassy local minima obtained with $a_0 = \sigma/4.6$ and $L = 15a_0$ at density $n^* = 0.95$. The solid (dashed) line shows the results for a minimum with dimensionless free energy per particle equal to $-2.12(-1.78)$. The correlation function for the minimum with the lower free energy clearly shows more structure. The separation r is measured in units of σ.

equations for the two hydrodynamic fields: the density $\rho(\mathbf{r}, t)$ and the current density $\mathbf{j}(\mathbf{r}, t)$. These equations are somewhat lengthy and will not be reproduced here. They are differential equations in time, but integrodifferential equations in space, because they involve an integral over the range of the direct correlation function.

Despite these complications, the Langevin equations can be numerically solved for samples large enough to avoid finite size effects, (typically one uses a system of size L^3 with $L = 15a_0$) and time ranges extensive enough for the decay of correlations to be studied over an extended density range. The system is kept from crystallizing by choosing values of a_0 and L which are incommensurate with the crystalline structure. These calculations are limited to densities $n^* \leq 0.93$. This limitation is not due to computational difficulties but rather to the Langevin dynamics becoming inadequate at higher densities where the system cannot be described as fluctuating about the liquid minimum.

Within this density range, the system reaches (local) equilibrium in the "supercooled" liquid state within computational times. One can then study [Lust *et al* (1993), Lust *et al* (1994)] the decay of the dynamic structure factor $S(\mathbf{q}, t)$. We work with its angular average $S(q, t)$. It is convenient to normalize this quantity by defining:

$$C(q,t) = \frac{S(q,t)}{S(q,t=0)}. \tag{4}$$

After obtaining the necessary numerical data, which requires an average over noise fields (i.e. over many different runs) and over time intervals consistent with equilibration having been achieved (see Lust *et al* (1993)), one begins the analysis of the density correlation results by attempting to characterize the decay of $C(q,t)$ by a single decay time. Thus we write a stretched exponential expression:

$$C(q,t) = \exp[-(t/\tau)^\beta], \tag{5}$$

with the characteristic time τ and the stretching exponent β being functions of q and n^*. One finds that τ is in fact a strong function of q, having a sharp maximum at the wavevector value where the static structure factor itself has its first and most prominent peak. This maximum value of τ, which we denote by τ^*, can be identified as the slowest time in the system. As a function of density, it was found [Lust *et al* (1993), Lust *et al* (1994)] to vary according to the Vogel-Fulcher law with the extrapolated divergence of τ^* occurring at $n_0^* = 1.23$. This is in good agreement with the results of MD simulation [Woodcock and Angell (1981)].

The simple stretched exponential form given above is, however, not a sufficiently good fit to $C(q,t)$ at higher densities, particularly for the slower decays occurring at wavevectors near the peaks of the static structure factor. A more general form that was found to be satisfactory for all densities $n^* \leq 0.93$ is the two-stage decay,

$$C(q,t) = (1-f)\exp[-(t/\tau_a)^\beta] + f\exp[-(t/\tau_b)], \tag{6}$$

which reduces to the form of Eq.(5) if $f = 0$ or if $\beta \approx 1$ and the two decay times coincide. This expression can also represent nonergodic behavior when the second decay time is longer than the computational time. Values of f are as large as $f = 0.5$ at $n^* = 0.93$ at relatively large wavevectors, while the smaller values of β are found for wavevectors near the peak of the static structure factor. A full discussion is given in the references mentioned above and further insights may be found in Valls and Dasgupta (1995).

An example of the results obtained and the fit to Eq.(6) is provided Fig.3. It corresponds to a density $n^* = 0.93$. The unit of time employed is, up to a factor of order unity, the Enskog collision time, and the system size was $L = 15a_0$ with $a_0 = \sigma/4.6$. Each curve, labeled by a "shell number" Q, represents an average of the data over a spherical shell of thickness π/L which includes the wavevector $(\pi Q/L, 0, 0)$.

The existence of a two-stage decay should be interpreted as representing first the combined effects of the very rapid phonon decay regime (coarse

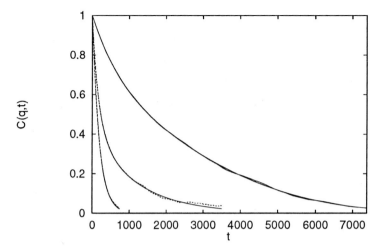

Fig. 3. Examples of the decay of the dynamic correlation function $C(q, t)$ at density $n^* = 0.93$. The solid curves are best fits to Eq.(6) and the dashed ones represent the numerical data. The three sets of curves correspond to $Q = 8$ (top), $Q = 13$ (middle), and $Q = 10$ (bottom). $Q = 8$ and 13 correspond respectively to the first and second peaks of the static structure factor.

grained by the Langevin dynamics) and the β-relaxation, followed by the second decay which we identify with the α-relaxation. These features of our numerical results are consistent with predictions of MC theories [Götze (1991)]. Throughout this density regime, the system is found to fluctuate in the vicinity of the uniform liquid minimum of the free energy. Thus, the dynamic behavior observed in this regime may be attributed unambiguously to nonlinear effects of interactions among small-amplitude density fluctuations about the uniform liquid minimum of the free energy.

5   Transitions Between Different Minima

At higher densities than those considered in the last section, the system must be studied, from the dynamical point of view, in a different way. The Langevin method becomes inadequate at higher densities because the system evolves towards a state characterized by large-scale spatial inhomogeneity of the density that cannot be described as arising from small-amplitude fluctuations about the liquid minimum. In practice, the Langevin method leads to equilibration at densities as high as $n_x^* \simeq 0.95$ [Dasgupta and Valls (1994)], but at higher densities, if one begins a Langevin simulation with nearly uniform initial conditions, the system does not equilibrate near the uniform liquid minimum. Instead, inhomogeneities in the density grow in time, indicating that the system has undergone a transition to the vicinity of some other free energy minimum, distinct of the liquid minimum. In other words, the inhomogeneous minima of the free energy become dynamically accessible from the neighborhood of the liquid minimum at densities higher than $n_x^* \simeq 0.95$.

The dynamics of the hard sphere system as characterized by the RY free energy can best be studied in this regime using a Monte Carlo technique developed in Dasgupta and Valls (1996). The density field (the only field that needs to be considered) is represented as before by the set of variables $\{x_i\}$. A Metropolis algorithm is implemented by sweeping the lattice sites i. Given a site i, one selects at random a neighboring site, j, among the sites within a hard sphere diameter from i. One then evaluates the quantity $s_{ij} = x_i + x_j$ and generates a random number p between zero and one. One then attempts to change the values of x_i and x_j to ps_{ij} and $(1-p)s_{ij}$ respectively, which of course conserves particle number. This change is accepted with probability 1 if $\Delta F \leq 0$, and with probability $\exp[-\lambda\beta\Delta F]$ if $\Delta F > 0$, where ΔF is the proposed change in the RY free energy, β is the inverse temperature, and $\lambda \equiv \sigma^3/(3n^*a_0^3)$ is the ratio between the computational and actual numbers of degrees of freedom. Since this algorithm obeys the principle of detailed balance, the Monte Carlo dynamics is expected to provide a correct description of the process of thermal activation over free-energy barriers.

We have used this dynamics to study the hard sphere system at densities $n^* \geq 0.94$. The initial conditions were chosen as follows: first, Langevin dynamics was used to follow the system from an initial time (with uniform density) until the point at which the free energy, measured relative to that at the liquid minimum, crosses zero. At that point the system is trying to reach a different minimum and Langevin dynamics must be stopped. We then used the density configuration at that time as input in a minimization routine [Dasgupta (1992)] that locates the minimum of F closest to the configuration then reached by the Langevin dynamics. This minimum obviously corresponds to a negative F. We find that the density configurations at the minima obtained this way are glass-like. These local minimum configurations are then taken as the initial conditions for the Monte Carlo routine described above. The objective, then, is to find out how long the system stays in the basin corresponding to the initial minimum as it evolves in time. We denote the time characterizing this evolution by τ_1. This time is found by proceeding with the Monte Carlo evolution and, at suitable periodic intervals, using the minimization routine to find which basin the system is in. When it no longer corresponds to the initial basin, a time τ_1 must have elapsed. Of course, such a time is actually obtained by averaging over a sufficient number of runs. The reader should consult Dasgupta and Valls (1996) for the technical details.

The results obtained from this procedure are shown in Fig.4. At the largest density studied, only one run out of five led to a transition to another basin

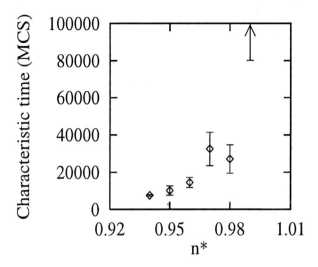

Fig. 4. The characteristic time τ_1 defined in the text, as a function of density n^*. The bottom of the arrow represents a lower bound to the value of τ_1 at $n^* = 0.99$. These results were obtained for $L = 15a_0$ and $a_0 = \sigma/4.6$.

within the time of computation (120,000 Monte Carlo steps (MCS) per site), so the result shown is only a lower bound. We see then that τ_1 grows very fast as the density $n_y^* \approx 0.99$ is approached from below. These results indicate that if the system is quenched to densities higher than n_y^*, then it would not be able to reach the equilibrium crystalline state during time scales accessible in simulations.

6 Conclusions

The overall picture that emerges from these numerical studies is as follows. Inasmuch as the hard sphere system is a prototypical simple liquid, the same picture should be qualitatively valid for other simple fragile liquids also. When the hard sphere system is compressed from a low-density liquid state to a density higher than n_f^*, it continues to fluctuate near the (metastable) uniform liquid minimum of the free energy over observational time scales if the density is lower than the first crossover density $n_x^* \simeq 0.95$. The dynamic behavior in this regime, governed by small-amplitude fluctuations of the density field about the uniform liquid minimum, is well-described by conventional MC theories. For values of $n^* > n_x^*$, the system evolves away from the liquid minimum within a computationally accessible time scale. The dynamic behavior in this regime is governed by thermally activated transitions between glassy free-energy minima. Numerical results suggest that $n_x^* < n_c^*$, the critical density at which the ideal glass transition of conventional MC theories is supposed to occur. Thus, the conventional MC description of the dynamics is expected to break down before the critical density is reached. The characteristic time scale for transitions between glassy free-energy minima increases very rapidly as a second crossover density $n_y^* \simeq 0.99$ is approached from below. Therefore, a system rapidly compressed from a low-density liquid state to densities higher than n_y^* is expected to remain stuck in a glassy free energy minimum over the time scales accessible in simulations. It is evident from this discussion that the two characteristic densities n_1^* and n_2^* mentioned in section 2 should be identified with the crossover densities n_x^* and n_y^*, respectively. The differences between the numerical values of n_1^* and n_2^* obtained in MD simulations and the values of n_x^* and n_y^* obtained in our work are to be attributed to our use of a discretized $C(r)$ which, as mentioned above, overestimates liquid-state correlations.

A question of obvious interest is whether the rapid growth of τ_1 near $n^* = n_y^*$ signals a true second order glass transition characterized by a divergence of τ_1 in the thermodynamic limit. Our limited investigation [Dasgupta and Valls (1996)] of the dependence of τ_1 on sample size does not show any evidence for such a divergence. It is interesting to note in this context that the results of a recent domain-wall scaling study [Ghosh and Dasgupta (1996)] suggest that simple liquids *do not* exhibit a thermodynamic glass transition at a non-zero temperature.

References

Angell C. A. (1988): J. Phys. Chem. Solids, **49**, 863.
Binder K., Young A. P. (1986): Rev. Mod. Phys. **58**, 881.
Das, S. P., Mazenko, G. F. (1986): Phys. Rev. **A 34**, 2265.
Dasgupta C. (1992): Eur. Lett. **20**, 31.
Dasgupta C., Ramaswamy R. (1992): Physica A **186**, 314.
Dasgupta C., Valls O. T. (1994): Phys. Rev. **E 50**, 3916.
Dasgupta C., Valls O. T. (1996): Phys. Rev. **E 53**, 2603.
Farrell J. E., Valls O. T. (1989): Phys. Rev. **A 40**, 7027.
Ghosh S. S., Dasgupta C. (1996): Phys. Rev. Lett. (in press).
Götze W. (1988): *Liquids, Freezing and the Glass Transition*, eds D. Levesque, J. P. Hansen and J. Zinn-Justin (Elsevier, New York).
Hansen J. P., MacDonald I. R. (1986): *Theory of simple liquids* (Academic, London).
Lust L. M., Valls O. T., Dasgupta C. (1993): Phys. Rev. **E 48**, 1787.
Lust L. M., Valls O. T., Dasgupta C. (1994): Phase Transitions, **50**, 47.
Ramakrishnan T. V., Yussouff M. (1979): Phys. Rev. **B 19**, 2275.
Valls O. T., Dasgupta C. (1995): Transp. Theor. and Stat. Phys. **24**, 1199.
Wolynes P. G. (1988): *Proceedings of International Symposium on Frontiers in Science, (AIP Conf. Proc. No. 180)*, eds. S. S. Chen and P. G. Debrunner (AIP, New York).
Woodcock L. V. (1981): Ann. N.Y. Acad. Sci. **371**, 274.
Woodcock L. V., Angell C. A. (1981): Phys. Rev. Lett. **47**, 1129.

Slow dynamics of glassy systems

Giorgio Parisi

Dipartimento di Fisica, Università *La Sapienza*, INFN Sezione di Roma I
Piazzale Aldo Moro, Rome 00185

Abstract. In these lectures I will study some properties that are shared by many glassy systems. I will shown how some of these properties can be understood in the framework of the mean field approach based on the replica method and I will discuss which are the difficulties which we have to face when we apply the replica method to realistic short range models. We finally present a partially successfully application of the replica method to soft sphere glasses.

1 Introduction

In these lecture I will study some systems which present some of the relevant characteristics of glasses. The aim is to construct simple microscopic models which can be studied in details and still behave in an interesting way. I will start from the simplest models, where only some of the observed characteristics are reproduced and we will later go to more complex systems.

The basic idea is to learn as much it is possible from mean field theory results and to try to understand the extent of the overlap among the rather complex phenomenology displayed by glasses and the exact results which are obtained for model with infinite range interaction. It is remarkable that the two cases have many points in common (with a few very interesting differences!) so that it seems to be a quite natural and reachable goal to construct a theory of glasses based on mean field tools. This construction is in progress and here I will review our present understanding.

This note is organized as follows: after this introduction we recall some of the main results which have been obtained in the framework of the mean field theory, both for systems with and *without* quenched disorder. In section III I will show which are the difficulties to extend these results to short range models, which properties are maintained and which are modified. Finally in the last section I will stress some of the properties of real glasses and I will describe a first tentative of doing esplicite computations for a soft sphere glass.

2 Mean field results

In these recent years there have been many progresses on the understanding of the behaviour of glassy systems in the mean field approximation. The mean field approximation is correct when the range of the interaction is infinite and this property allows us to write self consistent equations whose solution gives the solution of the model. The techniques that can be used are of various types, the replica method, a direct probabilistic method based of the cavity (Bethe) equations [2], [3] and the direct study of the dynamical equations [4], [5].

The main results have been the following:

- Model with random quenched disorder have been well understood both from the equilibrium and from the dynamical point of view.
- Some of the results obtained for systems with random quenched disorder have been extended to non disordered system. This step is crucial in order to have the possibility of extending these models to real glasses.

Let us see in details the results that have been obtained.

2.1 Disordered systems

In this case the thermodynamical properties at equilibrium can be computed using the replica method [2], [3]. A typical example of a model which can be solved with the replica method is a spin model with p spin interaction. The Hamiltonian we consider depends on some control variables J, which have a Gaussian distribution and play the same role of the random energies of the REM (i.e. the random energy model [6], [7]) and by the spin variable σ. For $p = 1, 2, 3$ the Hamiltonian is respectively

$$H_J^1(\sigma) = \sum_{i=1,N} J_i \sigma_i \qquad (1)$$

$$H_J^2(\sigma) = \sideset{}{'}\sum_{i,k=1,N} J_{i,k} \sigma_i \sigma_k \qquad (2)$$

$$H_J^3(\sigma) = \sideset{}{'}\sum_{i,k,l=1,N} J_{i,k,l} \sigma_i \sigma_k \sigma_l$$

where the primed sum indicates that all the indices are different. The N variables J must have a variance of $O(N^{(1-p)/2})$ if we want to have a non trivial thermodynamical limit. The variables σ are usual Ising spins, which take the values ± 1. ¿From now on we will consider only the case $p > 2$.

In the replica approaches one assumes that at low temperatures the phase space breaks into many valleys, (i.e. regions separated by high barriers in free energy). One also introduces the overlap among valley as

$$q(\alpha, \gamma) \equiv \frac{\sum_{i=1,N} \sigma_i^\alpha \sigma_i^\gamma}{N}, \qquad (3)$$

where σ^α and σ^γ are two generic configurations in the valley α and γ respectively.

In the simplest version of this method [8], [9] one introduces the typical overlap of two configurations inside the same valley (sometimes denoted by q_{EA}. Something must be said about the distribution of the valleys. Only those which have minimum free energy are relevant for the thermodynamics. One finds that these valleys have zero overlap and have the following probability distribution of *total* free energy of each valley:

$$P(F) \propto \exp(\beta m(F - F_0)), \tag{4}$$

where F_0 is the total free energy of the valley having lower free energy.

Indeed the average value of the free energy can be written in a self consistent way as function of m and q $(f(q, m))$ and the value of these two parameters can be found as the solution of the stationarity equations:

$$\frac{\partial f}{\partial m} = \frac{\partial f}{\partial q} = 0. \tag{5}$$

The quantity q is of order $1 - \exp(-A\beta p)$ for large p, while the parameter m is 1 at the critical temperature, and has a nearly linear behaviour al low temperature. The only difference is that m is no more strictly linear as function of the temperature.

The thermodynamical properties of the model are the same as is the Random Energy Model (indeed we recover the REM when $p \to \infty$): there is a transition at T_c with a discontinuity in the specific heat, with no divergent susceptibilities.

A very interesting finding is that if we consider the infinite model and we cool it starting at high temperature, there is a transition at a temperature $T_D > T_c$ [4], [5]. At temperatures less than T_D the system is trapped in a metastable state. The correlation time (not the equilibrium susceptibilities) diverges at T_D and the mode-mode coupling become exact in this region [10]

It was suggested some time ago [11], [12], [13] that these properties of the p-spin model strongly hint that this model may be considered a mean field realization of a glassy system and it should share with real glasses the physical origine of this behaviour.

It is interesting to note that although in the original approach the dynamical transition was found by using explicitly the equation of motions now we have techniques which allow the determination of the dynamical transition and of some of the properties of the exponentially large times needed to escape from a metastable state using only equilibrium computations [14], [15], [16], [17]. The main technical tool consists in computing the properties of a system with coupled replicas [18].

2.2 Model without quenched disorder

We could ask how much of the previous results can be carried to models without quenched results. It has been found in the framework of the mean field theory (i.e. when the range of the interaction is infinite), that there a the partial equivalence of Hamiltonians with quenched and random disorder. More precisely it often possible to find Hamiltonians which have the same properties (at least in a region of the phase space) of the Hamiltonian without disorder [19] - [26] . An example of this intriguing phenomenon is the following.

The configuration space of our model is given by N Ising spin variables [19]. We consider the following Hamiltonian

$$H = \sum_{i=1,N} |B_i|^2 - 1|^2, \tag{6}$$

$$\text{where} \quad B_i = \sum_{k=1,N} R_{i,k}\sigma_k. \tag{7}$$

Here R is an unitary matrix, i.e.

$$\sum_{k=1,N} R_{i,k}\overline{R_{k,m}} = \delta_{i,m}. \tag{8}$$

We could consider two different cases [19] :

− The matrix R is a random orthogonal matrix.
− The matrix R is given by

$$R(k,m) = \frac{\exp(2\pi i \; km)}{N^{1/2}} \tag{9}$$

In other words B is the Fourier transform of σ.

The second case is a particular instance of the first one, exactly in the same way that a sequence of all zeros is a particular instance of a random sequence.

The first model can be studied using the replica method and one finds results very similar to those of the p-spin model we have already studied.

Now it can be proven that the statistical properties of the second model are identical to those of the first model, with however an extra phase. In the second model (at least for some peculiar value of N, e.g. N prime [20], [19], [21]) there are configurations which have exactly zero energy. These configuration form isolated valleys which are separated from the others, but have much smaller energy and they have a very regular structure (like a crystal). An example of these configurations is

$$\sigma_k \equiv_{\text{mod}N} k^{(N-1)/2} \tag{10}$$

(The property $k^{(N-1)} \equiv 1$ for prime N , implies that in the previous equations $\sigma_k = \pm 1$). Although the sequence σ_k given by the previous equation is apparently random, it satisfies so many identities that it must be considered as an extremely ordered sequence (like a crystal). One finds out that from the thermodynamical point of view it is convenient to the system to jump to one of these ordered configurations at low temperature. More precisely there is a first order transition (like a real crystalization transition) at a temperature, which is higher that the dynamical one.

If the crystallisation transition is avoided by one of the usual methods, (i.e. explicit interdiction of this region of phase space or sufficient slow cooling), the properties of the second model are exactly the same of those of the first model. Similar considerations are also valid for other spin models [23], [24], [25] or for models of interacting particles in very large dimensions, where the effective range of the force goes to infinity [26], [27], [28], [29] .

We have seen that when we remove the quenched disorder in the Hamiltonian we find a quite positive effect: a crystallisation transition appears like in some real systems. If we neglect crystalization, which is absent for some values of N, no new feature is present in system without quenched disorder.

These results are obtained for long range systems. As we shall see later the equivalence of short range systems with and without quenched disorder is an interesting and quite open problem.

3 Short range models

The interest of the previous argument would be to much higher if we could apply them to short range models. In order to discuss this point it is convenient to consider two classes of models:

$$H = \sum_{x,y,z} J(x, y, z)\sigma(x)\sigma(y)\sigma(z)$$

$$H = \sum_{x,y,z} J(x - y, x - z)\sigma(x)\sigma(y)\sigma(z). \tag{11}$$

In both cases the sum is restricted to triplets spins which are at distance less than R (i.e. $|x-y| < R$, $|x-z| < R$, $|z-y| < R$). In the first case the model is not translational invariant while in the second case the Hamiltonian is translational invariant. Moreover in the second case the Hamiltonian depends only on a *finite* number of J also when the volume goes to infinity, so that the free energy density will fluctuates with J. The second model looks much more similar to real glasses than the first one. A careful study of the difference among these two models has not yet been done. In any case it is clear that the phenomenon of metastability which we have seen in the previous section for the infinite range models is definitely not present in these short range models. No metastable states with infinite mean life do exist in nature.

Let us consider a quite simple argument. Let us suppose that the system may stay in phase (or valleys) which we denote as A and B. If the free energy density of B is higher than that of A, the system can go from B to A in a progressive way, by forming a bubble of radius R of phase A inside phase B. If the surface tension among phase A and B is finite, has happens in any short range model, for large R the volume term will dominate the free energy difference among the pure phase B and phase B with a bubble of A of radius R. This difference is thus negative at large R, it maximum will thus be finite.

In the nutshell a finite amount of free energy in needed in order to form a seed of phase A starting from which the spontaneous formation of phase A will start. For example, if we take a mixture of H_2 and O_2 at room temperature, the probability of a spontaneous temperature fluctuation in a small region of the sample, which lead to later ignition and eventually to the explosion of the whole sample, is greater than zero (albeit quite a small number), and obviously it does not go to zero when the volume goes to infinity.

We have two possibilities open in positioning the mean field theory predictions of existence of real metastable states:

- We consider the presence of these metastable state with *infinite* mean life an artefact of the mean field approximation and we do not pay attention to them.
- We notice that in the real systems there are metastable states with very large (e.g. much greater than one year) mean life. We consider the *infinite* time metastable states of the mean field approximation as precursors of these *finite* time metastable states. We hope (with reasons) that the corrections to the mean field approximation will give a finite (but large) mean life to these states (how this can happen will be discussed later on).

Here we suppose that the second possibility is the most interesting and we proceed with the study of the system in the mean field approximation. We have already seen that in a short range model we cannot have real metastable states. Let us see in more details what happens.

Let us assume that $\Delta f < 0$ is the difference in free energy among the metastable state and the stable state. Now let us consider a bubble of radius R of stable state inside the metastable one. The free energy difference of such a bubble will be

$$F(R) = -\Delta f \, V(R) - I(R) \qquad (12)$$

where the interfacial free energy $I(R)$ can increase at worse as $\Sigma(R)$. The quantities $V(R) \propto R^D$ and $\Sigma(R) \propto R^{D-1}$ are respectively the volume and the surface of the bubble; σ is the surface tension, which can also be zero (when the surface tension is zero, we have $I(R) \propto R^\omega$, with $\omega < D - 1$).

The value of $F(R)$ increases at small R, reaches a maximum a R_c, which in the case $\sigma \neq 0$ is of order $(\Delta f)^{-1}$ and it becomes eventually negative at large R. According to enucleation theory, the system goes from the metastable to the stable phase under the formation and the growth of such bubbles, and the time to form one of them is of order (neglecting prefactors)

$$\ln(\tau) \propto \Delta f \, R^D \propto (\Delta f)^{(D-1)} \qquad (13)$$

$$\tau \propto \exp\left(\frac{A}{(\Delta f)^{(D-1)}}\right) \qquad (14)$$

where A is constant dependent on the surface tension. In the case of asymptotically zero surface tension we have

$$\tau \propto \exp(\frac{A}{\Delta f^l}) \tag{15}$$

$$l = \frac{D}{\omega} - 1 \tag{16}$$

This argument for the non existence of metastable states can be naively applied here. The metastable states of the the mean field approximation now do decay. The dynamical transition becomes a smooth region which separates different regimes; an higher temperature regime where mode mean field predictions are approximately correct and a low temperature region where the dynamics is dominated by barriers crossing. There is no region where the predictions of mean field theory are exact but mean field theory is only an approximated theory which describe the behaviour in a limited region of relaxation times (large, but not too large).

The only place where the correlation time may diverge is at that the thermodynamical transition T_c, whose existence seems to be a robust prediction of the mean field theory. It follows that the only transition, both from the static and the dynamical point of view, is present at T_c.

In order to understand better what happens near T_c we must proceed in a careful matter. The nucleation phenomenon which is responsible of the decay of metastable states is of the same order of other non-perturbative corrections to the mean field behaviour, which cannot be seen in perturbation theory. We must therefore compute in a systematic way all possible sources of non perturbative corrections. This has not yet been done, but it should not be out of reach.

One of the first problem to investigate is the equivalence of systems with and without random disorder. In systems with quenched disorder there are local inhomogeneities which correspond to local fluctuations of the critical temperature and may dominate the thermodynamics when we approach the critical temperature.

It is quite possible, that systems with and without quenched disorder, although they coincide in the mean field approximation, they will be quite different in finite dimension (e.g. 3) and have different critical exponents. The scope of the universality classes would be one of the first property to assess.

It is not clear at the present moment if the strange one order and half transition is still present in short range model or if it is promoted to a bona fide second order transition. If the transition remains of the order one and half (for example is conceivable that this happens only for systems without quenched disorder). It could also possible that there is appropriate version of the enucleation theory which is valid near T_c and predicts:

$$\tau \propto \exp(\frac{A}{(T - T_C)^\alpha}). \tag{17}$$

A first guess for α is $D - 2$ [13], [14] , although other values, e.g. 2/3, are possible. A more detailed understanding of the static properties near T_c is needed before we can do any reliable prediction.

4 Toward realistic models

4.1 General considerations on glasses

I would like now to shortly review the properties of real glasses, which, as we shall see, have many points in common with the systems that we have considered up to now. As usual we must select which of the many characteristics of glasses we think are important and should be understood. This is a matter of taste. I believe that the the following facts are the main experimental findings about glasses that a successfully theory of glasses should explain:

1. If we cool the system below some temperature (T_G), its energy depends on the cooling rate in a significant way. We can visualize T_G as the temperature at which the relaxation times become of the order of a hour.
2. No thermodynamic anomalies (i.e. divergent specific heat or susceptibilities) are observed: the entropy (extrapolated at ultraslow cooling) is a linear function of the temperature in the region where such an extrapolation is possible. For finite value of the cooling rate the specific heat is nearly discontinuous. Data are consistent with the possibility that the true equilibrium value of the specific heat is also discontinuous at a temperature T_c lower than T_G. This results comes mainly from systems which have also a crystal phase (which is reached by a different cooling schedule) under the very reasonable hypothesis that the entropy of the glass phase cannot be smaller than that of the crystal phase.
3. The relaxation time (and quantities related to it, e.g. the viscosity, which by the Kubo formula is proportional to the integral of the correlation function of the stress energy tensor) diverges at low temperature. In many glasses (the fragile ones) the experimental data can be fitted as

$$\tau = \tau_0 \exp(\beta B(T)), \tag{18}$$
$$B(T) \propto (T - T_c)^{-l}, \tag{19}$$

where $\tau_0 \approx 10^{-13}s$ is a typical microscopic time, T_c is near to the value at which we could guess the presence of a discontinuity in the specific heat and the exponent l is of order 1. The so called Vogel-Fulcher law [1] states that $\lesssim= 1$. The precise value of l is not too well determine. The value 1 is well consistent with the experimental data, but different values are not excluded.

Points 1 and 2 can be easily explained in the framework of the mean field approximation. Point 3 is a new feature of short range models which is not present in the mean field approximation. It is clearly connected to the non-existence of infinite mean life metastable states in a finite dimensional world. One needs a more careful analysis in order to find out the origine of this peculiar behavior and obtain quantitative predictions.

It is interesting to find out if one can construct approximation directly for the glass transitions in liquids. Some progresses have been done in the framework of mean field theory in the infinite dimensional cases. Indeed the model for hard spheres moving on a sphere can be solved exactly in the high temperature phase when the dimension of the space goes to infinity in a suitable way [26], [27], [29].

4.2 A first attempt to use the replica method for glasses

One of the most interesting results is the suggestion that the replica method can be directly applied to real glasses. The idea is quite simple [30]. We assume that in the glassy phase a finite large system may state in different valleys (or states), labeled by α. The probability distribution of the free energy of the valley is given by eq. (4). We can speak of a probability distribution because the shape of the valleys and their free energies depends on the total number of particles. Each valley may be characterized by the density

$$\rho(x)_\alpha \equiv < \rho(x) >_\alpha .$$ (20)

In this case we can define two correlation functions.

$$g(x) = \frac{\int dy < \rho(y)\rho(y+x) >>_\alpha}{V}$$

$$f(x) = \frac{\int dy < \rho(y) >_\alpha < \rho(y+x) >>_\alpha}{V}.$$ (21)

A correct description of the low temperature phase must take into account both correlation functions. The replica method does it quite nicely: g is the correlation function inside one replica and f is the correlation function among two different replicas.

$$g(x) = \frac{\int dy < \rho_a(y)\rho_a(y+x) >>}{V}$$

$$f(x) = \frac{\int dy < \rho_a(x)\rho_b(y+x) >>}{V},$$ (22)

with $a \neq b$.

The problem is now to write closed equation for the two correlation functions f and g. Obviously in the high temperature phase we must have that $f = 0$ and the non-vanishing of f is a signal of entering in the glassy phase.

The first attempt in this direction was only a partial success [30]. A generalized hypernetted chain approximation was developed for the two functions f and g. A non trivial solution was found at sufficient low temperature both for soft and hard spheres and the transition temperature to a glassy state was not very far from the numerically observed one. Unfortunately the value of the specific heat at low temperature is not the correct one (it strongly increases by decreasing the temperature). Therefore the low temperature behaviour is not the correct one; this should be not a surprise because an esplicite computation show that the corrections to this hypernetted chain approximation diverge at low temperature.

These results show the feasibility of a replica computation for real glasses, however they point in the direction that one must use something different from a replicated version of the hypernetted chain approximation. At the present moment it is not clear which approximation is the correct one, but I feel confident that a more reasonable one will be found in the near future.

References

[1] H. Vogel, Phys. Z, **22**, 645 (1921); G.S. Fulcher, J. Am. Ceram. Soc., **6**, 339 (1925).

[2] M.Mézard, G.Parisi and M.A.Virasoro, *Spin glass theory and beyond*, World Scientific (Singapore 1987).

[3] G.Parisi, *Field Theory, Disorder and Simulations*, World Scientific, (Singapore 1992).

[4] L.Cugliandolo, J.Kurchan, Phys. Rev. Lett.**71** (1993) 173.

[5] A. Crisanti, H. Horner and H.-J. Sommers, *The Spherical p-Spin Interaction Spin-Glass Model: the Dynamics*, Z. Phys. **B92** (1993) 257.

[6] B.Derrida, Phys. Rev. **B24** (1981) 2613.

[7] L. R. G. Fontes, Y. Kohayakawa, M. Isopi, P. Picco *The Spectral Gap of the REM under Metropolis Dynamics* Rome preprint (1996).

[8] D. J. Gross and M. Mézard, *The Simplest Spin Glass*, Nucl. Phys. **B240** (1984) 431.

[9] E. Gardner, *Spin Glasses with p-Spin Interactions*, Nucl. Phys. **B257** (1985) 747.

[10] J.-P. Bouchaud, L. Cugliandolo, J. Kurchan, Marc Mézard, cond-mat 9511042.

[11] T. R. Kirkpatrick and D. Thirumalai, Phys. Rev. Lett. **58**, 2091 (1987).

[12] T. R. Kirkpatrick and D. Thirumalai, Phys. Rev. **B36**, 5388 (1987).

[13] T. R. Kirkpatrick, D. Thirumalai and P.G. Wolynes, Phys. Rev. **A40**, 1045 (1989).

[14] G. Parisi, *Gauge Theories, Spin Glasses and Real Glasses*, Talk presented at the Oskar Klein Centennial Symposium, cond-mat 9411115.

[15] S.Franz and G.Parisi, J. Phys. I (France) 5(1995) 1401.

[16] E. Monasson Phys. Rev. Lett. **75** (1995) 2847.

[17] A. Cavagna, I. Giardina and G. Parisi, cond-mat 9611068.

[18] J. Kurchan, G. Parisi and M.A. Virasoro, J. Physique **3**, 18 (1993).

[19] E.Marinari, G.Parisi and F.Ritort, J.Phys.A (Math.Gen.) **27** (1994), 7615; J.Phys.A (Math.Gen.) **27** (1994), 7647.

[20] G. Migliorini, *Sequenze Binarie in Debole Autocorrelazione*, Tesi di Laurea, Università di Roma *Tor Vergata* (Roma, March 1994).

[21] I.Borsari, S.Graffi and F.Unguendoli, J.Phys.A (Math.Gen.), to appear and *Deterministic spin models with a glassy phase transition*, cond-mat 9605133.

[22] G. Parisi and M. Potters, J. Phys. A: Math. Gen. **28** (1995) 5267, Europhys. Lett. **32** (1995) 13.

[23] S. Franz and J. Hertz; Nordita Preprint, cond-mat 9408079.

[24] E. Marinari, G. Parisi and F. Ritort, J. Phys. **A**: Math. Gen. **28** 1234 (1995).

[25] E. Marinari, G. Parisi and F. Ritort J. Phys. **A**: Math. Gen. **28** 327 (1995).

[26] L.Cugliandolo, J.Kurchan, G.Parisi and F.Ritort, Phys. Rev. Lett. **74** (1995) 1012.

[27] L.Cugliandolo, J.Kurchan, E.Monasson and G.Parisi, Math. Gen. **29** (1996) 1347.

[28] G. Parisi, cond-mat 9701032.

[29] G.Parisi and F.Russo, work in progress.

[30] M. Mézard and G. Parisi cond-mat 9602002.

Classical and quantum behavior in mean-field glassy systems

Felix Ritort

Institute of Theoretical Physics, University of Amsterdam,
Valckenierstraat 65, 1018 XE Amsterdam (The Netherlands).
E-Mail: ritort@phys.uva.nl

In this talk I review some recent developments which shed light on the main connections between structural glasses and mean-field spin glass models with a discontinuous transition. I also discuss the role of quantum fluctuations on the dynamical instability found in mean-field spin glasses with a discontinuous transition. In mean-field models with pairwise interactions in a transverse field it is shown, in the framework of the static approximation, that such instability is suppressed at zero temperature.

1 Introduction

There is much current interest in the study of disordered systems. These are characterized by the presence of quenched disorder, i.e. disorder which is frozen at a timescale much larger than the typical observation time. In particular, a large amount of experimental and theoretical work has been devoted to the study of spin glasses[1]. These are alloys where magnetic impurities are introduced in the system. The magnetic impurities are frozen inside the host material and the interaction between them is due to the conduction electrons. This is the RKKY interaction which can be ferromagnetic or antiferromagnetic depending on the distance between the frozen impurities. Hence, the site disorder in the system leads to frustrated exchange interactions.

But there are also intrinsically non disordered systems which display glassy behavior as soon as they are off-equilibrium. The most well known examples are structural glasses [2], for instance dioxide of silica SiO_2. Structural glasses are characterized by the existence of a thermodynamic crystalline phase below the melting transition. Under fast cooling the glass does not crystallize at the melting transition temperature T_M and it stays in the supercooled state in local equilibrium. Instead if the temperature is slowly decreased this local equilibrium property is lost when the cooling rate is of the same order than the inverse of the relaxation time.

There are two main differences between glasses and spin glasses. The first difference has been already pointed out. It is that spin glasses are disordered systems while glasses are intrinsically clean. The second difference emerges from experimental measurements which show that in spin glasses there is a static quantity, the non linear susceptibility, which diverges at a critical temperature. This implies the existence of characteristic length scale or correlation length which diverges at the critical point. On the contrary in real glasses a divergence of a static susceptibility has not been observed. Hence, while in spin glasses there is common agreement that there is a true thermodynamic transition the situation for structural glasses is much less clear and such a thermodynamic transition (the so called ideal glass transition) is still only a theoretical concept.

The renewed interest in the connection between glasses and spin glasses comes from the observation that a certain class of mean-field microscopic spin-glass models seem to capture the essential features which are present in structural glasses. Furthermore, while spin glass mean-field models explicitly contain disorder it has been recently shown that this is not an essential ingredient and spin glass behavior is found even in microscopic models where disorder is absent. The connection appears then fully justified.

This connection between structural glasses and spin glasses at the classical level has been already emphasized in other talks in this conference (see the talks by S. Franz and G. Parisi). After reviewing recent developments in this direction I will address a different aspect of glassy behavior, this is the study of glassiness in the presence of quantum fluctuations. There are several reasons why quantum fluctuations are interesting [3]. The most compelling reason is that the low temperature behavior of a large variety of systems in condensed matter physics strongly depends on the presence of disorder, for instance the quantum Hall effect and the metallic insulator transition (see the talk by T. Kirkpatrick in these proceedings). Another reason is more theoretical and relies on the need for a better understanding of the role of randomness in quantum phase transitions in the regime where there is no dissipation. The main question we want to discuss in this proceedings concerns this last point, i.e. how the glassy behavior which emerges in the classical picture is modified when quantum fluctuations (i.e. non dissipative processes) are taken into account.

The talk is organized as follows. First I will review some recent work on the connections between glasses and spin glasses at the classical level, putting special emphasis in the difference between spin-glass models with continuous and discontinuous transitions. In section 3 I will discuss the relevance of quantum fluctuations in glassy phenomena and give a short reminder to the main results in the Sherrington-Kirkpatrick model in a transverse field. In section 4 I will show that a large class of exactly solvable models with discontinuous phase transition at finite temperature have a continuous phase transition at zero temperature. Particular results will be presented for the random orthogonal model. Finally I will discuss the implications of this result and present the conclusions.

2 Glasses v.s. spin glasses

It has been realized quite recently that systems without disorder can have a dynamical behavior reminiscent of spin glasses. While this suggestive idea has to be traced back to Kirkpatrick and Thirumalai [4] only recently has it been shown how this idea works in some microscopic models. In particular, Mezard and Bouchaud [5] studied the Bernasconi model [6] by mapping it onto a disordered model (the p-spin Ising spin glass with $p = 4$) and thus finding evidence for glassy behavior. It has also been shown in [7] how it is possible to map the Bernasconi model (with periodic boundary conditions) into a disordered model and solve it exactly by means of the replica method. Within this approach it is possible to show that both, the ordered and the disordered model, have the same high temperature expansion. While the thermodynamics of the models is different at low temperatures (the disordered model does not have a crystalline state) the ordered model reproduces all the features of the metastable glassy phase found in the disordered model including the existence of a dynamical singularity.

There have been several microscopic models such as the sine or cosine model [8], fully frustrated lattices [9], matrix models [10], the Amit-Roginsky model[11] and mean-field Josephson junction arrays in a magnetic field [12, 13] (see the talk by P. Chandra in this conference) where this approach has been succesfully applied. The main conclusion which emerges from these studies is that quenched disorder is not necessary to have spin glass behavior but it can be *self-generated* by the dynamics. Physically this means the following: the relaxation of the sytem becomes slower as the temperature is decreased and the local fields, acting on the microscopic variables of the system, can be considered as effectively frozen. It seems also that all mean-field models where this mapping is possible are those which show the existence of a dynamical singularity above the static transi- tion temperature. In fact, all models where this equivalence has been built up are characterized by a discontinuous transition. In the spin-glass language this corresponds to models with one-step replica symmetry breaking transition.

In what follows I will discuss the main results concerning mean-field spin glass models contrasting those with a continuous and discontinuous transition. The phase transition in spin glasses is described by an order parameter which is the Edwards-Anderson parameter (hereafter referred as EA order parameter). Because spin glasses are intrinsically disordered systems the magnetization is not a good order parameter since long range order is absent. In fact, below the spin-glass transition the spins tend to freeze in certain directions which randomly change from site to site. While spatial fluctuations of the local magnetization are large the temporal fluctuations are quite small and the parameter which measures the local spin glass ordering is given by $\langle \sigma_i \rangle^2$. The EA order parameter [14] is the average of this quantity over the whole lattice, $q_{EA} = \frac{1}{N} \sum_{i=1}^{N} \overline{< \sigma_i >^2}$. The EA parameter varies from zero to 1. If it is very close to 1 this means that the system is strongly frozen and thermal fluctuations are small. In spin-glass models with a continuous transition q_{EA} is zero above the spin glass transition T_g (in the paramagnetic phase) and continuously increases as the temperature is lowered below T_g (i.e. within the spin glass phase). In these models one finds $q_{EA} \simeq (T_g - T)$ which means that the critical exponent β is equal to 1, a typical value found in mean-field disordered systems. The simplest example of this class of models is the Sherrington-Kirkpatrick model [15] defined by

$$H = -\sum_{i<j} J_{ij}\sigma_i\sigma_j \qquad (1)$$

where the J_{ij} are random Gaussian variables with zero mean and variance $1/N$.

In models with a discontinuous transition q_{EA} is zero above T_g but discontinuously jumps to a finite value below T_g. Examples of models with a transition of this type are q-states Potts glass models with $q \geq 4$ [16] and p-spin glass models with $p \geq 3$ [17]. In the limit $p, q \to \infty$ both class of models converge to the random energy model of Derrida [18] which is characterized by an order parameter which is 1 just below T_g. Because this model is a particular limit where the energies of the configurations are randomly distributed and also because it can be fully solved without the use of replicas it is usually referred to as the simplest spin glass [19]. The two types of transitions are shown in figure 1.

Another example of a model with a discontinuous transition has been recently introduced [8]. This is the random orthogonal model (we will use the initials ROM in the rest of the paper to refer to this model) defined by eq.(1) where now the J_{ij} are matrix elements of a random orthogonal ensemble of matrices, i.e. $J_{ij}J_{jk} = \delta_{ik}$. In the ROM model q_{EA} jumps discontinuously to 0.9998 at the transition point [8]. This model can be considered as a very faithful microscopic realization of the random energy model of Derrida. It is important to note that both discontinuous and continuous spin glass transitions are continuous from the thermodynamic point of view. This is related to one of the main subtelities of spin glasses where order parameters are functions $q(x)$ in the interval [0:1] and thermodynamic quantities are integrals of moments of these functions [1]. The functional nature of the order parameter $q(x)$ is related to the existence of an infinite number of pure states in the spin-glass phase as shown by G. Parisi [1]. In discontinuous transitions the $q(x)$ has a finite jump for $x \to 1$ and all the moments $\int_0^1 q^p(x)dx$ remain continuous at the transition, hence there is no latent heat. The body of these results apply to mean-field models with long ranged interactions. Quite surprisingly it appeared that these mean-field models with a discontinuous transition are a nice realization of the entropy crisis theory proposed for glasses long ago by Gibbs and Di Marzio and later on by Adam and Gibbs in two seminal papers[20]. This is a heuristic theory which is based on the Kauzmann paradox[21] and proposes the collapse of the configurational entropy as the mechanism for a thermodynamic glass transition. The simplest example where this transition occurs is the random energy model of Derrida where the phase transition coincides with the point at which the entropy collapses to zero. This transition corresponds to what theorists refer to as the ideal glass transition and lies below the laboratory glass transition (see the talk by Angell in this proceedings) which is defined as the temperature at which the relaxation time is of order of a quite few minutes (more concretely the viscosity is 10^{13} Poise [2]).

It is important to note that the ideal glass transition is a thermodynamic transition where the configurational entropy collapses to zero but still the total

---

[1] The local order parameter inside one state is given by the EA order parameter which is given by the relation $q_{EA} = \max_x q(x)$.

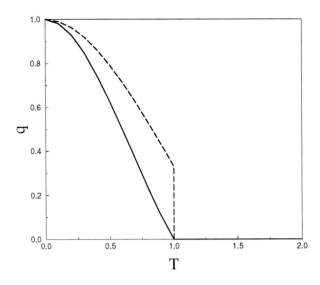

Fig. 1. EA order parameter as a function of T for a continuous transition (continuous line) and a discontinuous transition (long dashed line).

entropy can be finite since fluctuations inside a configurational state can still be present. In the random energy model, fluctuations in the low temperature phase are absent and the full entropy (which gets contributions from the configurational entropy and the entropy coming from local fluctuations inside one state) vanishes at the glass transition T_g (in what follows we will denote by T_g the ideal glass thermodynamic transition temperature). In the ROM the entropy at T_g is of order 10^{-4} and can be considered quite small (the entropy per free spin is $log(2)$).

The connection between structural glasses and spin glasses with a discontinuous transition would not be fully acomplished if dynamics is not taken into account. This was realized by Kirkpatrick, Thirumalai and Wolynes [22] who noted the existence of a dynamical transition T_d above the glass transition T_g in connection with an instability found in the mode coupling theory of glasses (MCT). S. Franz and J. Hertz have shown [11] that the dynamical equations of the Amit-Roginsky model (a model with pseudo-random interactions which, nevertheless, does not contain explicit disorder) can be mapped onto the dynamical equations of the p-spin spherical spin glass model with $p = 3$ [23]. At the dynamical transition a large number of metastable states (which grows exponentially with the size of the system) determines an instability in the relaxational dynamics of the system but does not induce a true thermodynamic transition.

In the region $T_g < T < T_d$ the free energy of the system is given by the para-magnetic free energy f_P but the dynamical response is fully determined by the presence of a very large number of metastable configurations $exp(NC^*)$ where C^* is the so called configurational entropy or complexity. Note that the free energy of these metastable states is higher than the free energy of the paramagnetic state and there is no thermodynamic transition at T_d. As the temperature is decreased the number of metastable configurations decreases and so does their free energy. When the free energy of the metastable states equals the paramag-netic free energy there is a phase transition. Because the number of metastable solutions with equilibrium free energy are not exponentially large with the size of the system the configurational entropy C^* also vanishes at this point.

The suspicious reader will find it extremely unclear how all these quantities (T_d, T_g, q_{EA}, C^*) can be analytically computed. Fortunately it is not necessary to fully solve the dynamics in order to find these quantities and there are pow-erful techniques to compute them. One of the simplest procedures [24] works for discontinuous transitions of the type described here and consists in expanding the free energy around $m = 1$ (m parametrizes the one step replica symmetry breaking and corresponds to the size of the diagonal blocks with finite order pa-rameter q in the Parisi ansatz [1]). The free energy is expanded in the following way,

$$\beta f(q) = \beta f_P + (m - 1)\mathcal{V}(q) \tag{2}$$

where f_P is the paramagnetic free energy (independent of q) and $\mathcal{V}(q)$ is a function called the potential [26]. Note in eq.(2) that $\mathcal{V}(q)$ plays the role of an entropy contribution to the paramagnetic free energy except for the factor $(m-1)$. The dynamical transition corresponds to an instability in the dynamics and is obtained by solving the equations,

$$\frac{\partial \mathcal{V}}{\partial q} = 0 \quad ; \quad \frac{\partial^2 \mathcal{V}}{\partial q^2} = 0 \tag{3}$$

which yield the transition T_d and the jump of the EA order parameter q_{EA}^d at the dynamical transition temperature. On the other hand, the static transition is obtained by solving the equations

$$\frac{\partial \mathcal{V}}{\partial q} = 0 \quad ; \quad \mathcal{V}(q) = 0 \tag{4}$$

and yield the transition T_g and the jump of the EA order parameter q_{EA}^g at the glass transition temperature. In the range of temperatures $T_g < T < T_d$ the complexity or configurational entropy C^* is given by the value of the potential $\mathcal{V}(q)$ in the secondary minimum (see figure 2) in the region $q_{EA}^d < q < q_{EA}^g$. For continuous transitions (such as that found in the Sherrington-Kirkpatrick model [15]) both temperatures (T_d and T_g) coincide and there is no discontinuous jump of q_{EA} at the transition temperature. The behavior of $\mathcal{V}$ as a function of q for different temperatures is shown in figure 2 for a discontinuous transition. Note that the behavior shown in figure 2 is quite reminiscent of a spinodal instability in first-order phase transitions. The behavior of the potential $\mathcal{V}(q)$ determines the phase transition and in particular the existence of an instability at T_d. The reason why a zero mode at T_d yields a divergent relaxation time in the dynamics is related to the one of the most prominent features of glassy systems: the dominance of an exponentially large of metastable states $(\exp(N\mathcal{C}^*))$ at that temperature[22, 25]. Note that in mean-field theory metastable states have an infinite lifetime, the time to jump from the metastable glassy phase to the paramagnetic state being equal to $exp(N\mathcal{B}^*)$ where $\mathcal{B}^* = \max_{0<q<q_{EA}^g} \mathcal{V}(q)$ is the height of the free energy barrier which separates the metastable and the paramagnetic phase. The extension of this approach to the computation of thermodynamic quantities below T_d in the metastable glassy phase has been considered in [26].

Summarizing, mean field spin-glass models with a discontinuous transition are good models to describe real glasses. The role of disorder is not essential and can be *self-generated* by the dynamics. These models show a thermodynamic transition (the ideal glass transition T_g) where the configurational entropy $\mathcal{C}^*$ collapses to zero and the EA order parameter jumps to a finite value. Concerning dynamics, these models are described by the mode coupling equations which are a good description of relaxational processes in real glasses at temperatures above but not too close to T_d. The instability found at T_d is a mean-field artifact which should be wiped out by including activated processes over finite energy barriers. In this sense, mode coupling equations are genuine mean-field dynamical equations. The interested reader can find more details in [27].

3 Ising spin glasses in a transverse field

There is much recent interest in the study of quantum phase transitions (see the talks by T. Kirkpatrick, R. Oppermann and H. Rieger in this proceedings). These transitions appear at zero temperature when an external parameter is varied. For a certain critical value of this parameter the system enters into the disordered phase. The general problem can be put in the following way. Let us consider the following Hamiltonian,

$$\mathcal{H} = \mathcal{H}_0 + \epsilon P \tag{5}$$

where $\mathcal{H}_0$ stands for the unperturbed Hamiltonian and P is a perturbation which does not commute with $\mathcal{H}_0$, i.e. $[\mathcal{H}_0, P] \neq 0$ and ϵ denotes the strength of the perturbation. Let us suppose that the system for $\epsilon = 0$ is in an ordered phase at $T = 0$. As the control parameter ϵ is varied and the strength of the perturbation P increases, the new ground state of $\mathcal{H}$ becomes a mixture of all the eigenstates of $\mathcal{H}_0$ and this tends to disorder the system. For a certain value of the control parameter ϵ the systems fully disorders. The effect of the control parameter ϵ in quantum phase transitions is rather similiar to the effect of temperature in classical systems. The difference is that now quantum effects are non dissipative while thermal effects are.

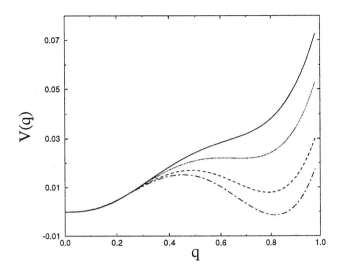

Fig. 2. Potential $\mathcal{V}(q)$ for a spin glass model with a discontinuous transition. The different regimes are: $T > T_d$ (continuous line), $T = T_d$ (dotted line), $T_g < T < T_d$ with complexity $\mathcal{C}^* = \min_{q>0} \mathcal{V}(q)$ (dashed line), $T = T_g$ with $\mathcal{C}^* = \mathcal{V}(q_{EA}^g) = 0$ (dot-dashed line)

In the realm of disordered systems it is essential to understand the role of disorder in quantum phase transitions. The rest of this talk will be devoted to discuss this problem in the framework of disordered mean-field spin glass models as presented in the previous section. Let us discuss how the classical glassy scenario is modified in the presence of quantum fluctuations. This problem has been already addressed in the literature, in particular the question wether replica symmetry breaking survives to the effect of quantum fluctuations. Large amount of work have been devoted to the study of the Sherrington-Kirkpatrick Ising spin glass in a transverse field [28, 29, 30, 31, 32]. The model is defined by,

$$\mathcal{H} = \mathcal{H}_0 + \Gamma P = -\sum_{i<j} J_{ij}\sigma_i^z\sigma_j^z - \Gamma\sum_i \sigma_i^x \qquad (6)$$

where σ_i^z, σ_i^x are the Pauli spin matrices and Γ is the the transverse field which plays the role of a perturbation. The indices i, j run from 1 to N where N is the number of sites. The J_{ij} are random variables Gaussian distributed with zero mean and $1/N$ variance. Note that for $\Gamma = 0$ this model reduces to the classical Sherrington-Kirkpatrick model which has a continuous transition to a replica broken phase with local spin-glass ordering in the z direction. The effect of the transverse field is to mix configurations and to supress the order in the z direction. For a critical value of Γ the ordering in the z direction is completely suppressed at the expense of ordering in the x direction. Note that the effect of a perturbation in the z direction in the form of a longitudinal magnetic field $P = -\sum_i \sigma_i^z$ has a quite different effect on the phase transition. The reason is that it commutes with the unperturbed Hamiltonian, hence it does not mix configurations.

The phase diagram of the model eq.(6) is reproduced in figure 3. There are two phases, the quantum paramagnetic (QP) and the quantum glass phase (QG). The transition is continuous all along the phase boundary and the QG phase resembles the classical one where replica symmetry is broken and a large number of pure states, with essentially the same free energy, contribute to the thermo-dynamics [2]. A large body of information on the SK model with a transverse field has been obtained from spin summation techniques [31] and perturbative expansions [30]. But only very recently has a full understanding of this model been achieved through seminal work by Miller and Huse[33] and in an independent way by Ye, Read and Sachdev [34]. They have been able to obtain the frequency response of the system as well as the crossover lines which separate regimes where thermal or quantum fluctuations are dominant. The values of the critical exponents as well as the nature of finite size corrections in the quantum critical point have been also numerically checked in [35].

In this framework one would like to understand the role of quantum fluctuations in systems with a discontinuous transition. In particular it is relevant to understand how quantum fluctuations could modify the dynamical instability T_d obtained in classical systems within the mode coupling approach. Note that at zero temperature the entropy must vanish and the idea of an entropy crisis as the transverse field Γ is varied is nonsense. On the other hand, since the statics and dynamics are inextricably linked in quantum phase transitions it is interesting to ask to what extent a dynamical instability (which is not related to any static singularity) can survive at zero temperature.

In this proceedings I want to discuss some recent results in a general family of solvable models which strongly suggest that the dynamical instability predicted in the mode coupling approach is completely suppressed at zero temperature when quantum fluctuations are taken dominant[36].

In particular we will show, always within this class of models, that the discontinuous transition becomes continuous at zero temperature. Particular results will be shown for the ROM (random ortoghonal model). The implications of this and other results will be also discussed.

3.1 Mean-field models with pairwise interactions

The family of exactly solvable models we are interested in are quantum Ising spin glasses with pairwise interactions in the presence of a transverse field. These are

---

[2] The extensive free energies of the different solutions only differ by finite - i.e. non extensive- quantities.

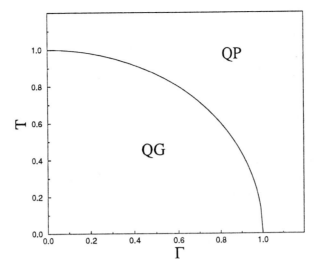

Fig. 3. Phase diagram for the SK model in a transverse field. The critical boundary separates the quantum paramagnetic phase (QP) from the quantum glass phase (QG).

described by the Hamiltonian,

$$\mathcal{H} = -\sum_{i<j} J_{ij}\sigma_i^z\sigma_j^z - \Gamma\sum_i \sigma_i^x \tag{7}$$

where σ_i^z, σ_i^x are the Pauli spin matrices and Γ is the the transverse field. The indices i, j run from 1 to N where N is the number of sites. The J_{ij} are the couplings taken from an ensemble of random symmetric matrices.

Details about how to analytically solve the quantum model (7) can be found in [36]. This are based on matrix theory techniques introduced in [7] to solve glassy models without disorder. Here we want we present some of the main results of the analysis of [36]. In practice the simplest way to solve the model eq.(7) is by means of the replica trick where we compute the average over the disorder of the n-th power of the partition function making the analytical continuation $n \to 0$ at the end,

$$\beta f = \lim_{n\to 0} \frac{\overline{Z_J^n} - 1}{n} \tag{8}$$

where

$$\overline{Z_J^n} = \int [dJ] \mathrm{Tr}\, exp(\sum_{a=1}^{n} \mathcal{H}^a) \tag{9}$$

and $\int[dJ]$ means integration over the random ensemble of matrices. This integral can be done using known methods in matrix theory [37, 8]. The final result of eq.(9) can be written in terms of a generating function $G(x)$ which depends on the particular ensemble of J_{ij} couplings via its spectrum of eigenvalues. For the two examples we will consider here we have $G_{SK}(x) = \frac{x^2}{2}$ (SK model) and $G_{ROM}(x) = \frac{1}{2}log(\frac{\sqrt{1+4x^2}-1}{2x^2}) + \frac{1}{2}\sqrt{1+4x^2} - \frac{1}{2}$ (ROM model).

By going to imaginary time and using the Trotter-Suzuki breakup we end up with a closed expression for the free energy. The final result is,

$$\overline{Z_J^n} = \int dQ\, d\Lambda exp(-NF(Q,\Lambda)) \tag{10}$$

where

$$F(Q,\Lambda) = -\frac{nC}{N} + \frac{1}{M^2}\mathrm{Tr}(Q\Lambda) - \frac{1}{2}\mathrm{Tr}G(AQ) - \log(H(\Lambda)) \tag{11}$$

where the constants A, B and C are given by $A = \frac{\beta}{M}$; $B = \frac{1}{2}\log(coth(\frac{\beta\Gamma}{M}))$; $C = \frac{MN}{2}\log(\frac{1}{2}sinh(\frac{2\beta\Gamma}{M}))$. The order parameters now depend on two set of indices: the replica index and the time indices corresponding to the imaginary time direction. The time indices t, t' go from 1 to M where M is the length of the discretized imaginary time direction. The order parameters are $Q_{ab}^{tt'}, \Lambda_{ab}^{tt'}$ (the Λ have been introduced as Lagrange multipliers in the saddle point equations) and the trace Tr is done over the replica and time indices. The term $H(\Lambda)$ is given by,

$$H(\Lambda) = \sum_{\sigma} exp(\sum_{ab}\frac{1}{M^2}\sum_{tt'}\Lambda_{ab}^{tt'}\sigma_a^t\sigma_b^{t'} + B\sum_{at}\sigma_a^t\sigma_a^{t+1}) \tag{12}$$

and the free energy is obtained by making the analytic continuation $\beta f = \lim_{n\to 0}\frac{F(Q^*\Lambda^*)}{n}$ where Q^*, Λ^* are solutions of the saddle point equations, $\Lambda_{ab}^{tt'} = \frac{AM^2}{2}\left(G'(AQ)\right)_{ab}^{tt'}$ and $Q_{ab}^{tt'} = \langle\sigma_a^t\sigma_b^{t'}\rangle$. The average $\langle(\cdot)\rangle$ is done over the effective Hamiltonian in (12). For $a = b$ we have translational time invariance and the order parameter becomes independent of the replica index, i.e $Q_{aa}^{tt'} = R(|t - t'|)$

Once we have written a closed expression for the free energy eq.(11) one can obtain the static and dynamical transition temperatures according to eq.(3,4). Such a solution always exists for models with a single quantum paramagnetic phase. In particular, using eq.(3), a closed expression for the dynamical instability can be obtained (see [36]). For a continuous transition this equation can be written in the simple form

$$\chi_0^2 G''(\chi_0) = 1 \tag{13}$$

where $\chi_0 = \beta\hat{R}_0$ is the longitudinal magnetic susceptibility and $\hat{R}_p = M^{-1}\sum_{t=0}^{M-1} e^{i\omega_p t}$ is the Fourier transformed order parameter $R(t)$ in terms of the Matsubara frequencies $\omega_p = \frac{2\pi p}{M}$.

Our main aim is to compute the order of the transition. We already know that some models within the family eq.(7) (for instance, the ROM) have a classical discontinuous transition with a dynamical instability above the static transition. Is the discontinuous nature of this transition changed in the presence of quantum fluctuations?

To answer this question we consider the static approximation introduced by Bray and Moore in the context of quantum spin glasses[28]. This approximation considers $R(t - t')$ to be constant which amounts to take into account only the zero frequency behavior $p = 0$ (small energy fluctuations) in the set of order parameters $\hat{R}_p$. We will later comment on the validity of this approximation.

Using this approximation one can write closed expressions for the paramagnetic free energy f_P and the complexity C. It is found [36] that at $T = 0$ the dynamical transition T_d and the static transition T_g coincide and the complexity vanishes. The transition then becomes continuous. The value of the critical field and all the thermodynamic observables at the critical point at zero temperature can be expressed in terms of the longitudinal susceptibility χ_0 which satisfies the following simple set of equations,

$$\chi_0^2 G''(\chi_0) = 1 \quad ; \quad \Gamma - \frac{1}{\chi_0} = G'(\chi_0); \tag{14}$$

Note that the first of eq.(14) can be obtained derivating the second of eq.(14)respect to χ_0. At the critical point the internal energy is given by $U = -\Gamma_c$ and the entropy $S = \frac{1}{2}(G(\chi_0) + 1 - \Gamma_c\chi_0 + log(\Gamma_c\chi_0))$. Above the critical point, always at zero temperature, the second of eq.(14) is still valid and yields the susceptibility as a function of Γ.

To go beyond the static approximation we should consider all the Matsubara modes $\hat{R}_p$ in the saddle point equations. The difficulty of this problem is similar to that found in strongly correlated systems where an infinite set of parameters has to be computed in a self-consistent way[38]. Nevertheless we expect the order of the transition to be correctly predicted. The essential idea is that for a continuous transition at zero temperature the gap vanishes. It would be quite surprising that higher frequency modes they could drastically modify the low frequency behavior. The order of the transition should not be determined by the decay of the correlation R_t in imaginary time but for its infinite time limit which is the EA parameter at the transition point [34].

In the next subsection we analyze our results for the particular case of the ROM model and compare them with those obtained in case of the SK model.

3.2 Results for the ROM

As has been already said, the ROM has a classical discontinuous transition at zero transverse field where $q_{EA}^d \simeq 0.962$ at the dynamical transition $T_d \simeq 0.134$ and $q_{EA}^d \simeq 0.9998$ at the static transition $T_g \simeq 0.065$. At the static transition the configurational entropy or complexity C^* vanishes and the total entropy is of order of 10^{-4}. The locations of the static and dynamical transitions can be evaluated within the static approximation to obtain the results shown in figure

4. Both transition temperatures decrease as a function of the transverse field merging into the same point at zero temperature as one would expect for a continuous transition. In figure 5 we show the EA order parameter $q = \overline{<\sigma^z>^2}$ as a function of Γ as we move along the static (q_{EA}^g) and dynamical (q_{EA}^d) phase boundaries. Note that both EA order parameters q_{EA}^d and q_{EA}^g vanish at zero temperature like $T^{\frac{1}{2}}$, hence the jump in the order parameter dissapears at zero temperature.

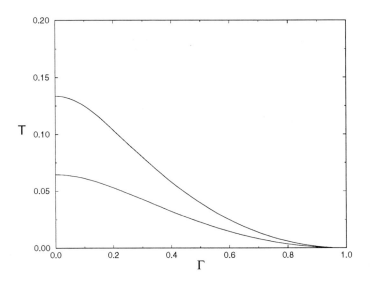

Fig. 4. Phase boundaries $T_g(\Gamma)$ (lower line) and $T_d(\Gamma)$ (upper line) in the ROM in the static approximation. At zero transverse field $T_g \simeq 0.0646, T_d \simeq 0.1336$.

By substituting the particular function $G(x)$ for the ROM model in the second of eq.(14) the susceptibility at zero temperature in the QP phase can be analytically obtained in the static approximation. One finds $\chi_0^{ROM} = \frac{\Gamma}{\Gamma^2-1}$ which diverges at the critical field $\Gamma_c = 1$. This is quite different to what is found in the SK model where $\chi_0^{SK} = \frac{\Gamma-\sqrt{\Gamma^2-4}}{2}$ and is finite at the critical point $\Gamma_c = 2$. Note that in the static approximation the critical field is given by the maximum eigenvalue of the coupling matrix J_{ij}.

Unfortunately, as we have said before, the static approximation gives incorrect results for the thermodynamic quantities. In particular, the entropy is finite

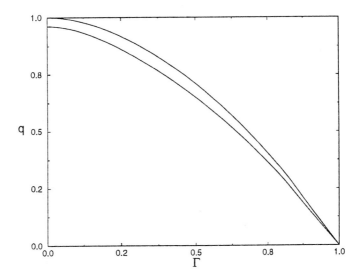

Fig. 5. EA order parameter q_{EA}^g (upper line) and q_{EA}^d (lower line) in the ROM on the static and dynamical phase boundaries boundaries as a function of the transverse field. At zero transverse field $q_{EA}^g \simeq 0.99983$, $q_{EA}^d \simeq 0.961$. Both q_{EA}^g and q_{EA}^d vanish linearly with $T^{\frac{1}{2}}$ at zero temperature.

at zero temperature in the SK model [32] and infinite in the ROM case. It is important to note that despite these failures of the static approximation some exact results can still be derived, in particular from eq.(13). When the transition is continuous equation (13) is exact. One finds for the ROM model that the longitudinal susceptibility χ_0 diverges at the critical point. This is different to what happens in the SK model where $\chi_0 = 1$ at the critical field. To see clearly the general implications of this result for continuous quantum phase transitions we observe that χ_0 is given by the decay in imaginary time of the correlation function $R(t)$ via the relation, $\chi_0 = \int_0^\beta R(t)dt$ (now the time t has become a continuous variable in the $M \to \infty$ limit). In the SK model the large time behavior of the $R(t)$ has been obtained [33, 34]. It is found that $R(t) \simeq t^{-2}$ which at zero temperature yields a finite value of the susceptibility. This decay is necessarily slower in the case of the ROM model where $\chi_0 = \infty$. Because the decay of $R(t)$ in imaginary time is related to the quantum critical exponents via the relation $R(t) \simeq t^{-\frac{\beta}{z\nu}}$ the z exponent is probably not a universal quantity and different mean-field models with continuous transitions may have different exponents. For the SK model $z = 2, \beta = 1, \nu = \frac{1}{4}$ yield the correct decay. These results suggest that the value of the exponent z is larger than 2 in the ROM.

The exact computation of the critical exponents and the full analysis of the problem beyond the static approximation in the ROM remains an interesting open problem.

4 Conclusions

In this talk I have reviewed some recent developments in the theory of spin glasses which allow for a comparison between glasses and spin glass models with a discontinuous transition. I have stressed that the two most common approaches to the glass transition, the thermodynamic approach based on the Adam-Gibbs theory and the dynamical approach based on mode coupling theory, appear quite naturally in the framework of mean-field spin glass models with a discontinuous transition. I have also stressed that disorder is not necessary to find spin-glass behavior and a large family of non disordered mean-field models indeed show spin-glass behavior. The connection between disordered spin glasses and structural glasses, at least at the mean-field level, appears to be fully justified. Going beyond the mean-field level is a major open problem. It is well accepted that the dynamical instability at T_d is an artifact of the mean-field theory but it is unclear if the entropy crisis survives in finite dimensions (see the talk of S. Franz in these proceedings).

We have also discussed the role of quantum fluctuations in glassy systems by studying the Ising spin glass in a transverse field. In quantum phase transitions statics and dynamics are inextricably linked. Then it is of relevance to understand the role of complexity in non relaxational quantum dynamics. We have shown that the classical glassy scenario with a dynamical transition above the thermodynamic transition is modified in the presence of quantum fluctuations. This result has been obtained in the framework of models with two spin interactions in the presence of a transverse field. In models with a discontinuous finite temperature transition we have shown, using the static approximation, that the transition becomes continuous at $T = 0$ and there is no room for a metastable glassy phase. We have argued in favour of this result even beyond the static approximation. Particular results have been presented for the ROM model where it has been shown that some critical exponents at the quantum phase transition should differ from the mean-field exponents derived in the case of the SK model. It is still too soon to understand the implications of this result which deserves further investigation. The removal of the instability at T_d, even in mean-field theory, could be a general consequence of the non dissipative nature of quantum processes. How general this result is for other type of models remains an interesting open problem. In this direction it would be very instructive to address the problem presented here within the approach developed in [33, 34] for the SK model as well as taking this research further by studying the zero temperature dynamical transition in quantum p-spin glass models [39] and Potts glass models [40].

Acknowledgements. I am very grateful to the following colleagues for fruitful collaborations in this and related subjects: J. V. Alvarez, S. Franz, D. Lancaster, E. Marinari, Th. M. Nieuwenhuizen, F. G. Padilla and G. Parisi. I acknowledge to the Foundation for Fundamental Research of Matter (FOM) in The Netherlands for financial support through contract number FOM-67596.

References

1. K. Binder and A. P. Young, *Spin Glasses: Experimental Facts, Theoretical Concepts and Open Questions* Rev. Mod. Phys. **58**, 801 (1986); M. Mézard, G. Parisi and M. A. Virasoro, *Spin Glass Theory and Beyond* (World Scientific, Singapore 1987); K. H. Fischer and J. A. Hertz, *Spin Glasses* (Cambridge University Press 1991); G. Parisi, *Field Theory, Disorder and Simulations* (World Scientific, Singapore 1992);

2. W. Gotze, *Liquid, freezing and the Glass transition*, Les Houches (1989), J. P. Hansen, D. Levesque, J. Zinn-Justin editors, North Holland; C. A. Angell, Science, **267**, 1924 (1995)

3. S. Sachdev in *Statphys 19* Ed. Hao Bailin, World Scientific 1996; S. L. Sondhi, S. M. Girvin, J. P. Carini and D. Shahar, *Continuous quantum phase transitions.* Preprint **cond-mat/9609279**.

4. T. R. Kirkpatrick and D. Thirumalai, J. Phys. A (Math. Gen.) (1989) L149

5. J. P. Bouchaud and M. Mezard, J. Physique I (Paris) 4 (1994) 1109.

6. J. Bernasconi, J. Physique **48** 559;

7. E. Marinari, G. Parisi and F. Ritort, J. Phys. A (Math. Gen.) **27** (1994) 7615;

8. E. Marinari, G. Parisi and F. Ritort, J. Phys. A **27** (1994) 7647;

9. E. Marinari, G. Parisi and F. Ritort, J. Phys. A (Math. Gen.) **28** (1995) 327;

10. L. F. Cugliandolo, J. Kurchan, G. Parisi and F. Ritort, Phys. Rev. Lett. **74** (1995) 1012;

11. S. Franz and J. Hertz, Phys. Rev. Lett **74** 2114 (1995);

12. G. Parisi, J. Phys. A (Math. Gen.) **27** (1994) 7555; E. Marinari, G. Parisi and F. Ritort, J. Phys. A (Math. Gen.) **28** (1995) 4481;

13. P. Chandra, L. B. Ioffe and D. Sherrington, Phys. Rev. Lett. **75** (1995) 713;

14. S. F. Edwards and P. W. Anderson, J. Phys. F (Metal. Phys.) **5** (1975) 965;

15. D. Sherrington and S. Kirkpatrick, Phys. Rev. Lett. **35**, 1792 (1975).

16. D. J. Gross, I. Kanter and H. Sompolinsky, Phys. Rev. Lett. **55** (1985) 304.

17. E. Gardner, Nucl. Phys. **B257** (1985) 747.

18. B. Derrida, Phys. Rev. **B24** (1981) 2613.

19. D. J. Gross and M. Mezard, Nucl. Phys. **B240** (1984) 431.

20. J. H. Gibbs and E. A. Di Marzio, J. Chem. Phys. **28** 373 (1958); G. Adams and J. H. Gibbs, J. Chem. Phys. **43** 139 (1965).

21. W. Kauzmann, Chem. Rev. **43** (1948) 219.

22. T. R. Kirkpatrick and D. Thirumalai, Phys. Rev. B **36**, (1987) 5342; Phys. Rev. B **36**, (1987) 5388; Phys. Rev. **B38** (1988) 4881; T. R. Kirkpatrick and P. G. Wolynes, Phys. Rev. B **36**, (1987) 8552.

23. A. Crisanti, H. Horner and H.-J. Sommers, Z. Phys. **B92** (1993) 257 ; L. F. Cugliandolo and J. Kurchan, Phys. Rev. Lett. **71** (1993) 173

24. G. Cwilich and T. R. Kirkpatrick, J. Phys A **22**, (1989) 4971; E. De Santis, G. Parisi and F. Ritort, J. Phys A **28**, (1995) 3025.

25. Th. M. Nieuwenhuizen, *Complexity as the driving force for dynamical glassy transitions.* Preprint cond-mat/9504059

26. R. Monasson, Phys. Rev. Lett **75** (1995) 2847; S. Franz and G. Parisi, J. Physique I **5** (1995) 1401

27. M. Mezard in *Statphys 19* Ed. Hao Bailin, World Scientific 1996; J. P. Bouchaud, L. F. Cugliandolo, J. Kurchan and M. Mezard; Physica **A226** (1996) 243.

28. A. J. Bray and M. A. Moore, J. Phys C. **13**, L655 (1980).

29. K. D. Usadel, Solid State Commun. **58**, 629 (1986).

30. H. Ishii and Y. Yamamoto, J. Phys. C **18**, 6225 (1985); J. Phys. C **20**, 6053 (1987);

31. G. Büttner and K. D. Usadel, Phys. Rev. B **41**, 428 (1990); Y. Y. Goldschmidt and P.Y. Lai, Phys. Rev. Lett. **64**, 2467 (1990).

32. D. Thirumalai, Q. Li, T. R. Kirkpatrick, J. Phys. A. **22**, 3339 (1989).

33. J. Miller and D. Huse, Phys. Rev. Lett. **70**, 3147 (1993);

34. J. Ye, S. Sachdev and N. Read, Phys. Rev. Lett. **70**, 4011 (1993); N. Read, S. Sachdev and J. Ye, Phys. Rev. B **52**, 384 (1995);

35. D. Lancaster and F. Ritort, *Solving the Schröedinger equation for the Sherrington-Kirkpatrick model in a transverse field.* Preprint cond-mat **9611022**.

36. F. Ritort, *Q uantum critical effects in mean-field glassy systems.* Preprint cond-mat **9607044**

37. C. Itzykson and J-B Zuber, J. Math. Phys. **21**, 411 (1980).

38. A. Georges. G. Kotliar, W. Krauth and M. J. Rozenberg, Rev. Mod. Phys. **68** (1996) 13

39. Y. Y. Goldschmidt, Phys. Rev. B **41**, 4858 (1990); V. Dobrosavljevic and D. Thirumalai, J. Phys. A **22**, L767 (1990);

40. T. Senhil and S. N. Majumdar, Phys. Rev. Lett. **76**, 3001 (1996).

Complexity as the Driving Force for Glassy Transitions

Th. M. Nieuwenhuizen

Van der Waals-Zeeman Laboratorium, Universiteit van Amsterdam
Valckenierstraat 65, 1018 XE Amsterdam, The Netherlands
e-mail: nieuwenh@phys.uva.nl

Abstract. The glass transition is considered within two toys models, a mean field spin glass and a directed polymer in a correlated random potential.

In the spin glass model there occurs a dynamical transition, where the the system condenses in a state of lower entropy. The extensive entropy loss, called complexity or information entropy, is calculated by analysis of the metastable (TAP) states. This yields a well behaved thermodynamics of the dynamical transition. The multitude of glassy states also implies an extensive difference between the internal energy fluctuations and the specific heat.

In the directed polymer problem there occurs a thermodynamic phase transition in non-extensive terms of the free energy. At low temperature the polymer condenses in a set of highly degenerate metastable states.

1 Introduction

The structural glass transition is said to occur at the temperature T_g where the viscosity equals 10^{14} Poise. The question why this transition occurs is often "answered" (more correctly: avoided) by saying that it is a dynamical transition. Surely, there is a continuum of time scales ranging from picoseconds to many years; at experimental time scales there is no equilibrium. Nevertheless, since some 20 decades in time are spanned, one would hope that equilibrium statistical mechanics can be applied in some modified way.

Crudely speaking, the observation time will set a scale. Processes with shorter timescales can be considered in equilibrium; processes with longer timescales are essentially frozen, as if they were random. To provide a (non-equilibrium) thermodynamic explanation of a model glassy transition will be the first subject of the present work.

Intuitively we expect that the resulting free energy is given by the logarithm of the partition sum, provided it has been restricted to those states that can be reached dynamically in the timespan considered. This non-equilibrium free energy will then differ from the standard case, and need not be a thermodynamic potential that determines the internal energy and entropy by its derivatives.

Experimentally one often determines the entropy S_{exp}, and thus the free energy $F_{exp} = U - TS_{exp}$, from the specific heat data by integrating C/T

from a reference temperature in the liquid phase down to T. As long as the cooling rate is finite there remains at zero temperature a residual entropy. In the limit of adiabatically slow cooling it vanishes.

Alternatively, a glass can be seen as a disordered solid. In this description the liquid undergoes a transition to a glass state with extensively smaller entropy. These states are sometimes called "states", "metastable states", "components" or, in spin glass theory, "TAP-states". As the free energy then becomes much larger, it is not so evident from thermodynamic considerations why the system can get captured in such a state with much and much smaller Gibbs-weight $\sim \exp(-\text{volume})$. The explanation is that the condensed system then has lost part of its entropy, namely the entropy of selecting one out of the many equivalent states. This part, $\mathcal{I}$, is called the *configurational entropy, complexity* or *information entropy* [1] [2]. Its origin can be understood as follows. When the Gibbs free energy $F_{\bar{a}}$ of the relevant state $\bar{a}$ has a large degeneracy $\mathcal{N}_{\bar{a}} \equiv \exp(\mathcal{I}_{\bar{a}})$, the partition sum yields $Z = \sum_a \exp(-\beta F_a) \approx \mathcal{N}_{\bar{a}} \exp(-\beta F_{\bar{a}})$, so $F = F_{\bar{a}} - T\mathcal{I}_{\bar{a}}$ is the full free energy of the system. The entropy loss arises when the system chooses the state to condense into, since from then on only that single state is observed. [3] As the total entropy $S = S_{\bar{a}} + \mathcal{I}_{\bar{a}}$ is continuous, so is the total free energy. For an adiabatic cooling experiment Jäckle has assumed that the weights p_a of the states a are fixed at the transition,[1] which implies that the free energy difference between the condensed phase and the liquid is positive and grows quadratically below T_c. This explains the well known discontinuity in quantities such as the specific heat. However, it may seem unsatisfactory that this higher free energy branch describes the physical state.

We shall first investigate these questions for the dynamical transition of a mean field spin glass model, and then for the static transition of a directed polymer in a correlated potential.

2   The p-spin glass

We first analyze these thermodynamic questions within a relatively well understood spin glass model, the mean field p-spin interaction spin glass. For a system with N spins we consider the Hamiltonian

$$\mathcal{H} = - \sum_{i_1 < i_2 < \cdots < i_p} J_{i_1 i_2 \cdots i_p} S_{i_1} S_{i_2} \cdots S_{i_p} \tag{1}$$

with independent Gaussian random couplings, that have average zero and variance $J^2 p!/2N^{p-1}$.

Kirkpatrick and Thirumalai[4] pointed out for the case of Ising spins that for $p = 3$ there is a close analogy with models for the structural glass transition, and that its properties are quite insensitive of the value of p as long as $p > 2$.

The spherical limit of this model, where the spins are real valued but subject to the spherical condition $\sum_i S_i^2 = N$, is very instructive. It has received quite some attention recently. The static problem was solved by Crisanti and Sommers. [5] There occurs a static first order transition to a state with one step replica symmetry breaking (1RSB) at a temperature T_g.

The dynamics of this spherical model was studied by Crisanti, Horner and Sommers (CHS) [6] and Cugliandolo and Kurchan [7]. Both groups find a sharp dynamical transition at a temperature $T_c > T_g$, which can be interpreted on a quasi-static level as a 1RSB transition. This dynamical transition is sharp since in mean field the metastable states have infinite lifetime. For $T < T_c$ one of the fluctuation modes is massless ("marginal"), not unexpected for a glassy state. At T_c^- there is a lower specific heat.

These dynamical approaches are the equivalent for the spin glass of the mode coupling equations for the liquid-glass transitions. temperature $T_c > T_g$ a dynamic phase transition has been reported. The presence of a sharp transition has been questioned, however. [9]

CHS integrate C/T to define the "experimental" entropy S_{exp} and the resulting free energy $F_{exp} = U - T S_{exp}$ exceeds the paramagnetic one quadratically. The interpretation of metastable states ("TAP-states") in this system is discussed in ref. [10] The statistics of those states was considered by Crisanti and Sommers (CS). [11] Assuming that the result of long time dynamics follows through being stuck in the metastable state of highest complexity, they reproduced the "experimental" free energy obtained of CHS. This confirms Jäckle's prediction of a quadratically higher free energy in the glassy state.

The long-time dynamics of 1RSB transitions fixes q_0 ($= 0$ in zero field), q_1 and x, which are just the plateau values and the breakpoint of a related Parisi order parameter function, respectively. In p-spin models they can simply be derived from a 1RSB replica calculation provided one fixes x by a marginality criterion for fluctuations on the q_1 plateau. The present author recently assumed that this is a very general phenomenon. expectation that a dynamical transition will automatically get trapped in a state with diverging time scale, if present. In a Potts model this then predicts a dynamical transition with marginal q_0 plateau and stable q_1 plateau.

As the replica free energy is minimized in this procedure, it lies near T_c^- *below* the paramagnetic value and has a larger slope. Though this is exactly what one expects at a first order phase transition, it is a new result for dynamical glassy transitions. It is the purpose of the present work to discuss the physical meaning of the mentioned free energies.

2.1 The replica free energy

At zero field the 1RSB replica calculation involves the plateau value q_1 and the breakpoint x. It yields the free energy [5]

$$\frac{F_{repl}}{N} = -\frac{\beta J^2}{4} + \frac{\beta J^2}{4} \xi q_1^p \tag{2}$$

$$-\frac{T}{2x}\log(1-\xi q_1)+\frac{T\xi}{2x}\log(1-q_1)$$

where $\xi = 1 - x$. The first term describes the paramagnetic free energy. Here and in the sequel, we omit the $T = \infty$ entropy. It is a constant, only fixed after quantizing the spherical model, [13] that plays no role in the present discussion. For the marginal solution q_1 is fixed by equating the lowest fluctuation eigenvalue to zero, which gives

$$\frac{1}{2}p(p-1)\beta^2 J^2 q_1^{p-2}(1-q_1)^2 = 1. \tag{3}$$

The condition $\partial F/\partial q_1 = 0$ then yields $x = x(T) \equiv (p-2)(1-q_1)/q_1$. This dynamical transition sets in at temperature $T_c = J\{p(p-2)^{p-2}/2(p-1)^{p-1}\}^{1/2}$ where x comes below unity. The same transition temperature follows from dynamics.

2.2 Components

A state, called a *component* by Palmer, [2] is labeled by $a = 1, 2, \cdots, \mathcal{N}$, and has a local magnetization profile $m_i^a = \langle S_i \rangle^a$. Its free energy F_a is a thermodynamic potential that determines the internal energy and the entropy by its derivatives. In the present model $F_a = F_{TAP}(m_i^a)$ is know explicitly. It is a minimum of the "TAP" free energy functional [14] [10] [11]

$$F_{TAP}(m_i) = -\sum_{i_1<\cdots<i_p} J_{i_1\cdots i_p} m_{i_1}\cdots m_{i_p} - H\sum_i m_i$$

$$-\frac{NT}{2}\log(1-q)-\frac{N\beta J^2}{4}(1+(p-1)q^p-pq^{p-1}) \tag{4}$$

where $q = (1/N)\sum_i m_i^2$ is the self-overlap. The state a occurs with weight p_a that is set by the type of experiment one describes. (In practice these weights are usually unknown.) Given the p_a's one can define the "component averages" such as $\overline{F} = \sum_a p_a F_a$, $\overline{C} = \sum_a p_a C_a$ and even the complexity [1] [2] $\mathcal{I} = -\sum_a p_a \ln p_a$. For any observable, the component overage is the object one obtains when measuring over repeated runs and averaging over the outcomes. According to the Gibbs weight the probability of occurrence is $p_a = \exp(-\beta F_a(T))/Z$, with $Z = \sum_a \exp(-\beta F_a)$.

The nice thing of the present model is that many questions can be answered directly. After setting $\partial F_{TAP}/\partial m_i = 0$, we can use this equation to express F_a in terms of q_a alone. This gives the simple relation $F_a = Nf(q_a)$ where

$$f(q) = \frac{\beta J^2}{4}[-1+(p-1)q^p-(p-2)q^{p-1}]$$

$$-\frac{Tq}{p(1-q)}-\frac{T}{2}\log(1-q) \tag{5}$$

The resulting saddle point equation for q_a coincides with the marginality for q_1 given below eq. (2). Since F_a only depends on the selfoverlap q_a, it is self-averaging. In the paramagnet one has $m_i = q = 0$, so both eqs. (4) and (5) reproduce the replica free energy $F = -N\beta J^2/4$. From the replica analysis we know that at T_c^- the value of $q_a \approx q_1$ is $q_1 = q_c \equiv (p-2)/(p-1)$. The component free energy $F_a = Nf(q_c)$ exceeds the free energy of the paramagnet by an extensive amount. As expected from experimental knowledge on glasses, the internal energy is found to be continuous. At T_c^- the free energy difference is solely due to the lower entropy, $S_a = -N\beta^2 J^2/4 - \mathcal{I}_c$, where

$$\mathcal{I}_c = N \left(\frac{1}{2} \log(p-1) + \frac{2}{p} - 1 \right) \tag{6}$$

is the value of complexity at the transition point.

This discussion supports the picture of the glass as a disordered solid, where the entropy of the component the system condenses in, and thus the component average $\overline{S}$, is much smaller that the entropy of the paramagnet. In real glasses this loss of entropy is due to the reduced phase space that arises by trapping of the atoms in a glassy configuration. In the quantized system $\overline{S}$ will vanish at $T = 0$. [13]

2.3   Value of the complexity

Kirkpatrick, Thirumalai and Wolynes [15] were the first to study the role of the complexity for Potts glasses in static situations in the temperature range $T_g < T < T_c$, see also [4] for Ising spin glasses. Statically (that is to say, on timescales $\sim \exp(N)$) the system condenses into a state with higher free energy but with complexity such that the total free energy is exactly equal to the paramagnetic free energy. Here we will investigate the role of the complexity on timescales $\sim N^\gamma$, relevant for the dynamical transition at T_c.

The free energies discussed for this problem are plotted in Figure 1.

A simple calculation shows that the 'experimental' free energy of CHS and CS, and the marginal replica free energy obtained from eq. (2) [12] have the following connection with the component average free energy $\overline{F} = Nf(q_1)$:

$$F_{exp} = \overline{F} - T\mathcal{I}_c \tag{7}$$

$$F_{repl} = \overline{F} - T\mathcal{I} = \overline{F} - \frac{T\mathcal{I}_c}{x(T)} \tag{8}$$

Since $x(T_c) = 1$ both expressions are at T_c equal to the paramagnetic free energy.

In order to trace back the difference between (7) and (8) we have decided to redo the analysis of the TAP equations. Hereto we consider the generalized partition sum

$$Z_u = \sum_a e^{-u\beta F_a(T)} \equiv e^{-\beta F_u} \tag{9}$$

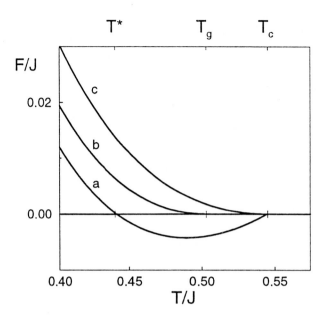

Fig. 1. Free energies of a spherical spin glass with random quartet couplings, after subtraction of the paramagnetic value. a) Marginal replica free energy b) Static replica free energy c) "Experimental" free energy, obtained by integrating C/T and by analysis of the degeneracy of the TAP states.

For $u = 0$ we thus calculate the total number $\mathcal{N}$ of TAP-states, while for $u = 1$ we consider their partition sum. The sum over the TAP states can be calculated using standard approaches. [16][14] A 1RSB pattern is assumed for the 6 order parameters. For instance, $q_{\alpha\beta} = (1/N) \sum_i m_i^\alpha m_i^\beta$ takes the values q_d for $\alpha = \beta$ and q_1 for $\alpha \neq \beta$ both inside a $\tilde{x} * \tilde{x}$ diagonal block of the 1RSB Parisi matrix, while vanishing outside these blocks. At fixed breakpoint $\tilde{x}$ the 12-dimensional saddle point can be found explicitly. For the long time limit of the dynamical approach the marginality condition should be taken, [12] in the form given in eq. (3). As expected, the above replica expression for q_1 is found back as solution of $\partial f(q_d)/\partial q_d = 0$ at $q_d = q_1$. The result $q_1 = q_d$ asserts that the mutual overlap between different states in the same cluster is equal to the selfoverlap. Like in the replica calculation of the ordinary partition sum, $\tilde{x}$ can still take any value. In analogy with the marginal replica calculation of eq (2), we expect $\tilde{x}$ to be fixed by the vanishing of a fluctuation eigenvalue. We have therefore analyzed that 12×12 longitudinal fluctuation matrix at marginality. For any value of $\tilde{x}$ it automatically has 3 zero eigenvalues, proving the marginality. Another eigenvalue vanishes for $\tilde{x} = 0$, $\tilde{x} = 1$ (twice) and for $\tilde{x} = x(T)/u$.

¿From this we infer that $\tilde{x} = x(T)/u$, and thus $\mathcal{I} = u\mathcal{I}_c/x(T)$. In our case

$u = 1$ it just implies that the calculated complexity is the replica value $\mathcal{I}$, and not $\mathcal{I}_c$, the one of CS. This leads to the conclusion that nothing went wrong in the replica calculation of the dynamical phase transition: the replica free energy is a generating function for the mean field equations, and its saddle point value is the logarithm of the partition sum.

This conclusion has been supported by a calculation for the spherical p-spin glass in a transverse field Γ. [17] In that extension of the model there again occurs a dynamical transition from the paramagnet to a 1RSB spin glass state, at transition temperature $T_c(\Gamma)$. The paramagnet of this model is non-trivial. There is a first order transition line (that we call *pre-freezing line*) separating regions with large and small ordering in the z-direction. This line intersects the dynamical PM-1RSB transition line at a point (T^*, Γ^*). Beyond this point there occurs a first order PM-SG transition with a finite latent heat. We expect the location of the transition line to follow from matching of free energies. The replica free energy is indeed suited for that, while the 'experimental' free energy does not lead to a meaningful match.

The free energy is F_{repl} is the physical one, in the sense that it takes into account the correct value of the complexity. Nevertheless the increase of complexity, $\mathcal{I} \sim 1/T$ for low T, remains to be explained.

2.4 Specific heat versus energy fluctuations

It would be nice to have a measurable quantity that probes the multitude of states. One object that should be accessible, at least numerically, is the specific heat. The standard expression $C = dU/dT = \sum_a d(p_a U_a)/dT$ is likely to differ from the component average fluctuations of the internal energy: $\overline{C} = \sum_a C_a = \sum_a p_a dU_a/dT = \beta^2 \sum_a p_a \langle \Delta U_a^2 \rangle$. question is whether their difference is extensive. Based on experience in a toy model, [19] we think it generally is in systems with 1RSB. Since in the present model the energy fluctuations are too small at $H = 0$, [20] it can only occur in a field. ¿From the internal energy in a small field we obtain

$$\frac{1}{N} C(T, H) = \frac{1}{2} \beta^2 J^2 (1 + (p-1)q^p - pq^{p-1})$$
$$- \beta^2 H^2 \frac{(p-1)^2(p-2)(1-q)^2}{p(pq+2-p)} \tag{10}$$

On the other hand, a short calculation shows that $C_a = -T d^2 F_a/dT^2$ remains only a function of q_a at the marginal point, which in the present model takes the field-independent value $q_1 = q_d$. This implies that $\overline{C}$ is field-independent as well, thus satisfying the Parisi-Toulouse hypothesis [21]. Interestingly enough, we find $C < \overline{C}$, whereas Palmer derives the opposite at equilibrium. Our reversed "dynamical" inequality is a new result that is due to the marginality.

The reversed dynamical inequality occurs due to non-equilibrium effects. We conjectured that it generally takes place outside equilibrium, for instance in cooling experiments above T_c in the three dimensional Edwards-Anderson spin glass. Some numerical support for this behavior was found, see [22]

3 Directed polymer in a correlated random potential

We introduce a new, simple model with a static glassy transition. Consider a directed polymer (or an interface without overhangs) $z(x)$ in the section $1 \leq x \leq L$ and $1 \leq z \leq W$ of the square lattice with unit lattice constant. In the Restricted Solid-on-Solid approximation the interface can locally be flat $(z(x+1) = z(x)$; no energy cost) or make a single step $(z(x+1) - z(x) = \pm 1$; energy cost $J)$; larger steps are not allowed. The polymer is subject to periodic boundary conditions $(z(0) = z(L))$ and we allow all values of $z(0)$.

Further there is a random energy cost $V(z)$ per element of the polymer at height z. Note that this is a correlated random potential, with energy barriers parallel to the x-axis.

3.1 The partiton sum

The partition sum of this system can be expressed in the eigenvalues of the tridiagonal transfer matrix T that has diagonal elements $\exp(-\beta V(z))$ and off-diagonal elements $\exp(-\beta J)$

$$Z = \text{tr} e^{-\beta \mathcal{H}} = \text{tr} T^L = \sum_{w=1}^{W} (\Lambda_w)^L \tag{11}$$

For a pure system $(V(z) = 0$ for all $z)$ at temperature $T = 1/\beta$ Fourier analysis tells that for small momentum $\Lambda(k) = \Re(1 + e^{-\beta J + ik})/(1 - e^{-\beta J + ik}) \approx \exp[-\beta f_B - Dk^2/(2\pi^2)]$ with bulk free energy density f_B and diffusion coefficient D that can be simply read off and are temperature dependent.

We shall consider the situation of randomly located potential barriers parallel to the x-axis. Hereto we assume binary disorder, so $V(z) = 0$ with probability $p = \exp(-\mu)$ or $V(z) = V_1 > 0$ with probability $1 - p$. Eq. (11) is dominated by the largest eigenvalues. It is well known that they occur due to Lifshitz-Griffiths singlarities. These are due to lanes of width $\ell \gg 1$ in which all $V(z) = 0$, bordered by regions with $V(z) \neq 0$. These dominant configurations are the "components", "TAP states" or "metastable states" of our previous discussion. The eigenfunction centered around z_0 has inside the lane the approximate form $\cos[\pi(z - z_0)/\ell]$ while it decays essentially exponentially outside due to the disorder. These states can thus be labeled by $a = (z_a, \ell_a)$. Since $k \to \pi/\ell$ the free energy of this state follows as

$$\beta F_\ell \equiv -L \ln \Lambda_\ell \approx \beta f_B L + \frac{DL}{2\ell^2} \tag{12}$$

The typical number of regions with ℓ successive sites with $V = 0$ is $\mathcal{N}_\ell = W(1-p)^2 p^\ell$. We now choose $W = \exp(\lambda L^{1/3})$ so the states with width ℓ have a configurational entropy or complexity $\mathcal{I}_\ell \equiv \ln \mathcal{N}_\ell \approx \lambda L^{1/3} - \mu \ell$.

3.2   The TAP-partition sum

For large L we may restrict the partition sum to these dominant states. We thus evaluate, instead of eq. (11), the 'TAP' partition sum

$$Z = \sum_\ell \mathcal{N}_\ell e^{-\beta F_\ell} \tag{13}$$

Note that it is obtained by simply omitting the contributions of states with low eigenvalue (high free energy). The total free energy

$$\beta F = -\ln Z = \beta f_B L - \lambda L^{1/3} + \mu \ell + \frac{DL}{2\ell^2} \tag{14}$$

has to be optimized in ℓ. The largest ℓ which occurs in the system can be estimated by setting $\mathcal{N}_\ell \approx 1$, yielding

$$\ell_{max} = \frac{\lambda L^{1/3}}{\mu} \tag{15}$$

It is a geometrical length, independent of T. Let us introduce $\tilde{D} = D\mu^2/\lambda^3$. The free energy of this state reads

$$\beta F = \beta f_B L + \frac{1}{2}\lambda L^{1/3}\tilde{D} \tag{16}$$

At low enough T the optimal length is smaller than ℓ_{max},

$$\ell = \left(\frac{DL}{\mu}\right)^{1/3} = \tilde{D}^{1/3}\ell_{max} \tag{17}$$

The free energy of this phase is

$$\beta F = \beta f_B L + \frac{1}{2}\lambda L^{1/3}(3\tilde{D}^{1/3} - 2) \tag{18}$$

For $\tilde{D} > 1$ ($T > T_g$) the interface is in an essentially non-degenerate state. For $\tilde{D} < 1$ it lies in one of the $\mathcal{N}_\ell \gg 1$ relevant states, which is reminiscent to a glass. So the model has a glassy transition at $\tilde{D} = 1$.

The internal energy of a state of width ℓ is

$$U_\ell = u_B L + \frac{L}{2\ell^2}\frac{\partial D}{\partial \beta} = u_B L + \left(\frac{\lambda L}{\tilde{D}^2}\right)^{1/3}\frac{1}{2}\frac{\partial \tilde{D}}{\partial \beta} \tag{19}$$

At $\tilde{D} = 1$ this coincides with the paramagnetic value, simply because $\ell \to \ell_{max}$. It is easily checked that the free energy (18) is a thermodynamic potential, and yields the same value for U. At the transition it branches off quadratically from (16). In the glassy phase the specific heat

$$C = \frac{dU}{dT} = c_B L + \frac{L}{2\ell^2}\partial_T\partial_\beta D + \frac{1}{3}(\lambda L\tilde{D}^{-5})^{1/3}(T\partial_T\tilde{D})^2 \tag{20}$$

exceeds the component averaged specific heat $\overline{C} = Lc_B + (L/2\ell^2)\partial_T\partial_\beta D$. In contrast to previous model, the specific heat it is larger in the glassy phase than in the paramagnet. This is because the free energy is lower.

3.3   On overlaps and hierarchy of phase space

In a given realization of disorder we define the 'overlap' of two states a and b, centered around z_a and z_b, respectively, as

$$q_{ab} = \lim_{t \to \infty} \langle \delta_{z(0),z_a} \delta_{z(t),z_b} \rangle \tag{21}$$

In the high temperature phase there is one non-degenerate state, so $P(q) = \delta(q - q_1)$. In the glassy phase we expect that $q_{ab} = q_1$ for all optimal states (a, b) at temperature T. The reason is that at thermodynamic equilibrium the whole phase space can be traversed, and negligible time is spent in non-optimal states. If so, then though there are many states, one still has $P(q) = \delta(q - q_1)$ and there is no replica symmetry breaking. This is standard for equilibrium situations without frustration.

   This puts forward the picture of replica symmetry breaking and hierarchy of phase space being a dynamical effect. At given timescale only some nearby states can be reached, "states within the same cluster". At larger times other clusters can be reached, and for times larger than the ergodic time of a large but finite system, all states are within reach. Only in the thermodynamic limit phase space splits up in truely disjoint sets. To investigate the validity of this picture in detail, one should solve the dynamics of the polymer.

3.4   The polymer model at $\tilde{T} = 1/T$

The comparison to the p-spin model is most direct when we compare the p-spin model at temperature T with the polymer model at temperature $\tilde{T} = 1/T$. In this interpretation, coming from high $\tilde{T}$, the polymer undergoes a gradual freezing into TAP states. This truely becomes relevant when the domain size is of order $\ell \sim L^{1/3}$, where the complexity starts to be smaller than $\log W = \lambda L^{1/3}$. This gradual freezing shows explicitly that the dynamical transition, as found in the mean field p-spin glass, is smeared in finite dimensions.

   For $\tilde{T}$ going down to $\tilde{T}_c$, the polymer gets captured in states with free energy closer and closer to the lowest free energy state available at that temperature.

   As in 1RSB spin glasses, the complexity also vanishes to leading order for $\tilde{T} \downarrow \tilde{T}_c$. In the low $\tilde{T}$ phase the complexity is no longer of order $L^{1/3}$. This is similar to the low T phase of the static p-spin model, where the complexity is non-extensive in the glassy phase.

   When considered as function of $\tilde{T}$, the specific heat makes a downward jump when cooling the system from large $\tilde{T}$ below $\tilde{T}_c$. The absence of a sharp dynamical transition and the vanishing of the complexity that occurs in this polymer model as function of $\tilde{T}$ are very analoguous to the expected behavior of realistic glasses.

Acknowledgments

The author thanks Yi-Cheng Zhang, H. Rieger and A. Crisanti for stimulating discussion. He is grateful to J.J.M. Franse for wise supervision.

References

[1] J. Jäckle, Phil. Magazine B **44** (1981) 533
[2] R.G. Palmer, Adv. in Physics **31** (1982) 669
[3] This sudden loss of entropy is reminiscent of the collaps of the wave function in the quantum measurement.
[4] T.R. Kirkpatrick and D. Thirumalai, Phys. Rev. Lett. **58** (1987) 2091
[5] A. Crisanti and H.J. Sommers, Z. Physik B **87** (1992) 341
[6] A. Crisanti, H. Horner, and H.J. Sommers, Z. Phys. B **92** (1993) 257
[7] L. F. Cugliandolo and J. Kurchan, Phys. Rev. Lett. **71** (1993) 173
[8] E. Leutheusser, Phys. Rev. A **29** (1984) 2765; U. Bengtzelius, W. Götze, and A. Sjölander, J. Phys. C **17** (1984) 5915
[9] R. Schmitz, J.W. Dufty, and P. De, Phys. Rev. Lett. **71** (1993) 2066
[10] J. Kurchan, G. Parisi, and M.A. Virasoro, J. Phys. I (France) **3** (1993) 1819
[11] A. Crisanti and H.J. Sommers, J. de Phys. I France **5** (1995) 805
[12] Th.M. Nieuwenhuizen, Phys. Rev. Lett. **74** (1995) 3463
[13] Th.M. Nieuwenhuizen, Phys. Rev. Lett. **74** (1995) 4289; ibid 4293
[14] H. Rieger, Phys. Rev. B **46** (1992) 14665
[15] T.R. Kirkpatrick and P.G. Wolynes, Phys. Rev. B **36** (1987) 8552; D. Thirumalai and T.R. Kirkpatrick, Phys. Rev. B **38** (1988) 4881; T.R. Kirkpatrick and D. Thirumalai, J. Phys. I France **5** (1995) 777
[16] A.J. Bray and M.A. Moore, J. Phys. C **13** (1980) L469; F. Tanaka and S.F. Edwards, J. Phys. F **10** (1980) 2769; C. De Dominicis, M. Gabay, T. Garel, and H. Orland, J. Phys. (Paris) **41** (1980) 923
[17] Th.M. Nieuwenhuizen, unpublished (1996)
[18] V. Dobrosavljevic and D. Thirumalai, J. Phys. A **23** (1990) L767R
[19] Th.M. Nieuwenhuizen and M.C.W. van Rossum, Phys. Lett. A **160** (1991) 461
[20] This happens since at $H = 0$ its (free) energy fluctuations are not $\mathcal{O}(\sqrt{N})$ but $\mathcal{O}(1)$, as can be seen by expanding the result for $\ln[Z^n]_{av}$ to order n^2. For $H = 0$ there appear no terms of order $n^2 N$; for $H \neq 0$ they do appear.
[21] G. Parisi and G. Toulouse, J. de Phys. Lett. **41** (1980) L361
[22] Such behavior has been observed for $T > T_g$ in a numerical cooling experiment in the 3d Edwards-Anderson model. (H. Rieger, private communication, April 1995)

A Solvable Model of a Glass

Reimer Kühn*

Institut für Theoretische Physik, Universität Heidelberg
Philosophenweg 19, D–69120 Heidelberg, Germany
e–mail: kuehn@hybrid.tphys.uni-heidelberg.de

Abstract. An analytically tractable model is introduced which exhibits both, a glass–like freezing transition, and a collection of double–well configurations in its zero–temperature potential energy landscape. The latter are generally believed to be responsible for the anomalous low–temperature properties of glass-like and amorphous systems via a tunneling mechanism that allows particles to move back and forth between adjacent potential energy minima. Using mean–field and replica methods, we are able to compute the distribution of asymmetries and barrier–heights of the double–well configurations *analytically*, and thereby check various assumptions of the standard tunneling model. We find, in particular, strong correlations between asymmetries and barrier–heights as well as a collection of single–well configurations in the potential energy landscape of the glass–forming system — in contrast to the assumptions of the standard model. Nevertheless, the specific heat scales linearly with temperature over a wide range of low temperatures.

1 Introduction

The present contribution is primarily concerned with the low–temperature properties of amorphous and glass-like materials. A prominent example of such an anomaly is the roughly linear temperature dependence of the specific heat at $T < 1\,\mathrm{K}$, which is in stark contrast to the T^3 behaviour known to originate from lattice vibrations in crystaline materials. Further anomalies are reported for the temperature dependences of the thermal conductivity and other transport properties.

To explain these anomalies, a phenomenological model — the so–called standard tunneling model (STM) [1], [2], [3] — has been introduced. It is based on two assumptions, which are plausible but until today are still lacking an analytic foundation based on microscopic modelling. First, it is assumed that even at temperatures well below the glass temperature, small local rearrangements of single atoms or of small groups of atoms are possible via tunneling between adjacent local minima in the potential energy surface of the system. Second, individual local double–well configurations of the potential energy surface are taken to be randomly distributed, and a specific assumption is advanced concerning the distribution $P(\Delta, \Delta_0)$ of asymmetries Δ, and tunneling–matrix elements Δ_0, viz., $P(\Delta, \Delta_0) \sim \Delta_0^{-1}$. The value of

* supported by a Heisenberg fellowship

Δ_0 is related with the barrier–height V between adjacent minima and the distance d between them. In WKB–approximation one has $\Delta_0 = \hbar\omega_0 \exp(-\lambda)$, with $\lambda = \frac{d}{2}\sqrt{2mV/\hbar^2}$. Here ω_0 is a characteristic frequency (of the order of the frequency of harmonic oscillations in the two wells forming the double well structure), m the effective mass of the tunneling particle, and d the separation between the two minima of the double well. In terms of Δ and λ one has $P(\Delta, \lambda) \simeq$ const.

The STM describes experimental data reasonably well at low temperatures, i.e. for $T < 1\,\mathrm{K}$ (for an overview, see e.g. [4]). At temperatures above $1\,\mathrm{K}$, one observes a (non–phonon) T^3 contribution to the specific heat and a plateau in the thermal conductivity which cannot be accounted for within the set of assumptions of the STM. To model these phenomena, alternative assumptions concerning the distribution of local potential energy configurations have been advanced, such as those leading to the so–called soft–potential model [5], [6], where it is assumed that locally the potential energy surface can be described by fourth order polynomials of the form $V(x) = u_0[u_2 x^2 + u_3 x^3 + x^4]$, with u_0 a fixed parameter and u_2 and u_3 independently distributed in a specific way. Under certain assumptions about these distributions, these systems also exhibit a collection of 'soft' (an)harmonic single–well potentials, supporting localized soft vibrations which can reasonably well account for both, the crossover to T^3–behaviour of the specific heat above $1\,\mathrm{K}$ as well as the plateau in the thermal conductivity.

On the other hand, simulations that tried to detect double–well potentials in quenched Lenard-Jones mixtures [7] produced results which did not fit well with the assumptions of the soft–potential model, but could be described by an ansatz that leads to a *generalized* soft–potential model, viz. $V(x) = w_2 x^2 + w_3 x^3 + w_4 x^4$, with all three coefficients w_α independently distributed in a specific way. Evaluations are, however, as yet based on rather moderate statistics.

For the time being, it is perhaps safe to say that both, the STM and the soft–potential model provide *phenomenological* descriptions, based on assumptions which — while plausible in many respects — are still lacking analytic support based on more microscopic approaches.

Here we propose a microscopic model inspired by spin–glass theory which exhibits both, a glass–like freezing transition at a certain glass-temperature T_g, and a collection of double–well configurations in its zero–temperature potential energy surface. Within this model, we shall not only be able to compute the full statistics of double–well configurations believed to be responsible for the low–temperature anomalies but also exhibit *relations* between low–temperature and high–temperature phenomena, e.g. between the low–temperature specific heat and the value of the glass–transition temperature itself.

Our line of reasoning is as follows. In Sec. 2, we propose an expression for the potential energy of a collection of particles forming an amorphous model system, which is taken to be *random* in its harmonic part, and which includes anharmonic on–site potentials to stabilize the system as a whole. We choose our setup in such a way that it can be analysed exactly within mean–field theory, and the replica method is used to deal with the disorder. By these means, the phase diagram of the model can be computed (Sec. 3), and we observe that it has a glass–like frozen phase at sufficiently low temperatures.

In mean–field theory, the system is described by an ensemble of effective single–site problems, characterized by single–site potentials which contain random parameters; replica theory in this context can be understood as a method to compute the distribution of these parameters self–consistently. The potential energy surface of the original model is thereby represented as the zero temperature limit of the aforesaid ensemble of independent single–site potentials, and we are able to identify regions of parameter space where some members of this ensemble have double–well form. The distribution of the parameters characterizing the double–well potentials (DWPs) — asymmetries and barrier heights — can be computed analytically within replica theory (Sec. 4). Taking these distributions as input of a tunneling model, we are able to *compute* the contribution of the tunneling states to various low–temperature anomalies. In the present paper we shall restrict our attention to the specific heat at low temperatures. Section 5 contains a summary of our achievements and an outlook on open problems.

2   The Model

Consider the following expression for the potential energy of a collection of N degrees of freedom (particles for short) forming an amorphous or glass–like system,

$$U_{\text{pot}}(v) = -\frac{1}{2} \sum_{i,j} J_{ij} v_i v_j + \frac{1}{\gamma} \sum_i G(v_i) \,, \qquad (1)$$

in which v_i $(1 \leq i \leq N)$ may be interpreted as the deviation of the i-th particle from some preassigned reference position. We propose to model the amorphous aspect of the system by taking the first, i.e. the harmonic contribution to $U_{\text{pot}}(v)$ to be *random*, so that the reference positions and thereby the entire system would quite generally turn out to be unstable in the harmonic approximation. This is why a set of anharmonic on–site potentials $G(v_i)$ is added to stabilize the system as a whole.

To be specific, we take the J_{ij} to be independent Gaussians with mean J_0/N and variance J^2/N. In order to fix the energy- and thus the temperature scale, we specialize to $J = 1$ in what follows. For the on–site potential we choose

$$G(v) = \frac{1}{2}v^2 + \frac{a}{4!}v^4 \,. \qquad (2)$$

That is, G also creates a harmonic restoring force, and by varying the parameter γ in (1) we can tune the number of modes in the system which are unstable at the harmonic level of description. Other forms of $G(v)$ may be contemplated; our method to solve the model does not depend on the particular shape of G. The only requirement is that it increases faster than v^2 for large v for the system to be stable.

The harmonic contribution to (1) is reminiscent of the SK spin–glass model [8], apart from the fact that we are dealing with continuous 'spins' here, without any local or global constraints imposed on them. Models of the type introduced above — albeit with different couplings and different choices for $G(v)$ — have, however, been studied in the context of analogue neuron systems [9], [10]. A model with Gaussian couplings, but different G has been considered by Bös [11].

The choice of quenched random couplings in (1) certainly puts our model outside the class of glass–models in the narrow sense. In view of recent ideas concerning the fundamental similarity between quenched disorder and so–called self–induced disorder as it is observed in glassy systems proper [13], it may nevertheless be argued that our choice should capture essential aspects of glassy physics at low temperatures.

To analyze the potential energy surface, we compute the (configurational) free energy of the system

$$f_N(\beta) = -(\beta N)^{-1} \ln \int \prod_i dv_i \exp[-\beta U_{\text{pot}}(v)] \tag{3}$$

and take its $T = 0$ limit, using replica theory to average over the disorder, so as to get *typical* results. Standard arguments [8] give $f(\beta) = \lim_{n \to 0} f_n(\beta)$ for the quenched free energy, with

$$n f_n(\beta) = \frac{1}{2} J_0 \sum_a m_a^2 + \frac{1}{4} \beta \sum_{a,b} q_{a,b}^2 - \beta^{-1} \ln \int \prod_a dv^a \exp\left[-\beta U_{\text{eff}}(\{v^a\})\right] . \tag{4}$$

Here

$$U_{\text{eff}}(\{v^a\}) = -J_0 \sum_a m_a v^a - \frac{1}{2} \beta \sum_{a,b} q_{ab} v^a v^b + \frac{1}{\gamma} \sum_a G(v^a) \tag{5}$$

is an effective replicated single–site potential, and the order parameters $m_a = N^{-1} \sum_i \overline{\langle v_i^a \rangle}$ and $q_{ab} = N^{-1} \sum_i \overline{\langle v_i^a v_i^b \rangle}$ are determined as solutions of the fixed point equations

$$m_a = \langle v^a \rangle \quad , \quad a = 1, \dots, n \tag{6}$$

$$q_{ab} = \langle v^a v^b \rangle \quad , \quad a, b = 1, \dots, n \ , \tag{7}$$

where angular brackets denote a Gibbs average corresponding to the effective replica potential (5), and where it is understood that the limit $n \to 0$ is eventually to be taken.

So far we have evaluated (4)–(7) only in the replica symmetric approximation by assuming $m_a = m$ for the 'polarization'-type order parameter, and $q_{aa} = \hat{q}$ and $q_{ab} = q$ for $a \neq b$ for the diagonal and off–diagonal entries of the matrix of Edwards-Anderson order parameters. These are determined from the fixed point equations

$$m = \langle\, \langle v \rangle \,\rangle_z \,,$$
$$\hat{q} = \langle\, \langle v^2 \rangle \,\rangle_z \,,$$
$$q = \langle\, \langle v \rangle^2 \,\rangle_z \,. \tag{8}$$

Here $\langle \ldots \rangle_z$ denotes an average over a zero-mean unit-variance Gaussian z while $\langle \ldots \rangle$ without subscript is a Gibbs average corresponding to the effective replica-symmetric single–site potential

$$U_{\mathrm{RS}}(v) = -[J_0 m + \sqrt{\hat{q}}\, z] v - \frac{1}{2} C v^2 + \frac{1}{\gamma} G(v) \tag{9}$$

with $C = \beta(\hat{q} - q)$. The replica symmetric approximation thus describes a Gaussian ensemble of independent single-site potentials $U_{\mathrm{RS}}(v)$, with parameters m, q and C which are determined self-consistently through (8).

3   Phase Diagram

The system described by (8)–(9) exhibits a glass-like freezing transition from an ergodic phase with $m = q = 0$ to a frozen phase with $q \neq 0$ at some temperature T_g depending on the parameters J_0 and γ of the model. If J_0 is sufficiently large, a transition to a macroscopically polarized phase with $m \neq 0$ may also occur.

The assumption of replica symmetry is not always correct. Spontaneous replica symmetry breaking (RSB) occurs at low temperatures (large β) and large γ. The precise location of the instability against RSB is given by the AT criterion [12]

$$1 = \beta^2 \left\langle\, (\, \langle v^2 \rangle - \langle v \rangle^2 \,)^2 \,\right\rangle_z = \frac{1}{q} \left\langle \left(\frac{\mathrm{d}}{\mathrm{d}z} \langle v \rangle \right)^2 \right\rangle_z \,, \tag{10}$$

the second expression being more appropriate for an evaluation in the $T = 0$-limit.

Phase boundaries between ergodic and non-ergodic phases are increasing functions of γ, diverging as $\gamma \to \infty$, and approaching zero for finite values of γ which can be read off from the $T = 0$ phase diagram of the model in Fig. 1.

Interestingly, the system can also exhibit a collection of DWPs in its zero-temperature potential energy surface. The following Sect. is devoted to extracting their statistics, and to analyzing their contribution to the low–temperature specific heat.

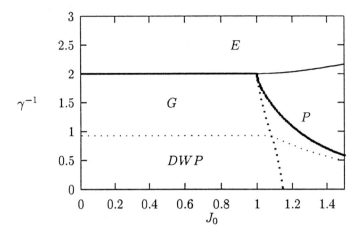

Fig. 1. $T = 0$ phase diagram. E denotes the $T = 0$–limit of the ergodic phase, G the glassy phase, and P a phase with macroscopic polarization. The bold line is the AT line. Below the (small) dotted line is the region with DWPs. The bigger dots separate the glassy phase G from the phase P with macroscopic polarization.

4 Double–Well Potentials and Specific Heat

We take the $T \to 0$ ($\beta \to \infty$) limit of the above set (8) of fixed point equations to compute the $T \to 0$ limit of the order parameters characterizing the ensemble (9) of effective replica symmetric single–site potentials, which in turn represents the potential energy surface of our model (in the RS approximation). For suitable values of external parameters, some of these potentials have DWP form. From (2),(9) it is clear that the condition for this to occur is $\gamma C > 1$; this region is marked DWP in Fig. 1. For not too large $|z|$ then, DWPs occur and the distribution of their characteristic parameters (barrier heights V, asymmetries Δ, and distance d between the wells, hence $\lambda = d\sqrt{mV/2\hbar^2}$) derives from the Gaussian distribution of z and can be computed *analytically*. For larger $|z|$, the $U_{\mathrm{RS}}(v)$ only exhibit (anharmonic) single well forms; see Fig. 2.

In contrast to the assumptions of the STM, we find that Δ and λ are strongly correlated random variables; in the RS approximation both are functions of one Gaussian random variable, viz. z, as shown in Fig. 3. The distributions $P(\lambda)$ and $P(\Delta)$ are depicted in Fig. 4. Both have singularities at their upper boundary, the former an integrable divergence, the latter a cusp singularity. A notable feature here is that upper and lower limits of the λ and Δ ranges are *given* within our approach, and the total mass under either distribution gives the fraction of degrees of freedom which 'see' DWPs.

Despite the differences in the shape of $P(\Delta, \lambda)$ from that assumed in the

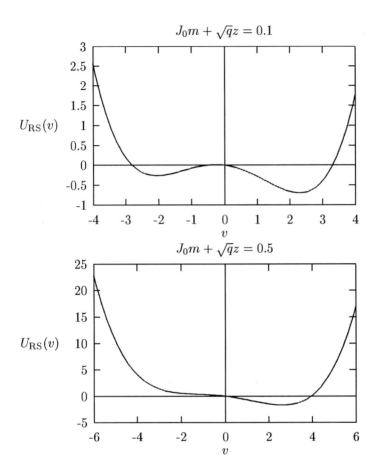

Fig. 2. Effective replica–symmetric single–site potential for two different values of $\tilde{h} = J_0 m + \sqrt{q}z$. **(a)** $\tilde{h} = 0.1$. **(b)** $\tilde{h} = 0.5$. In **(a)**, the asymmetry Δ is the difference between the two minima, the barrier height V is the difference between the maximum and the (lower) minimum, and d is the separation between minima on the v axis. In both cases we have $\gamma^{-1} = 0.5$ and J_0 such that $m = 0$.

STM, we find that the contribution of the tunneling states to the specific heat – taking the tunnel splitting to be given by $\epsilon = \sqrt{\Delta^2 + \Delta_0^2}$ [1], and ignoring higher excitations – exhibits an extended range of temperatures where it scales linearly with T, and we find this phenomenon to be more pronounced, as we move deeper into the DWP phase, i.e., deeper also into the glassy phase as it is described in our model (see Fig. 5). At very low temperatures exponential behaviour of the specific heat is observed which is due to the cutoff in the λ distribution, which in turn creates a cutoff in the density of states at low energies.

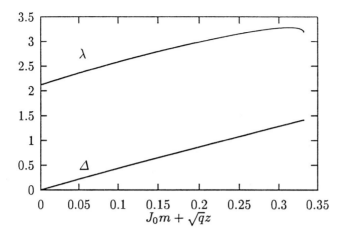

Fig. 3. Asymmetry Δ and λ as functions of $\tilde{h} = J_0 m + \sqrt{q}z$. The parameters γ and J_0 are as in Fig. 2.

5 Summary and Outlook

We have proposed and solved a simple model which exhibits both, a glass–like freezing transition, and a collection of DWPs in its zero temperature potential energy landscape. The latter are generally believed to give rise to a number of low-temperature anomalies in glassy and amorphous systems via a tunneling mechanism that allows particles to move back and forth between the wells forming the DWP structure at temperatures where thermally activated classical motion would still be rather unlikely. Within our model, we were able to compute the distribution of the parameters characterizing the DWPs analytically, and we found, in particular, strong correlations between asymmetries Δ and the parameter λ which determines the magnitude of the tunneling matrix element Δ_0. Nevertheless, we observe an extended range of low temperatures at which the contribution of the two-level tunneling systems to the specific heat scales linearly with temperature. The correlations between λ and Δ can be weakened (but most likeley *not* eliminated) by introducing *local* randomness, i.e., by introducing either a randomly i–dependent γ or by making other parameters of the on–site potentials i–dependent in a random fashion. It has been demonstrated elsewhere [10] that the model remains solvable with these modifications.

So far, we have evaluated only the replica symmetric approximation. At the same time we know that RSB is observed in the interesting region of the phase diagram where DWPs actually do occur. However, it can be shown that DWPs and even the correlation between asymmetries and tunneling matrix element persist at all finite levels of RSB [16]; the distribution of the parameters characterizing them may of course change in details. The one-

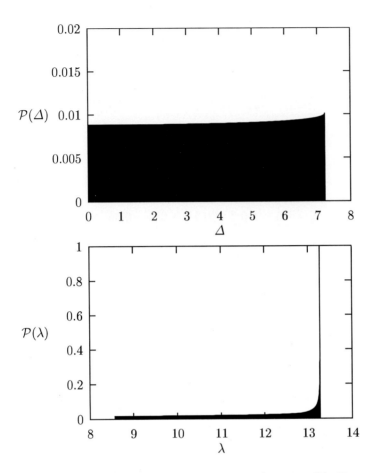

Fig. 4. Distributions $\mathcal{P}(\Delta)$ (for positive Δ), and $\mathcal{P}(\lambda)$. Here $\gamma = 4$, while J_0 is as in Fig. 2. The total mass under either distribution is 0.138 and gives the fraction of particles which 'see' DWPs.

and two-step RSB approximations [14] as well as Parisi's full RSB scheme [15] for this model are currently being evaluated [16].

Concerning the motivation of (1) via the idea of an expansion of the potential energy about a set of reference positions, it should be noted that in principle a linear random–field type term should be added to U_{pot}. While such a term does change details of the low-temperature properties of the system, we find that the main physics is left invariant [16].

Up to now we have not investigated *relations* between high–temperature and low–temperature properties of our system in any detail, but it should be obvious that they exist — all properties are determined from the two model–parameters γ and J_0 — and that they are within relatively easy reach

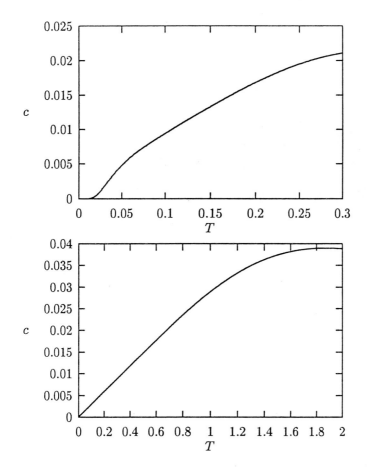

Fig. 5. Low T specific heat. **(a):** $\gamma = 2$, $J_0 = 0$. **(b):** $\gamma = 4$, $J_0 = 0$. Note the extended linear region in **(b)**. In both cases there is a region at very low T where the behaviour is exponential. In **(b)** this region is too small to be resolved.

of our approach; they are currently under study [16]. Moreover, we have as yet treated the DWPs only as two–level systems, which entails that their contribution to the specific heat levels off and eventually decreases to zero as the temperature is further increased. Clearly, the contribution of higher excitations as well as that of the single well configurations to physical quantities must finally be taken into account as well.

It should be pointed out that our model will also exhibit interesting *dynamical* properties at or near its glass transition temperature and very likely throughout the glassy phase, which are worth investigating. Indeed, *formally* such an investigation already exists [17], albeit for the Langevin dynamics of a model with G chosen differently, so as to describe Ising spins. It would be interesting to see which features of that model will survive in the present context, and which will turn out to be altered.

In the present paper we have chosen to quantize our system only *after* a mean–field decoupling. For ordered, translationally invariant systems, the validity of this procedure has been rigorously proven by Fannes et al. [18]. A corresponding proof for disordered systems is still lacking. Therefore, it would be interesting to study the system in a full quantum statistical context right from the outset, using imaginary time path integrals in conjunction with the replica method. It would seem feasible to solve the system along these lines at least within the so called static approximation. The clear connection to the DWP concept, to which we have access precisely through the mean–field decoupling scheme is, however, likely to be lost in a solution along these lines.

Acknowledgements It is a pleasure to thank C. Enss, H. Horner, U. Horstmann, S. Hunklinger, Th. Nieuwenhuizen, P. Neu, A. Würger, A. Verbeure, and in particular O. Terzidis for illuminating discussions and helpful advice.

References

[1] P.W. Anderson, B. Halperin, and S. Varma, Phil. Mag. **25**, 1 (1972)

[2] Phillips, J. Low Temp. Phys. **7**, 351 (1972)

[3] J. Jäckle, Z. Phys. **257**, 212 (1972)

[4] S. Hunklinger and W. Arnold, in *Physical Acoustics*, edited by W.P. Mason and R.N. Thurston (Academic, New York, 1976) Vol. XII, p. 155; S. Hunklinger and A.K. Raychauduri, in *Progress in Low Temperature Physics*, edited by D.F. Brewer (Elsevier, Amsterdam, 1986) Vol. IX, p. 267

[5] V.G. Karpov, M.I. Klinger, and F.N. Ignat'ev. Zh. Eksp. Teor. Fiz. **84**, 760 (1983) [Sovj. Phys. JETP **57**, 439 (1983)]

[6] U. Buchenau, Yu.M. Galperin, V.L. Gurevich, D.A. Parshin, M.A. Ramos, and H.R. Schober, Phys. Rev. B**46**, 2798 (1992)

[7] A. Heuer and R.J. Silbey, Phys. Rev. Lett. **70**, 3911 (1993)

[8] D. Sherrington and S. Kirkpatrick, Phys. Rev. Lett. **35**, 1792 (1975); S. Kirkpatrick and D. Sherrington, Phys. Rev. **B 17**, 4384 (1978); see also M. Mézard, G. Parisi, and M. A. Virasoro, *Spin Glass Theory and Beyond*, (World Scientific, Singapore, 1987)

[9] J. Hopfield, Proc. Natl. Acad. Sci. USA, **81**, 3088 (1984)

[10] R. Kühn, S. Bös, and J.L. van Hemmen, Phys. Rev. A **43**, 2084 (1991); R. Kühn and S. Bös, J. Phys. A **26** 831 (1993)

[11] S. Bös, PhD Thesis, Gießen (1993), unpublished

[12] J.R.L. de Almeida and D.J. Thouless, J. Phys. A **11**, 983 (1978)

[13] T. Kirkpatrick and D. Thirumalai, J. Phys. A**22**, L149 (1989); J.P. Bouchaud and M. Mézard, J. Phys. I (France) **4**, 1109 (1994): E. Marinari, G. Parisi, and F. Ritort, J. Phys. A**27**, 7615 (1994); — ibid. 7647 (1994); J.P. Bouchaud, L. Cugiandolo, J. Kurchan, and M. Mézard, preprint cond–mat/9511042

[14] G. Parisi, J. Phys. **A13**, L115 (1980)

[15] G. Parisi, J. Phys. **A13**, 1101 (1980)

[16] U. Horstmann and R. Kühn, in preparation

[17] H. Sompolinsky and A. Zippelius Phys. Rev. **B25**, 6860 (1982)

[18] M. Fannes, H. Spohn, and A. Verbeure, J. Math. Phys. **21**, 355 (1980)

On the long times, large length scale behaviour of disordered systems

Jean-Philippe Bouchaud[1] and Marc Mézard[2]

[1] Service de Physique de l'État Condensé, Centre d'études de Saclay, Orme des Merisiers, 91191 Gif-sur-Yvette Cedex, France
[2] Laboratoire de Physique Théorique de l'ENS ** , 24 rue Lhomond, 75231 Paris Cedex 05, France

Abstract. We discuss the large scale effective potential for elastic objects in the presence of a random pinning potential. In the static approach, converging analytical results show that the large scale, low temperature free energy landscape consists in a succession of parabolic wells of random depth, matching on singular points where the effective force is discontinuous. These parabolas are themselves subdivided into smaller parabolas, corresponding to the motion of smaller length scales, in a hierarchical manner. Consequences for the *dynamics* of these pinned objects are underlined, and compared to the mean field theory of aging effects.

Email: bouchaud@amoco.saclay.cea.fr, mezard@physique.ens.fr

1 Introduction

The physics of disordered systems has been a flourishing theme of modern statistical physics. The basic reason, already emphasized long ago by Anderson in the context of electron localisation, is that a dirty sample is not, in general, equivalent to a 'pure' one with renormalized parameters. In other words, an altogether new phenomenology appears on large length scales. This is the case of spin-glasses, whose behaviour is very different from usual (ordered) magnetic materials. From a theoretical point of view, however, the spin-glass has turned out to be a very difficult problem to tackle. A comparatively simpler problem, still containing much of the physics of spin-glasses, is that of elastic objects pinned by random impurities, which covers a vast domain of applications, some of which of technological importance: the pinning of flux lines in superconductors [1], [2], [3], of dislocations, of (Bloch) domain walls in magnets [4], [5], or of charge density waves [7], [8], controls in a crucial way the properties of these materials. In this case the theoretical apparatus is somewhat more luxuriant, and reliable results are available even though perturbation theory badly fails [9]. A general 'scaling' picture has emerged, including the generic presence of rare events and large fluctuations in, e. g.,

---

** Unité propre du CNRS, associée à l'Ecole Normale Supérieure et à l'Université de Paris Sud the response function [10], [11], [12]. However, scaling arguments cannot provide a detailed statistical description of the energy landscape seen by the large wavelength modes of the pinned objects. This information is very important from a physical point of view, since the effective *pinning force* induced by the impurities on the elastic object determines its fluctuations, and its response to an external driving force. In the example of flux lines in superconductors, the ohmic dissipation is controlled by the mobility of these lines; hence the detailed knowledge of the large scale pinning potential is crucial. There is actually one beautiful experiment by Porteseil et al. [5], where the statistics of the pinning potential acting on a *single Bloch wall* has been determined. In this context, the construction of the effective pinning potential seen by a domain wall was addressed a long time ago by L. Néel [6].

In a recent work together with L. Balents [13], our motivation was to understand the relation between two general approaches which have been proposed to describe the *statics* of these pinned objects. The first one is the variational replica method which combines a Gaussian trial Hamiltonian with 'replica symmetry breaking' (RSB) inspired from spin-glasses. This non-trivial analytical trick (see below) is needed to obtain a correct description of the low temperature, strongly pinned phase [14], [15], [16]. The second is the 'functional renormalisation group' (FRG) which aims at constructing directly the correlation function for the effective pinning potential acting on long wavelengths, using renormalisation group (RG) ideas [18]. These two methods are, at first sight, so remote that it is not obvious to compare them directly, in particular because the physical quantities on which one focuses are different. Working out the precise connections between these two approaches was fruitful: first of all, from a technical point of view, it helps understanding the precise assumptions on which these theories are based, and thus their limitations. Second, it provides hints on the generic statistical features of the large scale energy landscape, and show how these are encoded in the two theories.

The basic picture which emerges is the following: the effective, long wavelength pinning potential is a succession of *parabolic wells of random depth, matching on singular points* where the effective force (i.e. the derivative of the potential) is discontinuous. These discontinuities induce a singularity in the effective potential correlation function (singularity which is indeed found within the FRG); in the replica language, these singular points are a direct consequence of RSB. The replica calculation furthermore provides detailed information on the effective pinning potential, for example on the depth of the wells, which correspond physically to long-lived metastable configurations. As discussed below, this allows one to make motivated conjectures on the long-time dynamics of such pinned objects.

2 The Hamiltonian

We consider the general problem of pinned elastic manifolds described by the Hamiltonian:

$$\mathcal{H}(\{\phi(\mathbf{x})\}) = \int d^D\mathbf{x} \left[\frac{c}{2} \left(\frac{\partial \phi(\mathbf{x})}{\partial \mathbf{x}} \right)^2 + V_0(\mathbf{x}, \phi(\mathbf{x})) \right], \tag{1}$$

where $\mathbf{x}$ is a D-dimensional vector labelling the *internal* coordinates of the object (or manifold), and $\phi(\mathbf{x})$ an N-dimensional vector giving the position in physical space of the point labelled $\mathbf{x}$. Various values of D and N actually correspond to interesting physical situations. For example, $D = 3$, $N = 2$ describes the elastic deformation of a vortex lattice (after a suitable anisotropic generalisation of Eq. (1)), $D = 2$, $N = 1$ describes the problem of domain walls pinned by impurities in 3 dimensional space, while $D = 1$ corresponds to the well-known directed polymer (or single flux line) in a $N + 1$ dimensional space. The elasticity of the structure is characterized by the modulus c. [1]

The pinning potential $V_0(\mathbf{x}, \phi(\mathbf{x}))$ is a random function, which we shall choose to be Gaussian distributed with zero mean and a correlation function:

$$\overline{V_0(\mathbf{x}, \phi)V_0(\mathbf{x}', \phi')}_c = NW\delta^D(\mathbf{x} - \mathbf{x}')R_0\left(\frac{(\phi - \phi')^2}{N} \right), \tag{2}$$

where W measures the strength of the pinning potential. We shall here concentrate on the case where the correlation function is short ranged (although the long range case is also interesting), and we shall choose for convenience $R_0(y) = \exp(-\frac{y}{2\Delta^2})$, where Δ is the correlation length of the random potential.

3 Statics

The quantity on which previous studies have particularly focused on is the 'wandering exponent' ζ which characterize the geometrical conformation of the pinned object at low temperatures. This wandering exponent is also related to the 'energy exponent' (often called θ) relating the typical pinning energy to the length scale. These exponents are known to be non trivial below $D = 4$ internal dimensions, where a small disorder is a relevant perturbation. Approximate values for these exponents can be obtained from simple Imry-Ma like arguments, which are rather successful [4], and are complemented by numerical simulations, some exact results [9], and by the two approximate methods mentionned before [14], [18].

---

[1] Obviously for each specific case the elastic term should be adapted (for instance in vortex lattices one should introduce the three moduli C_{11}, C_{44}, C_{66})

A more ambitious question is to understand how the microscopic pinning potential affects the large length scale Fourier modes of the elastic manifold, with the hope that the resulting effective pinning potential has some universal features, which do not depend too much on the detailed statistics of the microscopic pinning. Both the FRG and the replica approach provide some indication on this effective potential. The small parameter in the FRG is $\epsilon = 4 - D$ (N arbitrary), while the replica variationnal approach becomes exact for $N = \infty$, but with D arbitrary. Therefore where both methods are simultaneously effective in the neighbourhood of $D = 4$, $N = \infty$.

3.1 The FRG method

The FRG method consists in writing down a recursion relation for the correlation function of the potential acting on 'slow' modes $\phi_<$, after 'fast' modes $\phi_>$ (corresponding to wavevectors in a high-momentum shell $[\Lambda/b, \Lambda]$) have been integrated out using perturbation theory (which requires for consistency that $\epsilon \to 0$), and after a proper coarse graining of the variables both in x space and in ϕ space [18]. Assuming that the effective random pinning potential *remains gaussian* for all scales ℓ, one can write a recursion relation for its two point correlation $R_\ell(y)$ – which thus characterizes completely the effective pinning potential. The iteration of the recursion equation from the 'initial' condition $R(y) = R_0(y)$ converges towards the *fixed point* disorder correlation $R^*(y)$, describing the long wavelength properties. As noticed initially by D. S. Fisher, an important property of $R^*(y)$ is singular behaviour for small y [18]

$$R^*(y) - R^*(0) = \epsilon[a_1 y - a_{3/2}|y|^{\frac{3}{2}}] + ..., \tag{3}$$

In terms of the effective *force* f acting on the manifold, (defined as minus the derivative of the effective potential with respect to ϕ), one finds that the force correlation function behaves as

$$\overline{[f^*(\phi) - f^*(\phi')]^2} = 12\epsilon a_{3/2}|\phi - \phi'|. \tag{4}$$

Since the effective potential is (by assumption) gaussian, the same is true of the force. Therefore, for $N = 1$, Eq. (4) means that the effective force acting on the manifold is a simple *random walk* in ϕ space.

3.2 The Replica approach

¿From a technical point of view, the replica method (in particular with RSB) looks rather mysterious, handling all along hierarchical 0×0 matrices. However, after rather a lot of work, first initiated within the 'cavity' approach to spin glasses [19], and later for the random manifold problem [14], [13], the physical content of these algebraic manipulations has emerged very clearly. Actually, the RSB scheme encodes in an extraordinarily compact way the full probabilistic construction of the effective disordered potential seen by the manifold. The price to pay is that it must be implemented in a variational way, which is exact only when $N \to \infty$. It can alternatively be seen as a Hartree resummation of the perturbation theory (the coupling constant being here the strength of the disorder), which is exact when $N \to \infty$.

Furthermore, the construction of the effective potential is somewhat intricate when replica symmetry is 'fully' broken [14]; the computation of its correlations is not a straightforward task. In the simpler case of a 'one-step' breaking scheme, these correlations were calculated first in [17], in the context of Burgers's turbulence, where the 'effective force' in the pinning problem can be identified with the velocity field in Burgers' equation. The small y behaviour of the velocity correlator was found to be precisely of the form given by (4); this result was then extended to the continuous RSB scheme (relevant to the case $D = 4 - \epsilon$) in [13] in a way which we now briefly describe.

One first isolates a particular, very slow mode $\mathbf{k}_0 \to 0$ of the manifold. The effective force acting on $\boldsymbol{\varphi}_0 \equiv \boldsymbol{\varphi}(\mathbf{k}_0)$ is defined as $f_\Omega^\mu(\boldsymbol{\varphi}_0) = -\frac{1}{\beta} \frac{\partial}{\partial \varphi_0^\mu} \ln \mathcal{P}_\Omega(\boldsymbol{\varphi}_0)$, where $\mathcal{P}_\Omega(\boldsymbol{\varphi}_0)$ is the probability to observe $\boldsymbol{\varphi}_0$ for a given realisation of the random pinning potential Ω. The log of this probability thus defines an effective potential; this interpretation is however only correct if all the other modes are thermalized. Then the correlation function of $\mathbf{f}$ can be expressed with replicas through the following identity:

$$\overline{f_\Omega^\mu(\boldsymbol{\varphi}_0) f_\Omega^\nu(\boldsymbol{\varphi}_0')} = \lim_{n \to 0} \frac{4}{n^2} \frac{\partial^2}{\partial \varphi_0^\mu \partial \varphi_0'^\nu} \overline{[\mathcal{P}_\Omega(\boldsymbol{\varphi}_0)]^{\frac{n}{2}} [\mathcal{P}_\Omega(\boldsymbol{\varphi}_0')]^{\frac{n}{2}}}. \qquad (5)$$

which is directly calculable within the Gaussian RSB Ansatz, which asserts that:

$$\overline{\prod_{a=1}^{n} \mathcal{P}_\Omega(\boldsymbol{\varphi}_0^a)} = \sum_\pi \exp[-\frac{\beta}{2} \varphi_0^{\pi(a)} G_{ab}^{-1}(\mathbf{k}_0) \varphi_0^{\pi(b)}], \qquad (6)$$

where n is the number of replicas, G is the optimal matrix determined *via* the variational equations (see [14], [17] for more details) and π denotes all the $n!$ permutations of the replica indices.

3.3 Results. The 'scalloped' effective potential

After an elaborate calculation [13], one finally finds that, in the limit $k_0 \to 0, D \to 4, N \to \infty$, the force-force correlation function behaves precisely as in Eq. (4). In particular, the overall ϵ factor and the $|y|^{\frac{1}{2}}$ singular term are recovered. Hence, at the level of the two point correlation function, the two methods indeed give the same result.

However, as stated above, the replica calculation actually contains much more information, in particular on the higher moments of the large scale potential (V_Ω^*) statistics. It turns out that this effective potential is highly non-Gaussian, with cusp like singularities separating parabolic regions. Let us explain how this comes about within the simpler case of 'one-step' RSB (which holds in the case $D = 1$). In this case, the effective pinning potential can be expressed as:

$$V_\Omega^*(\varphi) = -\frac{1}{\beta} \ln \left[\sum_\alpha e^{-\beta F_\alpha - \frac{(\varphi - \varphi_\alpha)^2}{u_c \Delta^2}} \right], \tag{7}$$

where α label the 'states', centered around φ_α and of free-energy F_α, both depending on the 'sample' Ω. The φ_α are uniformly distributed, while the free energies F_α are exponentially distributed:

$$\rho(F_\alpha) \propto_{F_\alpha \to -\infty} \exp(-\beta u_c |F_\alpha|). \tag{8}$$

A crucial point is that when replica symmetry is broken, the parameter u_c is less thAn one, which means that the weights $e^{-\beta F_\alpha}$ are very broadly distributed, in such a way that the sum, Eq. (7), is, for each value of ϕ, dominated by very few states α. More precisely, one finds that the potential has for $N = 1$ the shape drawn in Fig. 1 [17]: it is made of parabolas matching at points where the derivative (i.e. the force) is discontinuous in the limit of zero temperature (and smoothed over a certain scale at finite temperature). The singular behaviour of the force-force correlation function, Eq. (4), is precisely due to these cusps: with a probability proportional to the 'distance' $|\varphi_0 - \varphi_0'|$, there is a shock which gives a *finite* contribution to $\mathbf{f}(\varphi_0) - \mathbf{f}(\varphi_0')$. This means in particular that all the moments $\overline{|\mathbf{f}(\varphi_0) - \mathbf{f}(\varphi_0')|^p}$ grow as $|\varphi_0 - \varphi_0'|$ for $p \geq 1$, instead of $|\varphi_0 - \varphi_0'|^{\frac{p}{2}}$ as for Gaussian statistics. As mentionned above, this picture was obtained in the context of Burgers turbulence in [17]. In the case of continuous RSB, the construction of the effective potential is more complicated. Basically it is recursively constructed via a set of 'Matrioshka doll' Gaussians [14], [13]. It is schematically drawn in Fig 2 for the the transverse fluctuations $\phi(\ell) - \phi(0)$. The $|y|^{\frac{3}{2}}$ singular structure of the two point correlations $R(y)$ for small y however still holds.

The reason why the FRG succeeds in capturing correctly the singularity in $R(y)$ under the wrong assumption that the large scale potential remains gaussian is not totally clear. To get some idea about this issue, one must keep in mind that the effective potential calculated within the FRG procedure involves an extra step which we have not performed within the replica construction, which is a coarse graining of the ϕ variables. In the FRG calculation, one restricts to configurations which are such that ϕ is constant on scales ℓ, and scales as ℓ^ζ. The correct choice of ζ then ensures that there are only a few shocks on the scale ℓ. This is perhaps why the FRG can still be controlled in first order in ϵ (and probably only in first order), the departure from Gaussian statistics being in some sense (which we do not fully understand) 'weak'.

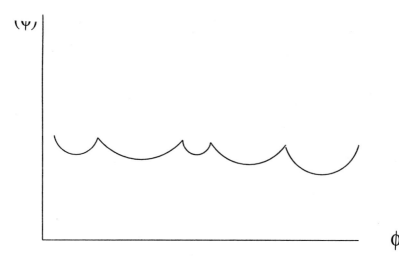

Fig. 1. Schematic view of the effective energy landscape as a succession of parabolic wells matching at singular point. This picture actually corresponds to a 'one-step' replica symmetry breaking scheme.

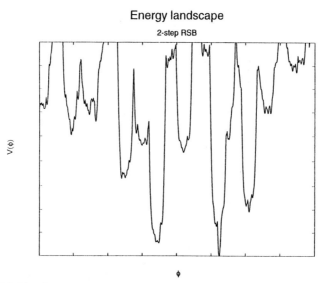

Fig. 2. Multiscale energy landscape corresponding to a full replica symmetry breaking scheme. In this case, the construction is that of parabolas within parabolas, in a hierarchical manner. The depth of the wells (and thus also the height of the barriers) typically grows as $|\phi - \phi'|^{\frac{\theta}{\zeta}}$. The figure actually corresponds to a two-step breaking scheme, with $u_1 = 0.5$ and $u_0 = 0.05$. The inset is a zoom on a particular region, showing the first level of Gaussians.

3.4 Relation with Burgers' equation

The relation with Burgers' equation can be considered from two points of view. The first one, studied in [17], is based on a direct mapping between Burgers' equation and the equation governing the dependence of the partition function of a 'polymer' ($D = 1$) in a random medium as a function of its length. Another, possibly more fundamental connection, comes from the renormalization. The FRG fast- mode elimination by which the renormalized effective potential is defined looks as follows (For simplicity, we keep $N = 1$ and we do not discuss the explicit x dependance; The details, including the reasons why the potential can still be considered as local in x space after proper rescaling, can be found in [18], [13]):

$$\beta V_R(\varphi_<) = -\ln\left[\int d\varphi_> e^{-\beta[\frac{c\Lambda^2}{2}\varphi_>^2 + V_0(\varphi_< + \varphi_>)]}\right].\tag{9}$$

The interesting remark is then that $V_R(\varphi_<)$ is precisely the Cole-Hopf solution of a Burgers equation [20]:

$$\frac{\partial V(\varphi,t)}{\partial t} = \frac{1}{2\beta c\Lambda^2}\frac{\partial^2 V(\varphi,t)}{\partial \varphi^2} - \frac{c\Lambda^2}{2}\left(\frac{\partial V(\varphi,t)}{\partial \varphi}\right)^2\tag{10}$$

with the microscopic pinning potential as initial condition:

$$V(\varphi,t=0) = V_0(\varphi)\qquad V_R(\varphi) = V(\varphi,t=1).\tag{11}$$

Now, it is well known [20], [21] that a random initial condition (here the bare pinning potential acting on φ) develops shocks which separates as 'time' (here length scales) grows, between which the 'potential' $V(\varphi)$ has a parabolic shape. Elimination of fast modes in a disordered system thus naturally generates a 'scalloped' potential, with singular points separating potential wells – the famous 'states' appearing in the replica theory. Furthermore, the statistics of $V(\varphi)$ at late 'times' (i.e. on large length scales) is known to be to a large degree *universal* with respect to the detailed nature of the initial conditions, with a few universality classes [21]. This suggests that it might be possible to characterize not only the two point function R^*, but also the *full fixed point functional distribution* $\mathcal{P}[V^*(\varphi)]$ within an extended RG treatment.

3.5 Experimental: Shocks in Bloch walls

¿From an experimental point of view, it is interesting to note that Porteseil et al. [5] have determined some statistical aspects of the effective pinning potential acting on the center of mass ($k_0 = 0$) of a Bloch wall. In particular, they have determined the power spectrum of the local pinning force acting on a single Bloch wall driven at a certain (small) velocity, and found that it behaves as ω^{-2}, i.e. precisely what one should expect from Eq. (4). The shocks can even be seen directly by monitoring the driving magnetic field, in such a way that the Bloch wall moves as slowly as possible. On the other hand, Néel's construction of the effective pinning potential started from the 'physical' assumption that this potential had no discontinuities The existence of these singularities is however allowed in the limit of long-wavelengths and small temperatures.

As a final remark, the RSB predicts that the local curvature of the 'wells' does not fluctuate. As explained in [22], this is an artefact of the large N limit; correspondingly, Porteseil et al. have experimentally determined the distribution of these local curvatures.

4 Dynamics

4.1 The 'trap' model

We believe that our construction of the effective pinning potential could turn out to be very useful in order to understand the *dynamics* of such objects at finite N. It suggests for instance to analyse the relaxation of one mode in terms of hops between the different wells which appear in the above construction ('traps'), corresponding to metastable configurations with a given long-wavelength conformation. Of course this approach is to some extent uncontrolled, since it assumes that the dynamics of one mode can be understood using the effective static potential determined above. This is correct only if all the higher modes are already equilibrated. The hope is that there is a sufficiently strict hierarchy of time scales such that smaller wavelength modes, even if they are 'slow', are much faster than the particular mode one is interested in. Although this problem needs to be clarified, it is interesting to discuss the main features of the dynamics within this simplified 'trap' model.

The main point (already emphasized in [23] in the context of spin-glasses – see also [25] in this volume) is that the trapping time distribution, which is controlled by the barrier heights (and hence by the tail of the distribution of F_α), is a broad distribution with a *diverging first moment*. More precisely, for a given Fourier mode k_0, the lifetime of each 'trap' is activated $\tau \simeq \tau_0 \exp(\beta \Delta E)$, where τ_0 is an 'attempt' time scale (presumably k_0 dependent and no longer microscopic if $k_0 \to 0$). From the exponential distribution of free-energies, one finds that the distribution of τ is (in the continuous RSB picture) a power-law:

$$\rho(\tau) \simeq_{\tau \gg \tau_0} \frac{\tau_0}{\tau^{1+u(k_0)}} \qquad (12)$$

where the exponent $u(k_0) \propto k_0^\theta$ depends on the 'size' of the jump, i.e. the mode k_0 involved in the change of conformation. (the energy exponent θ is close to 2 in dimension $D \sim 4$). In particular, small $u(k_0) < 1$ corresponds to large wavelengths, for which the average of τ calculated from Eq. (12) diverges. In this case, the dynamics becomes non stationary and aging effects appear at low temperatures and large wavelength modes. In particular, the time dependent a.c. response $\chi(k_0, \omega, t)$ of the mode k_0 is found to be [23]:

$$\chi(k_0, \omega, t) \propto (\omega t)^{u(k_0)-1} \qquad \text{for} \quad u(k_0) < 1 \tag{13}$$

(while it is independent of t for $u(k_0) > 1$). Interestingly, when the temperature is decreased, $u(k_0)$ also decreases. This means that a mode which is equilibrated at a certain temperature ($u(k_0) > 1$) can be out of equilibrium at a smaller temperature, and give an aging contribution to the total response. This is similar in spirit to the temperature cycling experiments in spin-glasses [24], [25] or in dipolar glasses [26].

4.2 Mean-field results: a brief survey

The above approach is motivated by the static energy landscape construction which we have discussed previously. However, it is still to some extent phenomenological. Another path is to try a direct analytic study of the dynamics of randomly pinned objects using mean field methods which hold for large N. Starting from the dynamical field theory approach [27], [28], one can write [29] in this limit some coupled equations for the correlation:

$$C(x, t; x', t') = < \frac{1}{N} \sum_\alpha \phi_\alpha(x, t)\phi_\alpha(x', t') > \tag{14}$$

and the response:

$$r(x, t; x', t') = < \frac{1}{N} \sum_\alpha \frac{\partial \phi_\alpha(x, t)}{\partial \eta_\alpha(x', t')} > . \tag{15}$$

These dynamical equations involve a memory kernel and an effective noise which are determined self consistently. The understanding of this mean field dynamics parallels that of spin glasses. For simplicity we shall discuss the case of a point particle ($D = 0$) in a random medium, rather than that of a manifold (see [33]).

At high temperature one finds a solution which is, at large enough time, both time translational invariant (TTI) and obeys the fluctuation dissipation theorem (FDT). At a critical temperature the relaxation time becomes infinite [29]. In the low temperature phase, as first recognised in spin glasses by Cugliandolo and Kurchan [30], the solution looses the two properties of TTI and FDT (Notice that this can be seen only keeping the initial time, and thus the age of the system, fixed, while sending $N \to \infty$). The solution of the dynamical equations at low temperature was developped in [31] for the case of long range correlated noise, and then in [33] for the case of short range correlations. Roughly speaking, the main features of the solution look as follows. There exist several different ways of sending the two times t and t' to infinity while keeping a non trivial dependance of the correlation and response. The usual asymptotic regime corresponds to having $t \to \infty$, $t' \to \infty$, with a fixed value of $\tau = t - t'$. Then the correlation and response go to their asymptotic forms $C_{as}(\tau)$ and $r_{as}(\tau)$, and represent the stationnary dynamics.

In the simplest case (corresponding to 'one-step' replica symmetry breaking solutions for the statics), the interesting aging regime is unique and corresponds to the domain in which $t \to \infty$, $t' \to \infty$, with a fixed value of $\lambda = \frac{h(t')}{h(t)}$, where $h(t)$ is an increasing function which has not yet been determined by the theory. A possibility could be that $h(t) \propto t$, as is the case for coarsening models or in the trap model alluded to above. In this aging regime, one has $C(t, t') \sim \hat{C}(\lambda)$, and similarly for the response. This implies a response 'anomaly', which means that the system develops a long term memory. Mathematically this can be found from the following relation:

$$\lim_{t_w \to \infty} \int_0^{t_w} ds \ r(t_w, s) \neq \int_0^\infty d\tau \ r_{as}(\tau) , \tag{16}$$

Furthermore in this aging regime the FDT is substituted by [30]:

$$Tr(t, t') = x \frac{\partial C(t, t')}{\partial t'} \tag{17}$$

where the fluctuation dissipation (FD) ratio x is smaller than one.

In more complicated cases (corresponding to continuous replica symmetry breaking solutions for the statics), there exist several aging regimes. For instance one could take the limits $t \to \infty$, $t' \to \infty$, with a fixed value of $\lambda = (t - t')/t^u$. In each such regime there will be a modified FDT as in (17), with a FD ratio which depends on the regime. (For some non understood reason the set of values of the FD ratio is related to the Parisi order parameter function in the replica solution of the statics, whenever there is a continuous RSB solution [31], [32]).

5 Discussion and Conclusion

The derivation of the aging effect from mean field dynamics, and the subsequent analytical progress, is an important breakthrough. Actually many features found within this framework turn out to be rather similar to those obtained within the trap model. Yet the physical mechanisms underlying aging in these two models are probably quite different (see the related discussion in [25]). In particular, in the simplest 'one-step' models, the dynamics is rather insensitive to temperature, and remains qualitatively the same from the dynamical transition temperature to zero temperature. Conversely, temperature is crucial for activated dynamics, underlying the 'trap' picture. Aging in 'one step' models is related to a diffusion in a very high dimensional phase space, where trapping and activated effects are absent (there is always a path to escape). In this respect, various possible scenarios have been proposed recently, including purely entropic barriers [34], [36], or diffusion along basin boundaries in high dimensional space [35]. In the case of a point particle ($D = 0$) in a random potential, it is clear that for any finite dimension N, there will be a crossover time (which diverges with N) separating this 'wandering' regime from a 'trapping' regime at large times, where activated effects become important. A truly challenging task is to be able to control the finite N corrections to the mean-field models, which contain these activated effects. Precisely the same problem appears within the 'Mode-Coupling' theory of glasses, where activated effects are discarded [38], [39], [40].

Continuous RSB models are even more intricate, and it would be very interesting to understand in more details what happens in this case. It thus seems necessary at this stage to develop new tools to understand the various types of aging which have been seen (a first tentative classification has been proposed in [37]).

To conclude, we have shown that the FRG and RSB techniques give compatible results when they can be compared. Both suggest quite an appealing physical picture: the phase-space of the system is, on large length scales, divided into 'cells' corresponding to favourable configurations where the potential is locally parabolic, and whose depth is exponentially distributed. This general shape can be understood through an analogy with Burgers' equation: elimination of the fast modes is a non linear operation which generates 'shocks' in the effective force, and 'laminar' regions between the shocks (corresponding to the cells). Perhaps surprisingly, this suggests that the large scale energy landscape is *universal*, in the same way as the velocity-field statistics generated the Burgers equation for large times.

Full replica symmetry breaking corresponds to the fact that these cells are themselves subdivided into smaller cells, etc... Each level of the hierarchy corresponds to a different length scale, finer details corresponding to smaller length scales. The replica variationnal approach has allowed one to extend ideas from spin glass mean field theory to some finite dimensional problems. A very interesting issue concerns the applicability of these ideas back to the harder problem of spin-glasses in finite dimensions. Actually, an early version of the picture developed in the present paper was proposed within the context of one dimensional spin-glasses in [41].

The dynamical picture which is naively inferred from this construction landscape is that of the (multi-level) trap model [23]. The precise link (or absence thereof) with mean field dynamics is not very well understood; however the 'trap' picture seems rather different from the diffusion like mechanism underlying aging within 'one-step' models.

Acknowlegments: We want to thank L. Balents, L. Cugliandolo, J. Hammann, J. Kurchan and E. Vincent for many important discussions on these matters.

References

[1] D. S. Fisher, M.P.A. Fisher, D. A. Huse, Phys Rev B 43, 130 (1991)

[2] G. Blatter, M. V. Feigel'man, V. B. Geshkenbin, A. I. Larkin and V.M. Vinokur, Rev. Mod. Phys. **66** (1994) 4.

[3] for a recent review, see e.g. M. Kardar, D. Ertas, in 'Scale Invariance, Interfaces and Non-equilibrium dynamics', NATO-ASI (1995), Kluwer.,M. Kardar, 'Lectures on directed paths in random media', Les Houches Summer School on 'Fluctuating geometries in Statistical Mechanics and Field Theory', in press.

[4] T. Nattermann and P. Rujan, Int. J. Mod. Phys. B3 (1989) 1597; T. Natterman and J. Villain, Phase transitions 11 (1988) 5

[5] R. Vergne, J. C. Cotillard, J.L. Porteseil, Rev. Phys. Appl. **16** 449 (1981).

[6] L. Néel, Cahier de la Physique **12** (1942) and **13** (1943)

[7] H. Fukuyama and P. A. Lee, Phys. Rev. **B 17**, 535 (1978).

[8] D. S. Fisher, Phys. Rev. **B 31**, 7233 (1985).

[9] T. Halpin-Healey and Y.C. Zhang; Phys. Rep. **254** (1995) 217

[10] G. Parisi, J. Phys. France **51** 1595 (1990)

[11] M. Mézard, J. Phys. France **51** 1831 (1990)

[12] T. Hwa, D. S. Fisher, Phys. Rev. **B 49** 3136 (1994)

[13] L. Balents, J.P. Bouchaud and M. Mézard, cond-mat/9601137, to appear in J. Physique (August 1996).

[14] M. Mézard, G. Parisi, J. Physique I **1** 809 (1991); J.Phys. **A23** L1229 (1990)

[15] J.P. Bouchaud, M. Mézard, J. Yedidia, J. Phys Rev B **46** 14 686 (1992)

[16] S. E. Korshunov, Phys. Rev. B **48**, 3969 (1993); T. Giamarchi, P. Le Doussal, Phys. Rev. **B 52** 1242 (1995)

[17] J.P. Bouchaud, M. Mézard, G. Parisi, Phys. Rev. **E 52** (1995) 3656

[18] D.S. Fisher, Phys. Rev. Lett. **56** (1986) 1964, L. Balents, D. S. Fisher, Phys. Rev. **B 48** (1993) 5949

[19] M. Mézard, G. Parisi, M.A. Virasoro, "Spin Glass Theory and Beyond", (World Scientific, Singapore 1987)

[20] J. M. Burgers, 'The Non-Linear Diffusion Equation', D. Reidel Pub. Co. (1974)

[21] S. Kida, J. Fluid. Mech. **93** (1979) 337

[22] J.P. Bouchaud, M. Mézard, preprint cond-mat 9607006

[23] J.P. Bouchaud, *J. Physique I (Paris)* **2** (1992) 1705; J.P. Bouchaud, D.S. Dean, *J. Physique I (Paris)* **5** (1995) 265. See also: C. Monthus, J.P. Bouchaud, preprint cond-mat 9601012, submitted to J. Phys. A.

[24] E. Vincent, J. Hammann, M. Ocio, p. 207 in "Recent Progress in Random Magnets", D.H. Ryan Editor, (World Scientific Pub. Co. Pte. Ltd, Singapore 1992)

[25] E. Vincent, J. Hammann, M. Ocio, J. P. Bouchaud, L. Cugliandolo, this volume.

[26] F. Alberici, P. Doussineau, A. Levelut, preprint.

[27] P.C. Martin, E.D. Siggia and H.A. Rose; Phys. Rev. **A8** (1978) 423; C. de Dominicis, L. Peliti, Phys. Rev. **B18** (1978) 353.

[28] H. Sompolinsky and A. Zippelius, Phys. Rev.Lett. **47** (1981) 359; Phys. Rev. A **25** (1982) 6860.

[29] H. Kinzelbach and H. Horner, *J.Phys. I France* **3**, 1329 (1993) ; *J.Phys. I France* **3**, 1901 (1993);

[30] L. F. Cugliandolo and J. Kurchan, Phys. Rev. Lett. **71** (1993) 173

[31] S. Franz and M. Mézard, Europhys. Lett. **26** (1994) 209; Physica **A209** (1994) 1

[32] L. F. Cugliandolo and J. Kurchan, J. Phys. **A27** (1994) 5749

[33] L. F. Cugliandolo and P. Le Doussal; Phys. Rev. **E53**, 1525 (1996). L. F. Cugliandolo, J. Kurchan and P. Le Doussal; Phys. Rev. Lett. **76**, 2390 (1996).

[34] A. Barrat and M. Mézard; J. Phys. I (France) **5** (1995) 941

[35] J. Kurchan and L. Laloux; J. Phys. **A29**, 1929 (1996)

[36] F. Ritort, Phys. Rev. Lett. **75** (1995) 1190; S. Franz and F. Ritort, preprint cond-mat/9508133 ; C. Godrèche, J.P. Bouchaud and M. Mézard, J. Phys. **A28** (1996) L603

[37] A. Barrat, R. Burioni and M. Mézard, J. Phys. **A29** (1996)1311

[38] For reviews, see W. Götze, in *Liquids, freezing and glass transition*, Les Houches 1989, JP Hansen, D. Levesque, J. Zinn-Justin Editors, North Holland. see also W. Götze, L. Sjögren, *Rep. Prog. Phys.* **55** (1992) 241

[39] S. Franz and J. Hertz, Phys. Rev. Lett. **74** (1995) 2114

[40] J-P Bouchaud, L. F. Cugliandolo, J. Kurchan and M. Mézard; Physica **A226**, 243 (1996)

[41] M. Feigel'man, L. Ioffe, Z. Phys. B **51** (1983) 237.

Hexatic Glass

Eugene M. Chudnovsky*

Department of Physics and Astronomy, Lehman College, Cuny, Bronx, NY 10468-1589

Abstract. This is a short review of the current state of theory and experiment on the structure of elastic lattices weakly interacting with underlying random background. Examples are flux lattices in superconductors, magnetic bubble lattices in ferromagnetic films, charge density waves in semiconductors, and atomic monolayers physisorbed on random surfaces. We shall argue that the orientational order in these systems persists at longer scales than the translational order. The law of the decay of both types of order will be analysed, with the emphasis on its practical implications for superconductors. The depinning transition will be discussed in terms of the mobility of the lattice.

*e-mail address: chudnov@lcvax.lehman.cuny.edu

1   Introduction

When it comes to real systems, no order is long range. Large monocrystals tend to break into polycrystals, permanent magnets break into domains, etc. In an ideal system, such a breakdown of a homogeneous order always results in the onset of a more sofisticated inhomogeneous long range order. For instance, in an ideal type II superconductor, a suffuciently large magnetic field destroys the homogeneity of the wafe function, leading to the onset of the long range periodic array of vortices. In practice, however, this kind of the long range order never survives the imperfectness of the system. Practical applications of solids require the precise knowledge of how the long range order is destroyed. Surprisingly, the answer to this question is not straightforward even for simplest models of disorder. One example, that received much attention in the last years, is an elastic lattice in a random background. It can be a Wigner crystal in a semiconductor with impurities [1], a charge density wave in a weakly disordered medium [2], an atomic monolayer on an imperfect crystal surface [3], a magnetic bubble lattice in a ferromagnetic film with defects [4], a vortex lattice in a disordered superconductor [5], etc. The latter problem has been studied most intensively because correlations in the vortex lattice are responsible for such properties of superconductors as resistivity and critical current. For certainty, in what follows, we shall have this problem in mind.

Motion of flux lines in superconductors generates, through Maxwell equations, electric fields and Ohmic losses. Understanding of the conditions, under which flux lines become mobile, is, therefore, crucial for applications. The flux lines (or vortices) are defects of the superconducting phase. In order to decrease the volume occupied by the normal phase, they tend to match with defects of the solid which are incompatible with superconductivity. The resulting pinning potential, no matter how weak, destroys the long range periodicity of the vortex lattice on a certain scale, L_c, that is inversely proportional to a some power of the strength of disorder [6], [7]. The length L_c, which is called the translational correlation length, determines the size of the vortex bundle that can be displaced independently from the rest of the vortex lattice by an external force. Such a force originates from the interaction between the vortex lattice and the superconducting current $\mathbf{j}$. In d dimensions, the energy of the interaction of the bundle with the current scales as jL_c^d. Equating it to the pinning energy, $U_{pin}(L_c)$, one obtains the dependence of the critical current on the translational correlation length, $j_c \propto L_c^{-d} U_{pin}(L_c)$, and on all parameters that determine L_c.

Pinning is a cummulative statistical effect of a large number of weak pins. Thus, the energy of the pinning of a piece of the lattice of size L must scale slower than L^d. Consequently, for any weak current j, the interaction with the current must exceed the pinning energy at a certain scale $L_j > L_c$. This determines the size of the critical nucleus for depinning. Still, as long as $j < j_c$, a finite energy, $\sim U_{pin}(L_j)$, is needed to form the nucleus. At a non-zero temperature this may happen as a thermal fluctuation and will result in the finite resistivity of the superconductor, $\rho \propto \exp\left[-U_{pin}(j)/T\right]$. An important question is whether this resistivity disappears in the limit of $j \to 0$. The answer to this question determines whether the system is a true superconductor at $T \neq 0$.

Besides the translational order, the lattice can be characterized by the orientational order, that is, by correlations in the orientation of locally defined crystallographic axes. In a perfect lattice the two kinds of order are trivially coupled with each other. In a disordered lattice, however, these are two independent kinds of order [18]. It has been demonstrated [10] that the orientational order is more robust with respect to quenched randomness than the translational order. Rather big patterns of flux lattices in which the translational order does not extend beoynd 4-5 lattice spacings, exhibit almost perfect orientational order [5] It is that orientational order that a human eye cathes first when analysing the lattice. This explains why many lattices appear as such even when translational correlations only exist between nearest neighbours. Besides this philosophical observation, the degree of the orientational order in the flux line lattice may be important for understanding whether the superconducting-to-normal transition is a true phase transition accompanied by the symmetry change.

High-temperature superconductors exhibit a transition from almost immobile pinned flux-line lattice to the lattice that is dragged by the current. This transition can be achieved by either increasing temperature or the magnetic field. It forms a continuous irreversibility line in the $B(T)$ phase diagram. Two scenarios have been suggested for the irreversibility line [17]. In one the vortex lattice melts, the shear modulus turns zero, and vortices become mobile, since the random background does not pin individual vortices. Alternatively, it is possible that the lattice is preserved on a large scale even above the irreversibility line, and that it simply depins and becomes mobile as a whole. In the last part of this article we shall argue that the latter scenario is quite plausible.

Throughout this article we shall consider lattices free of dislocations. The honest argument behind this assumption is that the problem without dislocations is already rather difficult. Adding dislocations makes it almost hopeless at least as far as the analytical theory is concerned. By circumsizing the problem that way, one can find some releif in the fact that pictures of vortex lattices obtained in decoration experiments show remarkably large areas free of dislocations.

2 Translational Order

Let $\mathbf{u}(\mathbf{r})$ be the displacement field in the vortex lattice and $V[\mathbf{r}, \mathbf{u}(\mathbf{r})]$ describe the pinning potential. Then, in the absence of the superconducting current, the energy of the lattice is

$$U = \int d^d r \left[\alpha_{iklm} \nabla_i \nabla_k u_l u_m + V(\mathbf{r}, \mathbf{u}) \right] \quad , \tag{1}$$

where the first term represents elasticity; α_{iklm} being the elastic moduli. In the absence of pinning $\mathbf{u} = 0$. Pinning deforms the lattice, leading to a certain static deformation field, $\mathbf{u}(\mathbf{r})$.

Consider first a small portion of the lattice. If the pinning is weak, the displacements within that small portion must be small compared to the lattice spacing, a. Then $V(\mathbf{r}, \mathbf{u})$ can be written as $-\mathbf{f}(\mathbf{r}) \cdot \mathbf{u}$, and Eq.(1) reduces to the random force problem,

$$U = \int d^d r \left[\hat{\alpha} \nabla u \nabla u - \mathbf{f}(\mathbf{r}) \cdot \mathbf{u} \right] \quad . \tag{2}$$

Writing $< \mathbf{f}^2 > \equiv f^2$ and $< \mathbf{u}^2 > \equiv u^2$, the statistical average of the elastic energy on a scale L can be estimated as

$$U_{el} \sim L^d \frac{u^2}{L^2} \quad , \tag{3}$$

while the pinning energy is

$$U_{pin} \sim [L^d < (\mathbf{f} \cdot \mathbf{u})^2 >]^{1/2} \sim L^{d/2} f u \quad . \tag{4}$$

Equating (3) and (4), one obtains [6], [7]

$$< \mathbf{u}^2 > \; \sim \; \frac{f^2}{\alpha^2} L^{4-d} \quad . \tag{5}$$

When this quantity becomes of the order of a^2, the translational order in the lattice is lost, which gives

$$L_c \propto f^{2/d-4} \tag{6}$$

for the translational correlation length, that is, $L_c \propto 1/f$ in two dimensions and $L_c \propto 1/f^2$ in three dimensions.

The energy of the interaction between the flux lattice and the superconducting current is

$$U_j = \int d^d r \, \frac{1}{c} (\mathbf{j} \times \mathbf{B}) \cdot \mathbf{u} \quad , \tag{7}$$

where $\mathbf{B}$ is the average magnetic field through the superconductor. The depinning potential is, thus, proportional to jV_c, where V_c is the volume of size L_c. As soon as this energy becomes greater than the pinning energy, $U_{pin}(V_c)$, the lattice is no longer pinned but is dragged by the current through a dissipative medium. The critical current, therefore, scales with V_c as

$$j_c \propto \frac{U_{pin}(V_c)}{V_c} \quad . \tag{8}$$

which is a decreasing function of V_c (increasing function of f).

Let us now turn to large scales where $|\mathbf{u}| > a$. For large scales the argument that led to equations (4)-(6) no longer applies as it fails to account for the periodicity of the lattice:

$$V(\mathbf{u} + \mathbf{a}) = V(\mathbf{u}) \quad . \tag{9}$$

A qualitative, though non-rigorous, argument goes as follows [19]. Let us write the pinning potential as

$$V(\mathbf{r}, \mathbf{u}) = v(\mathbf{r}) \cos\left[\mathbf{G} \cdot (\mathbf{u} - \mathbf{r})\right] \quad , \tag{10}$$

which catches the periodicity, as well as the fact that the potential can depend on $\mathbf{u}$ only through $\mathbf{u} - \mathbf{r}$. The pinning energy at a scale L can be then estimated as

$$\begin{aligned}
U_{pin} &\sim \int^L d^d r \, \frac{1}{2} < \frac{\partial^2 V}{\partial \mathbf{u}^2} >_L \mathbf{u}^2 \\
&\sim - \int^L d^d r \, \frac{1}{2} v(\mathbf{r}) \cos\left(\mathbf{G} \cdot \mathbf{u}\right) < \cos\left(\mathbf{G} \cdot \mathbf{u}\right) > \mathbf{u}^2 \\
&\sim f L^{d/2} \mathbf{u}^2 \exp\left(-\mathbf{u}^2/2\right) \quad ,
\end{aligned} \tag{11}$$

while the elastic energy is still given by Eq.(3). Comparing the two energies one obtains [19], [9]

$$< \mathbf{u}^2 > \; \propto \; (4 - d) \ln(L) \quad . \tag{12}$$

There is also a numerical evidence of a very slow decay of orientational correlations at large distances [22].

Formulas (3), (7), (11), and (12) allow one to establish the L-dependence of pinning and depinning energies for large L:

$$U_{pin} \propto L^{d-2}\ln^2(L)$$
$$U_j \propto jL^d \ln(L) \quad . \tag{13}$$

The comparison of these two energies shows that for $j < j_c$ the size of the critical depinning nucleus scales with the current as

$$L_j \propto \frac{1}{\sqrt{j}} \quad . \tag{14}$$

Correspondingly, the energy barrier for the nucleation scales with the current as

$$U_{pin}(j) \propto \frac{\ln^2(1/j)}{j^{d/2-1}} \quad , \tag{15}$$

that is, as $\ln^2(1/j)$ in two dimensions and as $1/\sqrt{j}$ in three dimensions. Consequently, the linear resistivity of a superconductor,

$$\rho \propto \exp\left[-\frac{U_{pin}(j)}{T}\right] \quad , \tag{16}$$

disappears in the limit of $j \to 0$. The disordered vortex state, sometimes called the *vortex glass*, is, therefore, a true superconducting state [8].

3   Orientational Order

Consider a triangular flux line lattice characterized by elastic moduli C_{11}, C_{44}, and C_{66}. The corresponding energy, in the presence of the random force, $\mathbf{f}(\mathbf{r})$, is

$$U = \frac{1}{2}\int d^d r \left[(C_{11}-C_{66})(\partial_\alpha u_\alpha)^2 + C_{66}(\partial_\alpha u_\beta)^2 + C_{44}(\partial_z u_\alpha)^2 - f_\alpha u_\alpha\right] \quad . \tag{17}$$

It is easy to show [6] that

$$< [\mathbf{u}(\mathbf{r}) - \mathbf{u}(0)]^2 > \propto r_\perp^2 \quad for \quad d = 2$$
$$< [\mathbf{u}(\mathbf{r}) - \mathbf{u}(0)]^2 > \propto \left(r_\perp^2 + \frac{C_{66}}{C_{44}}z^2\right)^{1/2} \quad for \quad d = 3 \tag{18}$$

where $\mathbf{r}_\perp$ is the coordinate perpendicular to the direction of the field z. This must apply to small distances (compare with Eq.(5)).

The rotational deformations of the lattice are described by

$$\theta(\mathbf{r}) = \frac{1}{2}\nabla_\perp \times \mathbf{u}(\mathbf{r}) \quad . \tag{19}$$

The corresponding correlation function is [10], [11]

$$
\begin{aligned}
&< [\theta(\mathbf{r}) - \theta(0)]^2 > \propto (1/r_\perp)^{9f^2/2\pi C_{66}^2} \quad for \quad d = 2 \\
&< [\theta(\mathbf{r}) - \theta(0)]^2 > = const \quad for \quad d = 3 \ ,
\end{aligned}
\tag{20}
$$

which, again must be true at small distances. A large scale analysis of a kind we used in the previous section yields [20] constant for this correlator in two dimensions as well. This shows that the orientational order is more robust with respect to the random potential than the translational order.

In real systems and at very large distances the orientational order must decay anyway. Indeed, a pair of dislocations in a close proximity with each other must create a random torque τ that acts directly on θ [12],

$$
U_{int} = -\int d^d r\, \tau \cdot \theta \ . \tag{21}
$$

One can show [13] that

$$
< \tau^2 > \propto < f^2 >^2 \ . \tag{22}
$$

Consequently, the orientational correlation length scales as [15], [14]

$$
\frac{L_{orient}}{a} \sim \left(\frac{L_{transl}}{a}\right)^2 \ , \tag{23}
$$

that is, as $1/f^2$ in two dimensions and as $1/f^4$ in three dimensions. It should, therefore, come at no surprise that many physical systems [5], [2], [4] exhibit extended orientational correlations combined with a very short range translational order. Such a state can be called a *hexatic glass*. Numerical simulations of discrete lattices on random background [16] support conclusions obtained analytically.

4   Mobility

We now want to address the question of how the vortex lattice undergoes a transition from the pinned to a mobile state [21]. Consider a two-dimensional triangular lattice of particles coupled by a harmonic, nearest-neighbor interaction, and subject to a periodic potential and a uniform external force. The potential energy of the system is

$$
U = \frac{\kappa}{2} \sum_{<i,j>} (r_{i,j} - a)^2 + \sum_{<i,j>} u_{HC}(r_{i,j}) + S \sum_i \cos k_x x_i \cos k_y y_i - F \sum_i x_i \ , \tag{24}
$$

Here $\mathbf{x}_i = (x_i, y_i)$ denote the position of particle i. The first two sums run over all nearest-neighbor pairs in the triangular lattice, $r_{i,j} = |\mathbf{x}_i - \mathbf{x}_j|$, and $u_{HC}(r)$ is a hard-core potential, infinite for $r < 1/2$, and zero otherwise. In our simulations $\kappa = 2$, $a = 1$, and $k_x = k_y = 40$. (The hard-core contribution is included to prevent severe distortions of the lattice under large driving force.) The third term in equation (24) represents a static periodic potential having the symmetry of a square lattice and the wavelength that is small compared to the elastic lattice spacing. Since the background potential is incommensurate with the unstrained lattice, well-separated regions of the latter experience essentially uncorrelated potentials. It seems, then, reasonable to expect that this model captures the main effects of a truly random potential. Our choice of parameters corresponds to the weak pinning regime in which the translational correlation length is large compared to the lattice constant.

Without the driving force, this model exhibits short range translational correlations and long range orientational correlations [16], in accordance with the arguments provided in the previous sections. In the presence of the driving force, we find a sharp depinning transition at a critical force $F_c(0)$ when temperature is zero. For finite temperatures, the mobility grows exponentially with driving force before saturating when $F > F_c(0)$, and shows Arrhenius temperature dependence. The very rapid increase in mobility with driving force F, and with temperature, permits one to define an effective depinning line, $F_c(T)$, below which the mobility appears to be zero on the time-scale of the simulations [21]. It should be stressed that this model can afford only a qualitative description of pinned vortex lattices, as it employs a harmonic lattice rather than a logarithmic interaction between vortices, and a short-wavelength periodic background, rather than a random pinning potential. The observed phenomenology is quite robust, however, and will persist for a wide variety of interactions. These results indicate that a depinning line similar to that observed in high-temperature superconductors can be found in a model free of defects.

Acknowledgements

This work was supported by the Department of Energy under Grant No. DE-FG02-93ER45487.

References

[1] E.Y. Andrei et al., Phys. Rev. Lett. **60**, 2765 (1988).

[2] H. Dai, H. Chen, and C.M. Lieber, Phys. Rev. Lett. **66**, 3183 (1991); H.Dai and C.M.Lieber, ibid. **69**, 1576 (1992).

[3] S.E. Nagler et al., Phys. Rev. **B32**, 7373 (1985); N.Greiser et al., Phys. Rev. Lett. **59**, 1706 (1987).

[4] R. Seshardi and R.M. Westervelt, Phys. Rev. Lett. **66**, 2774 (1991); Phys. Rev. **B46**, 5142 (1992).

[5] C.A. Murray et al., Phys. Rev. Lett. **64**, 2312 (1990); C.A. Bolle et al., ibid. **66**, 112 (1991); D.G. Grier et al., ibid. **66**, 2270 (1990).

[6] A.I. Larkin, Zh. Eksp. Teor. Fiz. **58**, 1466 (1970) [Sov. Phys. JETP **31**, 784 (1970)]; A.I. Larkin and Yu.M. Ovchinnikov, J. Low Temp. Phys. **34**, 409 (1979).

[7] Y.Imry and S. Ma, Phys. Rev. Lett. **35**, 1399 (1975).

[8] M.P.A. Fisher, Phys. Rev. Lett. **62**, 1415 (1989).

[9] T. Giamarchi and P. Le Doussal, Phys. Rev. Lett. **72**, 1530 (1994).

[10] E.M. Chudnovsky, Phys. Rev. **B40**, 11355 (1989);

[11] Phys. Rev. **B43**, 7831 (1991).

[12] M.C. Marchetti and D.R. Nelson, Phys. Rev. **B41**, 1910 (1990).

[13] J. Toner, Phys. Rev. Lett. **66**, 2531 (1991).

[14] J. Toner, Phys. Rev. Lett. **67**, 1810 (1991).

[15] E.M.Chudnovsky, Phys. Rev. Lett. **67**, 1809 (1991).

[16] R. Dickman and E.M. Chudnovsky, Phys. Rev. **B51**, 97 (1995).

[17] For a review, see G. Blatter, M.V. Feigel'man, V.B. Geshkenbein, A.I. Larkin, and V.M. Vinokur, Rev. Mod. Phys. **66**, 1125 (1994).

[18] B.I. Halperin and D.R. Nelson, Phys. Rev. Lett. **41**, 121 (1978); D.R. Nelson and B.I. Halperin, Phys. Rev. **B19**, 2457 (1979).

[19] T. Nattermann, Phys. Rev. Lett. **64**, 2454 (1990).

[20] J-P. Bouchaud, M.Mezard, and J.S.Yedidia, Phys. Rev. Lett. **67**, 3840 (1991).

[21] R.Dickman and E.M.Chudnovsky, J. Phys. C: Condens. Matter, to appear.

[22] E.M.Chudnovsky and R.Dickman, to be published.

Slow Dynamics and Aging in Spin Glasses

Eric Vincent, Jacques Hammann, Miguel Ocio, Jean-Philippe Bouchaud and
Leticia F. Cugliandolo

Service de Physique de l'Etat Condensé, CEA Saclay, 91191 Gif-sur-Yvette Cedex,
France

1   Introduction

A crucial feature of the behavior of *real* spin glasses is the existence of
extremely slow relaxational processes. Any field change causes a very long-
lasting relaxation of the magnetization and, the response to an ac excitation
is noticeably delayed. In addition, the characteristics of this slow dynamics
evolve during the time spent in the spin-glass phase: the systems *age*.

Aging effects in real spin glasses have been layed down by experiments
[1], [2], [3] at a time where there was already an intense theoretical activity
on the equilibrium properties of mean-field spin-glass models [4]. Experimentalists started comprehensive studies of the non-equilibrium dynamics, which
happened to bring very instructive surprises, while in the meantime theoreticians developed extremely sophisticated methods for progressing towards
solutions of the equilibrium mean-field problem, thence inventing incentive
tools for the statistical mechanics of disordered systems.

This early epoch was not the time for the most productive dialogue between both parts. The situation is very different now; experiment and theory
have had, during these last years, a fruitful interplay. On the one hand, the
problem of the non-equilibrium dynamics has now been theoretically addressed from very different points of view; scaling theories of domain growth
[5], [6], [7], [8], a phase-space approach motivated by the Parisi solution to
mean-field models [9], a percolation like picture in phase space [10], random
walks in phase space [11], [12], [13], mean-field treatments of some simplified
situations [14], [15], are now providing us with various (and sometimes con-
tradictory) lightings of the experimental results, together with impulsing a
thrilling debate on the sound nature of the spin-glass phase. On the other
hand, more and more complex experimental procedures [16], [17], [18] have
been conceived with the aim of evidencing the materialization of some abstract theoretical notions, like *e.g.* the ultrametric organization of states or
the chaotic dependence of the spin-spin correlation function on temperature.

In this paper, we recall some important experimental features of the spin
glass dynamics. Since we intend to picture some aspects of the present state
of the dialogue between experimentalists and theoreticians, we give a detailed description of several ways of scaling the data and of the connection

between these scalings and the theoretical predictions. We obviously give up any pretention of giving an exhaustive comparison of theory and experiment; we mainly focus here on a perspective of spin glasses which proceeds from mean-field results [14] (abundant discussions of the scaling theories can be found in the literature of the past few years). In several occasions, we use as a guideline for the description of the experimental results a probabilistic model that views aging as a thermally activated random walk in a set of traps with a wide distribution of trapping times [12].

The slow dynamics of spin glasses - and, as well, of structural glasses [19] and other disordered systems - has been often interpreted in terms of thermal activation over barriers. One likes to think of a complex free-energy landscape (due to frustration) with peaks and valleys of all sizes. This picture has been extensively used in the litterature; Refs. [9], [11], [12], [13], ?, [20] are examples of different ways of drawing conclusions from it. In fact, the experiments never directly probe free-energy valleys or mountains, but rather give access to relaxation rates at various time scales, which may then be interpreted in terms of thermal activation over free-energy barriers [18].

However, when trying to describe the slow dynamics and aging of real spin glasses with a "phase-space" viewpoint, it is worth noticing that phase-space is infinite dimensional irrespectively of the finite or infinite dimensionality of real space. The geometrical properties of the infinite-dimensional phase space may put at work a different (non-Arrhenius) mechanism for slow dynamics that leads to slowing-down and aging even in the absence of metastable states [21]. The particle point in phase space, representing the system configuration, slowly decays through almost flat regions. This mechanism seems to be the one acting in the dynamics of mean-field spin-glass models (with a single aging correlation scale) [14] as well as in domain growth. In these models one does not see neither a severe change of behavior when approaching the zero-temperature limit nor rapid barrier-crossings from trap to trap in numerical simulations [15], [21].

We describe below some mean-field predictions which compare rather well with the experiments at constant temperature. Whether one can describe more subtle experimental results such as temperature variation dependences, etc. with mean-field models and/or with the above non-Arrhenius phase space geometrical description is still an open question. The rather good agreement at constant temperature suggests that the phase-space dynamical mechanism at the base of the dynamics of spin glasses may be a combination of rapid activated processes and slow decay through flat regions [14], [21], [22]. We might thence be led to revise our "common sense understanding" of the slow dynamics in disordered systems.

2 Experimental Evidence for Non-stationary Dynamics

2.1 Magnetization Relaxation in Response to a Field Change

In a measurement of the relaxation of the "thermo-remanent magnetization" (TRM), the system is cooled in a small field from above T_g down to some $T_0 < T_g$; it then "waits" in the field at T_0 during a time t_w, after which the field is cut, and the subsequent decrease of the TRM from the field-cooled (FC) value is recorded as a function of t. Following an "immediate fall-off" of the magnetization (depending on the sample and on temperature, of the order of 50 to 90 %), a slow logarithmic-like relaxation takes place; it is believed to head towards zero, although never reaching an end at laboratory time scales.

These endless-like relaxation processes and, more crucially, the existence of "aging" phenomena [1], [2], [3] are a salient feature of spin-glass dynamics: for different values of the waiting time t_w, different TRM-decay curves are obtained, as is evidenced in Fig. 1.a.

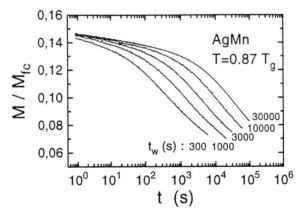

Fig. 1. a. Thermo-remanent magnetization M, normalized by the field-cooled value M_{fc}, vs. $t(s)$ ($\log_{10}$ scale) for the $Ag : Mn_{2.6\%}$ sample, at $T = 9K = 0.87T_g$. The sample has been cooled in a 0.1 Oe field from above $T_g = 10.4K$ to 9K; after waiting t_w, the field has been cut at $t = 0$, and the decaying magnetization recorded.

The dynamics depends on two independent time-scales, t ("observation time") and t_w ("waiting time"). This dynamics is *non-stationary*: the response at $t+t_w$ to an excitation at t_w depends on $t+t_w$ and t_w, and not only on t (breakdown of time-translational invariance). Qualitatively, one can see in Fig. 1.a that the longer the waiting time before cutting the field, the slower the overall response; the initial fall-off is smaller, the relaxation curve shows a slower decrease, the system has become "stiffer". Such aging phenomena have been early identified in the mechanical properties of glassy polymers [1] [23]; the slow strain following the application of a stress has been recognized to depend on the time spent in the glassy phase.

In spin glasses, the aging phenomena have initially been explored using the mirror experimental procedure of the TRM [1], in which the sample is cooled in zero-field, and after t_w a small field is applied. As far as the field remains low enough (usually, in the range 0.1-10 Oe), both procedures are equivalent; the relaxation of this "zero-field cooled magnetization" (ZFC) follows the same t *and* t_w dependence as the relaxation of the TRM [24], [25]. More precisely, it has been shown in [24] that, for all t, t_w values (*i.e.* all along the measured relaxations for various t_w), the sum of the ZFC-magnetization plus the TRM equals the field-cooled value. This is simply *linearity* in the response, since this experimental result shows that *the sum of the responses to different excitations* is equal to *the response to the sum of both excitations* (the response to a constant field being the field-cooled magnetization). That linearity holds for all t, t_w tells us that the presence of the (sufficiently small) field does not influence the aging process: waiting t_w in zero field and then applying a field during t (ZFC case) is equivalent, for the dynamics, to applying a field during t_w and then waiting t in zero field (TRM case). The only role played by the field in this context is to reveal the dynamic properties of the system. A recent study of the effect on the dynamics of increasing field values can be found in [20], [26], [27].

In the semi-log plot of Fig. 1.a, each curve shows an inflection point, and one first quantitative estimate of the t_w-effect on the relaxation is that this inflection point is located around $\log t \simeq \log t_w$. This fact has been noticed and given a physical meaning by Lundgren *et al.* [1]. The relaxations are slower than exponential; they do not correspond to a single characteristic response time τ, but are likely to be parametrized with the help of a wide distribution $g_{t_w}(\tau)$, which is defined hereby:

$$m_{t_w}(t) \equiv \frac{M(t + t_w, t_w)}{M_{fc}} = \int_{\tau_0}^{\infty} g_{t_w}(\tau) \exp(-\frac{t}{\tau}) d\tau \qquad (1)$$

where $\tau_0 \simeq 10^{-12} sec$ is a microscopic attempt time. $M(t + t_w, t_w)$ is the ZFC or TRM, depending on the experiment, and $M_{fc}(t + t_w)$ is the field-cooled value at time $t + t_w$. In the figures we abbreviate $M(t + t_w, t_w)/M_{fc}(t + t_w) = M/M_{fc}$. Lundgren *et al.* have pointed out that taking the derivative of (1) with respect to $\log t$ gives access to the distribution $g_{t_w}(\tau)$, since

---

[1] A wide class of materials like *e.g.* PVC, PS, Epoxy, or even bitumen, Wood's metal, amorphous sugar and cheese [23].

$$\frac{dm_{t_w}(t)}{d\log t} = -\int_{\tau_0}^{\infty} g_{t_w}(\tau)\frac{t}{\tau}\exp(-\frac{t}{\tau})d\tau \approx g_{t_w}(\tau = t) \ , \tag{2}$$

a rough approximation reflecting the sharp character of $\frac{t}{\tau}\exp(-\frac{t}{\tau})$ around $t = \tau$ (to be considered on a logarithmic scale, which is actually the scale which is suggested by the measurements). The plot of the relaxation derivatives shows bell-like shapes, with a broad maximum around $\log t = \log t_w$, and pictures $g_{t_w}(\tau)$ for various t_w [1]. Thus, in a first approximation, the aging phenomenon can be described as a *logarithmic shift towards longer times* of a wide spectrum of response times [2]. This shift is of the order of $\log t_w$, and therefore suggests that the dynamics be the same as a function of t/t_w.

Let us call "full aging" the pure t/t_w scaling, that is not far from being the correct one, as seen in Fig. 1.b where the data from Fig. 1.a is presented versus t/t_w. Most of the t_w-effect has been accounted for, though some systematic departures remain, and are worth being discussed (see Sect. 4.2).

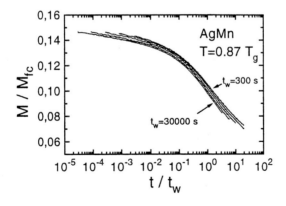

Fig. 1. b. Same TRM data as in Fig. 1a, presented as a function of t/t_w.

---

[2] Indeed, the spin-glass properties do not exactly depend on t_w, but rather on the *total* elapsed time $t_w + t$ (Sect. 4.2); they are evolving during the TRM measurement itself [3], [28]. The physical interpretation [1] of $g_{t_w}(\tau)$ therefore remains approximate.

2.2   Ac Susceptibility

The approximate t/t_w scaling of the TRM (or ZFC) curves is sufficient for a description of aging effects in ac experiments, where the in-phase and out-of-phase components of the response to a small ac excitation field at a frequency ω are measured. Aging is more visible (in relative value) in the out-of-phase component χ'' of the magnetic susceptibility, which represents dissipation. The *observation time*, corresponding to t in TRM experiments, is here constant, equal to $1/\omega$. When the sample is cooled from above T_g down to $T_0 < T_g$, the susceptibility does not immediately reach an equilibrium value, but shows a slow relaxation as time goes on. We denote t_a ("age") this time elapsed from the quench into the spin-glass phase; in the TRM experiment, the equivalent age is $t + t_w = t_a$. The *non-stationary* character of the dynamics, which appears in the TRM measurements as a dependence on the two independent time scales t and t_w, shows up in ac experiments as a dependence of χ'' on the two variables ω *and* t_a. This is clear in Fig. 2 where χ'' at various (low) frequencies ω is plotted as a function of ωt_a; applying a vertical shift, the curves can all be merged with respect to this reduced variable, which is equivalent to t/t_w in TRM experiments.

Thus, the approximate t/t_w scaling obtained from TRM and ZFC experiments can be fairly well transposed to an $\omega.t_a$ scaling of the ac susceptibility $\chi''(\omega.t_a)$. The vertical shift corresponds to accounting for the various "equilibrium values" ($\chi''_{eq}(\omega) = \lim \chi''(\omega, t_a \rightarrow \infty)$) at different frequencies. In Fig. 2, for technical reasons, the zero of the scale has not been measured and the shift is arbitrary. As an example of the relative orders of magnitude, let us mention that, at $\omega = 0.01 Hz$, the amount of the relaxing part is roughly equal to the equilibrium value; for χ' in the same conditions, it would be of the order of 10% of the equilibrium value. The frequency-dependence of χ''_{eq} has been determined in other studies [25], [28]; it can be represented by a power law with a very small exponent (or else a power law of a logarithm)

$$\chi''_{eq}(\omega) \propto \omega^\alpha , \tag{3}$$

where α increases in the range $0.01 - 0.1$ when approaching T_g from below. This is valid in the $10^{-2} - 10^5 Hz$ range which has been explored, and has been measured rather in insulating than in intermetallic spin glasses (due to eddy currents in metals). Both classes of samples have been found to present the same general spin-glass behavior [3], [28]. The $\omega.t_a$ scaling indicates that the smaller the frequency, the longer the time t_a during which a significant relaxation, characteristic of aging effects, can be found. Therefore, at higher frequencies ($\omega \geq 10 Hz$) aging disappears very rapidly, yielding almost instantaneously a stable value $\chi''_{eq}(\omega)$; conversely, at lower frequencies, the determination of $\chi''_{eq}(\omega)$ becomes problematic, implying measurements over tens of hours or days.

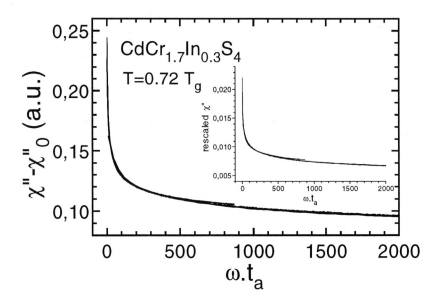

Fig. 2. Out of phase susceptibility $\chi(\omega, t_a)$ vs. $\omega \, t_a$ for the insulating $CdCr_{1.7}In_{0.3}S_4$ sample. The four curves, corresponding to $\omega = 0.01, 0.03, 0.1, 1$. Hz, have been vertically shifted (see text). t_a is the total time elapsed from the quench (age). The inset shows a scaling of the same data which follows from mean-field results (see Sect. 3.2).

2.3 Time Regimes in ac and dc Experiments

Let us summarize the conditions for observing either stationary or non-stationary (aging) dynamics in ac(χ") and dc (TRM or ZFC) measurements. We can define two distinct time regimes, which apply to both experiments.

- For $\omega.t_a \to \infty$, "equilibrium dynamics" is recovered in ac experiments, in the sense that only one time scale is needed: the dynamics is then *stationary*. This time regime corresponds, for TRM's, to $(t_w + t)/t = t_a/t \gg 1$ or equivalently $t \ll t_w$, that is the very beginning of the TRM-decay curves.

- If ωt_a is comparable to t_a ($\omega t_a = O(1)$) in the χ" experiments and, equivalently, t is comparable to t_w in the TRM experiments, one observes non-stationary dynamics.
  The χ" measurements are limited to $\omega t_a > 1$ since $\omega.t_a < 1$ cannot be experimentally realized (the harmonic response is not defined at times shorter than one period). Hence, with χ" we can only explore the beginning of the aging regime.
  In contrast, aging is predominant in TRM-measurements over the largest

part of the accessible time scale since, for TRM's, $(\omega.t_a) \equiv t_a/t = (t_w + t)/t$ becomes rapidly close to 1 as the observation time t elapses. Measuring the TRM decay we have access to a larger time-window in the aging regime.

Therefore, aging in χ'' can only be compared with the TRM decay at the beginning of the aging regime – that we call early epochs (also called "quasi-stationary regime" in [28]). This might explain why a full aging scaling seems to apply better to χ'' results than to the TRM (see Fig. 1.b and 2).

3 Aging Theories for Old Results

3.1 Scaling Theories

The different time scales in which *stationary* and *non-stationary* dynamics are occurring are likely to be mapped onto length scales in the real space of spins. When approaching T_g from above, the onset of a critical regime has been characterized by the diverging behavior of the characteristic time in ac experiments and of the non-linear susceptibility in dc studies, [29]. A thermodynamic phase transition would imply the divergence of a characteristic correlation length ξ when $T \rightarrow T_g$. In this equilibrium picture the spin-glass phase is believed to be an ensemble of randomly oriented spins, which are frozen due to infinite-range correlations corresponding to a long-distance "order".

However, aging shows that indeed equilibrium has not been established when crossing T_g. In several "scaling theories" [5], [6], [7], [8] of non-equilibrium phenomena in spin glasses, the spin correlations are considered to be limited to some *finite* range $\xi(t)$ (out-of-equilibrium situation); as time elapses, spin rearrangements yield a slow (due to frustration) extension of equilibrium correlations, towards the equilibrium situation of infinite range ($t \rightarrow \infty, \xi(t) \rightarrow \infty$). An ac experiment at frequency ω, as well as a TRM or ZFC relaxation at time t, can be viewed as probing the spin-glass excitations at a given length scale L which should be an increasing function of the *characteristic probe time* $1/\omega$ for χ'' and t for TRM. In the "droplet model" by Fisher and Huse [6], one has $L \propto \log^{1/\psi} t$ ($\psi \leq d - 1$), and in the "domain model" by Koper and Hilhorst [7] $L \propto t^{p/d}$ ($p \sim 0.5$). These theories do not aim at a microscopic description at the scale of spins, but make this mapping of *time* onto *length* scales quantitative, in terms of scaling laws. At least qualitatively, these models provide us with a convenient picture of aging phenomena, which is the following. For short probe times compared to the age ($\omega.t_a \gg 1$, or $t/t_w \ll 1$), short-ranged excitations are involved ($L \ll \xi(t)$), and the increase with time of $\xi(t)$ does not affect the dynamics, which is found to be stationary (no aging). Conversely, for longer probe times compared to the age ($\omega.t_a \sim 1$, or $t \geq t_w$), the characteristic length of the relevant excitations is of the same order of magnitude as $\xi(t)$, which is increasing due to aging, and the dynamic properties are strongly affected by aging (non-stationary dynamics).

The accurate quantitative agreement with the data is still to be discussed (see *e.g.* Ref. [13]). Some critical remarks to the simple scaling approaches, which rise up in view of other results (T-variation experiments), are discussed in Sect. 5.1.

3.2 Non-equilibrium Dynamics in Microscopic Theories

The Models. The classical "realistic" microscopic model of spin glasses is the 3-D Edwards-Anderson (3DEA) model [30]

$$H = -\sum_{\langle i,j \rangle} J_{ij} S_i S_j \,, \tag{4}$$

where J_{ij} are Gaussian or bimodal random variables, S_i are Ising spins and $\langle i,j \rangle$ represents a sum over first neighbours on a cubic 3D lattice. It is very difficult to obtain analytical results for the statics or the dynamics of 3DEA, in consequence, even after more than 20 years of research on the field of spin glasses, very few results are available. A lot of efforts have been devoted to the numerical study mainly of the equilibrium properties of the 3DEA. Again, the situation is still pretty unclear: basic questions as to the existence of a thermodynamic phase transition are still not answered [31].

The standard mean-field extension of the 3DEA model is due to Sherrington and Kirkpatrick (SK) [32] and corresponds to the same interactions as in (4) but with the sum extended to hold over all pairs of spins in the system. The study of the SK model - and of some other related mean-field models - had for a long time been confined to the search for equilibrium properties [4].

Numerical Simulations of Out-of-Equilibrum Phenomena. One may wonder whether aging, which did at first sight appear as some imperfection of the experiments, is really intrinsic to the Hamiltonian (4). We now know that the answer is yes. Only recently, attention has been paid to the study of the out of equilibrium dynamics of microscopic spin-glass models. Andersson *et al.* [33] and Rieger [34] reproduced in a numerical simulation the procedure of, *e.g.*, the TRM experience using the 3DEA model. The results show that it captures the main features of real spin glasses: both slow dynamics and aging effects. Later, numerical simulations of the large D "hypercubic" spin-glass cell in real space showed that also this model, that is expected to reproduce the SK model, when $D \rightarrow \infty$, captures the main characteristics of aging [35], thus confirming the previous analytical results that we describe in the following paragraphs.

Analytical Approach: Formulation and Definitions. Again, it only happened recently that *analytical* developments evidenced aging effects in mean-field spin-glass models [14], showing that these simplified models can describe, at least qualitatively, the phenomenology of real spin glasses. The idea in the case of mean-field spin-glass models is just to try to solve the *exact* dynamical equations derived for N, the number of dynamical variables in the model, tending to infinity. These equations are well-defined, have a unique solution and, in the absence of a magnetic field, only involve the correlation and the response functions (see Eqs.(5),(10) below for their definitions). The initial condition is chosen to be random so as to mimic the initial configuration just after the quench in the experimental situation. One then considers the large-time limits, but only after having already taken the thermodynamic limit $N \to \infty$ to obtain the asymptotic behavior of the solution.

It is important to notice that in this approach it is not necessary to assume *a priori* any particular structure of phase space - typically to say that there are many metastable states due to frustration separated by high barriers - to obtain the dynamical behavior of the problem. The solution can *a posteriori* be given a geometrical interpretation [12], [13], [21], [22].

The solution shows that the equations are self-consistently solved in the large-time limit by an *aging* solution. The reason why these equations can be solved is the *weak long-term memory* of the system [14]. The results fall into the *weak-ergodicity breaking scenario* previously proposed in [12], [13] within the trap model (see Sect. 3.3 below). When looking at the large-time dynamics of the system, the weak long-term memory property allows us to neglect the contribution of any finite time-interval after the quenching time. The system forgets what happens in "finite" time intervals with respect to the "infinite" observation time. It keeps, however, an averaged memory of its history. The weak-ergodicity breaking scenario tells us that the evolution of the system continues forever; the dynamics slows down as time elapses but the system is never completely stopped in its evolution. The waiting-time t_w gives us an idea of the age of the system.

In the following we shall be a bit more technical and describe the main features of the formalism and the solution.

The auto-correlation function is defined as

$$C(t + t_w, t_w) \equiv \frac{1}{N} \sum_{i=1}^{N} \overline{\langle s_i(t + t_w) s_i(t_w) \rangle} \,, \tag{5}$$

with the overline representing a mean over different realizations of the disorder and $\langle \rangle$ an average over different realizations of the thermal noise. We then define

$$C_F(t) \equiv \lim_{t_w \to \infty} C(t + t_w, t_w) \qquad C_F(0) = 1 \qquad \lim_{t \to \infty} C_F(t) = q_{EA} \,. \tag{6}$$

This allows us to separate the auto-correlation into two *additive* parts, a *stationary* term and an *aging* term C_A [14], [13]:

$$C(t + t_w, t_w) = C_F(t) - q_{EA} + C_A(t + t_w, t_w) \ . \tag{7}$$

In the absence of a magnetic field, the weak ergodicity breaking scenario [12], [13], [14] implies

$$\lim_{t \to \infty} C(t + t_w, t_w) = 0 \quad \forall \text{ fixed } t_w \ , \tag{8}$$

thus

$$\begin{array}{l} \lim_{t \to \infty} \lim_{tw \to \infty} C_A(t + t_w, t_w) = q_{EA} \\ \lim_{t \to \infty} C_A(t + t_w, t_w) = 0 \ . \end{array} \tag{9}$$

It will turn out that the two scales corresponding to these two limits are well-separated for mean-field models, in the sense that in the time-regime where C_F varies then C_A stays constant, and *viceversa*. In other words, one can think of q_{EA} as a value of the correlation separating different "correlation-scales", $C > q_{EA}$ and $C < q_{EA}$: when $t \ll t_w$, $C > q_{EA}$ and we have stationary dynamics, while when $t \gg t_w$, $C < q_{EA}$ and we have non-stationary dynamics and aging just as described in Sect. 2.3 for the general features of aging in spin glasses. [3]

In the same way, the response function can be equivalently separated into a *stationary* and a *non-stationary* term

$$R(t + t_w, t_w) \equiv \frac{1}{N} \sum_{i=1}^{N} \left. \frac{\partial \langle s_i(t + t_w) \rangle}{\delta h_i(t_w)} \right|_{h=0} = R_F(t) + R_A(t + t_w, t_w) \ . \tag{10}$$

with

$$R_F(t) \equiv \lim_{\substack{tw \to \infty \\ C(t+t_w,t_w) > q_{EA}}} R(t + t_w, t_w) \quad and \tag{11}$$

$$R_A(t + t_w, t_w) \equiv \lim_{\substack{tw \to \infty \\ C(t+t_w,t_w) < q_{EA}}} R(t + t_w, t_w) \ . \tag{12}$$

$R_F(t)$ satisfies the fluctuation-dissipation theorem (FDT) and $R_A(t + t_w, t_w)$ satisfies a generalized FDT [14]

$$R_F(t) = -\frac{1}{T} \frac{dC_F(t)}{dt} \qquad R_A(t+t_w, t_w) = \frac{X[C_A(t + t_w, t_w)]}{T} \left. \frac{\partial C_A(t + t_w, t')}{\partial t'} \right|_{t'=tw} \ , \tag{13}$$

---

[3] In some numerical works, the form $C(t + t_w, t_w) = t^{-x(T)} \Phi(t/t_w)$ has been often used to scale the data for all times t ([34], [35], [36], [37], see also [28] for a related discussion of the experimental data). It should be remarked that, though at first glance this scaling seems to be similar to the one following from the WEB scenario, it implies quite a different conclusion for the global behavior of the system. Note that if one takes the limit $\lim_{t \to \infty} \lim_{t_w \to \infty}$, that corresponds to exploring the *end of the stationary dynamics*, Eqs.(6) yield $\lim_{t \to \infty} \lim_{t_w \to \infty} C(t+t_w, t_w) = q_{EA}$, while $\lim_{t \to \infty} \lim_{t_w \to \infty} t^{-x(T)} \Phi(t/t_w) = \Phi(0) \lim_{t \to \infty} t^{-x(T)} = 0$. The stationary dynamics *and* the fact that the correlation decays to q_{EA} at the end of this time-regime, have a very clear geometrical interpretation [13], [14], [21].

with $0 \leq X[C_A] \leq 1$ a monotonically increasing function of $0 \leq C_A \leq q_{EA}$. ($X = 1$ corresponds to the usual FDT.)

With these definitions one can solve the asymptotic (large t_w) dynamics of several mean-field disordered models [14], [15], [38], [39], ?.

Analytical Approach: Stationary Regime. For all these models, when t is large, the *stationary* part of the correlation function $C_F(t)$ decays with a power law

$$C_F(t) \sim q_{EA} + c_\alpha \left(\frac{\tau_0}{t}\right)^\alpha , \tag{14}$$

where τ_0 is a microscopic time-scale and α has precisely the same meaning as the exponent in (3), describing the frequency dependence of the equilibrium out of phase susceptibility.

The temperature dependence of α depends on the model. For models [4] such as the p-spin spherical spin glass [40], [39] or the model of a particle moving in an infinite-dimensional random potential [41], [39], $\alpha = 1/2$ at $T = 0$ and it decreases when increasing the temperature. For the SK model, conversely, $\alpha = 1/2$ at $T = T_c$ and it decreases when decreasing the temperature [43]. Finally, for the mixed ($p = 2 + 4$) spherical model introduced in Ref. [44], α has a non-monotonic dependence on T; $\alpha(T = 0) = \alpha(T_c) = 1/2$. The value of the exponent α measured experimentally follows the tendency of the one holding for SK and the mixed ($p = 2 + 4$) spherical models close to the critical temperature, though the value of α from the experiments is considerably smaller ($\alpha \leq 0.1$ *vs* $\alpha \sim 0.5$).

Analytical Approach: Non-stationary Regime. Following then very general requirements, it has been argued in Ref. [14], [15], and explicitly checked on several pure and disordered models, that in the large-time limit only two situations with different dynamical behavior seem to exist:

- On the one hand, there are models with only one time-scale – or equivalently, correlation-scale – apart from the stationary one. C_A scales as in a domain growth process within the non-stationary time-scale, in the sense that:

$$C_A(t_w + t, t_w) = \jmath^{-1} \left(\frac{h(t + t_w)}{h(t_w)}\right) \tag{15}$$

  with $h(t)$ a monotonically increasing function – analogous to the domain length $L(t)$. $\jmath^{-1}(u)$ is a function characterized by another exponent which

---

[4] Though the aim of Refs. [40], [41] was to study the equilibrium dynamics à la Sompolinsky [42] - a different situation from the out of equilibrium occuring in experiments - the calculation of the exponent α obtained in these works applies to the experimental case when adequately reinterpreted [14]. Let us also note that the in this paper the α and β exponents are exchanged with respect to Refs. [39].

we call here $(1 - x)$ (and has been called β in [39] and α in [45]). Close to $u = 1$, *i.e.* for the early epochs of the aging regime, $\jmath^{-1}$ reads

$$\jmath^{-1}(u) \sim q_{EA} - c(1 - u)^{1-x} , \qquad (16)$$

c is a constant and $x < 1$ implying that $\jmath^{-1}$ is non-analytical in the neighbourhood of $u = 1$ (see (35) below).

In this case, the FDT-violating factor $X[C_A]$ is a constant $X < 1$. These models, when treated statically with the replica trick, are solved by a *one step replica symmetry breaking* ansatz [4]. An example is the p-spin spherical model [46]. We call them "single-scale models".

Certainly the functions h and $\jmath^{-1}$ do depend on the specific model. At the mean-field level we have succeeded in obtaining $\jmath^{-1}$ and X for several models. However, surprisingly, there are for the moment no analytical results available for the scaling function $h(t)$ [5]. The simplest possibility is that $h(t)$ is a pure power-law $(t/\tau_0)^a$, as found in the trap model [12] or standard coarsening models. This solution is particular in the sense that (15) is then *independent* of the microscopic time scale τ_0, which can be taken to zero; in this case C_A simply depends on the ratio t/t_w (full aging situation, a qualitative approximation of the experimental results, as explained in Sect. 2.1). This is not the case for more general functional forms. Two explicit choices which have been proposed so far are:

$$h(t) = \exp\left[\frac{1}{1-\mu}\left(\frac{t}{\tau_0}\right)^{1-\mu}\right] \qquad \text{or} \qquad h(t) = \exp\left[\ln^a(t/\tau_0)\right] . \qquad (17)$$

The form on the left was proposed to account for experiments in polymer glasses by Struik [23], then used in the first accurate analyses of aging effects in the TRM-decay [3], [28], and recently found in the exact solution of the asymmetric spherical SK model with $\mu = 1/2$ [47] (see also [38]). The second form is suggested by the numerical data from the "toy model" of a point particle in a random potential with infinite dimension [39]. Both will be used below to scale the data for the TRM and the out-of-phase susceptibility. Note that in the limit $\mu = 1$ or $a = 1$, one recovers full aging with a pure power-law behavior for h, while $\mu = 0$ corresponds to time translation invariance (no aging). When $\mu < 1$ and $a > 1$ we have "sub-aging" that we define as follows. Taking t fixed and, say, in the beginning of the aging regime, $h(t_w)/h(t + t_w) \sim 1 - (d\ln(h(t_w))/dt_w)t$. This defines a characteristic relaxation time $\tau(t_w)$. We say that we have sub-aging (super-aging) when $\tau(t_w)$ grows slower (faster) than t_w.

Interestingly enough, one can in general derive a relation between α, $(1 - x)$ and X (see (14), (16) and (13) for their definitions) [39], [45]:

---

[5] This technical difficulty is related to the introduction of a time re-parametrization invariance when studying the *exact* mean-field equations for large and widely separated times t_w and $t + t_w$.

$$X \frac{(\Gamma[1 + (1 - x)])^2}{\Gamma[1 + 2(1 - x)]} = \frac{(\Gamma[1 - \alpha])^2}{\Gamma[1 - 2\alpha]} , \qquad (18)$$

for single scale models. We shall use this equation to predict X at the beginning of the beginning of the aging regime in Sect. 4.2.

− On the other hand, there are models such as SK that have an infinite number of time-scales - correlation scales - apart form the stationary one [42], [15], [38]; mathematically, one has ultrametricity in time for all correlations such that $C_A < q_{EA}$, in the sense that $C_A(t_1, t_3) = \min(C_A(t_1, t_2), C_A(t_2, t_3))$, $t_1 > t_2 > t_3$ [15]. The decay is here infinitely slower than in the single-scale models. The FDT-violating factor $X[C_A]$ is a nontrivial function of C_A. These models are solved by a *full replica-symmetry breaking* ansatz when using the replica trick at the static level [4], and can be called "multi-scale" models.

It is important to notice that a scaling like (15) inside a correlation scale and ultrametricity between different correlation scales are expected to hold on very general grounds, in particular for more realistic finite dimensional models (in the limit of large-times), *provided that* the rather mild assumptions used in [14], [15] are satisfied. This justifies the fact that we shall use, in the following, a scaling-law like (15) to scale the data for real spin glasses, without refering to any particular model.

Connection with Measurable Quantities. If linear response theory holds, the TRM is just

$$M(t + t_w, t_w) = h \int_0^{t_w} ds \, R(t + t_w, s) . \qquad (19)$$

For large waiting-time t_w, this integral can be rewritten using the decomposition in stationary and non-stationary decays ((10) and (13)) and using (14), for $t \gg \tau_0$ we have

$$\frac{M(t + t_w, t_w)}{M_{fc}(t + tw)} - A \left(\frac{\tau_0}{t}\right)^\alpha \propto \int_0^{C_A} dC_A' \, X[C_A'] . \qquad (20)$$

where A is a constant. For single scale models as in (15) the scaling reads

$$\frac{M(t + t_w, t_w)}{M_{fc}(t + tw)} - A \left(\frac{\tau_0}{t}\right)^\alpha \propto \jmath^{-1}\left(\frac{h(t_w)}{h(t + t_w)}\right) . \qquad (21)$$

In the early epochs of the aging regime for the TRM, that should be compared to the non-stationary behavior of the $\chi''(\omega, t)$, one has

$$\frac{M(t + t_w, t_w)}{M_{fc}(t + tw)} - A \left(\frac{\tau_0}{t}\right)^\alpha \propto q_{EA} - c\left(1 - \frac{h(t_w)}{h(t + t_w)}\right)^{1 - x} , \qquad (22)$$

where we used (16).

The out-of-phase susceptibility can also be simply related to the correlation function (5). For high-frequencies, $\omega t \to \infty$, the aging term does not contribute (it is the integral of a slowly varying function $R_A(t, s)$ times a rapidly oscillating function). Using (14) to approximate the remaining integral, one finds the stationary part of the a.c. susceptibility:

$$\chi''(\omega, t) \to \chi''_{eq}(\omega) \propto \omega^\alpha , \qquad \omega t \to \infty . \tag{23}$$

Conversely, if $\omega t \geq 1$, *i.e.* for low frequencies, the aging part strongly contributes. For single-scale models we then have

$$\chi''(\omega, t) - \chi''_{eq}(\omega) \sim \frac{X}{T} h\omega \int_0^t ds \exp(i\omega s) \jmath^{-1}\left(\frac{h(s)}{h(t)}\right) , \qquad \omega t \geq 1 . \tag{24}$$

For ωt finite but large, one has in general:

$$\chi''(\omega, t) - \chi''_{eq}(\omega) \propto \jmath^{-1}(1) - \jmath^{-1}\left(1 - \frac{1}{\omega t} \frac{d \ln h(t)}{d \ln t}\right) \tag{25}$$

$$\propto \left(\frac{d \ln h(t)}{d \ln t} \frac{1}{\omega t}\right)^{1-x} \qquad 1 \ll \omega t < \infty , \tag{26}$$

where we introduced the power-law behavior of $\jmath^{-1}(u)$ in the vicinity of $u = 1$ defined in (16). Hence the conclusions for χ":

- If $h(t)$ is a simple power-law, then $\chi''(\omega, t) - \chi''_{eq}(\omega)$ scales as ωt, as obtained in [12], [13] (full aging).
- If $h(t) = \exp(1/(1 - \mu)t^{1-\mu})$ then $\chi''(\omega, t) - \chi''_{eq}(\omega)$ is a function of $\omega t \times (t/\tau_o)^{\mu-1}$. When $\mu \neq 1$ there is a correction to the pure ωt scaling (sub-aging for $\mu < 1$)).
- If $h(t)$ is of the form $\exp[\ln^a(t/\tau_0)]$, then the ωt scaling is corrected by a slowly varying factor $\ln^{(a-1)}(t/\tau_0)$ (sub-aging if $a > 1$). The χ" data are scaled within this assumption in the inset of Fig.2.

In Sect. 4, we apply the scaling relation (15) from single-scale models together with these proposals for $h(t)$ to the TRM and χ'' data.

The mean-field models which lead to the above results are very instructive: general statements and new ideas (like the violation of FDT) have emerged from their study. However, the physical mechanism underlying aging in these models is not yet very clear. The single-scale models, in particular, are very weakly sensitive to temperature, suggesting that no activated effects are involved, and that aging is rather related to large dimensional effects: the system wanders indefinitely in a large phase-space, without ever reaching a local minimum of the free energy [21]. Conversely, the trap model which we shall discuss now relies on activated effects to generate a broad distribution of time scales, which also leads to aging. However, this model is phenomenological, and does not emerge from a precise microscopic description - although some steps in this direction have recently been made [22], [48]. A tentative classification of the different models of aging has been proposed in [49].

3.3 The Trap Model

In the trap model [12], aging has been shown to naturally occur in a situation called "weak ergodicity breaking", which corresponds here to a statistical impossibility for the system to realize equilibrium occupation rates of the metastable states. This model has appeared as a fertile guideline for the analysis of the experiments; we therefore recall its main points, and come back to it later in Sect. 5.2. In the simplest version of the model [12], aging is sketched by a random walk in a collection of "traps" with random trapping times τ, all equally accessible. To each trap is associated a certain magnetization M and ac susceptibility $\chi_\tau(\omega)$. The properties of a real sample are obtained by averaging over an ensemble of decorrelated *subsystems*, corresponding to spins in different regions of space [6]. An important input is common to microscopic theories [4], and also to the more general problem of manifolds in random media [22], [48]; the distribution of trap depths is taken as an exponential. For thermally activated processes, this yields the following distribution of trapping times:

$$\psi(\tau) = \frac{x\tau_0^x}{\tau^{1+x}} \quad \text{(for } \tau \gg \tau_0) \ , \tag{27}$$

where x (from the distribution of barrier heights) is a temperature dependent parameter describing the structure of the phase space. In a comparable way, the "random energy model" (REM) of Derrida [50] involves $x = T/T_g$. The crucial point is that $x < 1$ in the spin-glass phase; in consequence, the mean value of $\psi(\tau)$ is divergent, that is the mean time needed to explore the whole set of traps (and thus to reach ergodicity) is infinite. This was called weak ergodicity breaking, in the sense that the equilibrium situation is never realized, although the system never gets trapped in a finite region, leading to an asymptotically zero correlation function – see (8). This is very different of the usual ergodicity breaking, where the system can reach rather quickly an equilibrium configuration, but remains in a restricted sector of the phase space. Since $x < 1$, the distribution (27) is very broad. After a random walk during t_w, the system has visited numerous short-life traps, but in a relatively small time compared with t_w; as usual with such broad distributions, the significant contributions arise from the largest - although rare - events. Thus, after t_w, the system has the largest probability to be found in a trap of characteristic time t_w itself; if a magnetic field is varied at t_w, most subsystems will need a time of order t_w before changing their magnetization. It has been shown [12] that the TRM-decay is then a function of t/t_w, and similarly that the ac susceptibility is a function of $\omega.t$. Thus, on the basis of a statistical description in the space of the metastable states, the main features

---

[6] At this stage, the spins do not enter directly. In later developments [26], however, the number of spins to be flipped for escaping from a trap (and hence the size of the subsystems referred to above) has been estimated from the influence of the field amplitude on the dynamics, as observed in the experiments.

of aging can be obtained. Further developments of this approach [13], [52] are discussed below at the light of various aspects of the experimental results. We now turn to a more detailed description of the combination of aging *and* stationary dynamics as seen in *both* TRM and χ" measurements.

4 TRM Experiments:
Aging and Non-aging Dynamics Disentangled

4.1 Departures from Full t/t_w Aging

In Fig.1.b, the TRM curves are presented as a function of t/t_w; it is clear that this is not exactly the correct reduced variable (failure of a full aging scaling). The same effect has been found in other samples [3], [28] and can be seen in results from other laboratories (*e.g.* [1]). It has been first identified in polymer mechanics [23]. Indeed, the χ" results recall us that stationary dynamics, namely $\chi"_{eq}(\omega) \propto \omega^\alpha$ (an additive contribution to the aging part, see Fig. 2), must intervene in the TRM decay, particularly in the $t \ll t_w$ regime. This frequency-dependent $\chi"_{eq}(\omega)$ is equivalent to an additive contribution of the form $t^{-\alpha}$ to the TRM.

The question of the departure from full aging has already been discussed in the past and the TRM's have been very accurately parametrized as the *product* of a $t^{-\alpha}$ factor times a (t, t_w) dependent factor ([28], see also the footnote[2] above). However, an additive combination of the stationary and aging parts arises naturally in the above theoretical approaches [14], [12], and we shall reanalyze these data in this light. We express the stationary part of the TRM, in units of the field-cooled value M_{fc} like in Fig. 1, as $A(\tau_0/t)^\alpha$. τ_0 is again a microscopic time, which allows homogeneity of the units, and then A is a non-dimensional constant, expected to be of order 1. We have taken the same TRM-data as in Fig. 1, and adjusted A and α in order to try to merge the aging parts $f(t_w, t)$

$$\frac{M}{M_{fc}} = A \left(\frac{\tau_0}{t}\right)^\alpha + f\left(\frac{t}{t_w}\right) \tag{28}$$

of all 5 curves of various t_w's as a function of t/t_w. We have no α values from χ"-measurements on the $AgMn$ sample, so we have simply kept α in the $0.01 - 0.1$ range obtained for the $CdCr_{1.7}In_{0.3}S_4$ sample in previous analyses [28]. Fig. 3.a shows the result; whatever the choice of parameters (also, trying a logarithmic decrease instead of a power law), it seems impossible to merge all 5 curves together in the *whole time regime*. Systematic discrepancies are always found in the large t/t_w range: no simple function of t alone can account for these deviations. However, for small t/t_w, the good quality of the scaling *is equivalent* to the simple $\omega.t$-dependence of $\chi"(\omega, t)$, which is actually measured in the *same time regime* (*i.e.* $\omega t > 1$).

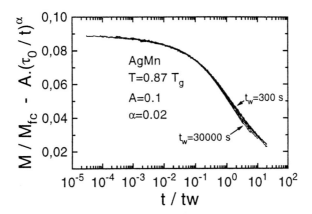

Fig. 3. a. Aging part of the TRM (Eq.28): the estimated stationary contribution has been subtracted from the full measured value. The data (same as in Fig. 1ab) is plotted *vs.* t/t_w.

In the long time $t/t_w > 1$ region, the discrepancies shown in Fig. 3.a present the same features - although somewhat weaker - as in previous analyses using a multiplicative stationary contribution [28]. In Fig. 3.a, as well as in Fig. 1.b (the same, without stationary dynamics subtraction), it is clear that a t/t_w-scaling slightly overestimates the aging effect (sub-aging); the "youngest" curve (shortest t_w) lies *above* the others, whereas it was *below* in the raw data (Fig. 1.a), and the effect is systematic for the 5 curves.

4.2   Sub-aging Scaling

A way to succeed in merging all TRM curves is to use a generalized scaling function of the form $h(t) = \exp[\frac{1}{1-\mu}(\frac{t}{\tau_0})^{1-\mu}]$ with $\mu < 1$ (one of both examples quoted above in (17)). This was proposed in the context of spin glasses in [3], [28], as a phenomenological scaling procedure inspired from polymer mechanics [23], which proved to account with great accuracy for the observed sub-aging situation. We refer the reader to [3], [28] for the arguments which lead to the above choice of $h(t)$, or rather, along the lines of [28], to the *effective time* λ defined as [7] :

---

[7] The careful reader will notice a change of λ-units when compared with [28]; within the present definition, λ is equal to $\lambda(\frac{\tau_0}{t_w})^\mu$ from [28].

$$\frac{h(t + t_w)}{h(t_w)} = \exp \frac{\lambda}{\tau_0} \quad . \tag{29}$$

As a function of λ, the 5 curves in Fig. 3.a recover precisely the same shape [3], [28], as displayed in Fig. 3.b. In other words, λ results from a change of variable which allows us to see the aging dynamics as *stationary*. It can also be obtained from

$$\frac{d\lambda}{\tau_0^\mu} = \frac{dt}{(t_w + t)^\mu} \quad , \tag{30}$$

which means that, since the age $t_w + t$ is varying during the TRM experiment, the effect of an elementary time interval dt depends on the elapsed time; due to sub-aging, the effective time scale associated to the age $(t_w + t)$ is $(t_w + t)^\mu$, with $\mu < 1$.

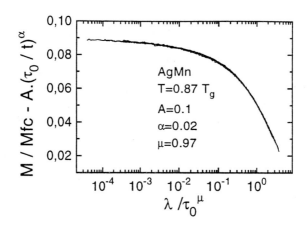

Fig. 3. b. Aging part of the TRM as in Fig. 3a, but *vs* the scaling variable λ/τ_o^μ defined in (29) (1st example in (17)). 5 curves, obtained for $t_w = 300, 1000, 3000, 10000, 30000 s$ (same data as in Fig. 1 and 3a), are superimposed onto each other.

This type of scaling has been successfully applied to various samples [3], [28]; μ is found in the $0.8 - 0.9$ range and it is almost independent of temperature in the $0.3 < T/T_g < 0.9$ range. This μ-trick is a very convenient way of parametrizing the sub-aging deviations from a full t/t_w-scaling, although not necessarily having a direct physical meaning. $\mu = 0$ corresponds to the case of no t_w-dependence (no aging), while $\mu = 1$ would yield a full t/t_w scaling. The intermediate values of μ (sub-aging) correspond to the fact that the apparent relaxation time scales sub-linearly (as $(t_w + t)^\mu$) with the age $t_w + t$.

The alternative choice of $h(t)$ proposed above in (17) can also be considered [53]. It is actually very close to the μ-case, in particular in the limit $\mu \to 1$. Again, for $a = 1$, full aging is recovered, while $a > 1$ describes a sub-aging situation, with an apparent relaxation time $\tau(t_w) = t_w/(a \ln^{(a-1)}(t/\tau_0))$. We have also applied this other sub-aging scaling to the non-stationary part of the relaxation displayed in Fig. 3a; the result is shown in Fig. 3.c, again the curves are fairly well superimposed onto each other.

Eq.(18), derived for single-scale mean-field spin-glass models, relates α, x and the FDT violating factor X defined in (13). According to (15) and (16), we have made a fit of the beginning of the TRM decay in Fig. 3.c (early epochs), and we have obtained $x \sim 0.96$ (see (16)). This is consistent with $X = 1$, which would suggest that FDT applies without corrections even at the beginning of the aging regime, if one accepts that (18) holds. This would be in accord with previous indications obtained through the comparison of χ'' and noise measurements made in the early epochs of the aging regime ("quasi-stationary regime") [28]. A careful and detailed study of the noise auto-correlation and response functions would help us knowing if and how FDT is violated, namely if X departs from 1 when going deeper in the aging regime. .

For the sake of comparison, we show in Fig. 3d how the μ-scaling applies to the full value of the TRM (the stationary contribution is neglected, and *not subtracted* from the magnetization). An almost acceptable scaling is obtained; the discrepancies are not much larger than the experimental uncertainties, but *they are systematic*: all curves are crossing in the middle of the figure. This corresponds to $\mu = 0.87$. Neglecting the influence of the stationary contribution thus enhances the deviations from $\mu = 1$.

The very good scaling of the 5 curves in Figs. 3.b, 3.c, involves 3 free parameters: here $A = 0.1, \alpha = 0.02, \mu = 0.97$ or $A = 0.1, \alpha = 0.02, a = 2.2$. They correspond to a full account of the results, with good coherence between TRM and χ'' experiments. Let us summarize our points:

– The out-of-phase susceptibility $\chi''(\omega, t)$ is well represented by the sum of a stationary contribution, which varies like ω^α (or $(\ln \omega)^{power}$), plus a non-stationary contribution which varies as a negative power $x - 1$ of $\omega.t$ (possibly with slowly varying corrections suggested by (26) above, but which are hardly visible on the χ'' data; see the rescaled χ'' in the inset of Fig.2).

– The TRM decay curves are the sum of a stationary contribution $\propto t^{-\alpha}$, plus an aging contribution. In the time range ($t \ll t_w$) which is common

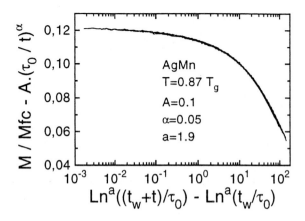

Fig. 3. c. Same as Fig. 3b, but as a function of a scaling variable which corresponds to the 2nd example in (17).

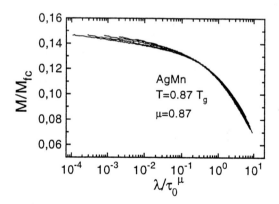

Fig. 3. d. Full measured value of the TRM (as in Fig. 1ab), but *vs* the scaling variable λ/τ_o^μ defined in (29).

to χ"-measurements, this aging part can be approximated (in agreement with χ") by a t/t_w-scaling. The long time regime $t \geq t_w$ seems however to prefer a sub-aging scaling of the type given in (17).

The origin of this weak $(1 - \mu = 0.03)$ but insistant sub-aging behavior is an interesting point, still uncompletely understood. As mentionned above, from a theoretical point of view this would mean that the $\tau_0 \rightarrow 0$ limit of the underlying model does not exist. From a more physical point of view, several scenarios which might explain this effect can be considered. The simplest one concerns the effect of the field amplitude, which indeed (for larger fields) is known to suppress progressively the t_w-effect [8](see [26] for details). However, the parameter μ seems to stick to a plateau value (< 1) for the explored low-field range of $10 - 0.1 Oe$ (systematic scaling analysis with *additive* (rather than multiplicative) stationary corrections are however needed to confirm this point).

This sub-aging behavior might also be interpreted as a sign that aging is actually "interrupted" beyond very long, but finite, times [52]. For example, if the trapping time distribution is cut-off beyond a certain ergodic time t_{erg} (which itself depends on the subsystem), then *for part of the subsystems* ergodicity will be realized within the time of the measurement: their dynamics will no more depend on t_w (interrupted aging). As shown in [52], this produces an effective sub-aging scaling very close to the $\mu < 1$ effect described here. The result is a value of a typical time t_{erg}, which must be understood as a *crossover time scale*, beyond which μ will further decrease to 0. In the analysis of [52], however, we had not properly taken into account the stationary contribution, which we have shown here to bring μ much closer to 1 (Figs. 3.b and 3.d). We had found t_{erg} of the order of $10^{6-7} sec$, which might therefore be underestimated.

5   Towards a Hierarchical Description of the Space of the Metastable States

5.1   Temperature Variation Experiments

Main Qualitative Features: ac Measurements. Until now, we have only presented results which are obtained from aging experiments *at constant temperature* after the quench into the spin-glass phase. In another class of experiments, the temperature is varied during aging. In terms of thermally activated processes in a mountaneous free-energy landscape, one may expect to explore in more detail the various scales of the free-energy barriers which are involved in the slow dynamics. Thermal activation should - at first sight

---

[8] In a similar way, μ is seen to decrease with the stress amplitude in polymer glasses [23]

- be able to speed up or slow down the aging evolution. The conclusions of these experiments have been surprisingly instructive.

An early result was obtained in [54]; measuring the aging relaxation of χ" in a CuMn spin glass, the authors observed that any step increase or decrease in temperature was causing an instantaneous increase of χ", followed by a slow decrease. We may call this effect "restart of aging", since the relaxation is renewed by the temperature change, which hereby produces a similar effect as obtained after the quench. This phenomenon reveals that the slow aging evolution towards equilibrium is significantly disturbed by relatively small temperature changes. Such a "chaotic dependence" of the equilibrium states on temperature has been predicted to occur as a consequence of frustration in [56], where it is argued that the relatively small free-energy of an overturned region of spins (droplet) results from large cancellations at the surface of the droplet. These cancellations should be very sensitive to temperature, hence the strong effect of a small temperature variation.

When studied in more detail, the temperature variation experiments do not simply show a restart of aging upon *any* temperature change. We present in Fig. 4 an experiment performed in such a way that the reaction of the χ"-relaxation to either a *decrease* or an *increase* in temperature is very different [16]. The sample is first quenched from above $T_g = 16.7$ K to 12 K; due to aging, χ" slowly relaxes. After 350 min, the temperature is decreased to 10 K. Despite the reduced thermal energy, the relaxation does not slow down, but restarts abruptly from a higher value, in agreement with [54]; this is the surprising chaotic-like effect, which looks as if the system was (at least partially) restarting aging from the quench. But, when after another 350 min at 10 K the sample is heated back to 12 K, the result is very different: χ" resumes its slow relaxation from the value which had been reached before the temperature variation (see the inset of Fig. 4), in a *memory-like* effect.

There is no contradiction with the results in [54]. The experimental conditions of Fig. 4 are chosen here as the most illustrative of this twofold effect. Indeed, for smaller temperature variations, more intricate situations can be found [55]; aging at the lower temperature may partly contribute to aging at the higher temperature, even with a temporary incoherent transient, as emphasized *e.g.* in [17]. What we want to stress here is that a restart of aging is always observed after a temperature decrease; in contrast, what is found upon a temperature increase is a memory of previous aging *at this higher temperature*. On the one hand, if a long time has been spent previously at the higher temperature, as is the case in Fig. 4, only a weak relaxation is found; on the other hand, if the system has been *directly* quenched to a given temperature, heating up afterwards to a higher temperature - for which no memory of previous aging exists - will produce a strong relaxation at this temperature (see examples of various situations in [55]).

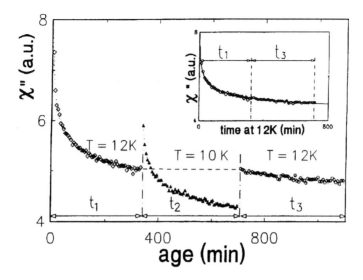

Fig. 4. Out of phase susceptibility $\chi''(\omega, t_a)$ of the $CdCr_{1.7}In_{0.3}S_4$ sample ($T_g = 16.7K$) during a temperature cycle. The frequency ω is 0.01 Hz, and t_a is the time elapsed from the quench. The inset shows that, despite the strong relaxation at 10 K, both parts at 12 K are in continuation of each other.

Same Effects in TRM Experiments. The effect is quantitatively confirmed by measurements of the TRM-relaxation [25], [16]. The comparison with χ'' requires some care; the χ''-relaxation directly shows *on-line* aging as a function of time, whereas in the TRM the effect of aging during t_w is considered *afterwards*, during the relaxation which follows the field cut-off at t_w. The TRM curve shows the relaxation processes in a wide time window, and thus yields more extensive information than χ'' at a given frequency ω (which mainly reflects the processes of characteristic time $\simeq 1/\omega$). In Fig. 5.a, a negative temperature cycling has been applied during the waiting time.

¿From the above χ'' results, one expects that aging processes be restarted when going to $T_0 - \Delta T$; but for sufficiently large ΔT this evolution will be erased when coming back to T_0. This is what can be checked in Fig. 5.a for $\Delta T = 1K$: the procedure yields a relaxation curve which is exactly superimposed onto that obtained after simply waiting $t_{w+} + t_{w+} = 30min$ at constant T_0, aging during $t_{w-} = 1000min$ at 11 K has not contributed. Note that the equivalent normal curve has $t_w = 30min = 2t_{w+}$, not $15min$ (see Fig. 5a); the memory of the first aging stage has indeed been preserved.

Intermediate ΔT values produce intermediate situations; for $\Delta T = 0.3K$, the resulting curve is the same as that obtained after waiting $t_{eff} = 100min$

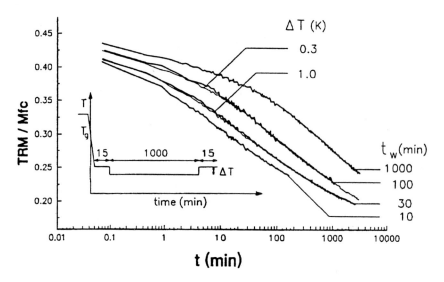

Fig. 5. a. Effect on the TRM relaxation of a *negative* temperature cycle ($CdCr_{1.7}In_{0.3}S_4$ sample, $T_g = 16.7K$). After waiting $t_{w+} = 15min$ at T_0 ($= 12K = 0.7T_g$), the sample is cooled to $T_0 - \Delta T$ for $t_{w-} = 1000min$, and then is heated back to T_0; after another $t_{w+} = 15min$, the field is cut and the relaxation measured (at T_0). These relaxations (thin lines) are compared with normal ones, measured after waiting $t_w = 10, 30, 100 or 1000min$ at constant T_0 (bold lines).

at constant T_0. From this example, we can work out a quantitative discussion of the effect. If we consider that the same characteristic free-energy barrier has been crossed by thermal activation during

    i) $t_{w+} + t_{eff} + t_{w+} = 100min$ at T_0 and

    ii) $(t_{w+}$ at $T_0) + (t_{w-}$ at $T_0 - \Delta T) + (t_{w+}$ at $T_0)$,

then we can write

$$(T_0 - \Delta T) \ln \frac{t_{w-}}{\tau_0} = T_0 \ln \frac{t_{eff}}{\tau_0} \quad , \tag{31}$$

which yields for the attempt time τ_0 the unpleasant value of $\sim 10^{-42}s$. Obviously, the *memory effect* cannot be explained by the only thermal slowing down of jumping processes over constant height barriers.

In addition, simple thermal effects cannot be expected to explain the restart of aging, which is again evidenced in the positive cycling procedure of Fig. 5.b (all temperatures remain below T_g).

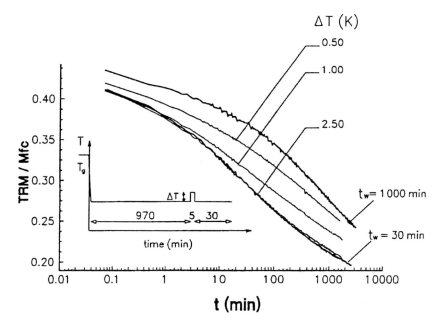

Fig. 5. b. Effect on the TRM relaxation of a *positive* temperature cycle ($CdCr_{1.7}In_{0.3}S_4$ sample, $T_g = 16.7K$). A short heating cycle is applied after 970 min of waiting time at T_0; then one still waits 30 min before cutting the field and measuring the relaxation at T_0 (thin lines). These relaxations are compared with normal ones, measured after waiting $t_w = 30$ or $1000min$ at constant T_0 (bold lines).

What appears in this procedure is the restart of aging due to the temperature decrease at the end of the cycle; for a sufficient ΔT ($= 2.5K$), the restart is so strong that the 970 min of previous aging are completely erased, as proved by the superposition of the $\Delta T = 2.5K$ curve with a normal $t_w = 30min$ one.

Again, for intermediate ΔT values, the effect is only partial; in its short-time part, the $\Delta T = 1K$ curve sticks to the young $t_w = 30min$-curve, whereas in its long-time part it goes closer to older ones. In contrast with the result in Fig. 5.a (negative T-cycling), the present procedure (positive T-cycling) yields curves which are not equivalent to a given waiting time at constant temperature. Clearly, the reason for that is that the positive temperature cycle ends by cooling down to T_0 (which produces a partial restart of aging at T_0), whereas the negative temperature cycle ends by heating back to T_0 (which lets the system retrieve the memory of the previous aging at T_0).

Thus, the same features of aging are observed in χ" and TRM experiments [16], [25]. A chaotic nature [56] of the spin-glass phase appears when the temperature is decreased, the restart of aging processes being very similar to what initially happens after the quench. On the other hand, a memory effect is found when the temperature is raised back, and this memory effect goes far beyond what can be expected from thermal slowing down.

A Hierarchical Sketch. These effects have been interpreted in terms of a *hierarchical organization* of the metastable states as a function of temperature [25], [16]. The empirical picture is sketched in Fig. 6.

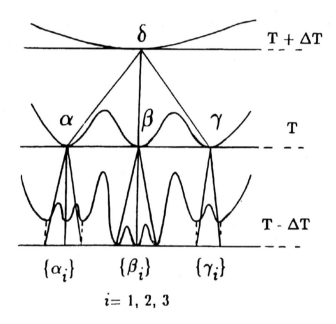

Fig. 6. Schematic picture of the hierarchical structure of the metastable states as a function of temperature.

During aging at a given T, the system samples the free-energy valleys (metastables states) at a given level of a hierarchical tree; the restart of aging upon lowering the temperature is figured out as a subdvision of the free-energy valleys into others, pictured as the nodes of a lower level of the tree, which thus develops multifurcating branches as the temperature decreases. The system is partially quenched since it must now search for equilibrium in a new, unexplored landscape, therefore aging (at least partially) restarts. The hierarchical picture naturally provides us with the observed memory effect; when the temperature is raised back, the newly born valleys and barriers merge back to the previous free-energy landscape. Thus, aging at $T - \Delta T$ may have *not contributed* to the evolution at T, as far as all subvalleys which could be explored at $T - \Delta T$ originate from unique valleys of the landscape at T (which is realized for large enough ΔT).

The interpretation of these effects in a picture where aging is seen as the growth of *compact*, independent domains (droplets [6], domains [7]) remains, in our opinion, difficult [9] . The droplets may be broken into smaller ones when the temperature is decreased, thus producing a restart of aging; but the memory effect implies that some information is kept somewhere about the *stage of aging which had been reached before decreasing the temperature*. Thus, large-scale correlations should be kept untouched, while on the other hand they should apparently be destroyed. A way to satisfy these requirements could be to consider that the droplets should be fractal (non space filling) [8], [51], with a non trivial internal structure, in such a way that large droplets may contain smaller ones which can be activated independently. These droplets inside droplets might lead, under some energetic conditions, to a space-transcription of the hierarchical organization of the states.

More Quantitatively: Rapid Growth of Free-Energy Barriers. The observed effect on aging of various thermal histories is in contradiction with thermal activation over *constant height* barriers, as evidenced above in (31). The picture can be made more quantitative along this same line (Fig. 5.a and 31), which we now further develop. In one experiment, the spin glass is aged during t_w^- at $T - \Delta T$, and brought back to T before cutting the field and measuring the relaxation. In another experiment, the spin glass is simply aged during t_w^0 at T, and the relaxation is measured at T. If both decay curves are the same in the whole time window, we may consider that the aging states reached in both procedures are the same, and that equivalent regions of both landscapes at $T - \Delta T$ and T have been explored. We can characterize each evolution by a typical height B of the maximum barrier which has been crossed in each case, namely

$$B(T - \Delta T) = (T - \Delta T) . \ln \frac{t_w^-}{\tau_0} \tag{32}$$

$$B(T) = T . \ln \frac{t_w^0}{\tau_0} . \tag{33}$$

Thanks to the identity of the TRM curves obtained in each procedure, one may consider that this quantifies the T-variation of the *same barrier*, which limits in both cases the *same region* of the phase space. The contradiction pointed out in (31) shows that $B(T - \Delta T) > B(T)$; in other words, the hierarchical picture of valleys bifurcating into valleys as the temperature decreases is supported by the observation of the growth of free-energy barriers.

---

[9] See [25], [16]. However, some experiments have been analyzed along this line, with the introduction of long-time effects for the breakup of the domains [7], [17].

An extensive series of TRM measurements on the AgMn sample, for multiple values of t_w^-, T and ΔT, has been performed [18]. In brief, the result is that the barriers are growing for decreasing temperatures below T_g, and in addition that their growth rate is so fast that *at any $T < T_g$* some of them should even *diverge* (see details in [18]). This provides us with an interesting link to the Parisi solution of the mean-field spin glass [4]. For decreasing temperatures, some barriers separating the metastable states are diverging, transforming the valleys into pure states in the sense of the Parisi solution; the hierarchical structure of the valleys, deduced from the experiments, can thus be related to that of the pure states. Also, the picture which emerges from these results is that of a *critical regime* at any $T < T_g$; in a sequence of micro-phase transitions starting at T_g, the spin-glass phase space continuously splits into nested and mutually inaccessible regions, within which non trivial dynamics takes place.

5.2 Aging as a Random Walk: Traps on a Tree

From a One-Level to a Multi-level Tree. In its first stage [12], the trap model deals with all-connected traps, which can be called a "one-level tree" (see Sect. 3.3). The basic quantity which is calculated is the spin-spin correlation function $C_A(t_w + t, t_w)$, which can be related to the relaxation function if the fluctuation-dissipation theorem holds (in non-equilibrium, generalized forms of FDT might nevertheless hold [14], see Sect. 3.2). The aging part M_A of the TRM-relaxation can thus be estimated from the decay of the aging correlation function, which is proportional (by a factor q_{EA}) to the probability $\Pi(t, t_w)$ that the system has not jumped out of a t_w-trap at $t_w + t$:

$$M_A(t + t_w, t_w) \sim C_A(t_w + t, t_w) = q_{EA}.\Pi(t, t_w) \quad . \tag{34}$$

The probability Π contains all the information on possible jumps from trap to trap, and thus represents the aging dynamics; it should tend to 1 when t_w goes to infinity, for any finite t. In this limit, equilibrium dynamics is recovered, since no jump occurs; the proportionality factor q_{EA} in (34) describes this "bottom of the traps" dynamics (*cf.* with (14) in Sect.3.2). From its definition, it represents the overlap between the various configurations which constitute the bottom of the traps; in a TRM experiment, it is this finite fraction of the initial magnetization which should be found in the ideal limit of infinite t_w.

Two asymptotic behaviors of $\Pi(t, t_w)$ (and thus of the TRM) are calculated in a "multi-level" version of the trap model [13]:

$$t \ll t_w \qquad \Pi(t, t_w) \sim A - \left(\frac{t}{t + t_w}\right)^{1 - x_M} \tag{35}$$

$$t \gg t_w \qquad \Pi(t, t_w) \sim \left(\frac{t}{t + t_w}\right)^{x_1} \quad ,$$

The predicted shape of the aging TRM is thus a constant minus a power law of exponent $1 - x_M$ for the short times $(t \ll t_w)$, and on the other hand a simple power law of exponent x_1 at long times $(t \gg t_w)$. In the simple one-level case of all-connected traps with trapping times distributed with a given x (27), one has $x_1 = x_M = x$, which does not yield a realistic TRM-decay shape. However, a satisfactory fit to the TRM measurements can be obtained with only two values of x [13], [52]; one finds from the initial part of the TRM (or from χ'' measurements) $x_M \sim 0.65 - 0.8$ and from its late part $x_1 = 0.05 - 0.35$ (for several samples and different temperatures). In Fig.7, we show the asymptotic behaviors corresponding to (35) for a typical TRM curve, $1 - x_M$ being equivalent to the exponent in (22) if $h(t)$ is a power law.

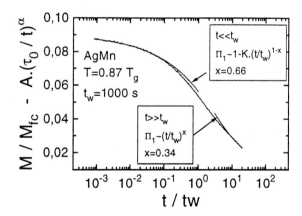

Fig. 7. Aging part of one of the above TRM curves (Fig. 1 and 3). Both asymptotic limits $t \ll t_w$ and $t \gg t_w$ have been fitted to (35). The variation of the effective index x along the different time regimes suggests a more complex than one-level tree structure.

Thus, if one thinks in terms of a one-level picture with no further assumption, it appears from the shape of the TRM itself that the effective x which labels the trapping time distribution (27) is *different* in the short and long-time regimes. Namely, x is closer to 1 for $t \ll t_w$ $(x = x_M)$, and decreases for $t \gg t_w$ $(x = x_1)$. This simple observation has been translated in terms of a hierarchical organization of the traps (traps inside traps); the model of all-connected traps [12] (one-level tree) has thus been generalized in [13], of which we now extract the main points.

The one-level tree is a distribution of traps of index say $x = x_1$ (< 1); following the construction of the Parisi tree of states [4], it can be extended into an n-level tree by multifurcating each trap into others of index $x_2 > x_1$, which themselves subdivide into others, until some final index $x_n > 1$ at the end-branches of the tree. At each level i, the x_i-distribution of trapping times corresponds [4], [12] to an x_i-dependent distribution of free energies, and this yields for the overlap between the corresponding states a value $q_i(x_i)$ which is the inverse of the Parisi order parameter $q(x)$ [4], thus $q_{i+1} > q_i$. Now, the states close to the end-branches (say the lower part of the tree) with $x > 1$ can all be visited in finite times, since for $x > 1$ the mean value of the distribution (27) is finite: the corresponding dynamics is *stationary*.

Aging phenomena appear when $x < 1$, i.e. above a given level c such that $x_c = 1$. The faster processes of the aging regime, which are seen in the short-time part of the TRM, occur among the states which are close to each other in the hierarchical geometry (large overlap), that is which are related by a tree-node at a level very cloes to c; therefore they correspond to x close to 1, as is indeed suggested by the exponent of the power law behavior at the beginning of the TRM. As time elapses, more distant states (of smaller overlap) can be explored, and the corresponding transitions imply passing higher nodes in the tree, of smaller index x, in agreement with the smaller exponent of the power law in the long-time part of the TRM.

The model can be solved for an arbitrary number of levels; but the simpler case of a *two-level* tree has been computed and fitted to the TRM experiments [13]. The fits are of a very good quality for both insulating and metallic samples: for a given t_w, the shape of the decay curve is quite well reproduced over the whole t scale. If the number of metastable states at each level is infinite, however, the predicted influence of t_w is precisely a t/t_w scaling (full aging), as in [12], and this does not satisfactorily account for the measured t_w-dependence, which corresponds to a sub-aging behavior (see the discussion in Sect. 4).

Effect of Temperature Changes. Now, the effect of the temperature on this tree-like organization of the metastable states can also be discussed. From the fit of the TRM-data [52], the x_M-index (e.g. from the short-time part) shows a tendency to increase towards 1 as T approaches T_g; this suggests that the spin-glass transition be closer to the REM scenario [50], where $x = T/T_g$, than to the mean-field equilibrium scenario [4] where the Parisi parameter x rather goes to zero at T_g. This leads us to a possible interpretation of the empirical *hierarchical structure versus temperature* in terms of the Parisi-like tree of traps [13].

T_g is characterized by a certain c-level of the tree where $x_c = 1$; the temperature dependence of x means that, when the temperature is lowered, this is now a lower (multifurcated) level of the tree which corresponds to $x = 1$ and limits the stationary dynamics. Thus, aging dynamics restarts among newly born states, as observed in the experiments and interpreted along the hierarchical scheme of Fig. 6. Conversely, when the temperature is raised back, the lower aging level comes back to equilibrium, and aging continues at the upper level, resuming from its previous stage of evolution. A more detailed comparison between theory and experiment is however desirable. Note finally that a hierarchical decoupling of time scales has also been argued on the basis of "second" noise spectra by Weissman and collaborators [58].

The influence of temperature is thus crucial in these pictures where activated processes play the dominant part in aging. Conversely, temperature is somewhat irrelevant for the aging dynamics of the mean-field models described above (at least those corresponding to single-scale dynamics). For example, the effect of temperature variations has been explicitly computed within the spherical $p = 2$ model [57], with the result that nothing comparable to experiments happens. In this respect, aging in these single scale models is again very similar to simple domain growth in a ferromagnet [21], which is rather insensitive to temperature. It would be very interesting to understand how temperature changes affect the dynamics of a multiscale model, such as the SK model.

6 Conclusions

In this paper, we have presented a survey of the experimental results concerning aging phenomena in spin glasses, focusing on the description of magnetization relaxation and low-frequency out-of-phase susceptibility measurements. We have discussed in some details how far two theoretical approaches of out-of-equilibrium effects, namely the trap model [12], [13], [52] and the microscopic approach [14], [15], can succeed in describing different aspects of the dynamics.

The relaxation of the out-of-phase susceptibility χ" towards a non-zero value indicates the presence of stationary dynamics. This same dynamics can be found, although less obviously, in the short-time part (compared to the waiting time t_w) of the decay of the thermo-remanent magnetization (TRM). In fact, an additive combination of a stationary and an aging regime in the auto-correlation (and thus also in the TRM and in χ") follows from the solution of some mean-field models [14], [15], [38], [39], [57]. This striking similarity of mean-field and real spin glasses raises the question of the nature of the slow dynamics in these systems. While one is used to think of thermal activation in mountaneous landscapes as the source of slow processes, the mean-field models now provide us with aging phenomena which are due to the flatness of large regions in the phase space [21], with no crucial role played by the temperature.

For the sake of a coherent description of both TRM *and* χ" data, one should therefore extract the same stationary contribution from all results, which we have made here in an additive way. The remaining *aging contribution* presents, for the TRM, systematic departures from a "full aging" situation of a pure t/t_w scaling ("sub-aging") [3], [28]. When the stationary dynamics of the TRM is additively accounted for, the departures from full aging become less pronounced, but still remain. For χ", whose aging regime only overlaps that of the TRM on a limited range, both full aging and sub-aging scalings remain compatible with the data. Using some recent analytical developments of the microscopic approach of spin-glass dynamics [15], we have shown that a sub-aging behavior will appear under some general conditions. Thus, the microscopic theory can now account for the scaling functions which had been postulated in the past by the experimentalists on phenomenological grounds [3], [28].

The experiments have shown that small temperature variations have a strong effect on aging phenomena [16], [17], [18]. A temperature decrease restarts the evolution, whereas raising back the temperature lets the system retrieve its previous stage of aging at this same temperature. These results have been interpreted in terms of a hierarchical organization of the metastable states as a function of temperature, in which the valleys of the free-energy landscape subdivide into others for decreasing temperatures [16], [18]. The mean-field models have not yet brought conclusive results on this question. The trap model [12], which provides us with a stochastic picture of aging, has recently been extended [13] in a way which sheds some light on the temperature variation experiments.

The simple picture of a random walk among all-connected traps (one-level tree of states) did not take into account the existence of stationary dynamics, which in this language is a "bottom-of-the-traps" dynamics. For this reason, traps must have an internal structure. On the other hand, the TRM shape is related to the index x of the trapping time distribution. The comparison of the experimental shapes with a one-level tree model shows that the "effective x" systematically varies along the curve; it is closer to 1 in the $t \ll t_w$ regime, and smaller in the $t \gg t_w$ regime [12], [52], [13]. Thus, the traps which are explored in the various time regimes are not connected in the same way.

These results are fairly well understood within a picture of hierarchically connected traps, in a multi-level tree geometry where x varies from one level to the other. The limit value $x = 1$ sets the tree level which is at the border of stationary and non-stationary dynamics. The tendency of x to increase when the temperature approaches T_g [52] indicates that this border $x = 1$ level changes with temperature; the resulting model of a tree-like organization where the limit of equilibrium dynamics changes with temperature [13] is now very close to the empirical picture of a hierarchy of metastable states as a function of temperature which had been drawn from the experiments [16].

This recent development, together with the predictions of slow dynamics and aging from microscopic models, have created a very stimulating atmosphere. They let us expect that an even tighter interaction among theoreticians and experimentalists will bring us closer to a more complete understanding of the physics of disordered systems.

Acknowledgements We would like to thank L. Balents, D. Dean, Vik. Dotsenko, M. Feigel'man, J. Kurchan, P. Le Doussal, M. Mézard, R. Orbach and G. Parisi for numerous stimulating discussions all along this work.

References

[1] L. Lundgren, P. Svedlindh, P. Nordblad and O. Beckman; Phys. Rev. Lett. **51** 911 (1983).

[2] R.V. Chamberlin; Phys. Rev. **B30**, 5393 (1984).

[3] M. Ocio, M. Alba and J. Hammann; J. Physique Lett. (France) **46**, L-1101 (1985).
M. Alba, M. Ocio and J. Hammann; Europhys. Lett. **2**, 45 (1986).

[4] K. Binder and P. Young; Rev. Mod. Phys. **58**, 801 (1986).
M. Mézard, G. Parisi and M. A. Virasoro; *Spin Glass Theory and Beyond*. World Scientific Lecture Notes in Physics Vol **9** (Singapore, 1987).
K. Fischer and J. Hertz; *Spin Glasses*, (Cambridge University Press, 1991).

[5] A. J. Bray and M. A. Moore; J. Phys. **C17**, L463 (1984), in Heidelberg Coll. in Glassy Dynamics, Springer-Verlag, 1986.

[6] D. S. Fisher and D. Huse; Phys. Rev. **B38**, 373 (1988).

[7] G. Koper and H. Hilhorst; J. Phys. France **49**, 429 (1988).

[8] M. Ocio, J. Hammann and E. Vincent; J. Magn. Magn. Mat. **90-91**, 329 (1990).

[9] Vik. Dotsenko, M. Feigel'man and Ioffe; *Spin Glasses and Related Problems*, Soviet Scientific Reviews **15**, (Harwood, 1990).

[10] H. Hoffmann and P. Sibani; Phys. Rev. **A38**, 4261 (1988).

[11] H. Hoffmann and P. Sibani; Phys. Rev. Lett. **63**, 2853 (1989); Z. Phys. **B80**, 429 (1990).

[12] J-P Bouchaud; J. Phys. I (France) **2**, 1705 (1992).

[13] J.P. Bouchaud and D.S. Dean; J. Phys. I (France) **5** (1995) 265.

[14] L. F. Cugliandolo and J. Kurchan; Phys Rev. Lett. **71**, 173 (1993); Phil. Mag. **71**, 501 (1995).

[15] L. F. Cugliandolo and J. Kurchan; J. Phys. **A27**, 5749 (1994).

[16] Ph. Refregier, E. Vincent, J. Hammann and M. Ocio; J. Phys. (France) **48**, 1533 (1987).
F. Lefloch, J. Hammann, M. Ocio and E. Vincent; Europhysics Lett. **18**, 647 (1992).

[17] P. Grandberg, L. Sandlung, P. Nordblad, P. Svendlindh, L. Lundgren; Phys. Rev. **B38**, 7097 (1988).

[18] M. Lederman, R. Orbach, J. M. Hammann, M. Ocio and E. Vincent; Phys. Rev. **B44**, 7403 (1991).
J. Hammann, M. Lederman, M. Ocio, R. Orbach and E. Vincent; Physica **A185**, 278 (1992).

[19] C. A. Angell; Science, **267**, 1924 (1995). C. A. Angell, P. H. Poole and J. Shao; Il Nuovo Cimento **16**, 993 (1994).

[20] Y.G. Joh, R. Orbach and J. Hammann; submitted to Phys. Rev. Lett. (This paper proposes a model for the time dependence of TRM and ZFC magnetization based on the same hierarchical picture as in Sect. 5.1, and a linear increase of the barrier heights with the Hamming distance).

[21] J. Kurchan and L. Laloux; J. Phys. **A29**, 1929 (1996).

[22] J.P. Bouchaud and M. Mézard, in this volume.

[23] L. C. E. Struik; *Physical Aging in Amorphous Polymers and Other Materials*, Elsevier, Houston (1978).

[24] P. Nordblad, L. Lundgren and L. Sandlund; J. Magn. Magn. Mat. **54**, 185 (1986).

[25] E. Vincent, J. Hammann and M. Ocio; in *Recent Progress in Random Magnets* p.207-236, editor D.H. Ryan, World Scient. Pub. Co. Pte. Ltd., Singapore 1992.

[26] E. Vincent, J-P Bouchaud, D. S. Dean and J. Hammann; Phys Rev **B52**, 1050 (1995).

[27] D. Chu, G.G. Kenning and R. Orbach; Philos. Mag. **B71**, 489 (1995).

[28] M. Alba, J. Hammann, M. ocio and Ph. Refregier; J. Appl. Phys. **61**, 3683 (1987).
Ph. Refregier, M. Ocio, J. Hammann and E. Vincent; J. Appl. Phys. **63**, 4343 (1988).
J. Hammann, M. Ocio and E. Vincent; ; in *Relaxation in Complex Systems and Related Topics* p.11-21,edited by I.A. Campbell and C. Giovanella, Plenum Press, New-York 1990.

[29] see e.g.: J. Souletie and J.L. Tholence; Phys. Rev. **32**, 516 (1985),
H. Bouchiat; J. Physique (France) **47**, 71 (1986),
E. Vincent and J. Hammann; J. Phys. **C20**, 2659 (1987),
and references therein.

[30] S. Edwards and P. W. Anderson; J. Phys. **F5**, 965 (1975).

[31] see, *e.g.*, P. W. Anderson and C. M. Pond; Phys. Rev. Lett. **40**, 903 (1978).
J. A. Hertz, L. Fleshmann and P. W. Anderson; Phys. Rev. Lett. **43**, 943 (1979).
E. Marinari, G. Parisi and F. Ritort; J. Phys. **A27**, 2687 (1994).
N. Kawashima and A. P. Young; Phys. Rev. **B53**, R484 (1996).

[32] D. Sherrington and S. Kirkpatrick; Phys. Rev. Lett. **35**, 1792 (1975).

[33] J. O. Andersson, J. Mattson and P. Svedlindh; Phys. Rev. **B46**, 8297 (1992); *ibid* **B49**, 1120 (1994).

[34] H. Rieger; J. Phys. **A26**, L615 (1993); J. Phys. **I** (France), 883 (1994); Annual Rev. of Comp. Phys **II**, 295 (1995).

[35] L. F. Cugliandolo, J. Kurchan and F. Ritort; Phys. Rev. **B49**, 6331 (1994).

[36] H. Yoshino; J. Phys. **A29**, 1421 (1996).

[37] G. Parisi, F. Ricci and J. Ruiz-Lorenzo; cond-mat/9606051.

[38] S. Franz and M. Mézard; Europhys. Lett. **26**, 209 (1994); Physica **A209**, 1 (1994).

[39] L. F. Cugliandolo and P. Le Doussal; Phys. Rev. **E53**, 1525 (1996).
L. F. Cugliandolo, J. Kurchan and P. Le Doussal; Phys. Rev. Lett. **76**, 2390 (1996), and in preparation.

[40] A. Crisanti, H. Horner and H-J Sommers; Z. Phys. **B92**, 257 (1993).

[41] H. Kinzelbach and H. Horner; J. Phys. I (France) **3**, 1329 (1993), *ibid*, 1901 (1993).

[42] H. Sompolinsky; Phys. Rev. Lett. **47**, 935 (1981).

[43] H. Sompolinsky and A. Zippelius; Phys. Rev. **B25**, 6860 (1982). P Biscari; J. Phys. **A23**, 3861 (1990).

[44] Th. Nieuwenhuizen; Phys. Rev Lett **74**, 4289 (1995), *ibid*, 4293.

[45] J-P Bouchaud, L. F. Cugliandolo, J. Kurchan and M. Mézard; Physica **A226**, 243 (1996).

[46] A. Crisanti and H-J Sommers; Z. Phys. **B87**, 341 (1992).

[47] L. F. Cugliandolo, J. Kurchan, P. le Doussal and L. Peliti; cond-mat/9606060.

[48] L. Balents, J-P Bouchaud and M. Mézard; cond-mat/9601137, J. Phys. **I** (France), to be published.

[49] A. Barrat, R. Burioni and M. Mézard, J. Phys. **A29** (1996)1311.

[50] B. Derrida; Phys. Rev. Lett. **45**, 79 (1980); Phys. Rev. **B24**, 2613 (1981).

[51] A similar concept has been proposed by J. Villain in *Europhys. Lett.* **2** (1986) 871.

[52] J.-P. Bouchaud, E. Vincent and J. Hammann; J. Phys. I (France) **4**, 139 (1994).

[53] L. F. Cugliandolo and J. Kurchan; unpublished.

[54] L. Lundgren, P. Svedlindh and O. Beckman, Journal of Magn. Magn. Mat. **31-34**, 1349 (1983).

[55] E. Vincent, J-P Bouchaud, J. Hammann and F. Lefloch; Phil. Mag. **71**, 489 (1995).

[56] D.S. Fisher and D.A. Huse; Phys. Rev. Lett. **56**, 1601 (1986). A.J. Bray and M.A. Moore; Phys. Rev. Lett. **58**, 57 (1987).

[57] L. F. Cugliandolo and D. S. Dean; J. Phys. **A28**, 4213 (1995).

[58] M. B. Weissmann, N.E. Isrealoff, G. B. Alers, Journal of Magn. Magn. Mat. **114**, 87 (1992).

Ultrametric Structure of Finite Dimensional Spin Glasses

Angelo Cacciuto[1], Enzo Marinari[1], and Giorgio Parisi[2]

[1] Dipartimento di Fisica and Infn, Università di Cagliari, via Ospedale 72, 09100 Cagliari, Italy
[2] Dipartimento di Fisica and Infn, Università di Roma *La Sapienza*, P. A. Moro 2, 00185 Roma, Italy

Abstract. We discuss numerical experiments that allow to detect an ultrametric structure of the phase space of the $4D$ spin glasses with quenched random couplings $J = \pm 1$. We discuss a constrained Monte Carlo method, and systematic problems like finite size effects. We compare our results to the ones we obtain for the mean field Sherrington Kirkpatrick model and by using unconstrained numerical simulations, and we exhibit a very coherent picture.

1 Introduction

Verifying the ultrametric structure of spin glass models by numerical simulations is a difficult task. Even for the SK model, where we know analytically what to expect, fully satisfactory numerical checks have not been yet obtained. Still, the question is very important: is the phase structure of finite D models reminiscent of the ultrametric organization of the mean field solution? Cacciuto et al. (1996) have discussed this issue in the $4D$ case, and found a positive evidence, that we will discuss in the following. The interested reader can read for example the introductions and discussions of Rammal et al. (1986), Parisi (1993): mean field techniques allow advanced computations about the ultrametric structure of the phase space (Franz et al. (1992a), Franz et al. (1992b)).

A good introduction to ultrametricity for physicists is Rammal et al. (1986). Here we just remind the reader that the usual triangular inequality

$$d_{1,3} \leq d_{1,2} + d_{2,3} \, , \tag{1}$$

is substituted in spaces endowed with an ultrametric distance by the stronger inequality

$$d_{1,3} \leq \max(d_{1,2}, d_{2,3}) \, . \tag{2}$$

One can for example define the squared distance among two spin configurations and relate it to the overlap q by

$$d_{\alpha,\beta}^2 \equiv \frac{1}{4 q_{\mathrm{EA}} V} \sum_{i=1}^{V} \left(\sigma_i^\alpha - \sigma_j^\beta \right)^2 = \frac{1}{2} \left(1 - \frac{q_{\alpha,\beta}}{q_{\mathrm{EA}}} \right) \, . \tag{3}$$

In an ultrametric space all triangles have at least two equal sides, that are larger than or equal to the third side. An hierarchical tree is an appropriate way of representing an ultrametric set of states. In the solution of the mean field spin glass theory one finds an exact ultrametric structure: states are organized on an hierarchical tree, and if we pick up three equilibrium configurations of the system and compute their distance we find an ultrametric triangle.

2 The 4D Constrained Monte Carlo Runs

The work of Cacciuto et al. (1996) is based on a constrained Monte Carlo procedure. One updates three replicas of the system (with the same set of couplings), and constrains the distance between replica one and replica two to a given value $q_{1,2}$, and the distance between replica two and replica three to $q_{2,3}$ (that can be equal to $q_{1,2}$). We have three replicas, two distances between them are fixed and we measure the third one, that we call q. For example if one fixes both $q_{1,2}$ and $q_{2,3}$ to some fraction of q_{EA} (in the case of Cacciuto et al. (1996) we used the value $\frac{2}{5}q_{EA}$) an ultrametric structure would imply that $q \geq \frac{2}{5}q_{EA}$, while the usual triangular inequality would only imply that $q \geq -\frac{7}{5}q_{EA}$. Obviously the choice of the constraint is crucial to obtain a sharp difference from the usual situation of an Euclidean metric.

It has been possible to thermalize lattices of up to 8^4. The computation turns out to be very successful, as we will see. The most serious problem turns out to be in the usual finite size effects: finite size effects are serious in spin glass models, and in this computation they appear clearly. In figure 1 the vertical line on the left, at $q \simeq -0.75$, depicts the bound given from the triangular inequality. The second vertical line, at $q \simeq .21$, depicts the ultrametric bound, while the vertical line on the right, at q_{EA}, is the upper bound for infinite volume. The probability distributions of the measured q value (the overlap among configuration C_α and configuration C_β, see before) for the 6 L values, from $L = 3$ to $L = 8$. The $L = 3$ $P(q)$ is the one with the smaller peak, far on the right, that end last on the left: $P(q)$ for increasing L values have higher peaks, and end more and more at q values close to zero. It is clear that already on small lattices the distribution is far from the triangular bound.

The probability for a measured distance q not to be ultrametric (i.e. one minus the normalized area S^U of the $P(q)$ integrated inside the ultrametric bound) decreases fast with the lattice size. On a $L = 8$ lattice half of the configurations are ultrametric (and indeed the big violation is from configurations with $q > q_{EA}$, which we expect from normal Monte Carlo runs to disappear in the continuum limit).

In order to be more quantitative we define the integral

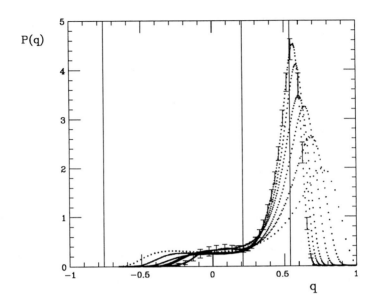

Fig. 1. $P(q)$ of the free overlap measured in the course of our constrained Monte Carlo as a function of q, for L values going from 3 (lowest curve, ending on the left closest to $q = 0$) to $L = 8$ (highest curve, ending on the left far from $q = 0$).

$$I^L \equiv \int_{-1}^{q_{\min}} dq \ (q(L) - q_{\min})^2 \ P(q) + \int_{q_{\max}}^{+1} dq \ (q(L) - q_{\max})^2 \ P(q) \ , \quad (4)$$

where $q_{\min}$ is the minimum q allowed (for us, for example, $q_{\min} = q_{1,2}$), and $q_{\max} = q_{\mathrm{EA}}$. I^L goes to zero if the system is ultrametric. We plot I^L in fig. (2) for the two choices of the constraint that have been discussed in Cacciuto et al. (1996).

For example in the case of two equal distances a very good best fit shown in the figure gives

$$I^L \simeq (-0.0001 \pm 0.0005) + (0.76 \pm 0.03) L^{-2.21 \pm 0.04} \ . \quad (5)$$

It is remarkable that the mean field computations of Franz et al. (1992a), Franz et al. (1992b) give an exponent of $\frac{8}{3} \simeq 2.67$, for the deviations from

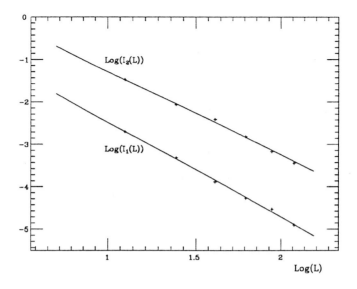

Fig. 2. The integral I^L as a function of L in a double log scale. The lower points are for the case where we have fixed $q_{1,2} = q_{2,3}$, the upper points where $q_{1,2} \neq q_{2,3}$ (see the text).

a pure ultrametric behavior in a finite system. Not only does one find the system converges, for large L, to an ultrametric behavior, but the rate of the convergence is very similar to the one computed in the mean field model. This is one of the quantitative agreements that make the relation of the mean field solution and the finite dimensional models clear and impressive.

3 Some Very Recent Results

We will report here about two sets of results that we have obtained recently (Cacciuto et al. (1997)).

The first point has been to verify the validity of the method. We have tried to reproduce the results of the former section, for the $4D$ spin glass, by using a free, non-constrained dynamics. This is important in order to get sure that the constrained method does not introduce some anomalous slowing down. Such free runs take a long time (since one has to investigate also a part of the phase space that is not relevant for the questions one wants to answer), but it turns out that a test can be done and that it gives a positive answer. One runs a long simulation of 3 replica's, and only sample the third value of q when the first two values are, for example, equal to $\frac{2}{5}q_{EA}$ (with a given precision, of the order of some percent). The runs have to be long since only some configurations triples satisfy the constraint. We have been able to check that for L up to 6 we get the some result that with the constrained approach.

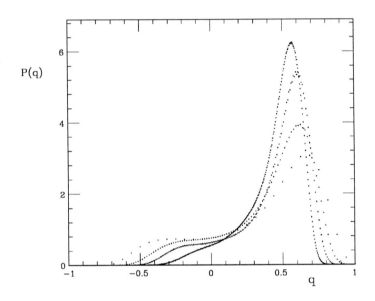

Fig. 3. As in figure 1, but for the SK model.

As a second piece of evidence we have analyzed the SK model, and applied the constrained method. Here we know analytically everything about the ultrametric nature of the replica broken solution. We find that the situation is very similar to the $4D$ case. Also here the main problem is related to strong finite size effects. In fig. 3 we show data for the SK model (from a constrained run), with $V = 64, 128, 256, 512$, at $\frac{T}{T_c} = 0.6$. The samples are 750 for the smallest lattice down to 27 for the largest one.

Th results of this study are very positive, since we have been able to detect clear signatures of ultrametricity in a finite dimensional spin glass. The similarity of the SK case and the $4D$ case is strong. The main problem seems to be able to thermalize large lattice sizes, where the feature of the infinite volume physics are more clear.

References

Cacciuto, A., Marinari, E., Parisi, G. (1996): preprint cond-mat/9608161, to be published on J. Phys. A

Cacciuto, A., Marinari, E., Parisi, G. (1997): in preparation

Franz, S., G. Parisi, G., Virasoro, M. A. (1992): Europhys. Lett. 17, 5

Franz, S., G. Parisi, G., Virasoro, M. A. (1992): J. Phys. I France 2, 1869

Parisi, G. (1993): J. Stat. Phys. 72, 857

Rammal, R., Toulouse, G., Virasoro, M. A. (1986): Rev. Mod. Phys. 58, 765

Entropy crisis in a short range spin glass

Silvio Franz

ICTP, Strada Costiera 11, P.O. Box 563, 34100 Trieste (Italy)
e-mail: *franz@ictp.trieste.it*

Abstract. We discuss the results of recent Monte Carlo simulations of a short range spin glass model with four-spin interactions, in connection with the physics of glasses. The model could be classified as a fragile glass, displaying stretched exponential relaxation with super-Arrhenius behavior of the relaxation time. The growth of this at low temperature appears to be associated to an entropy reduction according to the Adam-Gibbs equation. The correlation length appears to grow rather modestly compared with the relaxation time.

The entropy crisis scenario for the glass transitions makes its apparency a long time ago [1]. To explain the so-called Kauzmann paradox, Gibbs and di Marzio put forward the idea that a thermodynamic collapse of the configurational entropy was underlying the dynamical freezing of glasses. Soon after [2], in a refinement of the argument, Adam and Gibbs found a relation among relaxation time (τ) and configurational entropy (S_c) , namely, $\tau \sim e^{Const/TS_c(T)}$ that leads to the Vogel-Fulcher law under the hypothesis of linear vanishing of S_c at the Kauzmann temperature. Since then, few theoretical progresses have been made on this line. Very few models are known models displaying entropy crisis (and more in general glassy behavior), and being at the same time simple enough to allow for analytic comprehension. A remarkable exception in this landscape is represented by Mean Field Systems with Quenched Disorder. The simplest model in which the entropy crisis transition occurs is the celebrated Random Energy Model [4], and is a characteristic features of all models with "one step replica symmetry breaking transition" [5], in short 1RSBM. In the last years, the work of Kirkpatrick and Thirumalai has shown how the physics of these models is related to the one of structural glasses [6], and pointed out its strong connection with Mode Coupling Theory. Recently, there have been found models of interacting variables which lead to the same kind of physics even in absence of any quenched disorder [7]. This suggests that common mechanisms could underlly glassy phenomena both in disordered systems, and system like structural glasses with no quenched disorder. One of the most attractive features of 1RSBM is the existence, within a range of temperatures, of an exponentially large number of metastable states, leading to a very natural definition of the configurational entropy. As in all genuine mean-field theories "activated processes" that destabilize metastable states are neglected, and the metastable states have an infinite life time. Going from the mean field description to real finite dimensional systems it is necessary to reintroduce these processes. Hand waving arguments in this direction [8], result in a generalized Adam-Gibbs relation of the kind: $\tau \sim e^{Const/TS_c(T)^\gamma}$, with $\gamma = D - 1$. Unfortunately it is very difficult to check the validity of this formula by controlled analytical techniques. The natural testing ground for it are numerical simulations on finite dimensional models that reduce to 1RSBM for $D \to \infty$. In this talk, we report the results of Monte Carlo simulations on a short range spin glass with 4 spin interactions, obtained in collaboration with F.Ritort and D.Alvarez of the Universitad Carlos III of Madrid [9]. Simulations of finite dimensional analogous of 1RSBM are very rare, and we consider our work as just the beginning of a potentially very interesting field. Previous interesting numerical studies of disordered multi-spin interactions models appeared in [10].

The model concerns a set of Ising variables defined on the sites of a D dimensional hypercubic lattice, interacting via the Hamiltonian:

$$H = -\sum_{\square} J_{\square} \prod_{i \in \square} \sigma_i \tag{1}$$

where the sums runs over all the plaquettes $\square$ of the lattice (each plaquette has 4 spins). The "ferromagnetic" version of this model, with all $J_{\square} = 1$ has received recently attention in the context of random surface physics [11] and quantum gravity [12]. We concentrate here to the case where the coupling $J_{\square}$ are quenched random variables, drawn independently from the normal distribution with zero mean and unitary variance. For $D \to \infty$ the mean field limit is recovered, and we expect the physics of the model to reduce to the one of the "p-spin" spin glass model [13] for $p = 4$, a well representant of 1RSBM. The system has been studied through numerical simulations in two and three dimension. In two dimension, it can be seen that the thermodynamic corresponding to the HAmiltonian (1) is trivial [9]; accordingly we find that the relaxation to equilibrium is exponential, and the relaxation time follows a simple Arrhenius behaviour. More interesting results are obtained in $D = 3$[1]. The data for the energy as a function of the temperature for different cooling rates [9], show system undergoes an apparent calorimetric glass transition at a temperature $T_g \approx 0.9$. The relaxation time is estimated from the behaviour of the equilibrium spin-spin autocorrelation function. This can be fitted by a stretched exponential with temperature dependent exponent all along the decay, $C(t) = \exp[-(t/\tau(T))^{b(T)}]$. The dependence of τ on T (fig. 1) is not compatible with the Arrhenius law, and in the temperature range we can equilibrate the system is equally well fitted by the Vogel-Fulcher law $\tau = A \exp[B/(T - T_0)]$ and by the Ferry law $\tau = A \exp[B/T^2]$. The parameters of the fits are given in the caption of fig. 1.

---

[1] The result we present are sample averages of systems of 10^3 up to 20^3 spins. We checked the independence of the system size.

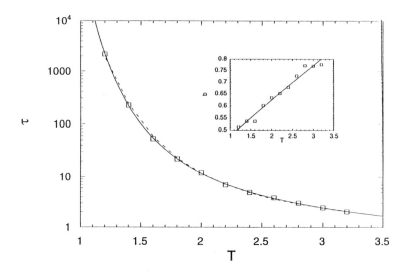

Fig. 1. The relaxation time as a function of the temperature together with the Vogel-Fulcher fit (dotted line) and the Ferry fit (slashes). The parameters are: $A = 0.27$, $B = 5.1$, $T_0 = .63$ for the Vogel-Fulcher fit and $A = .67$, $B = 11.5$ for the Ferry fit. The inset shows the temperature dependence of the exponent b.

In the light of the previous discussion, it is interesting to relate the increase of the relaxation time to the decrease of the configurational entropy. Unfortunately, we do not know how to disentangle the configurational contribution from the total entropy of the system. On that point we rely on mean field theory [6], where not too close to the transition, the configurational entropy dominates on the inter-state contribution. Simulations of the off-equilibrium dynamics [9] seem to confort this assumption. We then relate τ to the total entropy, obtained integrating the equilibrium internal energy data. In figure 2 we see that within the range of temperature we can equilibrate the system, the Adam-Gibbs relation is obeyed with an exponent $\gamma = 1$. At present, it is not clear if we have to attach to this value a fundamental meaning, or it is a just consequence of a spurious symmetry of the Hamiltonian (1) [9].

A major and unsolved issue in glass physics is whether the increase of the relaxation time is accompanied by large scale ordering, with a consequent growing correlation length. In order to investigate this possibility we have

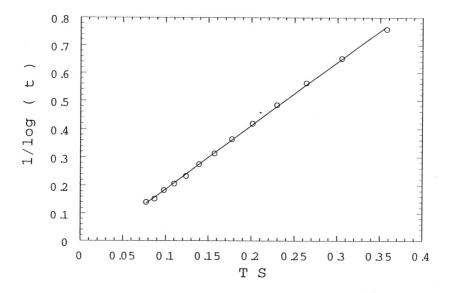

Fig. 2. The inverse of the logarithm of the relaxation time versus the $TS(T)$. The linear dependence is the one predicted by the Adam-Gibbs form.

considered the equal time correlation function[2]

$$G(x) = \overline{\langle \sigma_i \sigma_{i+e_\mu} \sigma_{i+xe_\nu} \sigma_{i+xe_\nu+e_\mu} \rangle^2} \sim \frac{\exp(-(\frac{x}{\xi}))}{x^{1+\eta}}, \qquad (2)$$

the angular brackets mean thermal average, and the bar means sample average. $\{e_\mu, \mu = 1, 2, 3\}$ stand for the three versors of the lattice and μ and ν are taken to be different in (2). Good fits are obtained supposing $\eta = -1$, i.e. with a simple exponential form for G. We then estimate the associated susceptibility $\chi = \sum_x G(x)$. Our data, displayed in figure 3 are in favour of of a slow increase of χ. The relation among among χ and τ is of the "activated form": $\tau \sim \exp(a\chi/T)$. Accordingly, the susceptibility can be well fitted by the finite temperature singularity $\chi = a + b/(T - T_0)$, and with the zero temperature one $\chi = 1 + c/T^2$. This last fit has to be preferred, as it gives the correct limit $\chi(T) \to 1$ for $T \to \infty$.

Summarizing, in this talk we have reported about one of the first investigations about the glassy nature of a finite dimensional analogous of 1RSBM. We find fragile glass behaviour with super-Arrhenius relaxation law. The relaxation time, if plotted against entropy, is seen to follow the Adam-Gibbs form on the temperature window we can explore. Our data are compatible

---

[2] This quantity is chosen according to symmetry considerations, and has been suggested to us by P. Young.

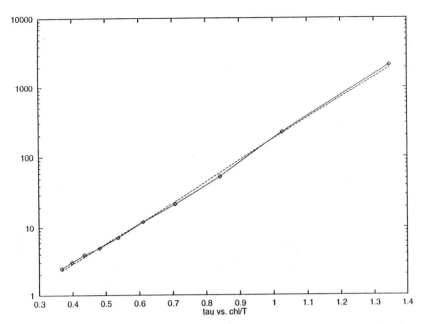

Fig. 3. Relaxation time versus susceptibility divided by temperature. The linear dependence demonstrate the "activated form" mentioned in the text. Note the modest increase of χ, that goes from $\chi = 1.1$ for $T = 3$ to $\chi = 1.6$ for $T = 1.2$.

both with a finite temperature singularity and with a progressive freezing with singularity at $T = 0$. In this respect the situation is similar to real experiments, where the severe growth of the relaxation time does not allow to reach equilibrium close enough to the expected singular point.

Unquestionably, the present model presents glassy behaviour with many phenomenological points in common with structural glasses. At this stage it is not clear what it just pathological in the model and what corresponds to basic mechanisms common to a wide class of systems. Systematic numerical study of finite dimensional 1RSBM are needed to clarify the picture. Steps in this sense have been undertaken recently by Parisi [14].

It is a pleasure to thank here my colleague and friend F. Ritort, with whom the ideas and the results presented here have been elaborated.

References

[1] J. H. Gibbs and E. A. Di Marzio, J. Chem. Phys. **28** 373 (1958)
[2] G. Adams and J. H. Gibbs, J. Chem. Phys. **43** 139 (1965)
[4] B. Derrida, Phys. Rev. **B24** 2613 (1981).
[5] M. Mézard, G. Parisi and M. A. Virasoro, *Spin Glass Theory and Beyond* (World Scientific, Singapore 1987); K. H. Fischer and J. A. Hertz, *Spin Glasses* (Cambridge University Press 1991);

[6] For a review see T. R. Kirkpatrick and D. Thirumalai Transp. Theor. Stat. Phys. **24** (1995) 927, and references therein.

[7] J. P. Bouchaud and M. Mezard, J. Physique I (Paris) **4** (1994) 1109; E. Marinari, G. Parisi and F. Ritort, J. Phys. A (Math. Gen.) **27** 7615 (1994); J. Phys. A (Math. Gen.) **27** 7647 (1994);

[8] T. R. Kirkpatrick and P. G. Wolynes, Phys. Rev. **B36** (1987) 8552; parisi Parisi, preprint cond-mat **9411115,9412004,9412034**

[9] D.Alvarez, S.Franz, F.Ritort, *Fragile-glass behavior of a short range p-spin model* preprint **cond-mat/9603200**, to appear in Phys. Rev. B

[10] H. Rieger, Physica **A184** (1992) 279, J. Kisker, H. Rieger and H. Schreckenberg, J. Phys. A (Math. Gen.) **27** (1994) L853

[11] A. Cappi, P. Colangelo, G. Gonnella and A. Maritan, Nucl. Phys. **B 370** (1992) 659

[12] G.K. Saviddy, K.G. Saviddy and F.J. Wegner, Nucl. Phys. **B 443** [FS] (1995) 565 and references therein

[13] D. J. Gross and M. Mezard, Nucl. Phys. **B240** 431 (1984); E. Gardner, Nucl. Phys. **B257** 747 (1985).

[14] G.Parisi work in progress.

Chiral and Spin Order in XY Spin Glass

Muriel Ney-Nifle

Laboratoire de Physique, Ecole Normale Supérieure, 46 allée d'Italie, 69364 Lyon cedex 7, France **

Abstract. We give a brief review on a family of two dimensional models with continuous symmetry and possibly regular frustration or frustration-type disorder. Next we focus on the random ± J bond XY model. There are two types of variables in this system and thus two candidates for an order parameter, the continuous spin variable and the discrete chiral variable. The question is : do the chiralities order more easily than the spins ? We present an analytical approach based on correlation function studies in one dimension and on domain wall scaling analysis in two dimension. Our results contradict recent interpretation of numerical simulations.

1 Introduction

This talk present some commonly accepted facts and also open questions about a family of models with continuous XY symmetry (specified by a phase) on the square lattice. We will start with the pure XY model and then add frustration and disorder and study the consequences for the transition. The pure XY model has a transition driven by point defects and we will focuss on these discrete degrees of freedom, especialy when frustration is introduced, and on the possible interplay with the continuous degree of freedom.

Next section gives a brief review on pure models, frustrated models and disordered models. In section 3 we present our analytical results on the XY spin glass. Section 4 is a summary.

There are several experimental realisations of two dimensional models with continuous symmetry [1].

∘ The superfluid and superconductors films have a complex order parameter $|\psi| \exp(i\theta)$. In case with inexpensive phase fluctuations one can describe the transtion by the phase-only Ginsburg-Landau free energy which integrates $|\nabla\theta|^2$ [1, 4].

∘ Melting of two dimensional solids such as smectic liquid crystal [3]. Layered high temperature superconductors (cuprates) in a perpendicular magnetic field seems to present several transitions within the mixed phase, in the magnetic field – temperature plane. The melting of the Abrikosov flux lattice into a liquid phase is a difficult issue when positional disorder of the flux lines comes into play, the relevance of the XY models is speculative.

∘ Spectacular realisations of XY models are the two dimensional lattices of superconducting islands weakly coupled by Josephson junctions [5]. The

---

** laboratoire associé au Centre National de la Recherche Scientifique

Josephson coupling between two superconductors with phases θ_i and θ_j is a cosine of the phase difference. Frustration and disorder occur when a prependicular magnetic field is turned on which is sensitive to the positional disorder of the superconducting grains.

o There are a family of layered magnets with classical XY spins of fixed length that are well descibed by XY models ($\mathbf{S}_i \cdot \mathbf{S}_j = \cos(\theta_i - \theta_j)$) [6, 7] .

2   The model

The family of systems under consideration is represented by models with continuous symmetry on two dimensional lattices, the prototype being the cosine hamiltonian on the square lattice. It describes angle variables $\{\theta_i\}$ on the sites of the lattice with nearest neighbor interactions J_{ij}

$$\mathcal{H}_{XY} = - \sum_{<i,j>} J_{ij} \ \cos(\theta_i - \theta_j) \tag{1}$$

This model is interesting because it can exhibit a phase transition driven by point defects (which are vortices or dislocations...) [1,8, 9]. The hamiltonian (1) can be decomposed in two parts, one for the singularities (such as topological point defects in the θ-field) and a singularity free part, the two contributions being coupled in general. One way to get rid of this coupling is to replace the cosine function by a Gaussian function with the same rotational symmetry, i.e. one keeps the 2π–periodicity. Then one gets a "reduced hamiltonian". For instance, in the ferromagnetic case ($J_{ij} = J > 0$), Villain replaces the cosine potential by [7]

$$e^{-\beta J(1-\cos(\theta_i - \theta_j))} \sim cst. \sum_{\{n_{ij}\}} e^{\beta - J_v (\theta_i - \theta_j - 2\pi n_{ij})^2} \tag{2}$$

where n_{ij} are intergers that ensure the 2π–periodicity. This model is known to be equivalent to the pure version of (1) and to reproduce properties of the phase transition. We shall see later how the replacement (2) can be generalised to any J_{ij}.

2.1   Pure XY model

The hamiltonian (1) with all $J_{ij} = J$ has the well known Berezinskii-Kosterlitz-Thouless (BKT) [8, 9] phase transition with the unbinding of point defects at T_{BKT}. At zero temperature, the ground state is colinear with the obvious rotational degeneracy. At higher temperature the vortices appear bound in pairs but come into play only close to T_{BKT} (see Fig.1 (a) showing ± 1 vortices on a unit cell of the square lattice). In addition to the vortex excitations one has the long-wavelength spin waves. Below T_{BKT}, the low temperature phase has what has come to be known as quasi-long-range order (the spin-spin correlation decay with a power law as the distance increases). We are at the lower critical dimension of the model.

The reduced hamiltonian of Villain (2) allows one to study accurately this transition [7, 1]. A standard transformation yields the explicit separation of the spin wave contribution from the vortex contribution. The latter is mapped onto a neutral Coulomb gas.

2.2 Regularly frustrated XY model

Frustration occurs when there are some antiferromagnetic (AF) bonds on the lattice. A plaquette of four bonds is frustrated when there is an odd number of AF bonds, that is, one can not minimise the energy of each bond individually (see Fig. 1).

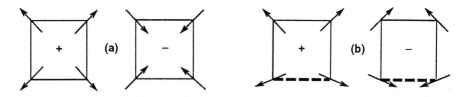

Fig. 1. (a) ± 1 vortices on non-frustrated plaquettes (b) $\pm\frac{1}{2}$ vortices, or "chiralities", on frustrated plaquettes (the dashed bold bond is AF).

The hamiltonian of a regularly frustrated XY system reads

$$\mathcal{H} = -J \sum_{<i,j>} \cos(\theta_i - \theta_j - \pi_{ij}) \tag{3}$$

where π_{ij} is 0 on a fixed regular set of ferromagnetic bonds $< ij >$ and π otherwise (AF bonds). Frustration introduces additional variables which generate a new symmetry in the hamiltonian : These variables are called "chiralities" and the new symmetry is the global reflection of the spins with respect to an arbitrary axis which is also the global flip of the chiralities. This can be illustrated on a the ground state of a single plaquette. Apart from the rotational degeneracy, there are two ground states that can not be related by a rotation of the spins but rather by a reflection (with respect to the vertical axis on Fig. 1 (b)). The chirality of the spins around a plaquette is $\pm\frac{1}{2}$. This is the result of the sum of the difference (restricted to $]-\pi, \pi[$) in angles θ_i around the plaquette. One gets from Fig.1 (fourth plaquette) $\left(-\frac{\pi}{4} - \frac{\pi}{4}\right) + \left(\frac{\pi}{4} - \frac{5\pi}{8}\right) + \left(\frac{5\pi}{8} + \frac{5\pi}{8}\right)_{]-\pi,\pi]} + \left(-\frac{5\pi}{8} + \frac{\pi}{4}\right) = -\frac{1}{2} * 2\pi$, that is, half a vortex of the previous section.

To study this system, one can go to the reduced hamiltonian by making the straightforward replacement of (3), similarly to (2), with the additional variables π_{ij}. Following standard tranformation mentioned above, one gets two contributions: the first one describes only the spin waves and the second one the vortices together with the chiralities. Again the latter can be mapped onto a neutral Coulomb gas [7] with the result

$$\mathcal{H} = \mathcal{H}_{SW} - 2\pi J \sum_{a,b} q_a \, q_b \, \log(r_{ab}) + E_c \sum_a q_a^2 \qquad (4)$$

The "charges" q_a lie on the dual lattice sites, i.e. the center of the plaquettes of the original lattice. They take *integer* values on non-frustated plaquettes as the vortices of the BKT theory, and *half-integer* values on frustrated plaquettes where lie the chiralities. The core energy E_c, introduced to describe the contribution of the inside of plaquettes, is a chemical portential, i.e. the number of thermally excited charges is not fixed.

At zero temperature the spins and also the chiralities are frozen. At higher temperature long-wavelength spin wave and short-wavelength vortex pair excitations superpose on the regular array of frozen frustrated plaquettes.

There are numerous studies of a particular case, namely the "fully frustrated model" where all the plaquettes are frustrated. There the ferromagnetic order of the non-frustrated model is highly perturbated but the ground state can still be determined and consists in the alternance of chiralities with opposite sign $(\pm\frac{1}{2})$. There are contradicting results [5,10,11] on :

▷ the number of transitions (one or two)
▷ the nature of the transitions (Ising, BKT,...)
▷ the order in which the transitions take place

2.3 Randomly frustrated XY model

One way to put some disorder in the model is to take randomly distributed π_{ij}. We quote briefly three possible distributions : the Gaussian, the uniform and the delta functions.

The Gaussain distribution for the π_{ij}, with variance σ and zero mean, leads to a family of models, from the pure XY system $(\sigma = 0)$ to the gauge glass model (see below) $(\sigma = \infty)$. For weak disorder, the quasi-long-range order of the BKT theory persists [12]. Very few is known on the disordered phase where the spin glass model lie that we are interested in.

The so-called gauge glass model occurs when the π_{ij} are uniformly distributed [13,14,15]. The critical temperature is zero in two dimension. It seems that it is non zero in three dimension but could vanish if one takes into account the screening of the Coulomb interaction [14,15].

A different disordered model is the $\pm J$ XY spin glass where the distribution of the π_{ij} is confined to two delta pics, i.e. there are ferromagnetic and AF bonds with equal probability. This models is in a different universality class than the gauge glass that is extensively studied. The plaquette frustration is maximum while it is not in the previous models where π_{ij} could take small values and lead to small deviations of the preferred angles around a plaquette. Here the hamiltonian has the chiral symmetry mentioned in the previous section.

3 $\pm J$ XY model

The question of what the lower critical dimension of the $\pm J$ XY spin glass is, is still open. Is there a transition in three dimension as presumably in the gauge glass and the Ising spin glass ? For the non-random ferromagnetic model, the lower critical dimension is $d_l = 1$, for Ising spins, and $d_l = 2$, for XY spins. With random $\pm J$ couplings, studies of spin correlation functions yield $2 < d_l \lesssim 3$, for Ising spins [16], and $3 < d_l \lesssim 4$, for XY spins [17].

Since discrete variables order usually more easily than continuous ones, one may ask the question whether the multi-spin chiralities are driving the transition to lower dimension than is believed upon considering only the continuous spin variables. This is the conjecture of T. Kawamura who claims that $2 < d_l < 3$ for the $\pm J$ XY model and that there is an Ising-like transition in three dimension.

In order to get an analytical insight on this issue, we studied the correlations length of one and two dimensional systems. Since $d < d_l$, there is no transition and the correlation functions diverge as the temperature vanishes with a power law giving the "correlation lenght exponent" ν. Since there are two variables in this model, one has the two correlation functions

$$\overline{\langle \mathbf{S_0} \cdot \mathbf{S_r} \rangle^2} \sim e^{-r/\xi_s} \quad and \quad \overline{\langle q_0 \, q_r \rangle^2} \sim e^{-r/\xi_c} \tag{5}$$

(where the overline denotes disorder average) yielding two exponents

$$\xi_s \sim T^{-\nu_s} \qquad \xi_c \sim T^{-\nu_c} \tag{6}$$

Another way to put the conjecture of Kawamura is to say that the chiralities order more easily than the spins in $d < d_l$ and thus

$$\xi_s/\xi_c \to \infty \quad when \quad T \to 0, \quad i.e. \quad \nu_s < \nu_c . \tag{7}$$

Some numerical studies give in two dimension

$$\nu_s \sim 1 \quad and \quad \nu_c \sim 2 \qquad (\pm 0.4) \tag{8}$$

with a large error bar depending on the simulation but in agreement with (7). These exponents are obtained by different methods, from the domain wall scaling analysis at zero temperature to suceptibility scalings with temperature, and different models are investigated (cosine [18,19] and reduced hamiltonians [20]).

3.1 Analytical results in quasi-unidimensional systems

We studied one dimensional systems of two kinds : the ladder lattice [21] and the tube lattice [22]. The first one consists of a chain of plaquettes (one plaquette in the transverse direction) and the second one has two plaquettes in the transverse direction with periodic boundary conditions.

We could perform an anlytical calculation of the exponents within the reduced hamiltonian (4). Whatever the dimension of the lattice, the spin-spin correlation is a product of a spin wave contribution and of a charge contribution. We showed that, on the ladder lattice, both are simply the spin wave–spin wave and the charge–charge correlations, and thus $\xi_s^{-1} = \xi_{sw}^{-1} + \xi_c^{-1}$. For both contributions we were able to calculate the exponents with the result $\nu_{sw} = 1$ and $\nu_c = 0.526...$ therefore $\xi_c/\xi_{sw} \to \infty$ when $T \to 0$, i.e. $\nu_s = \nu_c$. For the tube lattice the exponents were given by a domain wall analysis (see next subsection). In both cases we get

$$when \quad d = 1, \quad \nu_s = \nu_c. \tag{9}$$

in contrast to numerical results (8) in two dimension.

To answer to the question of whether the reduced hamiltonian of Villain was a good approximation of the original cosine model, M. Benakli [11] did some numerical simulations. His results based on domain wall scaling analysis and Monte Carlo determination of the ground state energies are in good agreement with ours (9). Thus, at least in one dimension, there does not seem to be any difference between both models.

3.2 Domain wall scaling analysis in two dimension

It is a difficult task to study analytically the two dimensional disordered cosine model. We do not know how to calculate the correlation functions (5) and, following the domain wall scaling analysis, it is unclear a priori which boundary conditions (b.c.) will give the two exponents and also what are the ground states.

To deal with these difficulties we did some approximations and finally obtained a relation between the exponents (without having determined there value).

Our approach is based on the following hypothesis [24] :
We start with the reduced hamiltonian which decoupled the spin wave and the charge (chirality and thermal vortex) contributions. The charge part maps onto a neutral Coulomb gas on a finite lattice with periodic b.c.

$$\mathcal{H}_{coulomb} = -\sum_{a,b} q_a \, q_b \, U_L(r_{ab}) \tag{10}$$

where U_L is the Coulomb potential including finite size effects (L is the linear dimension of the lattice) and q_a takes integer values on non-frustrated plaquettes and half-integer values otherwise. It is convenient and reasonable to work with the smallest values of the charges at zero temperature, that is, $q = 0$ or $q = \pm\frac{1}{2}$ (chiralities). There is one ground state configuration of (10) and its inverse by global chirality flip. This is analoguous to an Ising groung state but with AF long-range Coulomb interactions and diluted Ising variables (randomly distributed on the lattice).

One method to get the correlation length exponents is to study the L-dependence of the energy change when the b.c. of one side of the lattice are modified. This energy difference, at $T = 0$, scales as $L^{-1/\nu}$ with a negative exponent since there is no long range order in this system.

The question is now : how do we get both exponents ν_s and ν_c (which boundary conditions) ?

The spin exponent comes from the difference between *periodic* (P) and *antiperiodic* (AP) b.c. as usual. We determined the hamiltonian with both b.c. and get [24, 25], after having integrated over the spin wave part (which is Gaussian)

$$\mathcal{H}_{coulomb} + \left\{ \left(\delta^{\mathbf{P,AP}} + \frac{1}{L} \sum_a x_a q_a \right)^2 \right\} \; mod \; 1 \tag{11}$$

In addition to the Coulomb part (10) there is a positive term that takes values between 0 and 1. The difference in b.c. appears through the variable δ which is 0 or 1/2 for P or AP boundary conditions. Both terms in (11) describe chiralities and have "chiral excitations".

One finds the chiral excitation with lowest energy by minimising reversal costs of clusters of $\sim L^d$ spins within a given block of dimension L .(This is analoguous to the droplet theory of Ising spin glasses of [26]). The minimum energy costs scales as L^{-1/ν_c}. It is the chiral exponent that appears in this scaling since the reversal of a cluster of chiralities results from the change in b.c. from periodic to *reflecting* b.c. (that is, a reflection of the spins with respect to an arbitrary axis) and can be shown to result in a charge conjugation [18,24]. This is the analogue of the antiperiodic conditions but for the chiralities.

The ground state energy with P or AP b.c. differ from the Coulomb ground state by, at least, a chiral reversal that costs L^{-1/ν_c}. In conclusion the difference in energy between P and AP b.c., that gives the spin exponent by definition, is

$$\overline{(E^{AP} - E^P)^2}^{1/2} \equiv L^{-1/\nu_s} \geq L^{-1/\nu_c} \tag{12}$$

(where the overline denotes the disorder average), that is,

$$when \quad d = 2, \quad \nu_s \geq \nu_c \; . \tag{13}$$

Thus we get no evidence for chiral ordering extending on a longer scale than spin ordering .

4 Conclusions

4.1 Phase diagram of randomly frustrated XY system in d=2

Let us summarize some results that have been quoted in previous sections and sketch a phase diagram in the temperature vs. density of frustrated plaquettes (f) plane.

Without frustration, $f = 0$, one recovers the BKT transition with a whole line of critical points. With the introduction of weak disorder the quasi-long-range order seems to persist with apparently a non-reentrant shape [11]. On the other side of the diagram ($f = 1$) one gets the fully frustrated model with quasi-long-range spin order and also chiral order (there is an ongoing controversy on the number and the nature of the transitions [5, 11]). The effect of the disorder as one moves away from this line is still an open question with the possibility that the disorder is always relevant. Between these two limiting cases, at $f = 1/2$, lies the model that we were interested in and which has a spin glass ground state [27] and zero critical temperature.

4.2 $\pm J$ XY spin glass

There are two types of variables in this system and thus two candidates for an order parameter, the continuous spin variable and the discrete chiral variable. The question was : do the chiralities order more easily than the spins ? We presented an analytical appraoch based on correlation function studies in one dimension and on domain wall scaling analysis in two dimension. Our conclusions are that the correlation length of the chiralities, ξ_c, and of the spins, ξ_s, are equal in one dimension [21, 22] and that $\xi_s \geq \xi_c$ in two dimension [24]. We go one step further and speculate that there is only one length scale in any dimension below the lower critical dimension, $\xi_s = \xi_c$ when $d < d_l$.

This contradicts the results of several numerical studies [18, 19, 20] which give that the spin correlation length is smaller than the chiral one, $\xi_s \leq \xi_c$ (and possibly $2 < d_l < 3$ with a "chiral phase" in three dimension).

The reason for this discrepancy is still unclear. There are two possible issues that still have to be investigated carefully. First, the coupling between spin waves and chiralities could play a role that has been neglected in the reduced hamiltonian (used in analytical approaches). In the pure limit there is no essential difference while this coupling seems to be important in the fully frustrated limit [11]. On the numerical side, both models (cosine vs. Villain hamiltonian) verify the same inequality between chiral and spin length scales. Secondly, there are uncontrolled finite size effects in numerical simulations that are emphasized by the fact that there are long-range logarithmic interactions in the system. Even in the ferromagnetic case the finite size effects are important since both in simulations and in real experiments one does not get the asymptotic regime [6].

References

[1] Nelson D. R. (1983): *Phase Transition and Critical Phenomena*, **7**, Academic Press, London.

[2] Minhagen P. (1987): Rev. Mod. Phys. **59**, 1001.

[3] Strandburg J. (1988): Rev. Mod. Phys. **60**, 161.

[4] Bormann D., Beck H. (1994): J. Stat. Phys. **76**, 361.

[5] Teitel S., Jayaprakash C. (1983): Phys. Rev. B **27**, 598; see also S. Teitel conference in this volume.

[6] Bramwell S. T., Holdsworth P. C. W. (1993): J. Phys. C **5**, L53.

[7] Villain J. (1975): J. Physique **36**, 581; (1977): J. Phys. C **10** 4793.

[8] Berezinskii V. L. (1971): Phys. JETP **32**, 493.

[9] Kosterlitz J. M., Thouless D. J. (1973): J. Phys. C bf 6, 1181.

[10] Diep H. T. (1995): *Magnetic Systems with Competeting Interactions*, Wold Scientific.

[11] Benakli M. (1995) thesis; Benakli M., Zheng H., Gabay M. (1996): preprint cond-mat.

[12] Natterman T., Scheidl S., Korshunov S. E., Li M. S. (1995): J. Physique I **5**, 565; Tang L. H. (1996): preprint cond-mat.

[13] Fisher, M. P. A., Tokuyasu T. A., Young A. P. (1991): Phys. Rev. Lett. **66**, 2931.

[14] Bokil S., Young A. P. (1995): Phys. Rev. Lett. **74**, 3021.

[15] Wengel C., Young A. P. (1996): preprint cond-mat and poster in this volume.

[16] Marinari E., Parisi G., Ritort F., Ruiz-Lorenzo J. (1996): Phys. Rev. Lett. **76**, 843 ; Kawashima N., Young A. P. (1995): preprint cond-mat.

[17] Schwartz M., Young A. P. (1991): Europhys. Lett. **15**, 209; Ozeki Y., Nishimori H. (1992): Phys. Rev. B **46**, 2879; Jain S. (1996): preprint cond-mat.

[18] Kawamura H., Tanemura M. (1991): J. Phys. Soc. Japan **60**, 608; (1992): Phys. Rev. Lett. **68**, 3785; (1995): Phys. Rev. B bf 51, 12398.

[19] Ray P., Moore M. A. (1992): Phys. Rev. B **45**, 5361.

[20] Bokil S., Young A. P. (1996): preprint cond-mat.

[21] Ney-Nifle M., Hilhorst H. J., Moore M. A. (1995): Phys. Rev. B **48**, 10254.

[22] Thill M. J., Ney-Nifle M., Hilhorst H. J. (1995): J. Phys. A. **28**, 4285.

[23] Horiguchi, Morita T. (1990): J. Phys. Soc. Japan **59**, 888.

[24] Ney-Nifle M., Hilhorst H. J. (1995): Phys. Rev. B **51**, 8357.

[25] Vallat A., Beck H. (1994): Phys. Rev. B **50**, 4015.

[26] Fisher D. S., Huse D. A. (1988): Phys. Rev. B **38**, 386.

[27] Gawieck P., Grempel D. R. (1991): Phys. Rev. B bf 44, 2613.

A metal-insulator transition as a quantum glass problem

T.R. Kirkpatrick[1] and D. Belitzi[2]

[1]Institute for Physical Science and Technology and Department of Physics
University of Maryland, College Park, Maryland 20742
[2] Department of Physics and Materials Science Institute
University of Oregon, Eugene, Oregon 97403

Abstract. We discuss a recent mapping of the Anderson-Mott metal-insulator transition onto a random field magnet problem. The most important new idea introduced is to describe the metal-insulator transition in terms of an order parameter expansion rather than in terms of soft modes via a nonlinear sigma model. For spatial dimensions $d > d_c^+ = 6$ a mean field theory gives the exact critical exponents. For $d = 6 - \varepsilon$ the critical exponents are identical to those for a random field Ising model. Dangerous irrelevant quantum fluctuations modify Wegner's scaling law relating the conductivity exponent to the correlation or localization length exponent. This invalidates the bound $s \geq 2/3$ for the conductivity exponent s in $d = 3$. We also argue that activated scaling might be relevant for describing the AMT in three-dimensional systems.

1 Introduction

Metal-insulator transitions of purely electronic origin, i.e. those for which the structure of the ionic background does not play a role, are commonly divided into two categories. In one category the transition is triggered by electronic correlations, or interactions, and in the other it is driven by disorder. The first case is known as a Mott transition,[1] and the second one as an Anderson transition.[2] It is believed that for many real metal-insulator transitions both correlations and disorder are relevant. The resulting quantum phase transition, which carries aspects of both types of transitions, we call an Anderson-Mott transition (AMT).[3]

Until very recently virtually all approaches[3] studied the AMT only in the vicinity of two dimensions by generalizing[4] Wegner's theory[5] for the Anderson transition. Renormalization-group methods lead to a critical fixed point in $d = 2 + \varepsilon$ dimensions, and standard critical behavior with power-law scaling was found. However, the framework of these theories does not allow for an order parameter (OP) description of the AMT, and does not lead to a simple Landau or mean-field theory.[6] As a result, the physics driving the AMT remains relatively obscure in this approach, compared to standard theories for other phase transitions. An alternative line of attack has recently been explored by the present authors.[7], [8], [9], [10] We have shown that an

OP description of the AMT is possible with the tunneling density of states (DOS) as the OP. A simple Landau theory then yields the exact critical exponents, above the upper critical dimension, $d_c^+ = 6$. In this respect the AMT is conceptually simpler than the Anderson transition, which has no known simple OP description, and whose upper critical dimension may be infinite.

One of the most far-reaching implications of our approach is that the AMT is in some respects similar to magnetic transitions in random fields. Qualitatively, this can be understood as follows. Consider a model of an interacting disordered electron gas. In terms of anticommuting Grassmann fields, $\bar{\psi}$ and ψ, the action can be written,[3]

$$S = S_{\text{k}} + S_{\text{dis}} + S_{\text{int}} \quad , \tag{1}$$

with,

$$S_{\text{k}} = -\sum_\sigma \int dx \, \bar{\psi}_\sigma(x) \left[\partial_\tau - \frac{\nabla^2}{2m} - \mu \right] \psi_\sigma(x) \quad , \tag{2}$$

the kinetic or free part of S,

$$S_{\text{dis}} = -\sum_\sigma \int dx \, u(\mathbf{x}) \, \bar{\psi}_\sigma(x) \, \psi_\sigma(x) \quad , \tag{3}$$

the disorder part of S, and

$$S_{\text{int}} = -\frac{\Gamma}{2} \sum_{\sigma_1,\sigma_2} \int dx \, \bar{\psi}_{\sigma_1}(x) \, \bar{\psi}_{\sigma_2}(x) \, \psi_{\sigma_2}(x) \, \psi_{\sigma_1}(x) \quad , \tag{4}$$

denoting the interaction part of S. In these equations, $x = (\mathbf{x}, \tau)$ with τ denoting imaginary time, $\int dx \equiv \int d\mathbf{x} \int_0^{1/T} d\tau$, m is the electron mass, μ is the chemical potential, σ is a spin label, and for simplicity we have assumed an instantaneous point-like electron-electron interaction with strength Γ. $u(\mathbf{x})$ is a random potential which represents the disorder. For simplicity we also assume $u(\mathbf{x})$ to be δ-correlated, and to obey a Gaussian distribution with second moment

$$\{u(\mathbf{x})u(\mathbf{y})\} = \frac{1}{2\pi N_F \tau_{\text{el}}} \, \delta(\mathbf{x} - \mathbf{y}) \quad , \tag{5}$$

where the braces denote the disorder average, N_F is the bare DOS per spin at the Fermi energy, and τ_{el} is the bare elastic mean-free time. For future use we write S_{dis} as,

$$S_{\text{dis}} = -\sum_{n,\sigma} \int d\mathbf{x} \, u(\mathbf{x}) \, \bar{\psi}_{\sigma,n}(\mathbf{x}) \, \psi_{\sigma,n}(\mathbf{x}) \quad , \tag{6}$$

where a Matsubara frequency decomposition of $\bar{\psi}(\tau)$ and $\psi(\tau)$ has been used.

As mentioned above, the most obvious candidate for an OP for the AMT is the single particle DOS, N, at the Fermi level. In terms of Grassmann variables this quantity is proportional to the zero-frequency limit of the expectation value of the variable $\bar{\psi}\psi$:

$$N = \operatorname{Im} N(i\omega_n \to 0 + i0) \quad , \tag{7}$$

with,

$$N(i\omega_n) = \frac{-1}{2\pi N_F} \sum_\sigma \langle \bar{\psi}_{\sigma,n}(\mathbf{x})\, \psi_{\sigma,n}(\mathbf{x}) \rangle \tag{8}$$

where we have normalized the DOS by $2N_F$, and the brackets denote an expectation value with respect to the action S. Note that the so defined DOS is actually a local DOS, i.e. it depends on $\mathbf{x}$. Examining Eqs. (6) and (??), we see that the local OP for the AMT couples linearly to the random potential, and depending on the sign of $u(\mathbf{x})$ it will favor either an increasing or a decreasing DOS. Similarly, $S_{\text{int}} \sim -\Gamma N^2$, i.e. S_{int} always favors a decreasing DOS. We conclude that the interaction term in general frustrates the disorder term, just like in a random field (RF) magnet problem.

This conclusion has a number of important implications. For example, if conventional scaling exists at the AMT, then one expects hyperscaling to be violated due to a dangerous irrelevant variable (DIV), as it is in RF magnets.[11] As a consequence of this, we argue below that Wegner's scaling law relating the conductivity exponent s to the correlation length exponent ν is modified. Furthermore, if the AMT shares all of the features known to be induced in magnets by a random field, then one would expect glasslike features and unconventional or activated scaling similar to what has been predicted[12] and observed[13] in classical RF magnets.

The plan of this paper is as follows. In Section II we give a sketch of our order-parameter theory of the AMT. An explicit scaling theory near $d = 6$ is constructed. In the first part of Section III we give a general scaling theory of the AMT, assuming it is a conventional phase transition. In the second part of this section we review some aspects of an activated scaling theory for the AMT. We conclude in Section IV with a short discussion.

2 Formalism and Mean Field Theory

2.1 Formalism

Here we briefly review the formalism we have used to show that at least near $d = 6$, the AMT and the magnetic transition in a RF Ising model have many features in common. For details we refer to two recent papers.[9], [10]

Our starting point is the nonlinear sigma model (NLσM) that has been used to describe the AMT near two dimensions. The solution procedure we use near the upper critical dimension is closely analogous to the treatment of the $O(n)$ symmetric NLσM in the limit of large n.[14] The NLσM for the AMT is derived from Eq. (1) by assuming that all of the relevant physics can be expressed in terms of fluctuations of the particle number density, the spin density, and the one-particle spectral density. Technically, this is achieved by making long-wavelength approximations, and by introducing classical composite operators that are related to the Grassmannian variables mentioned above. The quenched disorder is handled by means of the replica trick. The resulting action reads,[3]

$$S[\tilde{Q}] = -\frac{1}{2G} \int d\mathbf{x} \; tr \left(\nabla \tilde{Q}(\mathbf{x}) \right)^2 + 2H \int d\mathbf{x} \; tr \left(\Omega \tilde{Q}(\mathbf{x}) \right)$$
$$-\frac{\pi T}{4} \sum_{n=s,t} \int d\mathbf{x} \left[\tilde{Q}(\mathbf{x}) \gamma^{(n)} \tilde{Q}(\mathbf{x}) \right] \quad , \tag{9}$$

where,

$$\left[\tilde{Q}(\mathbf{x}) \gamma^{(s)} \tilde{Q}(\mathbf{x}) \right] = K_s \sum_{n_1 n_2 n_3 n_4} \delta_{n_1+n_3, n_2+n_4} \sum_{\alpha} \sum_{r=0,3} (-1)^r$$
$$tr \left((\tau_r \otimes s_0) \, \tilde{Q}^{\alpha\alpha}_{n_1 n_2} \right) \; tr \left((\tau_r \otimes s_0) \, \tilde{Q}^{\alpha\alpha}_{n_3 n_4} \right) \quad , \tag{10}$$

and

$$\left[\tilde{Q}(\mathbf{x}) \gamma^{(t)} \tilde{Q}(\mathbf{x}) \right] = -K_t \sum_{n_1 n_2 n_3 n_4} \delta_{n_1+n_3, n_2+n_4} \sum_{\alpha} \sum_{r=0,3} (-1)^r \sum_{i=1}^{3}$$
$$tr \left((\tau_r \otimes s_i) \, \tilde{Q}^{\alpha\alpha}_{n_1 n_2} \right) \; tr \left((\tau_r \otimes s_i) \, \tilde{Q}^{\alpha\alpha}_{n_3 n_4} \right) \quad . \tag{11}$$

Here $\tilde{Q}$ is a classical field that is, roughly speaking, composed of two fermionic fields. It carries two Matsubara frequency labels, n and m, and two replica labels, α and β. The matrix elements are spin quaternions, with the quaternion degrees of freedom describing the particle-hole ($\tilde{Q} \sim \bar{\psi}\psi$) and particle-particle ($\tilde{Q} \sim \bar{\psi}\bar{\psi}$) channels, respectively. For simplicity we restrict ourselves to the particle-hole degrees of freedom. For this case the matrix elements can be expanded in a restricted spin-quaterion basis,

$$\tilde{Q}^{\alpha\beta}_{nm} = \sum_{r=0,3} \sum_{i=0}^{3} \, {}^i_r \tilde{Q}^{\alpha\beta}_{nm} (\tau_r \otimes s_i) \quad , \tag{12}$$

with $\tau_{0,1,2,3}$ the quaternion basis, and $s_{0,1,2,3}$ the spin basis ($s_{1,2,3} = i\sigma_{1,2,3}$ with $\sigma_{1,2,3}$ the Pauli matrices). The matrix Q is subject to the constraints,[3]

$$\tilde{Q}^2 = 1 \quad , \tag{13}$$

$$tr\,\tilde{Q} = 0 \quad, \tag{14}$$

$$\tilde{Q}^+ = C^T \tilde{Q}^T C = \tilde{Q} \quad, \tag{15}$$

where $C = i\tau_1 \otimes s_2$.

In Eq. (9), $G = 2/\pi\sigma$, with σ the bare conductivity, is a measure of the disorder, and $H = \pi N_F/2$ is a frequency coupling parameter. K_s and K_t are bare interaction amplitudes in the spin singlet and spin triplet channels, respectively, and $\Omega_{nm}^{\alpha\beta} = \delta_{nm}\,\delta_{\alpha\beta}\,\omega_n\,\tau_0 \otimes s_0$, with $\omega_n = 2\pi T n$, is a bosonic frequency matrix. Notice that $K_s < 0$ for replusive interactions.

The correlation functions of the $\tilde{Q}$ determine the physical quantities. Correlations of $\tilde{Q}_{nm}$ with $nm < 0$ determine the soft particle-hole modes associated with charge, spin and heat diffusion, while the DOS is determined by $< \tilde{Q}_{nn}^{\alpha\alpha} >$, i.e., $\tilde{Q}_{nm}$ with $nm > 0$. It is therefore convenient to separate $\tilde{Q}$ into blocks.

$$\tilde{Q}_{nm}^{\alpha\beta} = \Theta(nm)\,Q_{nm}^{\alpha\beta} + \Theta(n)\Theta(-m)\,q_{nm}^{\alpha\beta} + \Theta(-n)\Theta(m)\,(q^\dagger)_{nm}^{\alpha\beta} \quad. \tag{16}$$

Normally a NLσM is treated by integrating out the massive modes, i.e., the Q_{nm}, to obtain an effective theory for the massless modes, which are here the diffusion processes described by q and $q^\dagger$. However, since our goal is to obtain a field theory for the OP for the AMT, Q_{nn}, we instead integrate out the massless q-fields here.

Using standard techniques[14] the above program can be carried out. The resulting OP field theory for the AMT is,[9]

$$S[Q] = -\frac{1}{2G} \int d\mathbf{x}\, tr\, \left[(\nabla Q(\mathbf{x}))^2 + \langle \Lambda \rangle\,(Q(\mathbf{x}))^2\right]$$

$$+2H \int d\mathbf{x}\, tr\,(\Omega\,Q(\mathbf{x})) + \frac{u}{2G^2} \int d\mathbf{x}\, tr\, \left[(1-f)\,Q^2(\mathbf{x})\right]$$

$$-\frac{u}{4G^2} \int d\mathbf{x}\, tr\, Q^4(\mathbf{x}) - \frac{v}{4G^2} \int d\mathbf{x}\,\left(tr_+ Q^2(\mathbf{x})\,(tr_- Q^2(\mathbf{x}))\right)^2 + \cdots \quad, \tag{17}$$

where $tr_\pm$ denotes 'half-traces' that sum over all replica labels but only over positive and negative frequencies, respectively: $tr_+ = \sum_\alpha \sum_{n\geq 0}$, $tr_- = \sum_\alpha \sum_{n<0}$. $f = f(< \Lambda >)$ is a matrix with elements ${}_r^i f_{nm}^{\alpha\beta} = \delta_{ro}\,\delta_{io}\,\delta_{nm}\,f_n$ with $f_n > f_m > 0$ for $|n| < |m|$. f_n is an increasing function of disorder, G, and $|K_s|$. $< \Lambda >$ in Eq. (17) is proportional to $\Omega/ < Q >$, and u and v are finite constants, at least for $d > 4$. In giving Eq. (17) we have neglected terms that can be shown to be renormalization group (RG) irrelevant near the AMT.

2.2 Mean-Field Theory

Here we construct a mean-field or saddle-point (SP) solution of Eq. (17).[15], [14] We look for solutions, Q_{sp}, that are spatially uniform and satisfy,

$$\frac{i}{r}(Q_{sp})^{\alpha\beta}_{nm} = \delta_{ro}\,\delta_{io}\,\delta_{nm}\,\delta_{\alpha\beta}\,N^{(0)}_n \quad , \tag{18}$$

where the subscript (0) denotes the SP approximation. The replica, frequency, and spin-quaternion structures in Eq. (18) are due to the fact that $< \frac{i}{r}Q^{\alpha\beta}_{nm} >$ has these properties, and that in the mean-field approximation averages are replaced by the corresponding SP values.

In the zero-frequency limit, the SP equation of state obtained from Eq. (2.6) is,

$$\left(N^{(0)}_{n=0}\right)^2 = 1 - f_{n=0}(< \Lambda >) = t^{(0)} \quad , \tag{19}$$

or

$$N^{(0)}_{n=0} = \left(t^{(0)}\right)^{1/2} \quad . \tag{20}$$

Here $t^{(0)}$ is the mean-field value of the distance from the critical point, t. Equation (20) yields the mean-field value for the critical exponent β,

$$\beta = 1/2 \quad . \tag{21}$$

To obtain the remaining mean-field critical exponents we expand Q about its expectation value, which is proportional to N ,

$$\frac{i}{r}Q^{\alpha\beta}_{nm} = \delta_{ro}\,\delta_{io}\,\delta_{\alpha\beta}\,\delta_{nm}\,N_n + \sqrt{2G}\,\frac{i}{r}\varphi^{\alpha\beta}_{nm} \quad , \tag{22}$$

where the factor of $\sqrt{2G}$ has been inserted for convenience. The action S_G governing Gaussian fluctuations about the mean-field solution in the critical region then follows from Eq. (17) as,

$$S_G[\varphi] = -\int d\mathbf{x}\ tr\ \left[(\nabla\varphi(\mathbf{x}))^2 + \ell^{(0)}\varphi^2(\mathbf{x}) + \frac{2u}{G}\,t^{(0)}\varphi^2(\mathbf{x})\right]$$
$$-\frac{v}{2G}t^{(0)}\int d\mathbf{x}(tr_+\varphi(\mathbf{x}))(tr_-\varphi(\mathbf{x})) + O(\varphi^3) \quad . \tag{23}$$

Here

$$\ell^{(0)}_n = 2GH\omega_n/N^{(0)}_n \quad , \tag{24}$$

is the SP value of Λ. All remaining critical exponents can now be read off Eq. (23). Comparing the first and third terms on the r.h.s. yields the correlation length exponent $\nu = 1/2$. With $\ell^{(0)} \sim \omega/Q$, the first and second term gives the dynamical exponent $z = 3$. Finally, the $\varphi - \varphi$ correlation function near the transition has the standard Ornstein-Zernike form, which yields the critical exponents $\gamma = 1$ and $\eta = 0$. We thus have standard mean-field values for all static exponents,[15]

$$\beta = \nu = 1/2 \quad , \quad \gamma = 1 \quad , \quad \eta = 0 \quad , \quad \delta = 3 \quad , \tag{25}$$

and for the dynamical exponent we have,

$$z = 3 \quad . \tag{26}$$

Inspection of Eq. (23) shows that the AMT saddle point is a local minimum and therefore stable.

It is also possible to determine the critical behavior of the transport coefficients by directly computing the $q - q$ correlation functions and identifying the change, spin and heat diffusion coefficients (D_c, D_s, D_h). Near the mean-field AMT, all three of these coefficients behave in the same way and vanish like the OP,

$$D_a \sim N_{n=0}^{(0)}/GH \quad , \tag{27}$$

with $a = c, s, h$.

The next step in the standard approach for describing any continuous phase transition is to introduce RG ideas.[15] In the problem considered here, application of the RG method accomplishes three things. First, it generates all additional terms in the action that are consistent with the symmetry of the problem. Second, it enables us to prove that there exists an upper critical dimension, d_c^+, above which mean-field theory for the critical exponents is exact. Third, it enables us to do an ε − expansion below d_c^+. We begin by noting that Eq. (17) does not have the RF term we argued for in the Introduction. A Wilson-type RG procedure generates this term as well as others. It has the form,

$$S_{RF} = \frac{\Delta}{2} \int d\mathbf{x} \sum_{i=\pm} (tr_i \, \varphi(\mathbf{x}))^2 \quad . \tag{28}$$

In terms of the original fermion action, this contribution arises from a RF term of the form,

$$S_{RF} = \sum_{n,\sigma} \int d\mathbf{x} h_n(\mathbf{x}) \, \bar{\psi}_{\sigma,n}(\mathbf{x}) \, \psi_{\sigma,n}(\mathbf{x}) \quad , \tag{29}$$

where $h_n(\mathbf{x})$ is a random field with

$$\{h_n(\mathbf{x})h_m(\mathbf{x})\} = \theta(nm) \frac{\Delta}{4G} \delta(\mathbf{x} - \mathbf{y}) \quad . \tag{30}$$

All other terms generated by the renormalization process are irrelevant near the upper critical dimension, d_c^+.

Standard arguments imply that such a RF term yields $d_c^+ = 6$ instead of the usual $d_c^+ = 4$.[11] The same arguments also prove that the mean-field critical behavior quoted above is the exact critical behavior for $d > d_c^+ = 6$.

For $d = 6 - \varepsilon$, an ε–expansion of the critical exponents is possible. The main idea is that under renormalization, the disorder Δ scales to infinity while u, the coefficient of the quartic term in Eq. (17), scales to zero such that their product

$$g = u\,\Delta \quad , \tag{31}$$

scales to a stable fixed point value that is of $O(\varepsilon)$. In all of the other flow equations only the product g appears so that a stable critical fixed point is obtained. To first order in $\varepsilon = 6 - d$, the resulting critical exponents are,[8], [9]

$$\nu = \frac{1}{2} + \frac{\varepsilon}{12} + O(\varepsilon^2) \quad , \tag{32}$$

$$\gamma = \frac{1}{2} - \frac{\varepsilon}{6} + O(\varepsilon^2) \quad , \tag{33}$$

$$\delta = 3 + \varepsilon + O(\varepsilon^2) \quad , \tag{34}$$

$$\eta = 0 + O(\varepsilon^2) \quad , \tag{35}$$

$$z = 3 - \frac{\varepsilon}{2} + O(\varepsilon^2) \quad , \tag{36}$$

We finally mention that the RG flow properties, $\Delta \to \infty$, $u \to 0$, $g \sim O(\varepsilon)$, have an interesting physical interpretation. Δ represents the disorder, while u is a measure of the importance of quantum fluctuations about the SP. Because u determines physical quantities like the order parameter, this implies that quantum fluctuations are dangerously irrelevant near the OP driven AMT. This in turn modifies the standard hyperscaling equalities.

3 Scaling Descriptions of the Anderson–Mott Transition

Based on the known or suspected behavior of the random field Ising model, there are two distinct scaling scenarios one can imagine for the AMT. The first is a conventional one,[15] that takes into account in a general way the dangerous irrelevant variable discussed in Section II. The second, strikingly different one is new in the context of metal-insulator transitions, and is called the activated scaling scenario.[12]

3.1 Conventional Scaling Description of the AMT

There are standard ways to construct conventional scaling descriptions of either classical or quantum ($T = 0$) phase transitions. For the random field like transition considered here, one of the most important features is the presence of a dangerous irrelevant variable (DIV), namely u. Suppose that u is characterized by an exponent θ, defined so that $u(b) \sim b^{-\theta}$. One-loop perturbation theory gives $\theta = 2 + O(\varepsilon)$, but here we keep θ general. This adds a third independent exponent to the usual two independent static exponents. In addition, there is the dynamical scaling exponent z. For the case considered here it turns out that z is not independent, but rather it is equal to the scale dimension of the field conjugate to the OP. This is due to the fact that RF fluctuations are much more important than quantum fluctuations. The dominance of RF fluctuations compared to either thermal or quantum fluctuations is a general feature of RF problems. The net result, confirmed explicitly near $d = 6$, is,

$$z = y_h = \delta\beta/\nu \quad . \tag{37}$$

As for the classical RF problem, the DIV u, changes d in all scaling relations to $d - \theta$. For example, near the transition the OP obeys a scaling or homogeneity relation,

$$N(t, \Omega) = b^{-\beta/\nu} N(b^{1/\nu} t, b^z \Omega) \quad , \tag{38}$$

with β related to ν and η by the usual scaling law, but with $d \to d - \theta$ due to the violation of hyperscaling by the DIV,

$$\beta = \frac{\nu}{2} (d - \theta - 2 + \eta) \quad . \tag{39}$$

Similarly, the exponents δ and γ are given by,

$$\delta = (d - \theta + 2 - \eta)\nu/2\beta \quad , \tag{40}$$

$$\gamma = \nu(2 - \eta) \quad . \tag{41}$$

Next we consider the transport coefficients. The charge, spin, or heat diffusion coefficients, which we denote collectively by D, all scale like a length squared times a frequency, so that

$$D(t, \Omega) = b^{2-z} D(t\, b^{1/\nu}, \Omega\, b^z) = t^{\nu(z-2)} D(1, \Omega/t^{\nu z}) \quad . \tag{42}$$

Denoting the static exponent for the diffusion coefficient by s_D, defined by $D(t, \Omega = 0) \sim t^{s_D}$, we have found,

$$s_D = \nu(z - 2) = \beta - \nu\eta = \frac{\nu}{2}(d - 2 - \theta - \eta) \quad . \tag{43}$$

The behavior of the electrical conductivity σ, which is related to the charge diffusion coefficient by means of an Einstein relation, $\sigma = D_c\, \partial n/\partial\mu$, depends on the behavior of $\partial n/\partial\mu$. If $\partial n/\partial\mu$ has a constant contribution at the AMT, then $\sigma \sim t^s$ vanishes as D_c, so that

$$s = s_D = \frac{\nu}{2}(d - 2 - \theta - \eta) \quad . \tag{44}$$

Dimensionally, however, all of the thermodynamic susceptibilities scale like an inverse volume times a time, which implies a singular part $(\partial n/\partial\mu)_s$ of $\partial n/\partial\mu$ that scales like,

$$(\partial n/\partial\mu)_s (t, T) = b^{-d+\theta+z} (\partial n/\partial\mu)_s (tb^{1/\nu}, Tb^z) \quad . \tag{45}$$

If there is no constant, analytic, background term, then Eqs. (??) and (45) give,

$$s = \nu(d - 2 - \theta) \quad . \tag{46}$$

In either case, Wegner's scaling law $s = \nu(d - 2)$,[16] which previously had been believed to hold for the AMT as well as for the Anderson transition, is violated, unless Eq. (3.5) holds *and* $\theta = 2 - d - \eta$. Finally, we note that Eqs. (44) and (46) are identical if $\eta = \theta + 2 - d$, and that this result is consistent with Wegner scaling apart from the replacement $d \to d - \theta$, and with $\partial n/\partial\mu$ being noncritical across the AMT. However, Eq. (39) shows that this result is not consistent with a vanishing OP unless the theory has multiple dynamical scaling exponents.[3]

3.2   Activated Scaling Description of the AMT

An important characteristic of a glass transition is the occurrence of extremely long time scales. While critical slowing down at an ordinary transition means that the critical time scale grows as a power of the correlation length, $\tau \sim \xi^z$ with z the dynamical scaling exponent, at a glass transition the critical time scale grows exponentially with ξ,

$$\ln(\tau/\tau_0) \sim \xi^\psi \quad , \tag{47}$$

with τ_0 a microscopic time scale, and ψ a generalized dynamical scaling exponent. Effectively, Eq. (47) implies $z = \infty$. As a result of such extreme slowing down, the system's equilibrium behavior near the transition becomes inaccessible for all practical purposes. That is, realizable experimental time scales are not sufficient to reach equilibrium, and one says that the system falls out of equilibrium. It has been proposed[12] that the phase transition in classical RF magnets is of this type, and there are experimental observations that seem to corroborate this suggestion.

Here we speculate that the analogy between RF magnets and the AMT leads to such 'activated' scaling for the AMT as well. For this quantum phase transition one expects time and inverse temperature to show the same scaling behavior, irrespective of whether the critical slowing down follows an ordinary power law, or Eq. (47). Quantum mechanics thus makes it very difficult to observe the static scaling behavior, since it requires exponentially small temperatures. Thus from Eq. (47) we see that static zero temperature scaling will be observed only if

$$T < T_0 \exp\left(-\xi^\psi\right) \quad , \tag{48}$$

with T_0 some microscopic temperature scale on the order of the Fermi temperature $\sim T_F$. This is potentially a crucial point in the interpretation of experimental data.

Activated scaling, as described by Eq. (47), follows from a barrier picture of the system's free energy landscape. The physical idea we have in mind is that while a repulsive electron-electron interaction always leads to a decrease in the local DOS, the random potential can in general lead to an increase in the local DOS as well. The competition between these two effects leads to frustration and to, for example, large insulating clusters within the metallic phase. Delocalizing these large clusters requires energy barriers to be overcome, which are assumed to grow like ξ^ψ as the AMT is approached. A further notion of the barrier model is that the frequency or temperature argument of the scaling function is expected to be $\ln(\tau/\tau_0)/\ln(T_0/T)$, rather than τT as in, for example, Eq. (??) and (??). The reason is that one expects a very broad distribution of energy barriers. The natural, self-averaging, variable is therefore $\ln \tau$ rather than τ.

It makes physical sense to assume scaling forms only for self-averaging quantities. For a system with quenched disorder it is known that the free energy is self-averaging, while the partition function is not, and correlation functions in general are not, either. Therefore, all thermodynamic quantities, which can be obtained as partial derivatives of the free energy, are self-averaging. For a general thermodynamic quantity, Q, one therefore expects a homogeneity law[10]

$$Q(t,T) = b^{-x_Q} F_Q\left(t\, b^{1/\nu}, \frac{b^\psi}{\ln(T_0/T)}\right) \quad , \tag{49}$$

where x_Q is the scale dimension of Q, and F_Q is a scaling function. For example, for the DOS one expects,

$$N(t,T) = b^{-\beta/\nu} F_N\left(t\, b^{1/\nu}, \frac{b^\psi}{\ln(T_0/T)}\right) \quad , \tag{50}$$

with β still given by Eq. (39). Alternatively, Eq. (50) can be written,

$$N(t,T) = \frac{1}{[\ln(T_0/T)]^{\beta/\nu\psi}} G_N\left[t^{\nu\psi} \ln(T_0/T)\right] \quad . \tag{51}$$

The scaling function G_N is related to the function F_N in Eq. (43) by $G_N(x) = F_N(x^{1/\nu\psi}, 1)$, and has the properties $G_N(x \to \infty) \sim x^{\beta/\nu\psi}$, and $G_N(x \to 0) \to$ const..

Equation (51) makes a qualitative prediction that can be used to check experimentally for glassy aspects of the AMT: Measurements of the tunneling DOS very close to the transition should show an anomalously slow temperature dependence, i.e., N should vanish as some power of $\ln T$ rather than as a power of T. While in principle this should be straightforward, similar checks for the RF problem have shown that a very large dynamical or temperature range is needed to produce conclusive results.

Other thermodynamic quantities can be considered and are discussed in detail elsewhere. One chief result is the occurence of a 'Griffiths phase', where both the spin susceptibility and the specific heat expansion coefficient are singular away from the AMT.[10]

We conclude this subsection by considering the electrical conductivity. Let $\tilde{\sigma}$ be the unaveraged conductivity, and σ_0 a suitable conductivity scale, e.g., the Boltzmann conductivity. Since $\tilde{\sigma}$ is directly related to a relaxation time, we expect it not to be self-averaging, while its logarithm should be self-averaging. We define $\ell_\sigma = < \log(\sigma_0/\tilde{\sigma}) >$ and assume it is self-averaging and that it satisfies,

$$\ell_\sigma(t, T) = b^\psi F_\sigma \left(t\, b^{1/\nu}, \frac{b^\psi}{\ln(T_0/T)} \right)$$
$$= \ln(t_0/T)\, G_\sigma \left(t^{\nu\psi} \ln(T_0/T) \right) \quad . \tag{52}$$

Notice that the scale dimension of ℓ_σ is necessarily ψ, since ψ characterizes the free energy barriers near the AMT.

As a measure of the conductivity, let us define,

$$\sigma(t, T) \equiv \sigma_0 \exp(-\ell_\sigma) \quad . \tag{53}$$

One can argue on physical grounds that $G_\sigma(x \to \infty) \sim 1/x$. This yields,

$$\sigma(t, T = 0) \sim \exp(-1/t^{\nu\psi}) \quad , \tag{54}$$

and

$$\sigma(t = 0, T) \sim T^{G_\sigma(0)} \quad . \tag{55}$$

Note that at zero temperature, σ vanishes exponentially with t, and that at the critical point σ vanishes like a nonuniversal power of T.

4 Discussion

We conclude by briefly summarizing our order parameter description of the AMT and its relation to the random field magnet problem. We also make a few additional comments on the experimental situation.

The RF nature of the AMT was made plausible in the introduction. In order to derive this result it is necessary to have an OP description of the AMT. In Section II we illustrated how to obtain an OP field theory for the AMT. This is an important advance because an OP description is conceptually simpler, and physically more intuitive, than the standard sigma model description of the AMT.[3] Renormalization of this OP field theory then generates the expected RF structure, which for unknown reasons is not present in the bare theory. The upper critical dimension d_c^+ is found to be $d_c^+ = 6$. For $d > 6$, mean-field theory gives the exact critical behavior, and for $d < 6$, an $\varepsilon = 6 - d$ expansion for the critical exponents can be obtained. One of the important results is that hyperscaling is violated at the AMT due to a dangerous irrelevant variable. As a consequence, Wegner's scaling law near the metal-insulation transition is modified.

In Section III we reviewed two distinct scaling scenarios for the AMT. The first one was a conventional scaling theory, in the presence of a dangerous irrelevant variable. The second one introduced the idea that activated scaling might be relevant near the AMT. Physically, one of the main results in this second approach is that static or zero-temperature scaling is expected to set in only at exponentially low temperatures, and that for practical purposes it is inaccessible close to the AMT.

Electron-electron interactions are necessary in order for the AMT discussed here to exist, since for noninteracting electrons one has an Anderson transition with an uncritical DOS.[16] This point is correctly reflected by the theory since f_n in Eq. (19) vanishes for noninteracting systems, so that the critical point discussed here is never reached: For $K_{s,t} \to 0$ the critical disorder for the AMT increases without bound, $G_c \to \infty$. This suggests a number of distinct phase transition scenarios. The simplest one is that for sufficiently small interaction constants, or large G_c, the AMT discussed here gets preempted by some other transition, such as a pure Anderson transition. This scenario is particularly likely if $K_t = 0$, and if the electron-electron interactions are short ranged, since in this case K_s is irrelevant near the Anderson transition FP, at least near $d = 2$. For this case a likely phase diagram is shown in Fig. 1. A different possibility is that in the above picture the Anderson transition is replaced by an AMT of a different type than the one discussed here, possibly one that is related to the transition studied near $d = 2$ for the case when either K_s and K_t are nonzero, or the electron-electron interaction is of long range.[3]

In Section III.B we suggested that the AMT is a quantum glass transition.[10] Following this notion, our chief results are as follows: (1) The specific heat and spin susceptibilities are singular as $T \to 0$ even in the metallic phase.

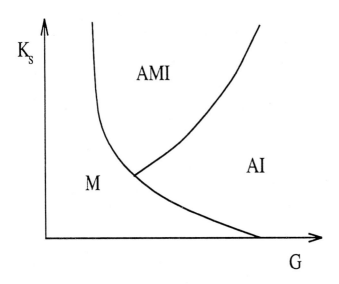

Fig. 1. Schematic phase diagram in the disorder (G) - interaction (K_s) plane proposed for a system with $K_t = 0$ and a short-ranged K_s. M, AI, and AMI denote a metal phase, an Anderson insulator, and an Anderson-Mott insulator, respectively. The transition from M to AI is an Anderson transition, while the one from M to AMI is an AMT.

These results are consistent with existing experiments, and the theory given here provides an alternative to the previous exploration in terms of noninteracting local moments. (2) The DOS is the order parameter for the quantum glass transition. At criticality it is predicted to vanish logarithmically with temperature. (3) The electrical conductivity $\tilde{\sigma}$ is so broadly distributed that it is not a self-averaging quantity, but log $\tilde{\sigma}$ is both self-averaging and a scaling quantity. This result may be relevant to explain the sample-to-sample fluctuations in the conductivity that are observed in Si:P at low temperature near the AMT.

Acknowledgments This work was supported by the NSF under grant numbers DMR-96-32978 and DMR-95-10185.

References

[1] N.F. Mott, *Metal-Insulator Transitions*, Taylor& Francis (London 1990).

[2] For a review, see, e.g., P. A. Lee and T. V. Ramakrishnan, Rev. Mod. Phys. **57**, 287 (1985).

[3] For a review, see, e.g., D. Belitz and T. R. Kirkpatrick, Rev. Mod. Phys. **66**, 261 (1994).

[4] A. M. Finkel'stein, Zh. Eksp. Teor. Fiz. **84**, 168 (1983) [Sov. Phys. JETP **57**, 97 (1983)].

[5] F. Wegner, Z. Phys. B **35**, 207 (1979).

[6] See A. B. Harris and T. C. Lubensky, Phys. Rev. **23**, 2640 (1981) for an order parameter description of the Anderson transition. Wegner's result[15] for the density of states in noninteracting systems proved that the transition studied in their theory is not realized.

[7] T. R. Kirkpatrick and D. Belitz, Phys. Rev. Lett. **73**, 862 (1994).

[8] T. R. Kirkpatrick and D. Belitz, Phys. Rev. Lett. **74**, 1178 (1995).

[9] D. Belitz and T. R. Kirkpatrick, Z. Phys. B **13**, 513 (1995).

[10] D. Belitz and T. R. Kirkpatrick, Phys. Rev. B **52**, 13922 (1995).

[11] G. Grinstein, Phys. Rev. Lett. **37**, 944 (1976).

[12] D. S. Fisher, Phys. Rev. Lett. **56**, 416 (1986); J. Villain, J. Phys. (Paris) **46**, 1843 (1985).

[13] For a review, see, D. P. Belanger and A. P. Young, J. Mag. Magn. Mat. **100**, 272 (1991).

[14] See, e.g., J. Zinn-Justin, *Quantum Field Theory and Critical Phenomena* (Clarendon, Oxford 1989).

[15] See, e.g., S.-K. Ma, *Modern Theory of Critical Phenomena* (Benjamin, Reading, MA 1976); and M. E. Fisher, in *Advanced Course on Critical Phenomena*, edited by F. W. Hahne (Springer, Berlin 1983), p.1

[16] F. Wegner, Z. Phys. B **25**, 327 (1976).

[16] F. Wegner, Z. Phys. B **44**, 9 (1981).

Quantum Spin Glasses

Heiko Rieger[1] and A. Peter Young[2]

[1] HLRZ c/o Forschungszentrum Jülich, 52425 Jülich, Germany
[2] Department of Physics, University of California, Santa Cruz, CA 95064, USA

Abstract. Ising spin glasses in a transverse field exhibit a zero temperature quantum phase transition, which is driven by quantum rather than thermal fluctuations. They constitute a universality class that is significantly different from the classical, thermal phase transitions. Most interestingly close to the transition in finite dimensions a quantum Griffiths phase leads to drastic consequences for various physical quantities: for instance diverging magnetic susceptibilities are observable over a whole range of transverse field values in the disordered phase.

1 Introduction

At very low temperatures the role of quantum fluctuations in any physical pure or disordered system become more and more important. As far as critical phenomena are concerned any finite temperature destroys quantum coherence of the lowest lying excitations determining the universality class of the transition, which remains therefore classical. However, if the transition takes place at strictly zero temperature, a new universality class and in particular new physics emerges. Obviously in the vicinity of such a transition even finite temperature properties are characterized by strong crossover effects between a quantum critical and classical regions. Thus the properties of such zero temperature transitions become experimentally accessible, which motivates the study these new universality classes.

The prominent feature of quantum phase transitions [1] is the fact that statics and dynamics are inextricably linked: the static features are defined by the Hamiltonian, which implies immediately the dynamics via the Schrödinger equation. This introduces an extra dimension, the (imaginary) time, into the problem and it is by no means guaranteed that this additional dimension is equivalent to one of the d space dimensions. In many *pure* systems it turns out to be so, which is the origin of the observation that "the correlation length is proportional to the inverse energy gap". This relation seems to be pretty robust even in the presence of a nontrivial anisotropy (c.f. the ANNNI model), for which reason it became folklore in strongly correlated systems.

However, in the presence of quenched (i.e. time-independent) *disorder*, this simple relation fails: Usually the randomness (modeling impurities etc.) is uncorrelated in space, but time-independence means a perfect correlation of the disorder in the time direction. This implies an *extreme* anisotropy, which manifests itself in a non-trivial relation between spatial correlation length ξ

and energy gap ΔE. For a generic second order phase transition scenario a dynamical exponent z can be introduced via $\xi \sim \Delta E^{-z}$, with z in general different from one.

Another drastic consequence of the perfect correlation of the disorder in the (imaginary) time direction are spectacular properties of physical observables within the so called Griffiths phase [2] surrounding the critical point itself. In contrast to the classical case one there may be a whole region of values for the parameter tuning the transition over which the zero-frequency susceptibilities *diverge*. Such a scenario has actually been described already a long time ago by McCoy [3] in the context of a one-dimensional model. It seems that this is indeed a generic feature of finite dimensional disordered systems with a quantum phase transition as we would like to demonstrate here.

In this article we focus on quantum spin glasses, in particular the transverse field Ising models defined by the Hamiltonian

$$H = -\sum_{\langle ij \rangle} J_{ij} \sigma_i^z \sigma_j^z - \Gamma \sum_i \sigma_i^x , \tag{1}$$

where σ_i are the Pauli spin matrices (modeling spin-$\frac{1}{2}$ degrees of freedom), $\langle ij \rangle$ nearest neighbor pairs on a d-dimensional lattice, Γ the transverse field and J_{ij} quenched random interaction strength obeying for instance a Gaussian distribution with zero mean and variance one. Here the quenched disorder mentioned above also produces frustration, which enhances the effect that the randomness has on critical and non-critical properties. For completeness we also include a discussion of the one-dimensional case, which is not frustrated but shares many features of the higher dimensional realizations. In the next section we report on the results for the critical properties and in section 3 we present the exciting features of the Griffiths phase in these models. Section 4 gives a summary and perspectives for future work.

2   Critical Properties

In any dimension d the quantum system described by eq. (1) has a second order phase transition at zero temperature that manifests itself in specific macroscopic properties of the ground state and low lying excitations. A critical value Γ_c for the transverse field strength separates a disordered or paramagnetic phase for $\Gamma > \Gamma_c$ from an ordered phase for $\Gamma < \Gamma_c$. This transition is characterized by a diverging length scale $\xi \sim |\Gamma - \Gamma_c|^{-\nu}$ and a vanishing characteristic frequency $\omega \sim \Delta E \sim \xi^{-z}$. The latter is the quantum analog of "critical slowing down" in the critical dynamics of classical, thermally driven transitions. Together with a third critical exponent, defining the anomalous dimension of the order parameter field, the thermal exponent ν and the dynamical exponent z give a complete description of the transition via a set of scaling relations for two and three dimensions. The one-dimensional case shows a somewhat richer scenario [4] and in $d > 8$ the violation of hyper-scaling adds another exponent [5]. We list the main features for the different dimensions:

2.1   d=1

Analytical [3], [4] and numerical [6], [7] investigations revealed the following picture: because of the logarithmically broad distribution of various physical quantities at criticality one is forced discriminate between average and typical properties. For instance the typical correlation length diverges with an exponent $\nu_{\mathrm{typ}} = 1$, whereas the average diverges with $\nu_{\mathrm{av}} = 2$. It turns out that it is not the energy gap but its logarithm that scales with system size and distance from the critical point, giving rise to an exponential rather than algebraic decrease of the energy gap with system size: $[\Delta E]_{\mathrm{av}} \sim \exp(-aL^{1/2})$. This means that $z = \infty$ and since the inverse gap corresponds to a characteristic relaxation or tunneling time it is reminiscent of an activated dynamics scenario in conventional spin glasses or random field systems.

For numerical work with finite size systems it is most useful to study the probability distribution of various quantities, such as the energy gap or local susceptibility $\chi^{(\mathrm{loc})}(\omega = 0)$. One finds at the critical point

$$P_L(\ln \chi^{(\mathrm{loc})}) \sim \tilde{P}(L^{-1/2} \ln \chi^{\mathrm{loc}}) \tag{2}$$

reflecting both features $\nu_{\mathrm{av}} = 2$ (since the system size comes with a power $1/\nu_{\mathrm{av}}$) and $z = \infty$ (since $\ln \chi^{(\mathrm{loc})}$ rather than $\chi^{(\mathrm{loc})}$ enters the scaling variable).

2.2   d = 2 and 3

Numerical investigation of the two- [8] and three-dimensional [9] quantum Ising spin glass model in a transverse field via quantum Monte Carlo simulations demonstrated the existence of a second order phase transition at a critical transverse field strength Γ_c. The results are compatible with the scaling predictions made by a droplet theory for these models [10].

In contrast to the one-dimensional case there is ample evidence that here the dynamical exponent z is finite, namely $z = 1.50 \pm 0.05$ in two and $z = 1.3 \pm 0.1$ in three dimensions. Also the subsequent study [11], [12] of the probability distribution of the logarithms of the local linear and nonlinear susceptibility at criticality gave no indication of a characteristic broadening with increasing system size that would indicate an infinite value for z.

Most interestingly, as mentioned in the introduction, the critical exponents for the quantum phase transition at strictly zero temperature are connected to the temperature dependence of various quantities for $\Gamma = \Gamma_c$, $T \to 0$. For instance the experimentally accessible nonlinear susceptibility *diverges* at $\Gamma = \Gamma_c$ quite strongly, i.e.

$$\chi_{\mathrm{nl}}(\omega = 0) = \left. \frac{\partial^3 M}{\partial h^3} \right|_{h=0} \propto T^{-\gamma_{\mathrm{nl}}/z} \tag{3}$$

with $\gamma_{nl}/z \approx 3.0$ in two and in three dimensions (M is the total magnetization and h a longitudinal magnetic field). A real space (Migdal-Kadanoff) renormalization group calculation [14] yields similar conclusions. Thus the strength of the divergence of the non-linear susceptibility is similar to the one at the classical transition. This is in sharp contrast to the results reported for recent experiments on $LiHo_x Y_{1-x} F_4$ [13]. Here the strength of the divergence seemed to decrease with decreasing temperature, and it has been speculated that at zero temperature, i.e. at the quantum phase transition, no divergence at all might be present (see next subsection). On the other hand, it might be possible that because of the long range nature of the dipolar interactions in $LiHo_x Y_{1-x} F_4$ the experiments might be better described by a model with appropriately chosen long range interactions.

2.3   Mean field (d=∞)

The quantum phase transition occuring in the infinite range model of the transverse Ising spin glass [15] and in a related quantum rotor spin glass [16], [5] can be handled analytically to a large extent [17]. The dynamical exponent is $z = 2$, the correlation length exponent $\nu = 1/4$ and the nonlinear susceptibility diverges with an exponent $\gamma = 1/2$. The latter again implies a divergence of the non-linear susceptibility. Although it is much weaker than for short range interaction models it should clearly be observable. However, the experiments [13] at small but finite temperatures yield an effective exponent γ_{eff} that is significantly smaller than $1/2$, and for $T < 25\,mK$ even one that is indistinguishable from zero. Therefore also mean-field theory does not cure the contradiction between experiment and theory for transverse Ising spin glasses.

Finally we would like to remark that the zero temperature phase transition in a number of mean field quantum spin glasses falling in universality classes different from the one considered here has been investigated recently [18].

3   Griffiths Phase

In a specific disorder realization one can always identify spatial regions that are more weakly or more strongly coupled than the average. The latter lead to a non-analytic behavior of the free energy not only at the critical point but in a whole region surrounding it, as already observed by Griffiths nearly 30 years ago [2]. However, a strongly coupled cluster (in spin glass terms one with a minor degree of frustration) tends to order locally much earlier, coming from the disordered phase, than the rest of the system. Upon application of an external field spins in this cluster act collectively and thus lead locally to a greatly enhanced susceptibility. However, one only gets an essential singularity in static properties of a classical system [2].

This effect is much stronger in dynamics than in statics. For classical disordered Ising system with heat bath dynamics close to a thermally driven phase transition it was shown [19] that these strongly coupled regions posses very large relaxation times for spin autocorrelations. By averaging over all sites one gets a so called enhanced power law

$$C_{\text{classical}}(t) = [\langle S_i(t)S_i(0)\rangle]_{\text{av}} \sim \exp\left\{-a(\ln t)^{d/d-1}\right\} \tag{4}$$

where $[\ldots]_{\text{av}}$ means an average over all sites and disorder realizations. However, this prediction, although rigorous to some extent (at least for the diluted ferromagnet [20]), could not be confirmed by extensive numerical investigations, neither for the spin glass [21] nor for the diluted ferromagnet [22], [23]. The main reason for this failure is most probably the extremely small probability with which big enough strongly coupled clusters that have an extremely large relaxation time occur in a finite size systems [24].

To make this point clear let us focus, for simplicity, on a diluted Ising ferromagnet below the percolation threshold, i.e. the site concentration $c < p_c$. Then a connected cluster of volume $V = L^d$ occurs with a probability $p = c^V = \exp(-\zeta L^d)$ with $\zeta = \ln 1/c > 0$, which is a very small number for large L. However, these spins within this cluster have a relaxation time τ that is exponentially large (because of the activation energy that is needed to pull a domain wall through the cluster to turn it over): $\tau \sim \exp(\sigma L^{d-1})$, with σ a surface tension. the combination of these two exponentials yield the result (4). Note that the occurrence of the exponent $d - 1$ in the relaxation time renders the final decay of the autocorrelation function *faster* than algebraic.

This is quite different in a random quantum system at zero temperature. Now the activated dynamics in the classical scenario has to be replaced by tunneling events. Let us stick to the above example of a diluted ferromagnet and add a transverse field at zero temperature. The probability for a connected cluster to occur is the same as before, but the relaxation time or inverse tunneling rate is now given by $\tau \sim \exp(\sigma' L^d)$. Remember that the classical ground state of the cluster under consideration would be ferromagnetic and twofold degenerate. The transverse field lifts this degeneracy for a finte cluster, but for Γ smaller than the critical field value, where the ferromagnetic order of the infinite pure system sets in, the energy gap is extremely small: it decreases exponentially with the volume of the cluster. This energy gap between ground state and first excited state sets the scale for the tunneling rate.

Now we see that in marked contrast to the classical system a d instead a $d-1$ appears in the exponent for the relaxation time. One might say that the same cluster relaxes faster via activated dynamics (although that is already incredibly slow) than by quantum tunneling. Therefore, by the same line of arguments that lead to (4), one now expects spin-autocorrelations to decay *algebraically* in the Griffiths phase of random quantum systems:

$$C_{\text{quantum}}(t) = \langle 0|\sigma_i^z(t)\sigma_i^z(0)|0\rangle \sim t^{-\zeta(\Gamma)/\sigma'(\Gamma)} \tag{5}$$

For simplicity, although we expect a similar form for real time correlations, we assume t to be imaginary time, i.e. the Operator $\sigma_i^z(t)$ to be given by $\exp(-Ht)\,\sigma_i^z\,\exp(Ht)$, H being the Hamiltonian. Apart from a different functional form (5) has drastic consequences for various quantities: For instance the local zero frequency susceptibility at temperature T is given by

$$\chi^{(loc)}(\omega = 0) = \int_0^{1/T} dt\,\langle C(t)\rangle \sim T^{-1+\zeta(\Gamma)/\sigma'(\Gamma)} \tag{6}$$

which *diverges* for $T \to 0$ if $\zeta(\Gamma)/\sigma'(\Gamma) < 1$ even if one is not at the critical point! Such a prediction should be experimentally measurable.

It should be noted that a diverging (surface) susceptibility in a whole region close to the critical point has already been found by McCoy nearly 30 years ago [3] for a classical two-dimensional Ising model with layered randomness, which is equivalent to the random transverse Ising chain. Actually we see now clearly that the non-analyticities reported by Griffiths [2] and the divergences calculated by McCoy [3] have a common origin: the existence of strongly coupled rare regions, which simply leads to a more drastic effect in a quantum system at low or vanishing temperatures than in a classical system.

3.1 d = 1

The random transverse Ising chain has been reinvestigated recently by D. Fisher [4] in a beautiful renormalization group treatment providing us with many astonishing analytical predictions. On the numerical side one can take advantage of the free fermion technique [25], by which the problem of diagonalizing the original quantum Hamiltonian is reduced to the diagonalization of a $L \times L$ or $2L \times 2L$ matrix. In this way one can consider very large system sizes ($L \sim 256$) with good statistics (i.e. number of disorder realizations $\geq 50,000$). However, we should note that some of the basic features of the model (e.g. $z = \infty$ at criticality, existence of a Griffiths phase, etc.) can already be seen for modest system sizes ($L \leq 16$) provided one studies the appropriate quantities which are sensible to the rare disorder configurations having strong couplings and/or small transverse fields. The probability distribution of the energy gap and/or various susceptibilities is such a quantity and in what follows we report the results of numerical investigations that support the above predictions.

In [7] we took a uniform distribution for bonds ($J_i \in [0, 1]$) and transverse fields ($h_i \in [0, h_0]$). The system is at its critical point for $h_0 = 1$ (since in this case the field- and bond distributions are identical and therefore the model self-dual) and in the paramagnetic (ferromagnetic) phase for $h_0 > 1$ ($h_0 < 1$). For each of 50,000 disorder realization for chains of length L (from $L = 16$ to $L = 128$) we calculated the energy gap Δ from which we calculated the probability distribution $P_L(\ln \Delta E)$. It turned out that *at* the critical point $h_0 = 1$ the probability distribution scales as described in section 2.1.

according to eq. (2). In particular the distribution gets broader even on a logarithmic scale with increasing system size.

Within the Griffiths phase for $h_0 > 1$, however, the distribution does not broaden any more, but gets simply shifted on a logarithmic scale with increasing system size. It turns out that now the distribution of gaps obeys

$$\ln\left[P(\ln \Delta E)\right] = \frac{1}{z}\ln \Delta E + \text{const.} \tag{7}$$

with a dynamical exponent that varies continuously throughout the Griffiths phase according to $1/z \sim 2\delta - c\delta^2$ (δ being the distance from the critical point $\delta = h - h_0$) and which diverges at the critical point $z \to \infty$ for $\delta \to 0$. The form (7) can be made plausible by an argument applicable to any dimension d given in the next subsection.

3.2 $d = 2$ and 3

For the numerical investigation of the astonishing features of the transverse field spin glass model defined in eq. (1) in 2 and 3 dimensions it is also most promising to focus on the probability distribution of the local linear and nonlinear susceptibility. The former is simply proportional to the inverse of the energy gap, whereas the latter is proportional to the third power of the inverse of the gap, hence the distribution of gaps $P(\ln \Delta E)$ plays again the crucial role. The form that we expect is similar to (7) because of the following argument:

From what has been said in the beginning of section 3 before eq. (5), by combining the probability of a strongly coupled cluster with its energy gap one gets a power law for the tail of the distribution of gaps: $P(\Delta E) \sim \Delta E^{\lambda-1}$ with $\lambda = \zeta(\Gamma)/\sigma'(\Gamma)$. Furthermore since the excitations that give rise to a small energy gap are well localized we assume that their probability is proportional to the spatial volume L^d of the system. Thus $P_L(\ln \Delta E) \sim L^d \Delta E^\lambda$, which can be cast into the conventional scaling form relating a time scale to a length scale via the introduction of a dynamical exponent: $P_L(\ln \Delta E) \sim (L\Delta E^{1/z})^d$ with $z = d/\lambda$. Therefore the distribution for the local susceptibility (because $\chi^{(\text{loc})} \propto 1/\Delta E$) should be given by

$$\ln\left[P(\ln \chi^{(\text{loc})})\right] = -\frac{d}{z}\ln \chi^{(\text{loc})} + \text{const.} \tag{8}$$

As a byproduct one obtains in this way also the form (7) for the one-dimensional case. The distribution of the local nonlinear susceptibility the factor d/z, which is actually the power for the algebraically detaying tail of $P(\chi^{(\text{loc})})$, should be replaced by $d/3z$ (since $\chi_{\text{nl}}^{(\text{loc})} \propto 1/\Delta E^3$):

$$\ln\left[P(\ln \chi_{\text{nl}}^{(\text{loc})})\right] = -\frac{d}{3z}\ln \chi_{\text{nl}}^{(\text{loc})} + \text{const.} \tag{9}$$

Thus one can characterize *all* the Griffiths singularities in the disordered phase by a *single* continuously varying exponent, z. This prediction is indeed confirmed by the numerical investigations [11], [12]. We see that the *average* susceptibility in the disordered phase ($\Gamma > \Gamma_c$) will diverge when $z > d$, and the nonlinear susceptibility when $z > d/3$.

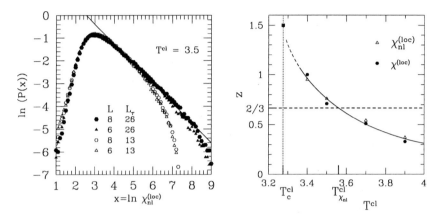

Fig. 1. Left: The probability distribution of the local non-linear susceptibility at a point in the disordered phase (the critical point is at $T_c^{cl} = 3.30$). The parameter T^{cl} is related to the transverse field strength via the Suzuki-Trotter mapping (see [9] for details). The straight line has slope -0.87 which gives via (9) $z = 0.76$. Since the slope is greater than -1, or equivalently $z > 2/3$, the average non-linear susceptibility does *diverge* at this point. **Right:** The dynamical exponent z, obtained by fitting the distributions of $\chi^{(loc)}$ and $\chi_{nl}^{(loc)}$ to Eqs. (8) and (9), is plotted for different values of T^{cl}. The estimates obtained from data for $\chi^{(loc)}$ are shown by the triangles and the estimates from the data for $\chi_{nl}^{(loc)}$ are shown by the hexagons. The two are in good agreement. The dotted vertical line indicates the critical point, obtained in Ref. [8] and the solid square indicates the estimate of z at the critical point. The dashed line is $z = 2/3$; the average non-linear susceptibility diverges for z larger than this, i.e. $T^{cl} > T_{\chi_{nl}}^{cl} \simeq 3.56$. The solid curve is just a guide to the eye.

From fig. 1 (taken from [11]) we see that in two dimensions this is indeed the case for the non-linear susceptibility in a range $\Gamma \in [\Gamma_c, \Gamma_{\chi_{nl}}]$. However, the numerical value of Γ_c and $\Gamma_{\chi_{nl}}$ might be non-universal and so we cannot make a precise prediction for what the range of field-values, in which χ_{nl} diverges, might be in an experiment.

In three dimensions [12] the situation is similar, although here the range of transverse field values where the nonlinear susceptibility diverges is much smaller than in 2d. This is plausible since for increasing dimension the spectacular effects of the Griffiths singularities, which have their origin in spatially isolated clusters, becomes weaker and seem to vanish in the mean field theory.

Finally we would like to emphasize that due to the divergent behavior of the linear or higher susceptibilities in part of the quantum Griffiths phase we expect this effect to be observable also in an experiment with a macroscopic but finite system. It has been argued by Imry [26] that the Griffiths singularity occuring in a *classical* diluted Ising magnet is an artifact of the thermodydnamic limit procedure. What is meant here is the essential singularity in a *static* quantity (like the magnetization or susceptibility). In essence the argument is that in order for these effects to be observable one needs a macroscopic cluster, which would occur only once in an astronomically large collections of finite samples. However, since in the *quantum* case, which we consider here, the statics is inextricably mixed with the *dynamics*, static quantities are much more sensible with respect to the existence of strongly coupled regions even of modest size. The distribution of cluster sizes n is still effectively cut off around $n_{max} \sim \log N$, where $N = L^d$ is the system size. However, such a *typical* cluster, which occurs with probability $\mathcal{O}(1)$ leads to a (local) nonlinear susceptibility of $\chi_{nl}^{(loc)} \sim N^{3\sigma'/\varsigma}$. For a macroscopic sample ($N \sim 10^{23}$) this is a huge number, which should render the divergence predicted for an infinite sample observable also in a macroscopic *typical* system (in contrast to the mere non-analyticities in the classical case).

4 Conclusions

The quantum phase transition occuring at zero temperature in transverse field Ising spin glasses can be described by a set three independent exponents that have been determined numerically. The dynamical exponent z, connecting time- or energy scales with a spatial diverging length is finite in $d > 1$ and infinite in the one-dimensional model. The critical point is surrounded by a quantum Griffiths phase, where various susceptibilities diverge. These divergences can be characterized by a single continuously varying exponent z. Its limiting value seems to be identical to the critical value, which is analytically established in 1d. The most important observation is that, although a Griffiths phase should be present in all random magnets, this is the first numerical verification of some of its theoretically predicted features. We would not be surprised if the divergences we found in the quantum Griffiths phase could also be measured in an appropriate experiment. It is interesting to speculate on whether the possible divergence of the non-linear susceptibility in part of the disordered phase might be related to the difference between the experiments [13] which apparently do not find a strong divergence of χ_{nl} at the critical point, and the simulations[8], [9] which do. Certainly more work in this direction has to be done.

Acknowledgement: The work of HR is supported by the Deutsche Forschungsgemeinschaft (DFG) and the work of APY by the National Science Foundation under grant No. DMR–9411964.

References

[1] S. Sachev, Z. Phys. B **94**, 469 (1994); Nucl. Phys. B **464**, 576 (1996); STAT-PHYS 19, ed. Hao Bailin, p. 289 (World Scientific, Singapore, 1996).

[2] R. B. Griffiths, Phys. Rev. Lett. **23**, 17 (1969).

[3] B. McCoy, Phys. Rev. Lett. **23**, 383 (1969).

[4] D. S. Fisher, Phys. Rev. Lett. **69**, 534 (1992); Phys. Rev. B **51**, 6411 (1995).

[5] N. Read, S. Sachdev and J. Ye, Phys. Rev. B **52**, 384 (1995).

[6] A. Crisanti and H. Rieger, J. Stat. Phys. **77**, 1087 (1994).

[7] A. P. Young and H. Rieger, Phys. Rev. B **53**, 8486 (1996).

[8] H. Rieger and A. P. Young, Phys. Rev. Lett. **72**, 4141 (1994).

[9] M. Guo, R. N. Bhatt and D. A. Huse, Phys. Rev. Lett. **72**, 4137 (1994).

[10] M. J. Thill and D. A. Huse, Physica A, **15**, 321 (1995).

[11] H. Rieger and A. P. Young, Phys. Rev. B **54**, *** (1996).

[12] M. Guo, R. N. Bhatt and D. A. Huse, Phys. Rev. B **54**, *** (1996).

[13] W. Wu *et. al*, Phys. Rev. Lett. **67**, 2076 (1991); W. Wu, D. Bitko, T. F. Rosenbaum and G. Aeppli, Phys. Rev. Lett. **71**, 1919 (1993).

[14] B. Boechat, R. dos Santos and M. Continentino, Phys. Rev. B **49**, 6404 (1994); M. Continentino, B. Boechat and R. dos Santos, Phys. Rev. B **50**, 13528 (1994).

[15] J. Miller and D. Huse, Phys. Rev. Lett. **70**, 3147 (1993).

[16] J. Ye, S. Sachdev and N. Read, Phys. Rev. Lett. **70**, 4011 (1993).

[17] The finite temperature features of this model are more delicate because of the issue of replica symmetry breaking. See Y. Y. Goldschmidt and P. Y. Lai, Phys. Rev. Lett. **64**, 2567 (1990) and B. K. Chakrabarti, A. Dutta and P. Sen: *Quantum Ising Phases and Transitions in Transverse Ising Models*, Lecture Notes in Physics **m41**, Springer Verlag, Berlin-Heidelberg-New York, (1996) for references.

[18] R. Oppermann and M. Binderberger, Ann. Physik **3**, 494 (1994); S. Sachdev, N. Read and R. Oppermann, Phys. Rev. B **52**, 10286 (1995); see also R. Oppermann in this volume. T. M. Nieuwenhuizen, Phys. Rev. Lett. **74**, 4289 (1995). F. Pázmándi, G. T. Zimányi and R. T. Scalettar, cond-mat/9602158 (1996).

[19] M. Randeira, J. P. Sethna and R. G. Palmer, Phys. Rev. Lett. **54**, 1321 (1985).

[20] A. J. Bray, Phys. Rev. Lett. **60**, 720 (1988), J. Phys. A **22**, L81 (1989); D. Dhar, M. Randeira and J. P. Sethna, Europhys. Lett. **5**, 485 (1988).

[21] A. T. Ogielski, Phys. Rev. B **32**, 7384 (1985); H. Takano and S. Miyashita, J. Phys. Soc. Jap. **64**, 423 (1995).

[22] S. Jain, J. Phys. C **21**, L1045 (1988); H. Takano, S. Miyashita, J. Phys. Soc. Jap. **58**, 3871 (1989); V. B. Andreichenko, W. Selke and A. L. Talapov, J. Phys. A **25**, L283 (1992); P. Grassberger, preprint (1996).

[23] Experimentally one finds qualitative indications for the existence of a classical Griffiths phase e.g. in diluted (anti)ferromagnets see: Ch. Binek and W. Kleemann, Phys. Rev. B **51**, 12888 (1995).

[24] For an investigation of the probability distribution of local relaxation times in classical spin glasses within the Griffiths phase see: H. Takayama, T. Komori and K. Hukushima, in *Coherent Approach to Fluctuations*, ed. M. Suzuki and N. Kawashima, p. 155 (World Scientific, Singapore, 1996); *ibid.* H. Takano and S. Miyashita, p. 217.

[25] E. Lieb, T. Schultz and D. Mattis, Ann. Phys. (NY) **16**, 407 (1961).

[26] Y. Imry, Phys. Rev. B **15**, 4448 (1977).

Fermionic Quantum Spin Glass Transitions

Reinhold Oppermann and Bernd Rosenow

Institut für Theoretische Physik, Universität Würzburg, D–97074 Würzburg

Abstract. This article reviews recent progress of the analytical theory of quantum spin glasses (QSG). Exact results for infinite range and one loop renormalisation group calculations for finite range models of either insulating or metallic type are presented. We describe characteristics of fermionic spin glass transitions and of fermionic correlations which are affected by these transitions and by spin glass order. Connections between tricritical thermal– and $T = 0$ QSG transitions are described. A general phase diagram with tricritical QSG transitions caused either by random chemical potential or by elastic electron scattering, and implying discontinuous $T = 0$–transitions in weak and in strong filling regimes, is also derived.

1 Introduction

Fermionic quantum spin glasses form a part of the overlap regime of interacting disordered fermion systems and spin glasses. Understanding such systems with frustrated or with general random interactions between fermionic spins touches many currently active research fields. The wellknown transverse field Ising spin glass can also be represented as a pseudofermionic quantum spin glass with imaginary chemical potential and thus be treated analogously to the ones discussed here which involve real fermion spins. One of the attractive features of fermionic spin glasses is the connection of spin and charge degrees of freedom. Even in a strongly localized Ising limit (with all spin and charge variables commuting) quantum statistics remains indispensable, since it governs the relative occupation of magnetic and nonmagnetic states. This provides a link between spin– and charge correlations, exerting a major influence on magnetic phase diagrams for example. Nonanalytic interplay is observed in the simplest fermionic spin glass, the SK–model extended to a four–state–per–site fermionic space (ISG_f). This phenomenon occurs at a thermal tricritical point, separating continuous spin glass transitions around half–filling from discontinuous regimes at either strong or weak filling. This structure of the phase diagram is compared with a similar one of classical spin 1 models including the clean BEG–model. Furthermore, model definitions on fermionic space generate a large set of correlation functions which display ubiquitous quantum dynamics of these models on the fermionic level, ie beyond the one seen in spin– and charge–correlations and caused by electronic transport. The fermionic Green's function and its density of states content is one of those quantities which probe the commutation properties of single fermion operators with the hamiltonian. The ISG_f shows such quantum-dynamics. Moreover, correlations of the statistical density of states fluctuations turn out to be related to the spin glass order parameter, which introduces Parisi replica symmetry breaking into the charge sector.

2 The Grand Canonical Fermionic Ising Spin Glass

In contrast to standard spin glass models like SK– or transverse field models, the models of interacting true fermion spins are naturally described in the grand canonical ensemble. Statistically fluctuating spin interactions leave only the choice between nonrandom local fermion concentration or nonrandom chemical potential. In any case however an effective spin dilution due to thermal redistribution among magnetic and nonmagnetic states is controlled by either μ or ν. This form of thermal spin dilution comprises at a special point the case of nonanalytically communicating charge- and spin–fluctuations. The simplest model displaying this type of behavior is the Ising spin glass on a fermionic space (ISG_f) with four states per site, defined by the hamiltonian $H = -\frac{1}{2}\sum_{i\neq j} J_{ij}\sigma_i\sigma_j - \mu\sum_i n_i$ with spins $\sigma_i = \Psi_{i,\alpha}^\dagger \sigma_{\alpha\beta}^z \Psi_{i,\beta}$, particle number operator $n_i = \Psi_{i,\alpha}^\dagger \Psi_{i,\alpha}$ and gaussian distributed exchange integrals J_{ij} [1]. The fermionic field operators obey the usual commutator relations $\{\Psi_{i\alpha},\Psi_{j\beta}\} = 0$ and $\{\Psi_{i\alpha}^\dagger,\Psi_{j\beta}\} = \delta_{ij}\delta_{\alpha\beta}$. Expressing the partition function with the help of Grassman integrals the disorder averaging of the free energy is performed by means of the replica trick $\beta[F]_{av} = [\log Z]_{av} = \lim_{n\to 0}\frac{1}{n}(1 - [\Pi_{\alpha=1}^n Z^\alpha]_{av})$. As a guide to the global phase diagram we study an exactly solvable infinite range version of the model but the formulae obtained in this subsection may equally well be considered as saddle point approximation for an interaction with finite range.

2.1 Phase Diagram

Details of the calculation are given in [1], here we just state the result for the replica symmetry broken saddle point free energy

$$\beta f = \frac{1}{4}\beta^2 J^2 \left[(1-\tilde{q})^2 - (1-q_1)^2 + q_1^2 - \int_0^1 dx\, q^2(x)\right] - \ln 2 - \beta\mu$$

$$- \lim_{K\to\infty} \int_{z_{K+1}}^G \ln\left[\int_{z_K}^G \left[\cdots\left[\int_{z_1}^G \left(\cosh(\beta\tilde{H})\right)\right.\right.\right. \tag{1}$$

$$\left.\left.\left.+ \cosh(\beta\mu)\exp(-\frac{1}{2}\beta^2 J^2(\tilde{q}-q_1))\right)^{m_1}\right]^{m_2/m_1}\cdots\right]^{m_k/m_{K-1}}\right]^{1/m_K}$$

Notice the appearance of the Parisi variables q_ν and the additional $\tilde{q}$ which lies at the heart of the following discussion. The Edwards - Anderson order parameter $q_{EA} = \lim_{t\to\infty} < S_i(t)S_i(0) >$ describes the freezing of spins in the spin glass phase and is given by the plateau height $q(1)$ of the order parameter function, whereas the replica diagonal $\tilde{q} = [< \sigma^a\sigma^a >]_{av}$ is related to the average filling factor $[\nu]_{av} = \frac{1}{N}\sum_i[n_i]_{av}$ by $[\nu]_{av} = 1+\tanh(\beta\mu)(1-\tilde{q})$. The last relation is exact even in the case of replica symmetry breaking. To obtain information about the phase diagram a replica symmetric approximation is sufficient, though. The symmetric saddle–point solutions are $q = \int_z^G \frac{\sinh^2[\beta\tilde{H}(z)]}{C_\mu^2(z)}$ and $\tilde{q} = \int_z^G \frac{\cosh[\beta\tilde{H}(z)]}{C_\mu(z)}$ with $C_\mu(z) = \cosh[\beta\tilde{H}(z)] + \cosh(\beta\mu)\exp[-1/2\beta^2(\tilde{q} - q)]$. Phase transitions are signalized by vanishing masses of the order parameter propagators which in the saddle point formalism are given by second derivatives of the free energy. Hence a positive mass for $\tilde{q}$ and a negative one for q guarantees stability. A similar system of coupled stability conditions was found for the BEG - model [2] and for a SK - model with crystal field [3]. Analyzing the stability limits one obtains curves of critical spin and charge fluctuations, respectively:

$$\mu_{c1}(T) = T\cosh^{-1}[(1/T - 1)\exp[1/(2T)]] \qquad (2)$$

$$\mu_{c2}(T) = T\cosh^{-1}[\frac{(1 \mp \sqrt{1 - 8T^2})^2}{8T^2}\exp[\frac{2}{1 \mp \sqrt{1 - 8T^2}}]] \qquad (3)$$

The two curves have a common tangent point at $\mu_{c3} = 1/3\cosh^{-1}[2\exp(3/2)] = 0.961056, T_{c3} = 1/3$. For smaller values of the chemical potential the paramagnetic solution becomes unstable with respect to spin fluctuations first; at the common tangent point both types of fluctuations become critical simultaneously and tricritical behaviour results.

2.2 Tricritical Point TCP: Exponents and Special Features

An expansion of the saddle point equations around the tricritical point yields in leading order

$$0 = gr_g - r_T\delta T^2 + 6\delta T\delta\tilde{q} - \frac{3}{4}\delta\tilde{q}^2 + 3q^2, \qquad 0 = 6q(\delta\tilde{q} - \delta T - q) \qquad (4)$$

where $\delta\tilde{q} \equiv \tilde{q} - \tilde{q}_{TCP}, gJ = \mu - \mu_{c3} + (\zeta^{-1}J - \mu_{c3})3\delta T$ as nonordering field, and $\delta T \equiv T - T_{c3}$. The constants are given by $r_g = \frac{2\zeta}{3}, r_T = 2(1 - \frac{3}{4}\zeta^{-2})$ with $\zeta \equiv tanh(\mu_{c3}/T_{c3}) \simeq 0.9938$. The average filling factor corresponding to μ_{c3} is evaluated as $[\nu_{c3}]_{av} \simeq 1.6625$. From (4) we get for q=0

$$\delta\tilde{q}_{dis} = 4(\delta T \pm |\delta T|\sqrt{1 + \frac{r_g g}{12\delta T^2} - \frac{r_T}{12}} = 4(\delta T \pm |\delta T|W) \qquad (5)$$

Only the solution with the - sign corresponds to a minimum of the free energy, in a region close to the line $\delta\mu = -3(1/\zeta - \mu_{c3})\delta T = -0.1354\delta T$ (tangent to both $\mu_{c1}(T_{c3})$ and $\mu_{c2}(T_{c3})$) g is of order δT^2 or smaller and usual critical behaviour results. However, if g is of order δT or larger the solution becomes to leading order $\delta\tilde{q} = \sqrt{\frac{r_g g}{12}}$ and thus displays a nonanalytical dependence on temperature and / or chemical potential. This type of crossover can also be seen from the scaling form of the free energy $f_{dis} = |\delta T|^{2-\alpha}\mathcal{G}(\frac{g}{\delta T^2})$ which allows for the identification of the specific heat exponent $\alpha = -1$ and the crossover exponent $\phi = 2$. The crossover function $\mathcal{G}(x)$ is regular for small x and has the asymptotic form $\mathcal{G}(x) \underset{x\to\infty}{\approx} (\frac{g}{\delta T^2})^{\frac{2-\alpha}{\phi}} (\mathcal{G}_\infty+\text{regular corrections})$. In the tricritical region the leading singularity in the free energy is given by $f_{TCP,sing} = \frac{4}{\sqrt{3}}(r_g g)^{\frac{3}{2}}$. For $\delta\mu = 0$ we have $g \sim \delta T$ and can read off the tricritical specific heat exponent $\alpha_3 = \frac{1}{2}$ from above.

Using $q = \delta\tilde{q} - \delta T$ one finds ordered solutions displaying the same crossover as discussed above, the transition from paramagnet to spin glass being continuous for positive δT and first order for negative temperature deviations. In the tricritical regime we find $q = \delta\tilde{q} = \frac{2}{3}\sqrt{-r_g g}$ which yields the tricritical order parameter exponent $\beta_3 = \frac{1}{2}$ and suggests that q and $\delta\tilde{q}$ act as order parameters simultaneously. From the fluctuation Lagrangian (see section 3) one reads off mass squared proportional to δT and hence $\gamma_3 = \beta_3 = \alpha_3 = \frac{1}{2}$.

3 Tricritical Landau Theory and the Parisi Solution of the Fermionic Ising Spin Glass

We derived a fluctuation theory for the tricritical and finite range ISG_f; for $T \neq 0$ a Lagrangian of the same structure is obtained for models including transport mechanism by integrating out dynamical degrees of freedom

$$
L = \frac{1}{t}\int d^d x [-\frac{3h^2}{2J}\sum Q^{ab} + \frac{r\kappa_1}{(\kappa_2)^2}\sum Q^{aa} + \frac{1}{2}\sum Q^{aa}(-\nabla^2 + u)Q^{aa}
$$

$$
+ \frac{1}{2}Tr'(\nabla Q^{ab})^2 - \frac{1}{t}\sideset{}{'}\sum Q^{aa}Q^{bb} - \frac{\kappa_1}{3}\sum(Q^{aa})^3
$$

$$
- \frac{\kappa_3}{3}Tr'Q^3 - \kappa_2 \sideset{}{'}\sum Q^{aa}Q^{ab}Q^{ba} + \frac{y_4}{4}\sideset{}{'}\sum(Q^{ab})^4], \tag{6}
$$

Here $4(\frac{\kappa_1}{t})^{(0)} = (\frac{\kappa_2}{t})^{(0)} = (\frac{\kappa_3}{t})^{(0)} = \frac{3^3}{2}$ and $u^{(0)} = 0$ denote the bare coefficients at tricriticality. One fourth order term relevant for replica symmetry breaking is kept. Replicas under $\sum'$ or Tr' are distinct. The $Q^{aa}Q^{bb}$–coupling is renormalization group generated as in the metallic quantum spin glass, its effects will be discussed in a subsequent section.

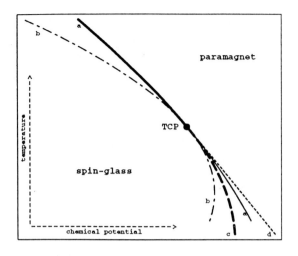

Fig. 1.
Vicinity of the tricritical point (TCP) for positive chemical potential. Continuous spin glass transitions occur on curve (a) above the TCP (thick unbroken line). Below the TCP first order thermodynamic transitions take place on curve (c). Curve (d), starting at the TCP, and curve (b) below the TCP limit the existence regime of ordered and disordered phases respectively.

Inserting the Parisi Ansatz for replica symmetrie breaking in the Lagrangian (6) the expansion of the replica symmetry broken saddle point free energy around the tricritical point reads

$$f - f_{TCP} = \mu - \mu_{c3} - \frac{3h^2}{2J}[\delta\tilde{q} - \int_0^1 q(x)dx] + J\{(\frac{3}{2}r_g g - r_\tau \tau^2)\delta\tilde{q}$$

$$+ \frac{3}{2}\tau[(\delta\tilde{q})^2 - \int_0^1 dx q^2(x)] - \frac{3}{2}[\int_0^1 dx[xq^3(x) + 3q(x)\int_0^x dy q^2(y)]$$

$$- 3\delta\tilde{q}\int_0^1 dx q^2(x) + \frac{1}{4}(\delta\tilde{q})^3] - \frac{y_4}{4}\int_0^1 dx q^4(x)\}, \tag{7}$$

Here variables and constants are defined as in (3), $\tau = \delta T/T_{c3}$. In contrast to crystal–field split spin glasses [3] the quartic coefficient y_4 of our free energy, Eq.(7), is nonzero and one obtains the Parisi solution $q(x) = \frac{9}{2y_4}x$ for $0 \leq x \leq x_1$ and $q(x) = q(1)$ for $x_1 \leq x \leq 1$. The plateau height is found to satisfy $q(1) = \delta\tilde{q} + O(\delta\tilde{q}^2)$. Consequently, plateau and breakpoint scale like $\sqrt{|\tau|} + O(\tau)$ at the TCP, while linear τ-dependence is reserved to $T_c > T_{c3}$. Adapting the notation of [4] we express our result for the irreversible response $q(1) - \int_0^1 q(x) \sim |\tau|^{\beta_\Delta}$ in terms of the exponent $\beta_{\Delta3} = 1$ for $T \to T_{c3}$ and $\beta_\Delta = 2$ for $T \to T_c > T_{c3}$. For the Almeida–Thouless line at tricriticality we find $\frac{H^2}{J^2} = \frac{80}{81}(\frac{2}{3}(1 - \frac{\mu_{c3}}{J}tanh(\frac{3\mu_{c3}}{J})))^{3/2}\tau_{AT}^{3/2} + O(\tau_{AT}^2)$ with $\tau_{AT} \equiv \frac{T_{c3} - T_{AT}(H)}{T_{c3}}$. Hence we obtain the critical exponent $\theta_3 = \frac{4}{3}$ near T_{c3}, while $\theta = \frac{2}{3}$ for all $T_c > T_{c3}$. These values do not satisfy the scaling relation $\theta_3 = \frac{2}{\beta_{\Delta3}}$ with $\beta_{\Delta3} = 1 + (\gamma_3 - \alpha_3)/2$. Along the lines described in [4], this problem of mean–field exponents will be resolved below by the renormalization group analysis of the coupling y_4 of the *finite–range and finite–dimensional* ISG_f.

4   Replica Symmetry Breaking for Fermions

The fermionic Ising spin glass allows for an exact evaluation of the bare fermion propagator $G_{ij,\sigma} = << \frac{1}{i\epsilon_n + \mu + \tilde{H}} >>$ at least in the disordered phase; here $\tilde{H}$ denotes the usual effective field and the double average refers to the replica–local and the Parisi block decoupling fields. The result can be written in the form

$$G_{ij,\sigma}(\epsilon_n) = -i\sqrt{\frac{\pi}{2J\tilde{q}}} \sum_{\lambda=0,\pm 1} \frac{(2-\lambda^2)e^{\frac{1}{2}(\beta J\lambda)^2 \tilde{q}} ch((1-\lambda^2)\beta\mu_\sigma)}{exp(\frac{1}{2}\beta^2 J^2 \tilde{q}) + ch(\beta\mu_\sigma)} \tag{8}$$

$$(1 - erfc(\frac{\epsilon_n - i\mu_\sigma + i\lambda\beta J^2 \tilde{q}}{J\sqrt{2\tilde{q}J}}))exp(\frac{\epsilon_n - i\mu_\sigma + i\lambda\beta J^2 \tilde{q}}{\sqrt{2\tilde{q}J}})^2 \delta_{ij},$$

where $\mu_\sigma \cong \mu + \sigma H$ includes a magnetic field H and $\epsilon_n = (2n+1)\pi k_B T/\hbar$. One easily extracts the disorder averaged electronic density of states [5]

$$< \rho(\epsilon) > = \frac{1}{\sqrt{2\pi\tilde{q}}J} e^{-\frac{(\epsilon+\mu)^2}{2J^2\tilde{q}}} \frac{ch(\beta\mu) + ch(\beta(\epsilon+\mu))}{ch(\beta\mu) + ch(\beta H)e^{\frac{1}{2}\beta^2 J^2 \tilde{q}}} \tag{9}$$

Below the freezing temperature $O(q^2)$–corrections occur in $< \rho >$. Statistical fluctuations $\delta\rho = \rho - < \rho >$ of the density of states are for example observable in $< \delta\rho_\sigma^a(\epsilon)\delta\rho_{\sigma'}^{a'}(\epsilon') >$. Taking Parisi symmetry breaking into account, this correlation becomes a function of the Parisi parameter x. Picking the one with $\sigma = \sigma'$ and $\epsilon = \epsilon'$ we obtain the (zero field) result

$$< \delta\rho_\sigma(\epsilon)\delta\rho_\sigma(\epsilon) > (x) = \frac{1}{4\pi J^2 \tilde{q}_c^5}(1 - \tilde{q}_c + \tilde{q}_c e^{-\frac{1}{2\tilde{q}_c}} cosh(\frac{\epsilon+\mu}{J\tilde{q}_c}))^2$$

$$(1 + \tilde{q}_c - \frac{(\epsilon+\mu)^2}{J^2})^2 e^{-\frac{(\epsilon+\mu)^2}{J^2\tilde{q}_c}} q^2(x) + O(q^3(x)). \tag{10}$$

This result shows that the fermionic density fluctuations reflect the irreversible magnetic response introduced by the spin glass order. It is valid for all fillings, one only has to insert the appropriate Parisi function and the filling–dependent value of the spin autocorrelation $\tilde{q}_c$ at the critical point; this implies that the scaling of the density correlator with $T - T_c$ is different in the second order regime (quadratic) and at the tricritical point (linear), while for the discontinuous regime a one–step RSB is expected. It is also clear that for models with transport mechanism the calculation of conductance fluctuations is of great interest. So far we have put aside the question of replica symmetry breaking of fermion propagators: by this we mean the possibility of a nonvanishing propagator between different replicas, ie $< \overline{\psi}^a \psi^{b \neq a} >$. In the light of the Mezard–Parisi instability of the random field Ising model we feel that this preferably would occur as a fluctuation effect, if at all.

5 Related Fermionic Spin Models

5.1 The Fermionic Ising Chain

Similarities between the phase diagrams of the clean BEG–model and of the fermionic Ising spin glass can be taken as indicative for the fact that rather spin dilution than disorder is the source of the tricritical crossover from continuous to discontinuous phase transitions. While it is complicated to solve 1D fermionic Ising spin glasses exactly, the clean fermionic Ising chain offers some simple exact solutions. Here we shall provide insight into the role of the chemical potential and moreover generalize known results into the complex μ–plane. Lee and Yang [7] derived the distribution of zeroes of the partition function of finite and infinite Ising chains within the complex magnetic field plane. Stimulated by the representation of conventional Ising chains by fermionic ones with special imaginary chemical potential, one may wish to extend the Yang Lee analysis to a fourdimensional space of complex (μ, ν). The transfer matrix $\mathbf{T_f}$ of the fermionic Ising chain reads

$$
T_f = e^{\beta\mu}
\begin{pmatrix}
e^{\beta\mu} & 1 & e^{\frac{1}{2}\beta(\mu+h)} & e^{\frac{1}{2}\beta(\mu-h)} \\
1 & e^{-\beta\mu} & e^{\frac{1}{2}\beta(h-\mu)} & e^{-\frac{1}{2}\beta(\mu+h)} \\
e^{\frac{1}{2}\beta(\mu+h)} & e^{\frac{1}{2}\beta(h-\mu)} & e^{\beta(J+h)} & e^{-\beta J} \\
e^{\frac{1}{2}\beta(\mu-h)} & e^{-\frac{1}{2}\beta(\mu+h)} & e^{-\beta J} & e^{\beta(J-h)}
\end{pmatrix}
\tag{11}
$$

The transfer matrices $\mathbf{T_f}$ and $\mathbf{T_s}$ of the standard $S = \pm 1$–chain and their eigenvalues do not map onto each other at $\mu = i\frac{\pi}{2}T$, while the partition functions obey $Z_f^{(N)} = TrT_f^N = (2i)^N Z_s^{(N)}(\mu = i\frac{\pi}{2}T)$ for any number N of sites. The largest eigenvalue determines the free energy of the infinite chain, while the second largest is required in addition to determine the correlation length. The eigenvalues for $h = 0$ are found as

$$
\lambda_{\pm} = e^{\beta\mu}[ch(\beta\mu) + ch(\beta J) \pm \sqrt{(ch(\beta\mu) + ch(\beta J))^2 + 4ch(\beta\mu)(1 - ch(\beta J))}]
$$
$$
\lambda_0 = 0 \quad , \quad \lambda_1 = 2e^{\beta\mu}sh(\beta J).
\tag{12}
$$

The correlation length is given by $\xi = 1/ln(\frac{\lambda_{\pm}}{\lambda_1})$. In the $T \to 0$–limit a transition arises at $\mu = J$ and due to the properties

$$
\lambda_1 \sim exp(\beta(\mu + J)), \quad \lambda_+ \sim
\begin{cases}
exp(\beta(\mu + J)), & \mu < J \\
exp(2\beta\mu), & \mu > J
\end{cases}
\tag{13}
$$

Thus ξ diverges only for $\mu < J$, since the energy for adding a fermion is larger than the gain from a magnetic bond if $\mu > J$. Hence

$$
\xi \sim
\begin{cases}
exp(\beta(2J - \mu)) & , 0 < \mu < J \\
exp(\beta J/2) & , \mu = J,
\end{cases}
\tag{14}
$$

while $\xi \sim T/(\mu - J)$ for $\mu > J$. The filling factor shows for $T \to 0$ that the system is completely filled for $\mu > J$ (empty for $\mu < -J$). Thus there is no physical $T = 0$–transition of this simple system. The correlation length diverges in the $T \to 0$ limit for all fillings ν. The zero–field partition function shows that Yang–Lee zeroes approach $\mu = \pm J$ for $T \to 0$. This means that for $(h = 0, T = 0)$ nonanalytic behaviour (as a function of the real chemical potential) can only occur at the values $\mu = \pm J$. It is instructive to consider $N = 2$ explicitly, which yields

$$Z_f^{N=2} = 4e^{2\beta\mu}[(ch(\beta\mu) + ch(\beta J))^2 + ch^2(\beta h)(e^{2\beta J} - 1) - sh(2\beta J)] \quad (15)$$

This almost trivial case already shows zeroes at $\mu_0 = \pm(J + (\frac{1}{2}ln2 \pm i\frac{\pi}{2})T)$, while allowing for finite complex magnetic field the first zero different from $\pm J$ in the $T \to 0$–limit becomes possible with $\mu = \pm(i\frac{\pi}{4} + 2im\pi)T$.More zeroes occur on the $T = 0$–axis as N is increased. For $N \to \infty$ a density function is expected in accordance with ξ diverging for any $|\mu| < J$.

5.2 Mapping the twodimensional Ising model with imaginary magnetic field $h = \frac{i\pi}{2}T$ into fermionic space

The complementary role of complex magnetic field and complex chemical potential can nicely be seen by recalling the exact solution for the 2d Ising model $h = i\frac{1}{2}\pi T$. This value corresponds to $\mu = i\frac{1}{2}\pi T$, which maps the fermionic Ising model onto the one above. Thus the exact solution of the 2d fermionic Ising model with $\mu = h = i\frac{1}{2}\pi T$ is known. Moreover this special model maps onto an interaction model of spinless fermions with a special species obeying bare Bose statistics but carrying along the minus signs of fermion interactions. The hamiltonian of this model can be written

$$H = -\sum_{ij} J_{ij}\sigma_i^z\sigma_j^z - \mu\sum_i(n_{i\uparrow} + n_{i\downarrow}) - h\sum_i\sigma_i^z \quad (16)$$

with $\mu = h = i\frac{\pi}{2}T$. This reduces to zero the imaginary field of one fermionic species, while the other field equals the distance between Bose– and Fermi–Matsubara energies. Setting $c = a_\uparrow$ and $d = a_\downarrow$ the hamiltonian reads

$$H = -\sum_{ij} J_{ij}[c_i^\dagger c_i c_j^\dagger c_j - c_i^\dagger c_i d_j^\dagger d_j - d_i^\dagger d_i c_j^\dagger c_j + d_i^\dagger d_i d_j^\dagger d_j] - i\pi T\sum_i d_i^\dagger d_i. \quad (17)$$

The new imaginary chemical potential $-i\pi T$ of d–fermions renders their Matsubara energies and single particle statistics bosonic. Thus anticommuting d–bosons interact with c–fermions. All vertices retain the fermionic minus–signs. In the fermionic path integral the c– and d–particles are described by the Grassmann fields $\psi_c(\tau)$ and by $\psi_d(\tau)$ respectively. The $i\pi T\hat{n}_d$–term can be absorbed by the phase transformation $exp(i\pi\tau)\psi_d(\tau) = \tilde{\psi}_d(\tau)$. The new anticommuting fields $\tilde{\psi}_d$ obey unusual **bosonic periodicity** $\tilde{\psi}_d(\tau) = \tilde{\psi}_d(\tau + \beta)$ on the imaginary time axis, while $\psi_c(\tau) = -\psi_c(\tau + \beta)$ remains fermionic. The bosonic feature of the anticommuting fields ψ_d's is of course seen in the bare statistics. Perturbatively the above exact conclusion maps the 2D Ising model with $h = i\pi T/2$ onto a coupled Fermi–Bose system (c–d) with additional (-1)–factors for each d–loop.

6 Metallic Quantum Spin Glass

So far we have discussed the magnetic phase diagram of fermionic lattice gases having in mind that the tricritical phenomenon and universal quantities are not changed by coherent hopping of the electrons in flat bands. On the other hand it is well known [8], [9] that both a random chemical potential and an increasing band width suppress the transition temperature continuously down to zero thereby producing a quantum phase transition (QPT). The field theory of such a QPT

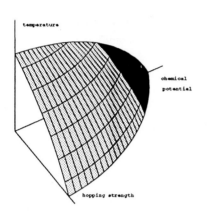

Fig. 2. *Phase diagram of fermionic spin glasses as function of chemical potential, hopping, and temperature. Above the marked area the system is in the paramagnetic, below in the spin glass phase. The shaded area corresponds to continuous transition, the black area to first order transitions. Tricritical points are located at the boundary of the two regions.*

shows surprising similarities with the thermal tricritical theory, above all the simultaneous occurence of critical charge and spin fluctuations [9]. However, electron hopping not only lowers the T_c of 2nd order transitions but also that of tricritical and 1st order transitions. For a metallic model with Gaussian random hopping and bandwidth $2E_0$ we found a quantum TCP at

$$E_F = (1 - (5/8)^{\frac{1}{2}})^{\frac{1}{2}} E_0, \qquad J_c = 3\pi E_0 [1 - E_F^2/E_0^2]^{-\frac{3}{2}}/32, \qquad (18)$$

Corrections to this Q - static approximation can be calculated by generalizing method [10]. The general phase diagram is sketched in figure 2.

7 Renormalisation Group Analysis

7.1 Tricritical Ising Spin Glass

We performed a 1–loop RG calculation for tricritical fluctuations. At each RG step the mass of charge fluctuations δQ^{aa} was shifted away. Introducing the anomalous dimensions $\tilde{\eta}$ and η for diagonal and offdiagonal fluctuations one finds at one loop level the following RG relations ($\epsilon = 8 - d$)

$$\frac{dr}{dl} = (\frac{d}{2} - 11\kappa_1^2 + 16\kappa_1\kappa_2 + 6\kappa_2^2)r - \kappa_2^2, \frac{du}{dl} = 2(1 - \kappa_1^2)u - 4\kappa_1^2 + 4\kappa_1\kappa_2,$$

$$\frac{d\kappa_1}{dl} = \frac{\epsilon}{2}\kappa_1 + 9\kappa_1^3, \frac{d\kappa_2}{dl} = (\frac{\epsilon}{2} + 6\kappa_2^2 - \kappa_1^2 + 16\kappa_1\kappa_2)\kappa_2, \frac{d\kappa_3}{dl} = (\frac{\epsilon}{2} + 9\kappa_2^2)\kappa_3,$$

Above $d = 8$ the RG flows towards the Gaussian fixed point with mean field exponents, for $d < 8$ there is no perturbatively accessible fixed point for real $\kappa's$. However, a preliminary analysis of the resulting strong coupling problem shows that there exists a solution with positive $\tilde{\eta}$ in contrast to the negative anomalous dimension typical of cubic field theories with imaginary coupling. The RG for the DIC y_4 showed that its long–distance behaviour is dominated by a κ^4–contribution (like in [4] but) for $d_c^{(u)} = 8 < d < 10$. This leads to the modified MF exponent $\theta_3 = 8/(d-4)$, which satisfies the scaling relation $\theta_3 = 2/\beta_{\Delta_3}$ in $d_{c3}^{(u)} = 8$ and reduces to the MF–result in 10 dimensions. The dimensional shift by 2 in comparison with [4] is due to coupling t.

7.2 Metallic Quantum Spin Glass

In the theory of quantum phase transitions time–dependent fluctuations are treated on an equal footing with spatial fluctuations [9]. While for the Lagrangian (7) it was sufficient to consider only the $\omega = 0$ - component of the Q - fields, in the quantum case all low energy fields must be kept as they are coupled via the quantum mechanical interaction $u \int d\tau (Q^{aa})^2$. However, the value $z = 4$ of the dynamical critical exponent renders the u - coupling dangerously irrelevant and allows for a perturbative mapping of the critical theory on a classical problem, the Pseudo - Yang - Lee edge singularity. This field theory has only one cubic coupling which corresponds to κ_1 in eq(7). The comparison of the metallic quantum case with the thermal tricritical theory allows one to understand the nature of the strong coupling RG fixed point: whereas the spin fluctuations in the thermal second order regime are governed by a perturbatively accessible fixed point, the TCP and the quantum case are charcterized by the combination of charge and spin fluctuations and a corresponding strong coupling problem.

References

[1] Oppermann R., Müller - Groeling A. (1993): Nucl. Phys. **B401**, 507
[2] Blume M., Emery V.J., Griffiths R.B. (1971): Phys. Rev. A **4**, 1071
[3] Mottishaw P. J., Sherrington D. (1985): J. Phys. C **18**, 5201
[4] Fisher D. S., Sompolinsky H. (1985): Phys. Rev. Lett. **54**, 1063
[5] Rosenow B., Oppermann R. (1996): Phys. Rev. Lett. to be published
[7] Lee T. D., Yang C. N. (1952): Phys. Rev. **87**, 410
[8] Oppermann R., Binderberger M. (1994): Ann. Phys. **3**, 494
[9] Sachdev S., Read N., Oppermann R, (1995): Phys. Rev. B **52**, 10286
[10] Huse D., Miller J. (1993): Phys. Rev. Lett. **70**, 3147

Polymer Winding Numbers
and Quantum Mechanics

David R. Nelson[1] and Ady Stern[2]

[1]Lyman Laboratory of Physics, Harvard University, Cambridge, MA 02138
[2]Department of Condensed Matter Physics, Weizmann Institute of Sciences, Rehovot 76100, Israel

Abstract. The winding of a single polymer in thermal equilibrium around a repulsive cylindrical obstacle is perhaps the simplest example of statistical mechanics in a multiply connected geometry. As shown by S.F. Edwards, this problem is closely related to the quantum mechanics of a charged particle interacting with a Aharonov-Bohm flux. In another development, Pollock and Ceperley have shown that boson world lines in $2 + 1$ dimensions with periodic boundary conditions, regarded as ring polymers on a torus, have a mean square winding number given by $\langle W^2 \rangle = 2n_s \hbar^2 / mk_B T$, where m is the boson mass and n_s is the superfluid number density. Here, we review the mapping of the statistical mechanics of polymers with constraints onto quantum mechanics, and show that there is an interesting generalization of the Pollock-Ceperley result to directed polymer melts interacting with a repulsive rod of radius a. When translated into boson language, the mean square winding number around the rod for a system of size R perpendicular to the rod reads $\langle W^2 \rangle = \frac{n_s \hbar^2}{2\pi m k_B T} \ln(R/a)$. This result is directly applicable to vortices in Type II superconductors in the presence of columnar defects. An external current passing through the rod couples directly to the winding number in this case.

1 Introduction

The study of the statistical mechanics of polymers in multiply connected geometries began many years ago with work by S.F. Edwards and by S. Prager and H.L. Frisch [1], [2]. The simplest nontrivial geometry consists of a polymer interacting with a repulsive rod, and the corresponding path integrals can be analyzed via an elegant analogy with the physics of a quantum mechanical particle interacting with a solenoidal vector potential [3]. The mean square winding number of the polymer around the rod, and other interesting quantities can be computed for this problem [4]. The physics bears a close mathematical relation to the famous Aharonov-Bohm effect for a real quantum mechanical particle interacting via its charge with a tube of magnetic flux [5].

More recently, Pollock and Ceperley have studied the winding numbers with respect to periodic boundary conditions for boson world lines in the Feynman path integral formulation of superfluidity in two dimensions [6]. The physics here is equivalent to *many* ring polymers interacting on the surface of a torus. The mean square winding number of the world lines around the torus can be expressed exactly in terms of the renormalized superfluid density of the equivalent boson system [7].

Winding numbers and the statistical mechanics of many polymer-like objects are also relevant to the physics of vortex lines in Type II superconductors [8]. Here, thermal fluctuations in the trajectory of a vortex defect in the superconducting order parameter can be described by a Feynman path integral for an elastic string. A collection of many such lines behaves like a directed polymer melt, with the added complication of a quenched random disorder potential [9]. Point-like disorder is always present to some degree due, for example, to quenched fluctuations in the concentration of oxygen vacancies in the high temperature cuprate superconductors. Drossel and Kardar have studied how the winding number distribution of a single directed polymer around various obstacles is affected by point-like disorder [10].

Dramatic improvements in the pinning efficiency of vortex lines have recently been achieved via the introduction of *columnar* damage tracks created by heavy ion irradiation. If the concentration of damage tracks (assumed to pass completely through the sample) exceeds the number flux lines, there is a low-temperature "Bose glass" phase, in which every vortex is trapped on a columnar defect [11]. At high temperatures, the vortices delocalize in an entangled flux liquid. If these vortex trajectories are viewed as the world lines of quantum mechanical particles, the physics in thick samples becomes equivalent to that of a low temperature boson superfluid in $2+1$ dimensions [9].

In this paper, we discuss how to compute polymer winding numbers from the Hamiltonian formulation of quantum mechanics, keeping in mind applications to vortex lines interacting with many columnar defects. A random distribution of columns maps onto a time-independent random potential in the quantum mechanical analogy [11]. We assume high temperatures, and samples which are clean before irradiation, so that point disorder is negligible and the usual Abrikosov flux lattice is either melted by thermal fluctuations or destroyed by the columnar disorder. An experimental realization of a multiply connected geometry for thermally excited flux lines is illustrated in Fig. 1. Superimposed on a dilute concentration of parallel columnar defects scattered randomly throughout a superconducting sample is a slender tube in which the columns are very dense. Such a "tube of columns" could be made by covering a sample with a mask containing a small submicron hole during irradiation with a strong dose of heavy ions. Imagine that this sample is first subjected to a very large magnetic field, such that the density of flux lines is approximately equal to the density of columns in the tube. It should be possible to choose the temperature such that the vortices in the tube are in the Bose glass phase, while those outside constitute a flux liquid. The many trapped vortices inside the tube should then present a virtually impenetrable barrier to the thermally excited vortex lines outside. The concentration of lines in the liquid outside this repulsive cylindrical obstacle could be varied by decreasing the magnetic field. Because flux creep in the Bose glass phase is quite slow [8], [11], the concentration of vortices in the tube should remain approximately constant as a field is turned down, leaving the barrier almost unchanged.

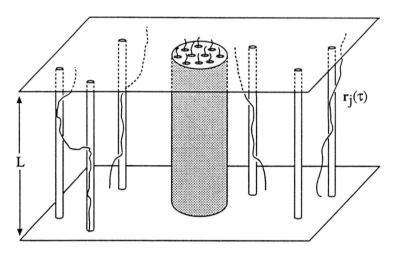

Fig. 1. Slab of Type II superconductor with a dense array of columnar pins confined to a cylinder with a dilute concentration of columnar defects outside. Vortices inside the cylinder are in the Bose glass phase, while those outside are free to move and entangle in a flux liquid

We shall explore the winding number fluctuations of the vortices with respect to the repulsive cylindrical tube in this experiment. As discussed below, a current passing through the tube couples directly to the net vortex winding number, and the mean square winding number of the unperturbed system gives the linear response of the net winding number to this current. (See Fig. 2) Such a current acts like an imaginary vector potential when this problem is mapped onto quantum mechanics [11]. Winding of vortices around a thin repulsive obstacle (or a set of such obstacles) could be probed via double-sided flux decorations [12] or indirectly by monitoring the flux flow resistivity in the plane perpendicular to the common direction of the applied field and the columns. The extra entanglements induced by the longitudinal current should impede vortex transport in this plane, and it would be especially interesting to look for changes in the in-plane resistivity as a function of the current through the tube [13]. This resistance should drop with increasing longitudinal current through the tubes, due to the enhanced vortex winding about the obstacle.

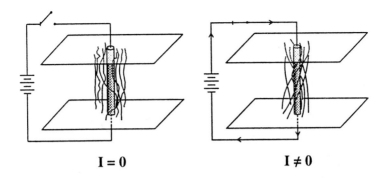

Fig. 2. Effect of a current through a cylindrical obstacle on a flux liquid. The lines wind around the obstacle with a preferred chirality when the current is on

In section II, we briefly review the Feynman path integral description of bosons in $2+1$ dimensions, and discuss the Pollack-Ceperley result for the winding numbers of bosons on a torus. We then describe the closely related statistical mechanics of directed polymer melts and vortex lines, and show how a current through a cylindrical obstacle couples to the winding number. In section III, we treat winding number statistics for isolated polymers both with and without columnar pin disorder. Results for *many* polymers or flux lines winding about an obstacle are presented in section IV.

2 Two-Dimensional Bosons and Directed Polymer Melts

2.1 Boson Statistical Mechanics

The partition function for a set of N nonrelativistic bosons interacting with pair potential $V(r)$ reads

$$\mathcal{Z} = Tr'\{e^{-\beta\hat{\mathcal{H}}_b}\}, \tag{1}$$

where $\beta = 1/k_B T$ and the boson Hamiltonian operator is

$$\hat{\mathcal{H}}_b = \sum_{j=1}^{N} \frac{-\hbar^2}{2m} \nabla_j^2 + \frac{1}{2} \sum_{i \neq j} V(|\mathbf{r}_i - \mathbf{r}_j|). \tag{2}$$

The prime on the trace means that only symmetrized boson eigenfunctions are to be included in the partition sum, and we shall focus on particles in two space dimensions. This trace can be rewritten in terms of a Feynman path integral by breaking up $\exp[-\beta\hat{\mathcal{H}}]$ into M pieces ($M \gg 1$),

$$e^{-\beta\hat{\mathcal{H}}_b} = e^{-\Delta\tau\hat{\mathcal{H}}_b} e^{-\Delta\tau\hat{\mathcal{H}}_b} \cdots e^{-\Delta\tau\hat{\mathcal{H}}_b} \qquad (M \quad \text{times}) \tag{3}$$

where $\Delta\tau M = \beta$. Upon inserting complete sets of position states between various terms in the product and taking the limit $M \to \infty$, the boson partition function $\mathcal{Z}_b$ may be expressed as an integral over a set of polymer-like trajectories $\{\mathbf{r}_j(\tau) = [x_j(\tau), y_j(\tau)]\}$ in imaginary time [14],

$$\mathcal{Z}_b = \frac{1}{N!} \sum_P \prod_{j=1}^{N} \mathcal{D}\mathbf{r}_j(\tau) \exp\left[- - -\frac{1}{2}\frac{m}{\hbar} \sum_{j=1}^{N} \int_0^{\beta\hbar} \left(\frac{d\mathbf{r}_j}{d\tau}\right)^2 d\tau \right.$$

$$\left. -\frac{1}{\hbar} \sum_{i>j} \int_0^{\beta\hbar} V(|\mathbf{r}_i(z) - \mathbf{r}_j(z)|) d\tau \right] \tag{4}$$

The normalized sum over permutations P insures that only boson eigenfunctions contribute to the sum, where the trajectories obey a type of periodic boundary condition,

$$\{\mathbf{r}_j(\beta\hbar\} = P[\{\mathbf{r}_i(0)\}], \tag{5}$$

and the operator P permutes the set of starting points $\{\mathbf{r}_i(0)\}$.

An approximate picture of the polymer statistical mechanics problem represented by (4) can be constructed as follows [14]: Provided interaction effects are not too large, each "polymer" simply diffuses in the imaginary time variable τ,

$$\langle|\mathbf{r}_j(\tau) - \mathbf{r}_j(0)|^2\rangle \approx 2\frac{\hbar}{m}\tau \tag{6}$$

where $\hbar/m$ plays the role of a diffusion constant. When the temperature is high, only the identity permutation contributes to Eq. (4). When projected onto the (x, y)-plane, the boson trajectories then behave like small *ring* polymers of typical transverse size given by the thermal deBroglie wavelength Λ_T,

$$\Lambda_T^2 \equiv 2\pi\hbar^2/mk_B T$$
$$\sim \langle|\mathbf{r}_j(\tau) - \mathbf{r}_j(0)|^2\rangle|_{\tau=\beta\hbar}. \tag{7}$$

At temperatures low enough so that $\Lambda_T \gtrsim n^{-1/2}$, where n is the particle number density, complicated cyclic permutations appear, as the trajectories coalesce to form much larger rings. Feynman suggested in 1953 that the lambda transition from a normal bulk liquid of He^4 to a superfluid is associated with a dramatic proliferation in the number and length of such cooperative ring exchanges [15].

Now, following Pollock and Ceperley, consider what happens when the excursions of the bosons in the xy-plane occur in a two-dimensional periodic box of size D [6], [7] (see Fig. 3). The permutation requirement (6) describing periodic boundary conditions in the τ-direction remains in effect. At high temperatures, virtually all ring polymers return to their initial positions $\{r_i(0)\}$ when $\tau = \beta\hbar$, and the spatial periodic boundary conditions are unimportant. In the low temperature limit, however, ring exchanges lead to huge composite trajectories which typically wrap completely around the torus embodied in the (x, y)-plane periodic boundary conditions. The collection of cyclic boson trajectories can be classified by a dimensionless topological invariant, the vector winding number $\mathbf{W}$,

$$\mathbf{W} = \frac{1}{\sqrt{\Omega}} \sum_{j=1}^{N} [\mathbf{r}_j(\beta\hbar) - \mathbf{r}_j(0)] \tag{8}$$

where $\Omega = D^2$ is the cross-sectional area of a square box with periodic spatial boundary conditions. In evaluating Eq. (8), we imagine that the $\{r_j(\tau)\}$ pass smoothly into neighboring periodic cells, without invoking the periodic boundary conditions (see Fig. 3). Consider the mean square winding number $\langle \mathbf{W}^2 \rangle$, where the angular brackets represent a path integral weighted by the exponential factor in Eq. (4) and divided by the boson partition function. Ring polymers which do not wrap completely around the torus make no contribution to Eq. (8). However, at temperatures low enough so that $\Lambda_T \gg n^{-1/2}$, virtually all trajectories belong to a cycle with a nontrivial winding number. We can then regard Eq. (8) as a normalized N-step random walk with typical step size Λ_T, and estimate that $\langle \mathbf{W}^2 \rangle \approx \frac{N\Lambda_T^2}{\Omega} = 2\pi n\hbar^2/mk_BT$, where $n = N/\Omega$ is the number density of bosons. Pollock and Ceperley showed that $\langle \mathbf{W}^2 \rangle$ is in fact *exactly* related to the *superfluid density* n_s of the equivalent boson system,

$$\langle \mathbf{W}^2 \rangle = 2n_s(T)\hbar^2/mk_BT$$
$$= 2n_s(T)(\hbar^2/m)\beta. \tag{9}$$

The implications of this remarkable connection between a *topological invariant* for a set of ring polymers on a torus and the *superfluid density* of the equivalent set of bosons is summarized in Fig. 4, where $\langle \mathbf{W}^2 \rangle$ is shown as a function of β, which is proportional to the number of "monomers" if the trajectories are viewed as ring polymers. Translational invariance of the boson system in the absence of disorder implies that $\lim_{T \to 0} n_s(T) = n$, [16], so we know that $\langle \mathbf{W}^2 \rangle$ diverges linearly with β as $\beta \to \infty$, with slope given exactly by $2n\hbar^2/m$. The existence of a sharp Kosterlitz-Thouless phase transition in superfluid helium films, moreover, implies that there must be a singularity in $\langle \mathbf{W}^2 \rangle$: This quantity *vanishes* below a critical value β_c, and the universal jump discontinuity in the superfluid density [17], $\lim_{T \to T_c^-} n_s(T)/T = \frac{2}{\pi} k_B \frac{m}{\hbar^2}$, implies the exact result [6]

$$\lim_{\beta \to \beta_c^+} \langle \mathbf{W}^2 \rangle = \frac{4}{\pi}. \tag{10}$$

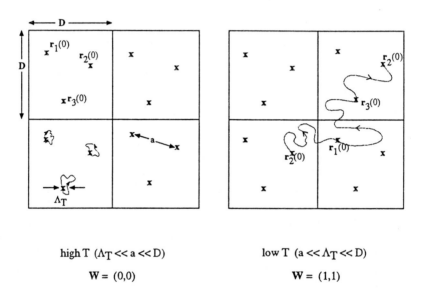

high T $(\Lambda_T \ll a \ll D)$ low T $(a \ll \Lambda_T \ll D)$

$\mathbf{W} = (0,0)$ $\mathbf{W} = (1,1)$

Fig. 3. Boson trajectories in $2+1$ dimensions projected onto the xy-plane. Three trajectories with characteristic interparticle distance a and four images of the periodic box of size D are shown. The periodic boundary conditions in the time-like direction mean that only small ring trajectories appear in the high temperature image at left. Trajectories which cross between the periodic cells and lead to nonzero winding numbers are shown in the low-temperature image on the right

Feynman hoped that his "cyclic ring exchange" picture of superfluidity could be used to understand the lambda transition in He^4 which was a quite mysterious phenomenon in 1953 [14], [15]. Here, the sophisticated understanding of vortex unbinding transitions in He^4 films developed in the past 25 years (using other methods) has been used to make a prediction about a topological quantity, the winding number.

Equation (9) allows winding number statistics extracted from computer simulations of bosons with periodic boundary conditions to be converted into measurements of the superfluid density [7]. In can be used, in particular, to probe the reduction in the superfluid density of helium films due to substrate disorder near $T = 0$. A disordered substrate potential maps onto randomness *correlated* along the imaginary time direction τ in the 2+1-dimensional world line picture of boson physics. The winding number fluctuations are reduced because this correlated randomness reduces the wandering of the boson world lines.

2.2 Directed Polymer Melts

One might think that the results described above would be applicable, at least in principle, to real ring polymers on the surface of a torus, whose fluctuations include fusing together to form large rings, similar to sulfer ring polymers in equilibrium [18]. Unfortunately, the boson model as applied to ring polymers is unrealistic: The individual "monomers" within a ring, indexed by the imaginary time coordinate τ in Eq. (4), are noninteracting, so there is no intra-chain self-avoidance. The only *inter*polymer interactions, moreover, occur between monomers with *same* imaginary time coordinate, which is also unrealistic for a melt of ring polymers.

The physics of directed polymer melts, on the other hand, is much closer to that of real bosons. Because the polymers are directed, interpolymer interactions at the same "imaginary time" coordinate dominate the physics. Vortex lines in superconductors provide an excellent physical realization of this system, but examples can also be found in polymer nematics in strong external magnetic or electric fields [19]. Consider a collection of N vortex lines or polymers in three space dimensions, labelled by x, y and τ (see Fig. 5). We assume that these lines are stretched out on average along the τ axis, so that they can be described by single-valued trajectories $\{\mathbf{r}_j(\tau)\}$. The partition function is a multidimensional path integral, similar to Eq. (4),

$$\mathcal{Z} = \prod_{j=1}^{N} \int \mathcal{D}\mathbf{r}_j(\tau) \exp[-F/T] \tag{11}$$

where

$$F = \frac{1}{2}g \sum_{j=1}^{N} \int_0^L \left(\frac{d\mathbf{r}_j}{dz}\right)^2 d\tau + \sum_{j=1}^{N} \int_0^L U(\mathbf{r}_j)d\tau + \sum_{i>j} \int_0^L V(|\mathbf{r}_i - \mathbf{r}_j|)d\tau. \tag{12}$$

Here, the mass m in the boson problem has been replaced by a line tension g, $\hbar$ has been replaced by the temperature T and the polymer system thickness L plays the role of $\beta\hbar$ for the bosons. Unless indicated otherwise, we shall henceforth use units such that $k_B = 1$. $V(r)$ is a repulsive interparticle potential and $U(\mathbf{r})$ represents the random potential due to columnar pins (not shown in Fig. 5). Both $V(r)$ and $U(\mathbf{r})$ are independent of τ. This approximation is quite reasonable for directed polymer melts and vortex lines in $2 + 1$ dimensions, provided $\langle |\frac{d\mathbf{r}}{dz}|^2 \rangle$, the mean square tilt away from the z axis, is small [11].

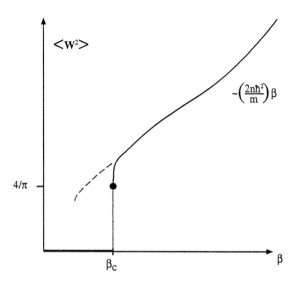

Fig. 4. Trajectories for directed polymer melts with "open" boundary conditions at the top and bottom oif a slab of thickness L

The statistical mechanics of directed polymers differs from that for bosons due to the absence of periodic boundary conditions in the "imaginary time" direction τ: we assume free boundary conditions in Eq. (11), i.e., we integrate freely over the starting and end points of the polymers. Hence there is no sum over permutations and no condition analogous to Eq. (5). This change, however, is less severe than one might think. Indeed, the directed polymer partition function can be written in a form similar to Eq. (1), [9]

$$\mathcal{Z} = \prod_{j=1}^{N} \int d\mathbf{r}'_j \prod_{j=1}^{N} \int d\mathbf{r}_j \langle \mathbf{r}'_1 \cdots \mathbf{r}'_N | e^{-\hat{\mathcal{H}}L/T} | \mathbf{r}_1 \cdots \mathbf{r}_N \rangle \tag{13}$$

where

$$\hat{\mathcal{H}} = -\frac{T^2}{2g} \sum_{j=1}^{N} \nabla_j^2 + \sum_{j=1}^{N} U(\mathbf{r}_j) + \sum_{i>j} V(|\mathbf{r}_i - \mathbf{r}_j|), \tag{14}$$

and the states $|\mathbf{r}_1 \cdots \mathbf{r}_N\rangle$ and $\langle \mathbf{r}'_1 \cdots \mathbf{r}'_N|$ describe entry and exit points for the polymers or vortices at the top and bottom of the system. The Hamiltonian $\mathcal{H}$ is identical to the Hamiltonian $\mathcal{H}_b$ in Eq. (2), provided we introduce a disorder potential $U(\mathbf{r})$ to model the effect of, say, a random substrate on the bosons. Because the initial and final states involve *symmetric* integrations over the entry and exit points, only boson eigenfunctions contribute to the statistical mechanics defined by Eq. (13). As, $L \to \infty$, the physics will be dominated by a bosonic ground state and bosonic low-lying excitations, just as in Eq. (1). Thus, the precise choice of boundary condition is irrelevant in the "thermodynamic limit" of large system sizes.

Can one define meaningful "winding number" problems for directed polymers with free ends? Because we are no longer dealing with ring polymers, the "winding number" $\mathbf{W}$ in Eq. (8) (with $\beta\hbar$ replaced by L) no longer has a precise topological meaning, even for periodic boundary conditions in the space-like directions. One can, however, still define and estimate the order of magnitude of the fluctuations in $\mathbf{W}$ as before. In terms of quantities appropriate for directed polymer melts, we have, up to constants of order unity

$$\langle \mathbf{W}^2 \rangle \approx nTL/g, \tag{15}$$

where n is the areal number density of polymers in a constant τ cross section. As illustrated (in the absence of a disorder potential) in Fig. 4, we expect that this quantity is nonzero for *all* values of $L \equiv \beta\hbar > 0$. Although the curves for directed polymers and bosons should agree when L and $\beta\hbar$ are large, there is no reason to expect a sharp Kosterlitz-Thouless phase transition for "bosons" with open boundary conditions [12]. The winding number variance for real bosons only vanishes identically for $\beta < \beta_c$ because the contribution to $\mathbf{W}$ from "ring polymer" trajectories which are not wrapped completely around the torus is exactly zero. There is no such condition for the open boundary conditions appropriate to a directed polymer melt.

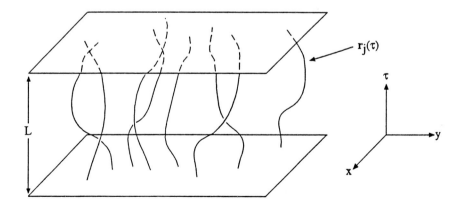

Fig. 5. Mean square winding number $\langle \mathbf{W}^2 \rangle$ for bosons in 2+1 dimensions a function of $\beta = 1/k_B T$. A Kosterlitz-Thouless transition occurs for $\beta = \beta_c$. Dashed line is the corresponding quantity for directed polymer melts

2.3 Winding Around an Obstacle

A more interesting case is the winding number distribution for many directed polymers in a disorder potential and interacting with a repulsive cylindrical obstacle. As discussed in the Introduction, this problem has a direct physical realization in the context of thermally excited vortex lines. The combination of a disorder potential and repulsive barrier is illustrated in Fig. 1. We model a repulsive obstacle by modifying the disorder potential in Eqs. (12) and (14),

$$U(\mathbf{r}) \to U(\mathbf{r}) + U_B[\mathbf{r}(\tau)], \tag{16}$$

where the barrier potential $U_B(\mathbf{r})$ (centered on the origin) is, e.g.,

$$U_B(r) = \begin{cases} \infty, & r < a \\ 0, & \text{otherwise.} \end{cases} \tag{17}$$

This problem is a natural generalization to many directed lines of the original multiply connected geometry problem for single polymers of Edwards [1] and of Prager and Frisch [2]. The (scalar) winding number with respect to the obstacle at the origin is now

$$\mathbf{W} = \sum_{j=1}^{N} \int_0^L d\tau \mathbf{A}[\mathbf{r}_j(\tau)] \cdot \frac{d\mathbf{r}_j}{d\tau}, \tag{18}$$

where the "vector potential" $\mathbf{A}(r)$ is [1]

$$\mathbf{A}(\mathbf{r}) = \frac{\hat{\tau} \times \mathbf{r}}{2\pi r^2}. \tag{19}$$

The boundary conditions along the τ direction can be periodic or open [3]. Periodic boundary conditions would imply integer winding numbers, but the noninteger winding numbers which characterize open boundary conditions are of direct physical interest for vortices, as we now explain.

A current in the τ direction through a cylindrical obstacle couples directly to the winding number of vortices outside in Type II superconductors. To see this, assume that the current is entirely confined to the cylinder, i.e., it cannot penetrate into the remainder of the sample. This "confined current approximation" should be applicable to a tube of columnar pins filled with pinned vortices in the bose glass phase, provided its conductivity is much higher than that of the flux liquid outside. Essentially infinite conductivity can be arranged in the tube because irradiation to produce columnar defects with matching field ~ 4 Tesla typically shifts the irreversibility temperature by 5–20°K. Choosing a temperature *above* the irreversibility line of the lightly irradiated region outside the tube but *below* the Bose glass transition inside insures that essentially all the current is confined to the tube. Under these conditions the additional free energy F of the vortices due the current is given by [21]

$$\delta F = -\frac{1}{4\pi} \int' d^2r \int_0^L d\tau \mathbf{H}_{\text{ext}}(\mathbf{r}) \cdot \mathbf{B}_\perp(\mathbf{r}, \tau) \tag{20}$$

where $\mathbf{H}_{\text{ext}}(\mathbf{r})$ is the field due to the extra current I through the wire

$$\mathbf{H}_{\text{ext}}(\mathbf{r}) = \frac{2I}{c} \frac{\hat{\tau} \times \mathbf{r}}{r^2}, \tag{21}$$

and the prime on the integration means that the region inside the obstacle is excluded. The perpendicular component of the magnetic field $\mathbf{B}_\perp$ due to the vortices obeys the anisotropic London equation [22]

$$\mathbf{B}_\perp(\mathbf{r}, \tau) = \phi_0 \sum_{j=1}^N \frac{d\mathbf{r}_j}{d\tau} \delta^{(2)}[\mathbf{r} - \mathbf{r}_j(\tau)] + \frac{M_\tau}{M_\perp}\lambda^2 \nabla_\perp^2 \mathbf{B}_\perp(\mathbf{r}, \tau) + \lambda^2 \partial_\tau^2 \mathbf{B}_\perp(\mathbf{r}, \tau), \tag{22}$$

where ϕ_0 is the flux quantum, $\frac{M_\tau}{M_\perp}$ is the mass anisotropy and λ is the screening length in the direction perpendicular to $\hat{\tau}$. Upon substituting Eqs. (21) and (22) into (20), integrating by parts and noting that $\left(\frac{M_\tau}{M_\perp}\lambda^2\nabla_\perp^2 + \lambda^2\partial_\tau^2\right)\mathbf{H}_{\text{ext}}(\mathbf{r}) = 0$ in the domain of integration, we find that

$$\delta F = -\frac{I\phi_0}{c} \sum_{j=1}^N \int_0^L d\tau \frac{\hat{\tau} \times \mathbf{r}_j}{2\pi r_j^2} \cdot \frac{d\mathbf{r}_j}{d\tau}$$
$$\equiv -\gamma W, \tag{23}$$

where we have defined a winding number coupling constant γ,

$$\gamma = I\phi_0/c. \tag{24}$$

Consider now the mean winding number $\langle W \rangle$ induced by the current for vortices outside the obstacle. Upon incorporating an additional factor $e^{-\delta F/T}$ into the Boltzmann weight in (11), we have, upon expanding $\langle W \rangle$ to leading order in the current,

$$\langle W \rangle = \frac{\phi_0}{2\pi cT} \langle W^2 \rangle|_{I=0} I. \tag{25}$$

Thus the equilibrium winding number *fluctuations* $\langle W^2 \rangle|_{I=0}$ determine the linear response of the net winding number to an external current.

3 Winding Number and Quantum Mechanics: Single Polymer in a Cylindrical Shell

We now study the winding number fluctuations of a single directed polymer, moving on average along the τ axis, and interacting with a repulsive cylindrical obstacle of radius a centered on the origin of an (x, y) plane perpendicular to $\hat{\tau}$. As discussed above, many of the results are directly applicable to vortex filaments in superconductors. We use the original method of Edwards [1], which maps the generating function for winding numbers onto the quantum mechanics of particles interacting with a solenoidal vector potential. Although winding of a single polymer is simple and relatively well understood [1], [2], [3], [4], [10], we review the basic results to illustrate the method and discuss effects due to columnar pins outside the obstacle. We use the *directed* polymer notation discussed at the end of section II because the neglect of self-avoidance is justified in this case. The single line results would, however, apply in principle to a "phantom" polymer without self-avoidance in two dimensions interacting with a repulsive disk and a random substrate potential. We defer new results for *many* directed polymers to section IV. We shall, for simplicity, usually impose periodic boundary conditions in the τ direction. This choice should not affect the results for $\langle W^2 \rangle$ in the limit of large sample thicknesses L. For a systematic study of the effects of different boundary conditions in the context of the boson mapping, see ref. [23].

3.1 Winding Number Formalism

Consider the winding number (18), specialized to the case of a single directed polymer with trajectory $\mathbf{r}(\tau)$,

$$W = \int_0^L d\tau \mathbf{A}[\mathbf{r}(\tau)] \cdot \dot{\mathbf{r}}(\tau) \tag{26}$$

with $\mathbf{A}(\mathbf{r}) = \frac{\hat{\tau} \times \mathbf{r}}{2\pi r^2}$ and $\dot{\mathbf{r}}(\tau) = d\mathbf{r}/d\tau$. Note that $\mathbf{A}(\mathbf{r})$ is the electro-magnetic vector potential that would have been present had we replaced the cylindrical obstacle by a solenoid enclosing one unit of magnetic flux. If we then regard τ as time, our expression for W becomes the quantum mechanical Aharonov-Bohm phase accumulated by a particle of unit charge traversing a path $\mathbf{r}(\tau)$, due to its interaction with the solenoid. As noted by Aharonov and Bohm [5], this phase is indeed the winding number, which is an integer for periodic boundary conditions, but can assume arbitrary values otherwise.

Our interest is, of course, not in the winding number corresponding to a single trajectory, but rather in the statistics of winding numbers. More concretely, we wish to study the distribution of W, averaged over all paths, with the appropriate Boltzmann weight. The winding number variance for a single polymer with line tension g is

$$\langle W^2 \rangle = \frac{\int \mathcal{D}\mathbf{r}(\tau) \left[\int d\tau \, [\dot{\mathbf{r}} \cdot \mathbf{A}(\mathbf{r}(\tau)]^2 \, e^{-\int_0^L \mathcal{L}_0[\mathbf{r}(\tau),\dot{\mathbf{r}}(\tau)]d\tau/T} \right]}{\int \mathcal{D}\mathbf{r}(\tau) e^{---\int_0^L \mathcal{L}_0[\mathbf{r}(\tau),\dot{\mathbf{r}}(\tau)]d\tau/T}}, \qquad (27)$$

where the free energy density

$$\mathcal{L}_0 = \frac{1}{2} g \left(\frac{d\mathbf{r}}{d\tau} \right)^2 + U(\mathbf{r}) \qquad (28)$$

plays the role of a Lagrangian in the imaginary time path integral formulation of quantum mechanics. Here, $U(\mathbf{r})$ includes the potential due to the columnar (i.e. τ-independent) disorder and a part representing the excluded volume interaction with the cylinder, as in Eq. (16).

Upon defining a new function

$$\mathcal{L}_\nu[\mathbf{r}(\tau), \dot{\mathbf{r}}(\tau)] = \mathcal{L}_0 - - - iT\nu \frac{d\mathbf{r}}{d\tau} \cdot \mathbf{A}[\mathbf{r}(\tau)], \qquad (29)$$

we have

$$\langle W^2 \rangle = -\frac{\partial^2}{\partial \nu^2} \ln \mathcal{Z}(\nu)|_{\nu=0}, \qquad (30)$$

where

$$\mathcal{Z}(\nu) = \int \mathcal{D}\mathbf{r}(\tau) \exp \left[- - -\frac{1}{T} \int_0^L \mathcal{L}_\nu[\mathbf{r}(\tau), \dot{\mathbf{r}}(\tau)]d\tau \right]. \qquad (31)$$

It is straightforward using standard path integral manipulations to express $\mathcal{Z}(\nu)$ in terms of $\hat{\mathcal{H}}(\nu)$, the Hamiltonian operator associated with $\mathcal{L}_\nu$,

$$\hat{\mathcal{H}}(\nu) = \frac{T^2}{2g} \left[\frac{1}{i}\nabla + \nu \mathbf{A}(\mathbf{r}) \right]^2 + U(\mathbf{r}), \qquad (32)$$

specifically,

$$\mathcal{Z}(\nu) = \text{Tr} \left\{ e^{-\hat{\mathcal{H}}(\nu)L/T} \right\}, \qquad (33)$$

where the trace means we have now imposed periodic boundary conditions. For a system with cross-sectional area Ω the mean square winding number itself is

$$\langle W^2 \rangle = \frac{L}{\Omega} \frac{\partial}{\partial \nu} \int d\mathbf{r} \langle j_\nu(\mathbf{r}) \rangle \cdot \mathbf{A}(\mathbf{r})|_{\nu=0}, \qquad (34)$$

$$\langle W^2 \rangle = \frac{L}{T} \left\langle \frac{\partial^2 \hat{\mathcal{H}}(\nu)}{\partial^2 \nu} \right\rangle_{\nu=0} - - -\frac{1}{T^2} \int_0^L d\tau \int_0^L d\tau' \left\langle \hat{T} \left[\frac{\partial \hat{\mathcal{H}}}{\partial \nu}(\tau) \frac{\partial \hat{\mathcal{H}}}{\partial \nu}(\tau') \right] \right\rangle_{\nu=0}, \qquad (35)$$

where

$$j_\nu(\mathbf{r}) \equiv -\frac{1}{T} \frac{\partial \hat{\mathcal{H}}}{\partial \mathbf{A}} = \frac{T}{g} \left[\frac{1}{i}\nabla + \nu \mathbf{A}(\mathbf{r}) \right] \qquad (36)$$

is the current operator, $\langle \hat{O} \rangle \equiv \frac{1}{Z}\mathrm{tr}\,\{\hat{O}e^{-\beta\hat{\mathcal{H}}}\}$, $\hat{T}$ is the time ordering operator, and time dependence of operators is defined according to the Heisenberg picture in imaginary time $\hat{O}(\tau) \equiv e^{-\hat{\mathcal{H}}\tau}\hat{O}e^{\hat{\mathcal{H}}\tau}$.

Equations (34,35) are central to our discussion. On the left-hand side they both have the fluctuations in the winding number. On the right-hand side, the first equation has the derivative of the persistent, thermodynamic, current due to a fictitious quantum particle flowing in a multiply connected geometry threaded by an Aharonov-Bohm flux. The derivative is taken with respect to the particle "charge" ν, and the current is weighted by the vector potential. Equation (35) expresses this derivative in terms of one- and two-time particle correlation functions. An analogous equation for *many* directed polymers in terms of boson correlation functions will be given in section IV. Information known from studies of quantum mechanical particles and many particle boson systems regarding these correlation functions will allow us to determine $\langle W^2 \rangle$.

Equations (34,35) are expressions for the variance of the winding number W. We now briefly discuss the probability that the winding number W equals n, denoted by $P_W(n)$. This probability is,

$$P_W(n) = \frac{\mathcal{P}_W(n)}{\sum_n \mathcal{P}_W(n)}, \tag{37}$$

where

$$\begin{aligned}
\mathcal{P}_W(n) &= \int D\mathbf{r}(z)\delta\left[\int d\tau \mathbf{A}[\mathbf{r}(z)]\cdot\dot{\mathbf{r}} - n\right]e^{-\int_0^L \mathcal{L}_0[\mathbf{r}(z),\dot{\mathbf{r}}(z)]dz/T}\\
&= \frac{1}{2\pi}\int d\nu \int D\mathbf{r}(\tau)e^{-\frac{1}{T}\int_0^L \mathcal{L}[\mathbf{r}(\tau),\dot{\mathbf{r}}(\tau)]d\tau + i\nu\int_0^L d\tau\mathbf{A}[\mathbf{r}(\tau)]\cdot\dot{\mathbf{r}} - i\nu n}\\
&= \frac{1}{2\pi}\int d\nu\,\mathcal{Z}(\nu)e^{-i\nu n} \tag{38}
\end{aligned}$$

with $\mathcal{Z}(\nu)$ given by (31). Equation (37) is again a mapping of the statistical property we are interested in onto a quantum mechanical problem, albeit a less transparent one: the probability $P_W(n)$ is proportional to the Fourier transform, with respect to ν, of the quantum mechanical partition function of a particle of charge ν on an Aharonov-Bohm ring threaded by a flux $1/2\pi$. For a related discussion, including an earlier derivation of Eq. (38), see Ref. [24].

The normalization of the probability distribution function $\mathcal{P}_W(n)$ is a matter of some subtlety. In the second line of Eq. (38) we used the identity,

$$\delta\left[\int d\tau\mathbf{A}[\mathbf{r}(\tau)]\cdot\dot{\mathbf{r}} - n\right] = \frac{1}{2\pi}\int d\nu e^{i\nu\left(\int_0^L d\tau\mathbf{A}[\mathbf{r}(\tau)]\cdot\dot{\mathbf{r}} - n\right)}. \tag{39}$$

The limits of the ν integration are determined by the allowed values of the winding number $\int d\tau\mathbf{A}[\mathbf{r}(\tau)]\cdot\dot{\mathbf{r}}$. Since we assume periodic boundary conditions, the allowed trajectories are closed ($\mathbf{r}(0) = \mathbf{r}(L)$), the winding number is an integer, and the integral over ν can be taken between 0 and 2π. However, if the allowed trajectories are not closed, the winding number can be non-integer, and the integral over ν should be taken between $-\infty$ and $+\infty$. For periodic boundary conditions, the integration domain $\nu \in [0, 2\pi]$ leads to $\sum_n \mathcal{P}_W(n) = 1$. Although we have focused on periodic boundary conditions, in the limit $L \to \infty$ one expects the difference between periodic and open trajectories to vanish.

3.2 A Single Polymer on a Thin Cylinder

We now specialize to a polymer, whose motion in the $x - y$ plane is confined to a ring of radius a, but is stretched out along the τ axis. The polymer's Lagrangian is given by Eq. (28) and we add to $U(\mathbf{r})$ a piece which confines the particle to an *annulus* just outside the cylindrical barrier of radius a. Our previous discussion maps its random walk around the ring onto the thermodynamics of a quantum charged particle on an Aharonov-Bohm ring. The solenoid's vector potential is then $\mathbf{A} = \frac{\hat{\theta}}{2\pi a} = \text{constant}$, where $\hat{\theta}$ is a unit vector around the ring.

In the absence of disorder, we easily reproduce the expected results. According to Eq. (35) the winding number variance is

$$\langle W^2 \rangle = \frac{LT}{g(2\pi a)^2}[1 - 2L\langle E(\nu = 0) \rangle] \qquad (40)$$

where $\langle E(\nu = 0) \rangle$ is the average kinetic energy of a chargeless particle on a ring. For a very large L the average energy is, to exponential accuracy, the energy of the ground state, i.e., 0, and $\langle W^2 \rangle = \frac{LT}{(2\pi a)^2 g}$. As expected for a random walk, the variance of the winding number is proportional to the number of steps taken (i.e., proportional to L). The probability distribution $P_W(n)$ is a Gaussian in this limit, as expected [10]. When $L \to 0$, $\langle E \rangle \to \frac{1}{2L}$, and $\langle W^2 \rangle \to 0$.

The presence of a disorder potential $U(\mathbf{r})$ modifies these results in an interesting way. The effect of disorder on

$$\langle j_\nu \rangle = \frac{1}{T}\left\langle \frac{\partial \hat{\mathcal{H}}}{\partial A} \right\rangle$$

$$= \frac{-2\pi a}{L}\frac{\partial}{\partial \nu} \ln \mathcal{Z}(\nu), \qquad (41)$$

i.e., the persistent current in a mesoscopic ring threaded by an Aharonov-Bohm flux, is well known: the disorder introduces a length scale, ξ, the quantum mechanical localization length of the ground state wave function. As $L \to \infty$,

$$\mathcal{Z}(\nu) \approx \exp[-L\epsilon_0(\nu)/T], \qquad (42)$$

where $\epsilon_0(\nu)$ is the ground state energy. As a crude, but illuminating model of localization, consider a *single* narrow trap of depth U_0 running up the side of the cylinder. The periodic boundary conditions around the cylinder then lead to a tight-binding model type result for the ground state energy as a function of ν,

$$\epsilon_0(\nu) = -U_0 - 2t_1 \cos(\nu) - 2t_2 \cos(2\nu) - - - \cdots \tag{43}$$

where the couplings t_n represent the matrix elements for tunneling around the cylinder n times, $t_n \sim e^{-2\pi an/\xi}$. For a square well of depth U_0 and size b with $U_0 \gg T^2/2gb^2$ we have (see, e.g. [11]) $\xi \approx T/\sqrt{2gU_0}$ and $t_1 \approx \frac{T^2}{gb^2} e^{-2\pi a/\xi}$. Upon using only the first two terms to evaluate (41), we find

$$\langle j(\nu) \rangle = \frac{4\pi a t_1}{T} \sin \nu, \tag{44}$$

and it follows from Eq. (34) that the mean square winding number behaves as

$$\langle W^2 \rangle = 2t_1 L/T$$
$$\sim e^{-2\pi a/\xi} L. \tag{45}$$

Although the variance is still proportional to L, the coefficient $2t_1/T$ which multiplies L vanishes *exponentially* fast as $a \to \infty$.

3.3 A Single Polymer on a Thick Cylinder

In the previous section we considered the statistics of winding numbers associated with the motion of a polymer on a thin cylinder or, equivalently, a random walker on a ring. For a more general cylinder of outer radius R, there are two extreme cases, depending on the ratio of the cylinder width $R - a$ and the typical transverse distance the polymer crosses in "time" L, namely $\sqrt{LT/g}$. In the previous section this ratio was zero. Suppose now that this ratio to be small, but nonzero, i.e. $0 < \frac{g(R-a)^2}{TL} \ll 1$. When using eigenfunctions of $\hat{\mathcal{H}}$ to evaluate winding numbers, this condition implies that the partition function has significant contributions only from one radial mode, the one contributing the lowest energy. In this limit, the density profile in the radial direction can be approximated by a constant, and an analysis similar to that following Eq. (40) leads, for a disorder-free sample in the limit $L \to \infty$, to

$$\langle W^2 \rangle = \frac{LT}{g\pi(R^2 - a^2)} \log \frac{R}{a}. \tag{46}$$

The opposite limit, in which the cylinder width is essentially infinite, was considered, also in the absence of disorder, in Refs. [4] and [10]. Here, we confine ourselves to a simplified discussion of $\langle W^2 \rangle$ using the quantum formalism. According to Eq. (27), the variance $\langle W^2 \rangle$ can be written,

$$\langle W^2 \rangle = \int_0^L d\tau \int_0^L d\tau' \left\langle \frac{dr_i(\tau)}{d\tau} \frac{dr_j(\tau')}{d\tau'} A_i[\mathbf{r}(\tau)] A_j[\mathbf{r}(\tau')] \right\rangle. \qquad (47)$$

Upon approximating the average in the integrand,

$$\left\langle \frac{dr_i(\tau)}{d\tau} \frac{dr_j(\tau')}{d\tau'} A_i[\mathbf{r}(\tau)] A_j[\mathbf{r}(\tau)] \right\rangle \approx \left\langle \frac{dr_i(\tau)}{d\tau} \frac{dr_j(\tau')}{d\tau'} \right\rangle \langle A_i[\mathbf{r}(\tau)] A_j[\mathbf{r}(\tau')] \rangle$$

$$\approx \frac{2T}{g} \delta(\tau - \tau') \langle A^2[\mathbf{r}(\tau)] \rangle, \qquad (48)$$

we have

$$\langle W^2 \rangle = \frac{2T}{g} \int_0^L d\tau \int d^2r \mathcal{P}(\mathbf{r}, \tau) A^2(\mathbf{r}), \qquad (49)$$

where $\mathcal{P}(\mathbf{r}, \tau)$ is the probability of finding the random walker at position $\mathbf{r}$ at height τ. For an infinitely thick cylinder, we have $P(\mathbf{r}, \tau) = \frac{g}{2\pi T\tau} e^{-gr^2/2T\tau}$ and $\langle W^2 \rangle \propto (\log L)^2$. The thermal averages in Eq. (34,35) now include many eigenstates.

The effect of disorder can also be understood by a similar analysis. Again, disorder introduces a localization length ξ, characterizing the eigenstates of the Hamiltonian $H(\nu)$ As long as ξ is larger than the cylinder's size, the effect of the disorder is weak. When ξ is smaller than the cylinder's outer radius, but larger than $2\pi a$, its inner circumference, the effective outer radius of the cylinder becomes ξ. Points on the plane whose distance from the obstacle is larger than ξ support mostly eigenstates that are insensitive to ν. Thus, they do not contribute to the winding number. Particles that start their way at such points never make it to the hole. Therefore, the winding number for such a system becomes,

$$\langle W^2 \rangle \propto \begin{cases} (\log L)^2 & \text{for} \quad \frac{TL}{g} < \xi^2, \quad \xi > 2\pi a, \\ \frac{L}{\xi^2} \log \left(\frac{\xi}{a} \right) & \text{for} \quad \frac{TL}{g} > \xi^2, \quad \xi > 2\pi a \end{cases} \qquad (50)$$

Finally, when $\xi < 2\pi a$ the contribution to the winding number comes from a 1-D strip around the hole, and the 1-D result applies, namely $\langle W^2 \rangle \propto e^{-2\pi a/\xi} L$. To summarize, for an infinitely wide cylinder in the long-time limit disorder makes the winding number increase, for particles that start their random walk close enough to the hole. In the absence of disorder the particles diffuse away from the hole, and thus end up with a small rate of winding. Disorder prevents them from diffusing away from the hole, and increases the winding number.

4 Winding Statistics of Many Polymers

We now turn to discuss a system with many polymers (or flux lines). As discussed in section 2, the two-dimensional Hamiltonian which describes the physics is that of interacting bosons, interacting both mutually and with randomly placed pinning centers. The pinning centers we discuss here are assumed to result from columnar defects, and thus create a τ-independent potential.

The Hamiltonian is a generalization of Eq. (14), namely,

$$\hat{\mathcal{H}} = \frac{T^2}{2g} \sum_{j=1}^{N} \left[\frac{1}{i} \nabla_j + \nu \mathbf{A}(\mathbf{r}_j) \right]^2 + \sum_{j=1}^{N} U(\mathbf{r}_j) + \sum_{i>j} V(|\mathbf{r}_i - \mathbf{r}_j|) \qquad (51)$$

where V is the interaction potential and U includes both disorder and the interaction with a repulsive cylinder centered on the origin. Note that neither V nor U depend on ν, which determines only the coupling of the bosons to the Aharonov-Bohm flux. The phase diagram of the Hamiltonian (51) includes several phases, most notably a superfluid phase and an insulating bose glass phase in the limit $L \to \infty$. The properties of these phases are reflected in the winding number distributions. The winding number for many directed lines is given by Eq. (18), and the many particle generalization of Eq. (35) is

$$\langle W^2 \rangle = \frac{TL}{g} \int d^2r \langle n(\mathbf{r}) \rangle A^2(r) - \int_0^L d\tau \int_0^L d\tau' \int d^2r \int d^2r' \langle \hat{T}[j_\alpha(\mathbf{r},\tau) j_\beta(\mathbf{r}',\tau')] \rangle_{\nu=0}$$
$$\times A_\alpha(\mathbf{r}) A_\beta(\mathbf{r}'). \qquad (52)$$

Several comments are in place regarding Eq. (52). First, the density operator $n(\mathbf{r}) = \sum_i \delta(\mathbf{r} - \mathbf{r}_i)$ describes the *total* density of particles (rather than, e.g., the superfluid density). Second, the current-current correlation function should be calculated for $\nu = 0$, i.e., the current operator is $j_\alpha(\mathbf{r}) = (1/2g) \sum_i [\delta(\mathbf{r} - \mathbf{r}_i) p_\alpha + p_\alpha \delta(\mathbf{r} - \mathbf{r}_j)]$. Finally, the interpretation of the second term becomes clearer upon Fourier transformation. After Fourier analysis, the imaginary-time current-current correlation function appearing on the right-hand side of Eq. (52) becomes the linear response function $\chi_{\alpha\beta}(\mathbf{q}, \omega = 0)$ defined by $j_\alpha(\mathbf{q}, \omega = 0) = \sum_\beta \chi_{\alpha\beta}(\mathbf{q}, \omega = 0) A_\beta(\mathbf{q}, \omega = 0)$, i.e., the function describing the current response to a weak, time-independent vector potential ?.

The current-current correlation function $\chi_{\alpha\beta}(\mathbf{r}, \tau) = \langle \hat{T}[j_\alpha(\mathbf{r}, \tau) j_\beta(0, 0)] \rangle$ in Eq. (52) may be decomposed in longitudinal and transverse parts. The corresponding Fourier decomposition reads

$$\chi_{\alpha\beta}(\mathbf{r}, \tau) = \frac{1}{\Omega L} \sum_{\mathbf{q}, \omega} e^{i\mathbf{q} \cdot \mathbf{r} - i\omega\tau} \left[\chi_\parallel(\mathbf{q}, \omega) \frac{q_\alpha q_\beta}{q^2} + \chi_\perp(\mathbf{q}, \omega) \left(\delta_{\alpha\beta} - \frac{q_\alpha q_\beta}{q^2} \right) \right].$$
$$(53)$$

However, the vector potential $\mathbf{A}(\mathbf{r}) = \frac{\hat{\tau} \times \mathbf{r}}{2\pi r^2}$ for winding around an obstacle is purely transverse, so only $\chi_\perp(\mathbf{q}, \omega)$ contributes to the winding number. Upon passing to the limit of large system dimensions in the xy-plane, and rewriting the second term of (52) in Fourier space, we have

$$\langle W^2 \rangle = \frac{TL}{g} \int d^2r \langle n(\mathbf{r}) \rangle A^2(\mathbf{r}) - L \int \frac{d^2q}{(2\pi)^2} \chi_\perp(\mathbf{q}, \omega = 0) |\hat{\mathbf{A}}(\mathbf{q})|^2 \qquad (54)$$

where $\hat{\mathbf{A}}(\mathbf{q})$ is the Fourier transform of $\mathbf{A}(\mathbf{r})$.

$$\hat{\mathbf{A}}(\mathbf{q}) = \frac{-i\hat{\tau} \times \mathbf{q}}{|q|^2}. \qquad (55)$$

Suppose the directed polymers are confined in an annulus of outer radius $R \gg a$, where a is the radius of the cylindrical obstacle. As $R \to \infty$, we can neglect any distortion of the line density due to the obstacle and approximate $\langle n(r) \rangle$ by its average value, n, far from the inner boundary. The first term of Eq. (54) then behaves like $\frac{TL}{2\pi g} n \ln(R/a)$, i.e., it diverges logarithmically with coefficient proportional to the average density of lines. Because $|\hat{\mathbf{A}}(\mathbf{q})|^2 = 1/q^2$, the integral in the second term is dominated by the behavior of $\chi_\perp(q, \omega = 0)$ in the limit $q \to 0$. Upon imposing upper and lower Fourier cutoffs of a^{-1} and R^{-1}, we $again$ find a logarithmically diverging contribution to the winding number, and the mean square winding number in the large R limit is

$$\langle W^2 \rangle = \frac{TL}{2\pi g} n \ln(R/a) - \frac{L}{2\pi} \left[\lim_{q \to 0} \chi_\perp(q, \omega = 0) \right] \ln(R/a). \qquad (56)$$

However, the transverse response function $\chi_\perp(q, \omega)$ is well known to be related to the $normal$ density n_n of the equivalent system of interacting bosons [16], [23]

$$\lim_{q \to 0} \chi_\perp(\mathbf{q}, \omega = 0) = \frac{T}{g} n_n. \qquad (57)$$

Upon defining the superfluid number density $n_s = n - n_n$, we are led to our final result, namely

$$\langle W^2 \rangle = \frac{TL}{2\pi g} n_s \ln(R/a). \qquad (58)$$

Equation (58) is an exact relation, valid in the limit $R \gg a$, between winding number fluctuations and the superfluid density of the equivalent boson system. When reexpressed in terms of the boson parameters of section 2 it reads

$$\langle W^2 \rangle = \frac{n_s \hbar^2}{2\pi m k_B T} \ln(R/a), \qquad (59)$$

which is similar in some respects to the Pollock-Ceperley result (9) for boson world lines winding around a torus [26]. Unlike the Pollock-Ceperley formula, however, Eq. (58) applies directly to a real physical system, namely "directed polymer melts" composed of vortex lines in Type II superconductors with columnar defects. As discussed in section 2, $\langle W^2 \rangle$ determines the linear response of the net winding number to a current through the obstacle.

According to Eq. (58) the winding number fluctuation around an obstacle for many interacting directed polymers is predicted to diverge linearly with the length of the obstacle and logarithmically with the cross-sectional area, as one might guess by multiplying the result (46) for a single polymer in a moderately thick cylinder by N. The meaning of the coefficient becomes clearer if we first define a "thermal deBroglie wavelength" Λ_T for the polymers by

$$\Lambda_T^2 = \frac{2\pi T L}{g}, \tag{60}$$

i.e., the directed polymer analogue of Eq. (7) with the usual identifications $\hbar \to T$, $\hbar/T \to L$ and $m \to g$. This length is a measure of the transverse wandering distance of the polymers as they traverse the sample. Equation (58) then becomes

$$\langle W^2 \rangle = \left(\frac{\Lambda_T}{2\pi} \right)^2 n_s \ln(R/a), \tag{61}$$

showing that only a part of the "superfluid fraction" n_s of lines, i.e., those within a transverse wandering distance of the obstacle, contribute to the winding number fluctuations. Correlated disorder along the $\hat{\tau}$ axis should decrease the winding number fluctuations. Equation (61) shows that this reduction is given entirely by the corresponding reduction in the superfluid density. The reduction in n_s for directed lines subjected to various kinds of correlated disorder is calculated explicitly in Ref. [23].

acknowledgements It is a pleasure to acknowledge helpful conversations with D. Bishop and D. Ceperley. Work by DRN was supported by the National Science Foundation, primarily by the MRSEC program through Grant DMR–9400396 and in part through Grant DMR–9417047.

References

[1] S.F. Edwards, Proc. Phys. Soc. **91**, 513 (1967) J. Phys. A**1**, 15 (1968).

[2] S. Prager and H.L. Frisch, J. Chem. Phys. **46**, 1475 (1967).

[3] For a review, see A. Grossberg and A. Khoklov, in Advances in Polymer Science **106**, 1 (1993).

[4] J. Rudnick and Y. Hu, J. Phys. A. Math. Gen. **20**, 4421 (1987).

[5] Y. Aharonov and D. Bohm, Phys. Rev. **115**, 485 (1959).

[6] E.L. Pollack and D.M. Ceperley, Phys. Rev. B**36**, 8843 (1987).

[7] For a review, see D.M. Ceperley, Rev. Mod. Phys. **67**, 279 (1995).

[8] G. Blatter, M.V. Feigel'man, V.B. Geshkenbein, A.I. Larkin, and V. Vinokur, Reviews of Modern Physics **66**, 1125 (1994).

[9] D.R. Nelson and S. Seung, Phys. Rev. B**39**, 9153 (1989), D.R. Nelson and P. Le Doussal, Phys. Rev. B**42**, 10113 (1990).

[10] B. Drossel and M. Kardar, Phys. Rev. E53, 5861 (1996); and cond-mat/96100119.

[11] D.R. Nelson and V. Vinokur, Phys. Rev. B48, 13060 (1993).

[12] Z. Yao, S. Yoon, H. Dai, S. Fan and C.M. Lieber, Nature 371, 777 (1995).

[13] We assume currents so small that the structure of the Bose glass state inside the tube is unaffected, due to the transverse Meissner effect[11]. For strong enough currents, the Bose glass in the tube may deform slightly, thus forming a repulsive *chiral* defect, similar to that discussed by Drossel and Kardar[10].

[14] R.P. Feynman, *Statistical Mechanics* (Benjamin, Reading, MA, 1972).

[15] R.P. Feynman, Phys. Rev. 91, 1291 (1953).

[16] P. Nozieres and D. Pines, *The Theory of Quantum Liquids*, vol. 2 (Addison-Wesley, Reading 1990).

[17] D.R. Nelson and J.M. Kosterlitz, Phys. Rev. Lett. 39, 1201 (1977).

[18] See, e.g., R. Bellissent, L. Descotes and P. Pfeuty, Journal of Physics C6, Suppl. A211 (1994).

[19] For reviews, see D.R. Nelson, Physica A177, 220 (1991); and D.R. Nelson in *Observation, Prediction and Simulation of Phase Transition in Complex Fluids*, edited by Ml. Bans et al. (Kluwer, The Netherlands, 1995).

[20] See, e.g., M.P.A. Fisher and D.H. Lee, Phys. Rev. B39, 2756 (1989).

[21] P.G.. deGennes, *Superconductivity of Metals and Alloys* (Addison-Wesley, Reading, MA 1989), chapter 2.

[22] See., e.g., V.G. Kogan, Phys. Rev. B24, 1572 (1981).

[23] U. Tauber and D.R. Nelson, submitted to Physics Reports.

[24] A. Comtet, J. Desbois and S. Ouvry J. Phys. Math Gen. 23, 3563 (1990); A. Comtet, J. Desbois and C. Monthus J. Stat. Phys. 73, 433 (1993); B. Houchmandzadeh, J. Lajzerowicz and M. Vallade J. de Physique 1 2, 1881 (1992).

[25] .D. Mahan, *Many-Particle Physics* (Plenum Press, New York 1981), sections (3.3) and (3.7).

[26] By taking $A(r) =$ const. in Eq. (52), one easily derives the Pollock-Ceperley formula (9) for bosons on a torus.

Localized Flux Lines and the Bose Glass

Uwe C. Täuber

University of Oxford, Department of Physics – Theoretical Physics, 1 Keble Road, Oxford OX1 3NP, U.K.

Abstract. Columnar defects provide effective pinning centers for magnetic flux lines in high–T_c superconductors. Utilizing a mapping of the statistical mechanics of directed lines to the quantum mechanics of two–dimensional bosons, one expects an entangled flux liquid phase at high temperatures, separated by a second–order localization transition from a low–temperature "Bose glass" phase with infinite tilt modulus. Recent decoration experiments have demonstrated that below the matching field the repulsive forces between the vortices may be sufficiently large to produce strong spatial correlations in the Bose glass. This is confirmed by numerical simulations, and a remarkably wide soft "Coulomb gap" at the chemical potential is found in the distribution of pinning energies. At low currents, the dominant transport mechanism in the Bose glass phase proceeds via the formation of double kinks between not necessarily adjacent columnar pins, similar to variable–range hopping in disordered semiconductors. The strong correlation effects originating in the long–range vortex interactions drastically reduce variable–range hopping transport.

1   Pinning of Flux Lines to Columnar Defects

The remarkably rich phase diagram of magnetic flux lines in high–T_c superconductors, especially when subject to point and / or extended disorder, has attracted considerable experimental and theoretical interest (for a recent review, see Ref. [1]). For the purpose of applying type–II superconductors in external magnetic fields, an effective vortex pinning mechanism is essential, in order to minimize dissipative losses caused by the Lorentz–force induced movement of flux lines across the sample. This issue becomes even more important for the high–T_c superconductors, because the strongly enhanced thermal fluctuations in these materials render the Abrikosov flux lattice unstable, and one therefore expects a first–order melting transition leading to a normal–conducting *flux liquid* phase with well below the upper critical field $H_{c_2}(T)$, at least for ideally pure systems [2]. It is very difficult to pin such a vortex liquid consisting of fluctuating flexible lines by just intrinsic point disorder (e.g., oxygen vacancies); yet asymptotically a truly superconducting low–temperature "vortex glass" phase has been predicted [3].

Therefore, in addition to point defects, the influence of extended or *correlated* disorder, promising stronger pinning effects, on vortex transport properties has been considered. Experimentally, linear damage tracks have been produced in oxide superconductors by irradiation with high–energy ions. These *columnar defects* serve as strong pinning centers for the flux lines, thus significantly increasing the critical current and shifting the irreversibility line upwards (for references, see Ref. [1], Sec. IX B).

In a long–wavelength description in the spirit of the London approximation, one may consider the following model free energy for N flux lines, defined by their trajectories $r_i(z)$ as they traverse the sample, interacting with each other and with N_D columnar defects which are aligned along the magnetic–field direction $\hat{z}$ [4], [5]:

$$F_N = \int_0^L dz \sum_{i=1}^N \left\{ \frac{\tilde{\epsilon}_1}{2} \left(\frac{dr_i(z)}{dz} \right)^2 + \frac{1}{2} \sum_{j \neq i}^N V[r_{ij}(z)] + \sum_{k=1}^{N_D} V_D[r_i(z) - R_k] \right\} . \quad (1)$$

Here $r_{ij}(z) = |r_i(z) - r_j(z)|$, and $V(r) = 2\epsilon_0 K_0(r/\lambda)$ denotes the repulsive vortex interaction; the modified Bessel function $K_0(x) \propto -\ln x$ for $x \to 0$, and $\propto x^{-1/2} e^{-x}$ for $x \to \infty$. Thus the London penetration depth λ defines the interaction range and the energy scale $\epsilon_0 = (\phi_0/4\pi\lambda)^2$. For low fields $B \leq \phi_0/\lambda^2$, i.e.: $\lambda \lesssim a_0 = (4/3)^{1/4}(\phi_0/B)^{1/2}$, the vortex line tension is $\tilde{\epsilon}_1 \approx \epsilon_0$ (a_0 is the triangular lattice constant in the corresponding pure system). Finally, the pinning energy is a sum of N_D z–independent potential wells $V_D(r)$ with average spacing d centered on the positions $\{R_k\}$, whose diameters are typically $b_0 \approx 100 \, \text{Å}$, with a variation of $\delta b_k/b_0 \approx 15\%$, caused by the ion–beam dispersion. This induces some distribution P of the pinning energies U_k, which may be determined from the formula $U_k \approx (\epsilon_0/2) \ln[1 + (b_k/\sqrt{2}\xi)^2]$, where ξ denotes the coherence length [5]. The thermodynamic properties of the model (1) are to be obtained by evaluation of the grand–canonical partition function

$$Z_{gr} = \sum_{N=0}^{\infty} \frac{1}{N!} e^{\mu NL/T} \int \prod_{i=1}^N \mathcal{D}[r_i(z)] e^{-F_N[\{r_i(z)\}]/k_B T} , \quad (2)$$

with the chemical potential $\mu = (H_{c_1} - H)\phi_0/4\pi$, which changes sign at the lower critical field $H_{c_1} = \phi_0 \ln(\lambda/\xi)/4\pi\lambda^2$.

As for pure systems [2], one can map this statistical problem of directed flux lines interacting with columnar pins, onto the quantum mechanics of bosons in two dimensions subject to static point disorder via the identification of the vortex trajectories $r_i(z)$ with the particle world lines in imaginary time [4], [5], [6]. In Table 1, the corresponding quantities for bosons and magnetic flux lines are listed (notice that the thermodynamic limit $L \to \infty$ for the directed polymers corresponds to a zero–"temperature" quantum problem), and the ensuing phase diagram is sketched schematically in Fig. 1. At high temperatures, one thus finds an *entangled liquid* of unbound flux lines (corresponding to a boson *superfluid*), separated by a sharp second–order transition (see Ref. [7] and references therein) from a low–temperature phase of localized vortices. This *Bose glass* phase is characterized by an infinite tilt modulus c_{44}, and turns out to be stable over a certain range of tipping angles of **B** away from the z direction [5], [8]. At least for $\lambda \ll d$, the theory also suggests a low–temperature *Mott insulator* phase, when $B = B_\phi$, i.e., the vortex density exactly matches that of the columnar pins. The Bose glass and the Mott insulator are distinct thermodynamic phases, for the latter should be characterized by infinite tilt *and* compressional moduli [5].

Table 1. Boson analogy applied to vortex transport.

Bosons	m	$\hbar$	$\beta\hbar$	$V(r)$	n	μ	ρ_s	**E**	**j**
Vortices	$\tilde{\epsilon}_1$	T	L	$2\epsilon_0 K_0(r/\lambda)$	B/ϕ_0	$(H - H_{c_1})\phi_0/4\pi$	$\rho^2 c_{44}^{-1}$	$\hat{z} \times \mathbf{J}/c$	$\mathcal{E}$

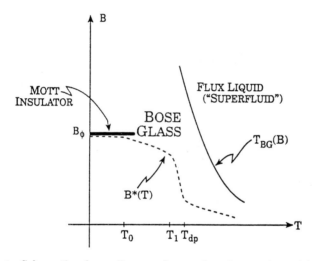

Fig. 1. Schematic phase diagram for vortices interacting with columnar defects aligned along the magnetic–field direction.

The formal analogy with a boson superfluid can be further exploited to investigate the effect of disorder on density and current correlation functions in the flux liquid phase within the Bogoliubov approximation [9]. Thus one may compute the corresponding transport coefficients, e.g., the boson superfluid density ρ_s whose flux line analog is the inverse tilt modulus of the vortex array (see Table 1). For parallel columnar defects of strength U_0 one eventually finds [10]:

$$c_{44}^{-1}(T) \approx (n\tilde{\epsilon}_1)^{-1}\left[1 - (T_{\mathrm{BG}}/T)^4\right], \quad \text{with } T_{\mathrm{BG}}(B) \approx T^*[\phi_0/(4\pi\lambda)^2 B]^{1/4} \quad (3)$$

and $k_{\mathrm{B}}T^* = b(\tilde{\epsilon}_1 U_0)^{1/2}$, in accord with estimates obtained in the Bose glass itself [5]. The critical behavior at the boson localization transition is not yet fully understood, although quantum Monte Carlo simulations [11] seem to support the scaling theory [7]. Direct simulations for the critical dynamics of flux lines near the Bose glass transition have also been performed [12].

2   Structural Properties in the Bose Glass Phase

Recently, the positions of flux lines and parallel columnar pins, aligned along the magnetic–field direction, were simultaneously measured in a BSCCO sample in the Bose glass phase at low magnetic fields using a combined chemical etching and magnetic decoration technique [13]. Figure 2 depicts the thus obtained positions of N_D columnar defects and N flux lines in a two–dimensional cross section perpendicular to $\mathbf{B}$ in a case where $f = N/N_D = B/B_\phi \approx 0.24$ and $\lambda \approx 0.45\, a_0$. While the columnar defects are to a good approximation randomly distributed in space, the flux lines clearly form a highly *correlated* configuration; correspondingly, the two–dimensional vortex structure function $S(q)$ displays a distinct peak at wave vector $q \approx 2\pi/a_0$, resembling an amorphous material (see Fig. 3). This correlated structure must obviously be the result of the mutual repulsive interactions V_{int} between the vortices which dominate over any statistical fluctuations in the pinning potentials δV_D; on the other hand, all the flux lines are attached to a defect, and therefore $V_D > V_{int}$ [14].

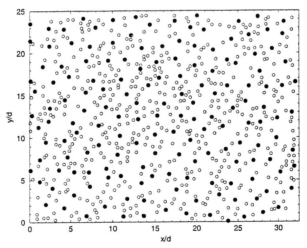

Fig. 2. Positions of empty columnar defects (open circles), and pins occupied by flux lines (filled circles), as obtained from a decoration experiment ($f \approx 0.24$).

In order to model this experimental situation, we can fortunately simplify the free energy (1) substantially: (i) As long as we limit ourselves to "low" temperatures (an estimate for BSCCO actually yields $T \lesssim T_1 = 0.9\, T_c$)

we can neglect thermal wandering of the flux lines, which is furthermore significantly suppressed by the confining line defects, we may drop the tilt–energy $\propto \tilde{\epsilon}_1$ and consider an effectively "classical" ($k_B T \sim \hbar$ in the boson analogy) two–dimensional system. (ii) For magnetic fields smaller than the crossover field $B^*(T)$ in Fig. 1, the vortices are sufficiently dilute to ensure that their repulsive interactions have no influence on the localization length, and in fact the "boson" statistics becomes irrelevant (for BSCCO, $B^*(T_1) \approx 0.7 B_\phi$). Having thus disposed of the entropic corrections, we may consider the following effective Hamiltonian and its grand–canonical counterpart [14], [15],

$$H = \frac{1}{2} \sum_{i \neq j}^{N_D} n_i n_j V(r_{ij}) + \sum_{i=1}^{N_D} n_i t_i \,, \quad \tilde{H} = H - \mu \sum_{i=1}^{N_D} n_i \,, \tag{4}$$

where $i = 1, \ldots, N_D$ denote the defect sites, randomly distributed in the x–y plane, $n_i = 0, 1$ is the corresponding site occupation number, and the t_i are random–site energies (originating in the varying pin diameters), whose distribution of mean zero and width $w \approx 0.1 l \epsilon_0$ we for simplicity choose to be flat: $P(t_i) = \Theta(w - |t_i|)/2w$.

This classical model is still far from trivial, however, due to the interplay of long–range interactions and disorder. Fortunately models of the form (4), albeit with a Coulomb potential replacing the screened logarithmic interaction $V(r)$, have been extensively studied in the context of disordered semiconductors, which allows us to adapt qualitative arguments and numerical simulation procedures from the literature (see, e.g., Refs. [16], [17]). Thus, in order to infer the (approximate) ground–state properties in the Bose glass phase as functions of the filling f and interaction range λ/d we employ a zero–temperature Monte Carlo algorithm that essentially minimizes the total energy (4) with respect to all possible one–vortex transfers; i.e., the ground state stability is checked by demanding that all single–particle hops from occupied sites i to empty sites j require an energy $\Delta_{i \to j} = \epsilon_j - \epsilon_i - V(r_{ij}) > 0$, where $\epsilon_i = t_i + \sum_{j \neq i} n_j V(r_{ij})$ are the interacting site energies. For $\lambda \gtrsim d$ and not too large $f \lesssim 0.6$ the resulting distribution of flux lines among the columnar pins closely resembles the experimental data in Fig. 2. More quantitatively, we can investigate the amorphous peak at $q a_0 \approx 2\pi$ in the vortex structure factor $S(\mathbf{q}) = \sum_{i,j} e^{i\mathbf{q} \cdot (\mathbf{r}_i - \mathbf{r}_j)}/N$, averaged over all directions in reciprocal space. Whereas the dependence on the interaction range is rather weak as long as $\lambda \gtrsim d$, the peak in Fig. 3 vanishes when $f \gtrsim 0.4$; this reflects that at higher fillings the system has to increasingly accomodate with the underlying randomness, and therefore the correlations induced by the interactions disappear. Furthermore, a triangulation of the vortex positions shows that typically only about 50 % of the sites are sixfold coordinated, and thus the structure formed by the flux lines rather resembles an amorphous state (justifying the name "Bose glass") rather than a weakly disordered triangular lattice [15].

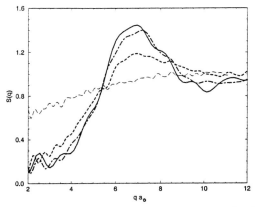

Fig. 3. Dependence of the vortex structure factor peak on the filling f for $\lambda/d = 2$; solid: $f = 0.1$, dot–dashed: $f = 0.2$, dashed: $f = 0.4$, and long–dashed: $f = 0.8$.

3   The Distribution of Pinning Energies

The above Monte Carlo algorithm can also be employed to determine the distribution of pinning energies $g(\epsilon)$, with the vortex interactions taken into account. As in the boson picture this is nothing but the single–particle density of states, $g(\epsilon)$ is readily obtained by sampling the site energies ϵ_i for a number of disorder realizations and then performing a quenched average. Following the mean–field arguments of Efros and Shklovskii (1984) for a system with long–range repulsive interactions $V(r) \propto r^{-\sigma}$, with $0 < \sigma < D$ (here, the space dimension is $D = 2$), one expects that the density of states vanishes at the chemical potential according to a power law,

$$g(\epsilon) \propto |\epsilon - \mu|^{s_{\text{eff}}} , \tag{5}$$

with a *gap exponent* $s_{\text{MF}} = (D/\sigma) - 1$. This means that as a consequence of the correlations induced by the long–range interactions, the probability of finding a low–energy empty site ($\epsilon \gtrsim \mu$) for introducing an additional particle is greatly reduced, as is the density of states of high–energy filled sites ($\epsilon \lesssim \mu$) for the corresponding process of removing a particle from the system. While the qualitative prediction (5) remains correct, the actual gap exponent is influenced by subtle fluctuation effects [17] and therefore differs from the mean–field predictions (for the currently most accurate results for Coulomb interactions, $\sigma = 1$, see Ref. [18]).

Figure 4 depicts two characteristic examples for the distribution of pinning energies $g(\epsilon)$, as obtained from our simulations [15], and normalized according to $\int g(\epsilon) d\epsilon = 1/d^2$. In the case of an infinite–range logarithmic potential, the Ewald summation method was used, and before taking the limit $\lambda \to \infty$, a term $\epsilon_i' - \epsilon_i = -2\pi f(\lambda/d)$ was subtracted from the site energies. The filled and empty states are separated by a wide soft gap centered at the

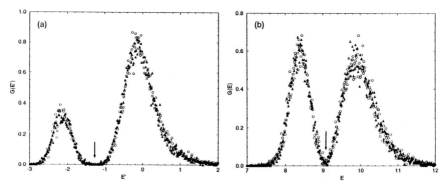

Fig. 4. Normalized distribution of pinning energies $G(E) = 2\epsilon_0 d^2 g(\epsilon)$ as function of the single–particle energies $E = \epsilon/2\epsilon_0$, with (a) $\lambda \to \infty$, and $f = 0.2$, (b) $\lambda/d = 2$, and $f = 0.4$. The location of μ is marked by the arrow.

chemical potential, with an effective gap exponent $s_{\rm eff} \approx 3$ for low filling. Remarkably, this characteristic double–peak structure and the marked depletion of states persists even for finite–range repulsion with $\lambda \geq d$ (Fig. 4b), and although $g(\mu)$ does not strictly vanish any more, the distribution of pinning energies for $\epsilon \approx \mu$ can be fitted with a power law (5); e.g., $s_{\rm eff} \approx 1.2$ for $\lambda/d = 1$ and $f = 0.1$. Upon increasing f, the gap closes quickly due to the stronger influence of the underlying randomness (compare Fig. 3), but for $f \lesssim 0.2$ the correlation effects only disappear when $\lambda < d$.

4   Variable–Range Hopping Transport of Flux Lines

An in–plane current $\mathbf{J} \perp \mathbf{B}$ induces a Lorentz force $\mathbf{f}_{\rm L} = \phi_0 \hat{\mathbf{z}} \times \mathbf{J}/c$, acting on the vortices. Accordingly, a term $\delta F_N = -\int_0^L dz \sum_i \mathbf{f}_{\rm L} \cdot \mathbf{r}_i(z)$ has to be added to the free energy (1). In the spirit of the thermally assisted flux–flow (TAFF) model of vortex transport, we write the superconducting *resistivity* (i.e.: the *conductivity* in the boson representation) as $\rho = \mathcal{E}/J \approx \rho_0 \exp\left[-U_{\rm B}(J)/T\right]$, where ρ_0 is a characteristic flux–flow resistivity, and $U_{\rm B}$ represents an effective barrier height. We shall see that in the Bose glass phase $U_{\rm B}(J)$ actually diverges as $J \to 0$,

$$\mathcal{E} \approx \rho_0 J \exp\left[-E_0/k_{\rm B}T\right](J_0/J)^{p_{\rm eff}}] \ . \tag{6}$$

Driven by an intermediate external current $J_1 < J < J_c$, a flux line will start to leave its columnar pin by detaching a segment of length z into the defect–free region, thereby forming a half–loop of transverse size r (Fig. 5a). The free energy corresponding to this saddle–point configuration is readily estimated to be $\delta F_1(r, z) \approx \tilde{\epsilon}_1 r^2/z + U_0 z - f_{\rm L} rz$, and optimizing successively

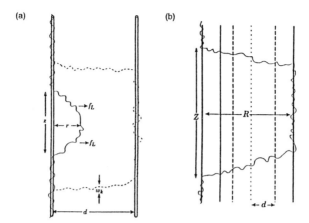

Fig. 5. Saddle–point configurations for vortex transport in the Bose glass phase. (a) Vortex half–loop. (b) Double–superkink in the variable–range hopping regime.

with respect to z and r one finds $z^* \approx r^* \sqrt{\tilde{\epsilon}_1/U_0} \approx c\sqrt{\tilde{\epsilon}_1 U_0}/\phi_0 J$. Hence the current–voltage characteristics assumes the form (6), with $E_0 = E_{\mathrm{K}} = d\sqrt{\tilde{\epsilon}_1 U_0}$, $J_0 = J_1 = cU_0/\phi_0 d$, and the effective *transport exponent* $p_{\mathrm{eff}} = 1$ in the *vortex half–loop regime* [5]. For $J \to J_1$, the flux line will extend a double kink of width $w_{\mathrm{K}} = d\sqrt{\tilde{\epsilon}_1/U_0}$ and energy $2E_{\mathrm{K}}$ to an adjacent columnar defect, and therefore for $J < J_1$ neighboring pins must be taken into account.

Thus, at low currents one enters another regime, where the flux line emits a pair of kinks to a possibly distant defect which happens to provide a favorable pinning energy (Fig. 5b). This is the vortex analog of variable–range charge transport in disordered semiconductors (see Ref. [16] and references therein). The cost in free energy for such a configuration of transverse size R and extension Z along the magnetic–field direction will consist of the double–kink energy $2E_{\mathrm{K}}(R) = 2E_{\mathrm{K}}(d)R/d$ stemming from the elastic term, and the difference in pinning energies of the highest–energy occupied site, $\epsilon_i \approx \mu$, and the empty site at distance R with $\epsilon_j = \mu + \Delta(R)$, where the distribution of pinning energies enters through the condition $\int_\mu^{\mu+\Delta(R)} g(\epsilon)\mathrm{d}\epsilon = R^{-D}$. Minimizing the saddle–point free energy $\delta F_{\mathrm{SK}} \approx 2E_{\mathrm{K}}R/d + Z\Delta(R) - f_L RZ$ in the regime where $g(\epsilon)$ has the form (5), one arrives at (6) with $E_0 = 2E_{\mathrm{K}}$ and the effective transport exponent [15]

$$p_{\mathrm{eff}} = (s_{\mathrm{eff}} + 1)/(D + s_{\mathrm{eff}} + 1) . \tag{7}$$

For short–range interactions, $g(\mu) > 0$, i.e., $s_{\mathrm{eff}} = 0$, and the Mott variable–range hopping exponent $p_{\mathrm{M}} = 1/(D + 1)$ is recovered [5]. But as the interaction range λ becomes larger than the average vortex separation $a_0 = d/\sqrt{f}$, the emerging correlations reduce the phase space for single–vortex hopping drastically, and $p_{\mathrm{eff}} < p_{\mathrm{M}}$; in addition, as the normalized distribution of pinning energies is broadened by the interactions, also the

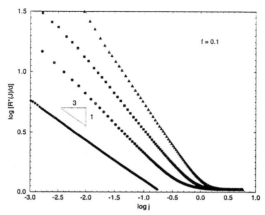

Fig. 6. Log–log plots (base 10) of the activation energy $U_B(J) \propto R^*(J)/d$ for variable–range hopping vs the reduced current $j = J\phi_0 d/2\epsilon_0 c$ for $f = 0.1$. Diamonds: $\lambda \to 0$, circles: $\lambda/d = 1$, squares: $\lambda/d = 2$, triangles: $\lambda/d = 5$.

current scale J_0 in (6) is increased. The result of the minimization procedure using the numerical data for $g(\epsilon)$ in the case $f = 0.1$ is shown in Fig. 6 as a function of the interaction range [15]. Assuming that flux line transport in the low–current regime is still dominated by single–vortex hops, this implies that long–range repulsive interactions reduce the resistivity by several orders of magnitudes with respect to the non–interacting situation. E.g., for $\log j \approx 1.5$ the exponent in (6) is about ten times smaller for $\lambda/d = 5$ as compared to $\lambda \to 0$ (see Fig. 6), and hence ρ is reduced by a factor of $\approx 10^{-5}$. In accord with the relation (7), the effective transport exponent assumes values upto $p_{eff} \approx 0.7$ for $\lambda \to \infty$ and $f = 0.1$, while, e.g., for $\lambda/d = 1$ and $f = 0.1$ we find $p_{eff} \approx 0.55$. We remark that the former value for for the variable–range hopping exponent of logarithmically interacting particles in two dimensions is consistent with an analysis based on the very different assertion that actually *collective* many–particle hops yield the lowest energy barriers [19]. Finally, a recent experiment on a BSCCO sample with $f \approx 0.15$ and $\lambda/d \approx 1.6$ reported a measured $p_{eff} \approx 0.57$ in the variable–range hopping regime across about half a decade in the current–voltage characteristics [20]; this apparently agrees well with our results, although due caution needs to be applied for this comparison as in this experiment the magnetic field distribution was in fact very inhomogeneous. Yet further experiments should clearly be capable to test the above predictions quantitatively.

In summary, experiments have established that columnar defects provide very effective pinning centers for flux lines in high–T_c superconductors, and demonstrated that in certain cases the repulsive vortex interactions lead to remarkable spatial correlations. As a consequence, a soft "Coulomb gap" emerges in the distribution of pinning energies, and vortex transport in the variable–range hopping regime should be drastically reduced.

Acknowledgments. The work presented here was done in delightful collaboration with David Nelson, whose ideas and insights have been invaluable. I would also like to thank H. Dai and C.M. Lieber for fruitful cooperation and their sharing of data with us prior to publication. This research was supported by the Deutsche Forschungsgemeinschaft under Contract Ta. 177/1, and by the National Science Foundation, in part by the MRSEC Program through Grant DMR-9400396, and through Grant DMR-9417047. Financial support from a Royal Society Conference Grant and from the Engineering and Physical Sciences Research Council through Grant GR/J78327 is gratefully acknowledged.

References

[1] G. Blatter, M.V. Feigel'man, V.B. Geshkenbein, A.I. Larkin, and V.M. Vinokur, Rev. Mod. Phys. **66**, 1125 (1994).

[2] D.R. Nelson, Phys. Rev. Lett. **60**, 1973 (1988); D.R. Nelson and H.S. Seung, Phys. Rev. B **39**, 9153 (1989).

[3] M.P.A. Fisher, Phys. Rev. Lett. **62**, 1415 (1989); D.S. Fisher, M.P.A. Fisher, and D.A. Huse, Phys. Rev. B **43**, 130 (1991).

[4] D.R. Nelson and V.M. Vinokur, Phys. Rev. Lett. **68**, 2398 (1992).

[5] D.R. Nelson and V.M. Vinokur, Phys. Rev. B **48**, 13060 (1993).

[6] I.F. Lyuksyutov, Europhys. Lett. **20**, 273 (1992).

[7] M.P.A. Fisher, P.B. Weichman, G. Grinstein, and D.S. Fisher, Phys. Rev. B **40**, 546 (1989).

[8] Further details on this "transverse Meissner effect" and vortex localization in transverse magnetic fields may be found in the contribution to this volume by D.R. Nelson.

[9] D.R. Nelson and P. Le Doussal, Phys. Rev. B **42**, 10113 (1990).

[10] U.C. Täuber and D.R. Nelson, Preprint (1996).

[11] M. Wallin, E.S. Sørensen, S.M. Girvin, and A.P. Young, Phys. Rev. B **49**, 12115 (1994).

[12] M. Wallin and S.M. Girvin, Phys. Rev. B **47**, 14642 (1993).

[13] H. Dai, S. Yoon, J. Liu, R.C. Budhani, and C.M. Lieber, Science **265**, 1552 (1994).

[14] U.C. Täuber, H. Dai, D.R. Nelson, and C.M. Lieber, Phys. Rev. Lett. **74**, 5132 (1995).

[15] U.C. Täuber and D.R. Nelson, Phys. Rev. B **52**, 16106 (1995).

[16] B.I. Shklovskii and A.L. Efros, *Electronic properties of doped semiconductors*, Springer–Verlag (New York, 1984).

[17] J.H. Davies, P.A. Lee, and T.M. Rice, Phys. Rev. Lett. **49**, 758 (1982); Phys. Rev. B **29**, 4260 (1984).

[18] A. Möbius, M. Richter, and B. Drittler, Phys. Rev. B **45**, 11568 (1992).

[19] M.P.A. Fisher, T.A. Tokuyasu, and A.P. Young, Phys. Rev. Lett. **66**, 2931 (1991).

[20] M. Konczykowski, N. Chikumoto, V.M. Vinokur, and M.V. Feigel'man, Phys. Rev. B **51**, 3957 (1995).

Structural Studies of Magnetic Flux Line Lattices Near Critical Transitions

Uri Yaron[†]

*Bell Laboratories, Lucent Technologies. 600 Mountain Ave., Murray Hill,
NJ 07974, USA*

Small Angle Neutron Scattering (SANS) studies are an essential tool for studying vortices in the bulk of type II superconductors, providing deep understanding of their three dimensional microscopic structure in a wide range of fields and temperatures. This talk will summarize detailed studies of flux lattices in the vicinity of two critical transitions: (1) flux lattices in the vicinity of the critical current in $2H\text{-}NbSe_2$. We find a clear evidence for a two step depinning process: as a function of increasing driving force three regimes are observed - first, no motion; then disordered, plastic motion; and finally at high velocities a coherently moving crystal. (2) flux lattices in the vicinity of the peak effect, below the upper critical field, H_{c2}, in Nb. Our studies reveal drastic structural disordering, characterized by complete loss of positional and orientational correlations, whereas the lines remain well correlated along their length.

The nature of the mixed state in type II superconductors remains an issue of great theoretical and practical interest[1]. The physical properties of the Flux Line Lattice (FLL) in the mixed state, are strongly linked to its structural properties[2]. It is therefore not surprising that a variety of imaging techniques were developed to study the FLL in the mixed state. Among the most common techniques are: Bitter decoration[3], magneto-optics[4], STM[5] and the scanning Hall bar technique[6]. All these are surface probes, that provide either good spatial resolution in static studies or dynamical information, but not both. Neutron scattering[7-9] is a bulk technique that provides information on the three dimensional properties of both static and moving FLL in a wide range of fields (~ 0.02 T up to several Teslas, a theoretically interesting regime in which vortices interact strongly) and temperatures. The method relies on the interaction between the magnetic moment of the neutron and the magnetic field modulation of the vortex structure, and provides a way of measuring the penetration depth, λ, the coherence length, ξ, the detailed shape of the magnetic field distribution and the FLL structure.

The experimental setup for Small Angle Neutron Scattering (SANS) in the horizontal field configuration, employed in our experiments, is shown in Fig. 1a. Our experiments were performed in the cold neutron guide hall of the Risø DR3 reactor. A superconducting magnet in the persistent mode was used to apply a horizontal field parallel within 1° to both the neutron beam and either the **c** axis (study of a single crystal $2H\text{-}NbSe_2$) or the cylinder axis (study of a single crystal Nb cylinder). The edges of the crystals and the electrical contacts were masked with Cd foil to reduce specular

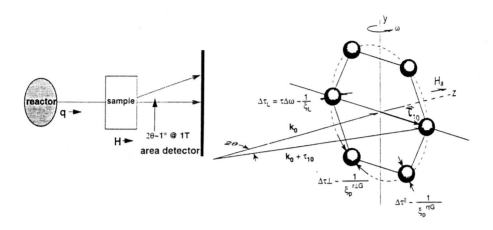

Fig. 1 (a) real and (b) reciprocal space descriptions of the scattering geometry.

reflections, scattering from a distorted FLL near the surface and small angle background from the contacts. We have used an incident neutron wavelengths of $\lambda_n \sim 9.9$ - 11.8 Å and a bandwidth $\Delta\lambda_n/\lambda_n \sim 18\%$, with an incident beam divergence $\sim 0.10°$ - $0.18°$ FWHM. The diffracted neutrons were counted by an area detector, located at the end of a 6 m evacuated chamber.

The reciprocal space description of the scattering geometry is shown in Fig. 1b. In principle, three correlation lengths could be extracted[9]. The radial and azimuthal widths of the six first order bragg peaks yield information about two positional correlation lengths, perpendicular to the flux lines, $\xi r\|G$ and $\xi r\bot G$, in directions **r** parallel and perpendicular to the reciprocal lattice vector **G**. These two lengths are related to compressional and shear displacements of the lattice, respectively. In addition, the azimuthal width is related to orientational order of the lattice. Well defined Bragg peaks are indicative of long range orientational order throughout the illuminated volume. In the lack of long range orientational order, as in an amorphous FLL, a ring of scattering would be expected. The last length which can be extracted is the longitudinal correlation length, ξ_L. This measure of correlations along the flux lines is extracted from the widths of the rocking curves. A rocking curve is a plot of the scattered intensity in the Bragg peaks as the sample and the magnet are rotated through the Bragg condition around an axis perpendicular to both the incoming neutron beam and the scattering wavevector. The horizontal field configuration optimizes the measure of ξ_L, as the resolution $\Delta\tau_L/\tau_L \sim 2$ x 10^{-3}, is roughly two orders of magnitude better than that for $\xi r\|G$ and $\xi r\bot G$.

Structural Evidence for a Two-Step process in the depinning of the FLL

An applied transverse transport current imposes a Lorentz force on the FLL, but motion of the lattice will be inhibited by pinning. Beyond the critical current the lattice can break free of its pins and flow. The microscopic nature of this process is still poorly understood. In particular, little is known about the detailed structure of the lattice as it begins to depin and flow.

Our experiments were performed on a large ($9 \times 5 \times 1$ mm^3) single crystal of 2H-NbSe$_2$. x-ray and neutron diffraction confirmed the 2H polytype with c-axis mosaic $< 0.1°$ FWHM. Magnetization and transport studies show a sharp (<40mK width) superconducting transition at T=7.4 K. Shown in Fig. 2 are the FWHM of the rocking curves and the measured voltage as a function of current, at T=4.8K following a field-cooled process with no applied current.

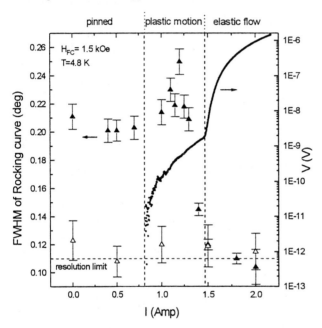

Fig. 2 The FWHM of the rocking curves for a sample cooled to 4.8K in an applied magnetic field H_{FC}=1.5 kOe ≈ 0.07 H_{c2} as a function of current I for both increasing (filled triangles) and decreasing (open triangles) currents. Also shown is the measured voltage V as a function of current. The vertical dashed lines separate the three regimes discussed in the text: pinned lattice, plastic motion and flowing crystal.

The combination of SANS and sensitive transport measurements suggests that the transition from a pinned lattice to a flowing crystal is by two steps. At low applied currents, I < 0.8A, the lattice is pinned. The lattice generated by a FC process is rather disordered and the FWHM of the rocking curves is significantly broader than the limit of resolution, implying $\xi_L \sim 10$ μm. It is believed that the disorder that characterizes this regime was frozen in upon crossing the highly disordered regime just below H_{c2}. In this regime we observe no voltage above our noise floor (~ 30 pV). Between ~ 0.8A and ~ 1.45A the voltage is consistent with activated flux creep, $V = V_0 \text{Exp}[(J/J_c-1)U/kT]$, with

$U/kT=7.4\pm0.5$, where U is an activation barrier. The motion in this regime is highly disordered as is evident from the broadening of the rocking curves, above their I=0 value, and from the extended tails of the non-Lorentzian rocking curves observed only in this regime, implying disorder on all length scales, down to the FLL constant, a_0. The motion in this regime is termed plastic as it is composed of bundles of flux lines that move in channels past pinned neighbors. Only at I ~1.45A a uniform flow sets in, and the FWHM of the rocking curves collapses to the limit of resolution, indicating that $\xi_L > 60$ μm. At this current a sharp kink is observed in the voltage response, above which the voltage follows a power law from, $V=V_0(J/J_c-1)^\alpha$, with $\alpha=1.7\pm0.1$. A remarkable feature is that the lattice disorder does not reappear upon decreasing the applied current back to I=0 (open triangles in Fig. 2), and the peaks remain resolution limited[10]. Cycling the current to above the critical current provides therefore a dynamical annealing mechanism. The transport measurements, on the other hand, are completely reversible, with no hysteresis. This suggests that the changes that occur in the FLL are on length scales which are larger than the pinning-related lengths discussed in the Larkin-Ovchinnikov Theory[1,11]. At higher fields, $H>H_{c2}/2$ hysteresis opens up in the I-V curves, in agreement with previous workers[12]. Unfortunately, at these fields SANS does not have the resolution to measure ξ_L. Transport and noise measurements in this regime show clear evidence of plastic motion[13]. Plasticity of the motion in this regime is considered to be related to the fast softening of the FLL as H_{c2} is approached, resulting a highly disordered FLL. Our work provides a clear evidence that even well below H_{c2}, the onset of motion of the FLL, driven to move against random pinning forces, is highly disordered. This observation is in agreement with recent theoretical calculations[14-15].

Structural Studies of the FLL Through the *Peak Effect* in Type II Superconductors

The *peak effect* - a dramatic increase in the critical current below H_{c2}, before it collapses to zero at H_{c2}, is one of the remaining puzzles from the earliest days of the study of type II superconductors[16]. The material of choice for studying the FLL in the vicinity of this effect is Nb, as the reflectivity from its FLL is large due to the short penetration depth. The study was performed on a very high quality, annealed Nb single crystal which was a right circular cylinder, 0.4 cm in diameter and 0.67 cm long with the cylinder axis nearly aligned with the [110] axis of the sample. Neutron diffraction studies showed the mosaic spread to be ~0.2° and transport, magnetization and specific heat studies showed a sharp superconducting transition, less than 40 mK wide at 9.16 K, and RRR~1000.

Complete sets of data as a function of field following a FC process were taken at two different temperatures, T=4.6K and T=6.4K. Shown in Fig. 2 are four false color contour plots taken at T=6.4K with relatively coarse resolution, after cooling in four different fields. Fig. 3 a, b, and c taken at 1.2 kOe, 1.35 kOe and 1.45 kOe show six clear first order Bragg peaks. In all cases, we find that the FLL Bragg peaks are oriented by the Nb crystal lattice. Fig 3d shows the image at H=1.48 kOe, ~ 50 Oe below $Hc2 \approx 1.53$ kOe. The observed ring of scattering is the signature of complete loss of long range orientational order.

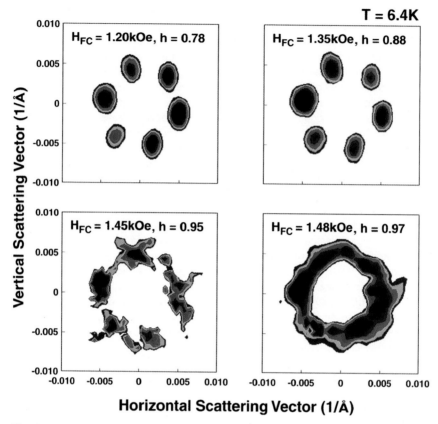

Fig. 3 Two dimensional false color intensity profiles of the scattered intensity vs. scattering vector for T=6.4K. As the field increases from (a) H=1.2 kOe (reduced field, h=0.78) to (b) H=1.35 kOe (h=0.88) to (c) H=1.45 kOe (h=0.95) and finally to (d) H=1.48 kOe (h=0.97) the Bragg peaks broaden into a ring of scattering.

Fig. 4 summarizes the data at 6.4K showing $\xi^{r\|G}$, the azimuthal width, ξ_L and the critical currents, $J_c^{transport}$ (registered at a constant-voltage criterion of 6nV/cm, measured on a 0.35x0.35x3.5 mm^3 piece cut from the large sample used for the SANS studies) and J_c^{mag} (calculated from magnetization measurements using the Bean model) for $0.65 < h < 1$, where h is H/H_{c2}. For fields below h~0.5 the sample is in the intermediate-mixed state. Initial broadening of the peaks, in the radial and azimuthal directions, from their resolution limited values start at h~0.75. In the region 0.75<h<0.90 the positional correlation length is short ranged, ~ 2a$_0$, while the orientational correlations

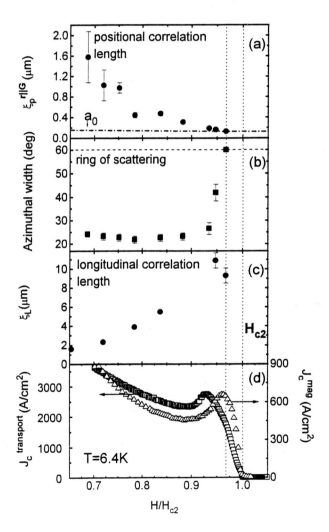

Fig. 4 Field dependence at T=6.4K of: (a) $\xi^{r\|G}$, (b) the azimuthal widths of the peaks, (c) ξ_L and (d) J_c^{mag} and J_c^{trans}.

remain longer than the dimensions of the sample. For h>0.93 there is a rapid increase in the azimuthal width and eventually at h ~ 0.97 (~ 50 Oe below H_{c2}) a ring of scattering is observed. This together with $\xi^{r\|G} \sim a_0$ is indicative of a completely disordered system. Throughout this regime ξ_L remains long, one to two orders of magnitude larger than $\xi^{r\|G}$. The overall structural observation is therefore that for fields above h~0.93 the FLL is characterized by complete loss of positional and orientational correlations, whereas the lines remain well correlated along their length.

Local and global magnetization measurements following a zero field cooled process by sweeping the field up to above H_{c2} and down to zero are irreversible up to H_{c2} and show a *fish tail* feature, leading to the observed peak in J_c^{mag}. The transport critical current shows two features - a peak at h~0.93 and a shoulder at h~0.97. Various surface treatments clearly show that the peak in J_c^{mag} and the shoulder in J_c^{trans} are a bulk response whereas the peak in J_c^{trans} is surface related.

In conclusion, at h ~ 0.97 we see changes in five quantities. $\xi^{r\|G}$ drops to ~a_0, the azimuthal width broadens into a ring of scattering, ξ_L begins to drop, there is a peak in J_c^{mag} and a shoulder in J_c^{trans}. Both J_c^{mag} and J_c^{trans} remain finite all the way up to H_{c2}, indicative of a glassy state. We therefore see clear evidence for an evolution of disorder with increasing field from a lattice to an orientationally ordered vortex glass to a vortex glass and then to the normal state. A more detailed presentation of the experimental results and discussion will be published soon[17].

† in collaboration with P. L. Gammel, D. A. Huse, R. N. Kleiman, C. S. Oglesby, E. Bucher, A. M. Chang, R. Ruel, L. Pfeiffer B. Battlog, G. D'Anna, D. J. Bishop, K. Mortensen, M. R. Eskildsen and K. N. Clausen.

References

1. for a review see G. Blatter, M. V. Feigel'man, V. B. Geshkenbein, A. I. Larkin and V. M. Vinokur, Rev. Mod. Phys. **66**, 1125 (1994).

2. for a review see D. J. Bishop, P. L. Gammel, D. A. Huse and C. A. Murray, Science **255**, 165 (1992).

3. C. A. Bolle *et al.*, Phys. Rev. Lett **72**, 4039 (1993).

4. C. A. Duran *et al.*, Nature **357**, 474 (1992).

5. H. F. Hess *et al.*, Phys. Rev. Lett. **69**, 2138 (1992).

6. A. M. Chang *et al.*, Appl. Phys. Lett. **61**, 1974 (1992).

7. D. Cribier, B. Jacrot, L. M. Rao and B. Franoux, Phys. Lett. **9**, 106 (1964); E. M. Forgan *et al.*, Nature **343**, 735 (1990); B. Keimer *et al.*, Science **262**, 83 (1993).

8. U. Yaron *et al.*, Phys. Rev. Lett. **73**, 2748 (1994).

9. U. Yaron *et al.*, Nature **376**, 753 (1995).

10. P. Thorel, R. Kahn, Y. Simon and D. Cribier, J. Phys. **34**, 447 (1973).

11. A. I. Larkin and Yu. Ovchinnikov, J. Low Temp. Phys. **34**, 409 (1979).

12. R. Wördenweber, P. Kes and C. C. Tsuei, Phys. Rev. **B33**, 3172 (1986).

13. S. Bhattacharya and M. J. Higgins, Phys. Rev. Lett. **70**, 2617 (1993); S. Bhattacharya and M. J. Higgins, Phys. Rev. **B49**, 10005 (1994); A. C. Marley, M. J. Higgins and S. Bhattacharya, Phys. Rev. Lett. **74**, 3029 (1995).

14. A. E. Koshelev and V. M. Vinokur, Phys. Rev. Lett. **73**, 3580 (1994).
15. T. Giamarchi and P. Le Doussal, Phys. Rev. Lett. **76**, 3408 (1996).
16. C. S. Tedmon Jr., R. M. Rose and J. Wulff, J. Appl. Phys. **36**, 829 (1965).
17. U. Yaron *et al.*, submitted to Science (1996).

Phase Diagram, Vortex Dynamics and Dissipation in Thin Films and Superlattices of 1:2:3 Superconducting Cuprates

M. Velez, J. I. Martin, E. M. Gonzalez, and J. L. Vicent

Departamento de Física de Materiales, Facultad de Físicas, Universidad Complutense, E-28040 Madrid, Spain

Abstract. The mixed state of High Temperature Superconductors (HTS) is a new field that allows us to study very interesting and new phenomena related with glassy behavior. There are many topics that could be addressed, but in this paper we are going to deal with two topics. The first one concerning to the single vortex dynamics, and the second one related with the dissipation mechanisms of the vortices and the so-calle d lock-in transition. The former will be studied using the anomalous behavior of the Hall effect in the mixed state with c-axis oriented 1:2:3 films the latter using as an appropriate tool the critical current, the pinning force and the resistivity (dissipation) in the m ixed state of 1:2:3 a-axis oriented superconductor / non-superconductors multilayers.

1   Introduction

The High Temperature Superconductors (HTS in the following) open a new phenomenological world in the realm of the Superconductivity with a clear projection in many fields of Physics. Since, the new experimental effects, that have been found in these materials, could be used as an ideal test to clarify controversial aspects mostly in Condensed Matter Physics and in Statistical Mechanics. These HTS are, from the experimental point of view, typical type II superconductors and they could be studied in the framework of the Ginzburg-Landau theory of second order phase tra nsitions.

The special properties of the HTS mixed state could be the adequate tools to understand and explore many phenomena in almost ideal experimental conditions, the experimental study of fluctuations is a clear example, fluctuations are accesible in a larger interval in HTS than in the tiny region in low temperature superconductors (LTS). The luc ky combination of several properties of the HTS is the main reason to be able to address new topics. Among these properties we could underline, of course, high transition temperature, above the liquid Nitrogen temperature; another important property is that these compounds are very anisotropic from the structural point of view. Therefore, the superconducting coherence length is small and anisotropic. Small coherence lengths and temperatures above the liquid Nitrogen lead to small vortex core and a mixed state behavior controlled, in some way, by thermal dissipation. The pinning of the vortices is one of the main problems from the applied and basic point of views. Movement of vortices means dissipation and therefore, the superconducting properties disappear. The so-called irreversibility line (Muller et al. (1987)) could move the real superconducting region in the (B,T) phase diagram to very low temperature values; for instance, if we applied a magnetic field of 5 T to a HTS sample of the cuprate family based on Bi (the so-called 2212 family) a critical current (J_c) different of zero could only be reached below T = 0.4 T_c. It is worth a while, before we go on, to mention that recently, Wu et al. (1995) have reported a method to obtained critical current densities (J_c) in HTS tapes (1:2:3 family) above 10^6 A/cm^2 at 77 K and with applied magnetic fiels of 5 T. This result is more than a strong indication that HTS are ready and suitable to be used in quite a few technologies, as the LTS are already from many years ago, but HTS with a clear advantage, since in HTS the complex and ex pensive liquid He world (needed in LTS applications) changes to the easier to handle and very cheap liquid Nitrogen world. HTS SQUID devices could be another clear example of the giant improvement that have been accomplished during the ten years of HTS life. The HTS SQUID noise has been lowered 8 orders of magnitude since 1987, and now there are very good quality HTS SQUID with the same performance than LTS SQUID, but working at 77 K.

One of the most interesting aspects of the HTS is the very complex mixed state behavior, the (H,T) phase diagram is not well understood yet. Fig. 1 shows the new HTS phase diagram. In any case, we have to point out that

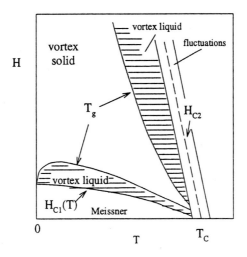

Fig. 1. Mixed state phase diagram in HTS

some of the new phenomena, which appear in HTS, could be found in some of the LTS; see for instance, the irreversibility line behavior reported by Guimpel et al. (1991) in LTS oxides. Recently, Blatter et al. (1994) have publish ed a very complete review of vortices in HTS. The interested reader could find there an exhaustive state of the art of this subject. Unfortunately or fortunately, the combination of high temperature, small coherence length, anisotropy and disorder (for ex ample due to the Oxygen content) lead to very complex glassy behavior. There are still many controversial aspects, but at the same time, the new phenomenology introduces a lot of exciting new possibilities for researchers in many experimental as well as t heoretical fields. Some of these new effects are related with the high transition tempera-tures; for example, increasing the temperature a perfect and ideal Abrikosov vortex-lattice could become a vortex-fluid. Another example could be the new phases as vo rtex-glass and vortex-liquid phases, the former with the resis-tivity going to zero when the current density is going to zero and the latter reaching a finite value of the resistivity when the current density vanishes, both phenomena related with the possible different types of vortex motion barriers and therefore creep behavior with a regimen of thermally assisted flux flow. The disorder could also play an important role. The disorder , in a first approach could transform the Abrikosov vortex-lattice in a vortex-glass. No disorder leads to a vortex-crystal with a first order transition (Safar et al. (1992)), ramdom disorder leads to vortex-glass (second order transition, Fisher et al. (1991)) and co rrelated disorder (Civale et al. (1991)) leads to Bose-glass second order transition, (Nelson et al. (1992)). The anisotropy and small coherence lengths lead to the so-called intrinsic lock-in transition (Tachiki et al. (1989)) and dimensionality problems (Lopez et al. (1991)), as f or instance, the interplay between the Abrikosov vortices and the coreless Josephson vortices, 2D vortex-pancakes (Clem (1991)), etc.

Actually a new world of phenomena in the frame of phase diagrams has risen from the HTS: vortex glass, vortex liquid, melting transition, thermally assisted flux flow, giant creep, Bose glass, splayed glass, lock-in transition and so on are only some of the topics that can be studied in the mixed state of HTS. In this paper we are going to focus, from the experimental point of view, on two relevant and not well understood subjects. The single vortex dynamics and the dissipation and lock-in transition in the anisotropic HTS. The former using the Hall effec t in the c-axis oriented films of the 1:2:3 cuprates and the latter using the critical currents and the tails of the resistivity transitions in a-axis oriented superlattices of 1:2:3 cuprates.

The natural superconducting anisotropy of these HTS compounds is due to the anisotropic crystalline structure with weakly superconducting areas between the CuO_2 planes. The paper will deal with films and superlattices based on the 1:2:3 cuprate family ($YBa_2Cu_3O_7$, YBCO in the following). The films could be grown either with the CuO_2 planes parallel to the substrates, the so-called c-axis oriented films or with the CuO_2 planes perpendicular to the substrates (a-axis films). Thin films with the former orientation (c-axis) have been grown from the beginning of the HTS, (Chaudari el al. (1987)). There are many interesting topics which have been addressed on this kind of films, but the latter orientation (a-axis) requires more critical growing conditions (Eom et al. (1990)) and, so far we know, very few groups are usually working on a-axis films. These a-axis oriented films show a very peculiar microstructure when they are grown on the usual cubic substrates, for example (100) SrTiO (STO). In this case, the films present the CuO_2 planes perpendicular to the substrates building microdomains and boundaries with a 90° misorientations of the c-axis lattice planes. The average size of these 90° microdomains is around 20 nm; therefore they are smaller than the superconducting penetration depth. In the mixed state these boundaries do not show any weak link behavior, see Velez et al. (1994).

The natural anisotropy of HTS could be enhanced, and in general modified, by artificially fabricated multilayers with alternately superconducting / non-superconducting layers. $PrBa_2Cu_3O_7$ (PBCO) is the most common insulator that is used in the HTS superlattice fabrication (Triscone et al. (1990)); although PBCO is a very special insulator with coexistence of metallic and nonmetallic electrical transport (Lee et al. (1995)), but the good matching and structural characteristics of these YBCO / PBCO superlattices are crucial to study many basic and applied subjects in the HTS field. In the case of studying the effects due to the PBCO layers in these 1:2:3 superlattices, the appropiate texture of the multilayered system is very important. In comparison with c-axis oriented superlattices, a-axis superlattices are the ideal system to study the contribution of the multilayered structure to the anisotropy. Since a-axis oriented multilayers allow us to separate the contributions of the natural anisotropy (CuO_2 planes) and the artificially induced anisotropy (PBCO layers), see Martin et al. (1995), because in this texture the CuO_2 planes run perpendicular to the PBCO layers. While in the case of c-axis oriented multilayers both anisotropies, the intrinsic (CuO_2 planes) and the artificial (PBCO layers) are acting at the same time, the insulating layers and the CuO_2 planes are parallel. Otherwise, c-axis oriented single films are the appropiate system for Hall effect measurements, because in comparison with a-axis films and multilayers, c-axis films have much lower normal state resistivity than a-axis systems. Therefore, in the latter system, the high background signal has to be carefully compensated, and more important the Hall effect signal is one order of magnitude lower in a-axis oriented than in c-axis oriented films.

2 Experimental Methods and Sample Characterization

Thin films of $RBa_2Cu_3O_7$ (R = Eu, Y, Ho) and superlattices of $EuBa_2Cu_3O_7$ / $PrBa_2Cu_3O_7$ have been grown by dc magnetron sputtering on single crystal substrates of (100) MgO, (100) $SrTiO_3$ (STO) and (100) $LaAlO_3$ (LAO) using targets of stoichiometric compositions. A commercial sputtering system was used with on-axis substrate-target geometry. The samples were obtained using the so-called not-aligned-chopped-power-oscillatory technique (Nakamura et al. (1992)). The multilayers were grown in the same system by alternately depositing $EuBa_2Cu_3O_7$ and $PrBa_2Cu_3O_7$ layers using two independent targets and stopping the substrates in front of the $EuBa_2Cu_3O_7$ (EBCO) and $PrBa_2Cu_3O_7$ (PBCO) cathodes by a computer controlled stepping motor. The total thickness of the samples were 250 nm. The substrate temperature and the partial Oxygen pressure during deposition allow us to obtain pure c-axis (Colino et al. (1991)) and pure a-axis (Velez et al. (1994)) oriented samples.

The structural and superconducting characterization of the films and superlattices has been reported elsewhere, see Martin et al. (1996). In summary, the films have been characterized by X-ray diffraction (XRD, Siemens D-5000) high resolution transmission electron microscopy (HRTEM, JEOL 4000), electron microscopy diffraction (EMD), scanning electron microscopy (SEM, Hitachi S-800), atomic force microscopy (AFM, Digital), resistivity, intensity-voltage (I,V) characteristics, critical current (J_c) and magnetotransport measurements up to magnetic fields of 9 Teslas using a commercial cryostat and temperature controller. A rotatable sample holder computer controlled by a stepping motor allows us to take angular dependence measurements. The rotation axis is perpendicular to the applied magnetic field direction. The films and superlattices were patterned via wet etching, the mask has two parts, one of them with a 50 μm bridge to measure critical currents and the other is a strip of width 500 μm and the adequate geometry for Hall effect and magnetoresistance measurements. A standard dc technique was used for the transport measurements. The Hall voltage was obtained from the antisymmetric part of the transverse voltage under magnetic-field reversal. The critical current density is defined using the 1 μv / mm criterion. The applied magnetic fields are always perpendicular to the electrical current direction.

3 Hall Effect in the Mixed State

The Hall effect is one of the most interesting topics in the mixed state; since it seems to be an adequate tool to study vortex dynamics. Josephson (1965) showed that vortex motion perpendicular to the current direction produces a dissipative electric field along the current direction and vortex motion parallel to the direction of the current produces an electric field perpendicular to both the current and the applied field direction. Therefore, the Hall effect seems to be an ideal tool to study vortex dynamics in the mixed state of type II superconductors. But, since the pioneer work of Reed et al. (1965), the Hall effect has remained an elusive topic. The single vortex motion was modeled by Bardeen and Stephen (1965) and by Nozieres and Vinen (1966), but both models do not explain the experimental Hall effect data.

The Hall effect shows an anomalous sign reversal in the mixed state in many systems, either HTS or LTS, see for example Harris et al. (1993). The origin of this anomaly cannot be understood in the frame of the classical vortex dynamics models. Many different approaches have been taken to figure out the origin of this anomaly, but this subject still remains very controversial, see for instance Li et al. (1996) and Clinton et al. (1995). In 1:2:3 cuprates, the c-axis films are the best choice to study this anomaly from the experimental point of view. Recently, the anomaly has been studied in c-axis oriented YBCO / PBCO superlattices (Qiu et al. (1995)) too.

Our experimental aim is not to test the different models that have been proposed, but we are trying to explore deeper some experimental facts, for example to study in detail, all the magnetic field range from low applied magnetic field to the magnetic field value where the Hall signal recovers the normal state sign. A very useful experimental technique, to get better insight, is to measure the (I, V) characteristics in longitudinal (ρ_{xx}) and transverse (ρ_{xy}) resistivities. Fig 2 (a) shows typical Hall effect data in the mixed state of a c-axis oriented film, taken with current density of J = 80 A / cm^2 . Figs. 2 (b) and 2 (c) show ρ_{xx} and ρ_{xy} resistivities versus current densities J. These Fig. 2 (b) and Fig. 2 (c), have been extracted from (I, V) data; each one of the curves has been recorded keeping constant the magnetic field from very low magnetic field values to magnetic fields close to the sign crossover in the Hall effect data. In this way, we can sweep the (H,T) phase diagram with each J value. Two important facts arise from these data, first of all they show the well known and clear correlation between both resistivities, not only in the vortex liquid region also in the vortex glass regime. Second and more important, the Hall effect is negative in the ohmic regime, as we already know, and in the non-ohmic regime too. The sign reversal happens from thermally assisted flux flow, flux creep up to the flux flow region (from the non-ohmic to the ohmic regimes).

Another interesting aspect of the Hall effect in the mixed state is the role played by intrinsic parameters in this anomaly. Hagen et al. (1993) have found that, in many systems of LTS and HTS, the Hall anomaly appears when the mean free path l and the BCS coherence length ξ_0 are similar, l/ξ_0 being between 1 and 5. They reported a value of 4.5 in the case of 1:2:3 cuprates. Following this result, Colino et al. (1994) have found a sample dependent empirical correlation between the maximum value of the Hall resistivity anomaly (ρ_{xy} max) and the parameter l/ξ_0. The recent models of Vinokur et al. (1993) and Khomskii et al. (1995) could be a hint to understand this peculiar Hall effect sign reversal. Hence, it is worth a while to study again the behavior of (ρ_{xy} max) vs. l/ξ_0. We have taken into account the different sample resistivities and a rough estimation of l have been done using a simple free-electron approximation (we have taken $\xi_0 = 1.2$ nm, the same coherence length value for all the samples). We have analyzed 14 different c-axis oriented samples, including data from the literature (YBCO/PBCO superlattices too), and different samples from this work.

The maximum value of ρ_{xy} vs H, in the region of the sign reversal, is

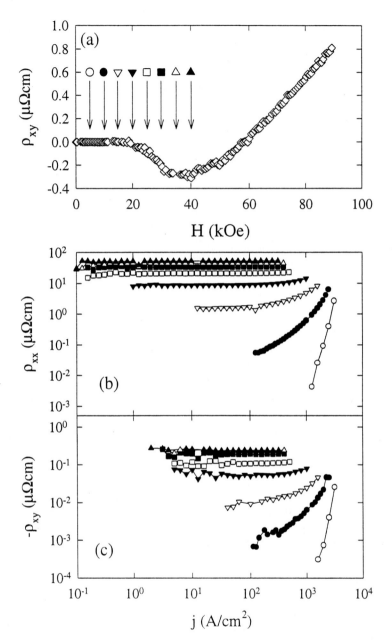

Fig. 2. (a) Hall resistivity vs magnetic field at 0.94 T_c of a c-axis oriented EBCO film. The arrows indicate the magnetic fields at which the (I,V) characteristics have been measured (from 5 kOe to 40 kOe in 5 kOe steps). (b) Longitudinal resistiv ity and (c) Hall resistivity vs current density. The symbols of (b) and (c) correspond to those indicated in (a).

temperature dependence and usually the highest value of a sample occurs around $T/T_c = 0.95$. Plotting these maxima vs l/ξ_0 we can see that samples with l around twice ξ_0 show the maximum sign reversal effect (see Fig. 3). That could mean that the ratio between the size of the vortex core (given by the coherence length ξ_0) and the mean free path, l, of the normal electron could be a hint to enhance the effect. The correlation between these intrinsic

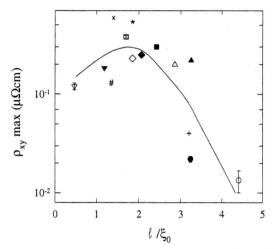

Fig. 3. Maximum absolute value of the Hall anomaly for each sample vs l/ξ_0 in several thin films and superlattices. $\diamondsuit$ Li et al. (1996); x Qiu et al. (1995); + Iye et al. (1990); $*$ Hagen et al. (1990); $\sharp$ Iye et al. (1990). The rest of the sy mbols are data from 1:2:3 films with different rare earths (Y, Eu, Ho) and growth on different substrates (STO, MgO, LAO) from this work. Solid line is a guide to the eye.

parameters and the anomalous Hall effect could be well understood in the frame of the models proposed by Vinokur et al. (1993) and Khomskii et al (1995). Following these authors, the transverse conductivity (σ_{xy}) could be written, after a straightforward calculation, as

$$\sigma_{xy} = \frac{nec}{B}(\frac{\Delta}{E_f})^2[\frac{1}{\pi^2}(\frac{l}{\xi_o})^2 - sign(\delta n)], \tag{1}$$

where E_f is the Fermi energy, Δ the superconducting gap, n the carrier density inside the core and $\delta(n)$ being the different between the carrier density inside the vortex core and outside the vortex core; since, according to Khomskii et al. (1995) could exist a difference between the density of carriers in the vortex core and outside the vortex core. This difference is due to a shift in the chemical potential in the superconducting state. When the first term, in expression (1), is large enough, therefore high values of l/ξ_0 (clean limit), the Hall effect in the mixed state and in the normal state have the same sign, but when the samples are between the clean and dirty limit, the Hall effect will show a sign reversal, since (Khomskii et al. (1995)) sign $\delta(n) = +1$.

Martin et al. (1994) have measured the Hall effect in the mixed state in the same sample with different Oxygen content. These results show that the value of the sign reversal could be changed and the effect could finally vanish with the variation of the Oxygen content. Wuyts et al. (1993) have shown that the decrease of Oxygen content decreases the carrier density n. All of these experimental results could be now relevant taking into account the models proposed by Vinokur et al. (1993) and Khomskii et al. (1995). Therefore, changing the value of n we will be able to tune the value of the Hall effect anomaly. This means that lower carrier density leads to lower value of the anomaly. This has to be considered in the models which deal with the role of this intrinsic parameter (n) in the explanation of the Hall effect anomaly.

In summary, we can conclude the following experimental results: 1) The sign reversal effect occurs in different vortex motion regimes. 2) This anomaly effect is enhanced for samples which are between the dirty and clean limit. 3) The effect decreases when the carrier density decreases from the optimum doping level. 4) last but not least, a subtle balance between intrinsic parameters (carrier density, mean free path and coherence length) has to be taken into account to model the single vortex motion. Finally, we have to point out, again, that this Hall effect sign reversal has been observed in HTS as well as LTS, but HTS allow us to study easier and deeper this anomaly than in LTS. In HTS we have a broader temperature intervals and several vortex dynamics regimes experimentally accesibles. The study of the single vortex dynamics behavior using the Hall effect anomaly is a clear example of the new possibilities open by the HTS.

4   Anisotropy and Dissipation in the Mixed State

The 1:2:3 HTS are anisotropic superconductors with a moderate anisotropy parameter, in comparison with other HTS families (Miu et al. (1995)), and also with anisotropic LTS, for instance intercalated transition-metal dichalcogenides (Vicent et al. (1980)). In anisotropic superconductors a possible dimensional crossover from three- (3D) to two-dimensional (2D) behavior could be driven by the applied magnetic field, the temperature (Coleman et al. (1983)), and al so with the modulation length in the case of artificially fabricated multilayers (Chun et al. (1984)). The usual Lawrence-Doniach model could be used as a starting point (Blatter et al. (1994)) for HTS. That is, the HTS could be modeled by mean of a stacking of superconducting slabs coupled by Josephson effect through the weakly superconducting region between the CuO_2 planes. Besides, these weakly superconducting areas could act as intrinsic pinning centers for the vortices (Tachiki et al. (1989)). They are the origin of the so-called lock-in transition, which enhances the J_c values when the applied magnetic field is parallel to the CuO_2 planes.

In this section, we will explore the possible influence of the layered structure on the anisotropy behavior, the lock-in transitions and the dissipation mechanisms in a-axis oriented EBCO / PBCO superlattices. We are going to focus on superlattices with thickness of the PBCO layers a few unit cells, clearly in the long range coupling regime, with long correlation length of vortices along the a axis direction and with the superconducting EBCO layers strongly coupled from the point of view of the vortex state (Suzuki et al. (1994)). Although the system is only expected to change its anisotropy, recent reports (Velez et al. 1996) have shown that even in this strong coupling regime the PBCO layers could modify the pinning forces behavior and the dissipation mechanisms; therefore, we can expect some effect on the anisotropy and scaling laws behavior too.

First of all, we have to define the geometry of the experiments. The angle θ, which defines the applied magnetic field direction, is measured from the substrate normal, that is $\theta = 0^o$ means that the applied magnetic field (H) is parallel to the CuO_2 planes. The films are patterned with the 90° boundaries making 45^0 with the direction of the applied current, (see inset in Fig. 4); The applied current is along the [013] axis. If we call θ_i the angle between the CuO_2 planes and H we have

$$sin\theta_i = sin\theta sin45^o \qquad (2)$$

This experimental geometry allows us to average the possible geometrical contribution coming from the microstructure. Expression (2) is taken into account in the scaling of the experimental data. In Fig. 4, the J_c angular dependence is shown for an a-axis oriented EBCO film. The enhancement of J_c at $\theta = 0^o$ is due to the natural lock-in transition (pinning of the vortices by the region between the CuO_2 planes). The same behavior is found in single crystal, Kwok et al. (1991), and c-axis oriented films, Iye et al. (1990).

The a-axis oriented EBCO/PBCO multilayers are an ideal system to study the interplay between the natural (CuO_2 planes) and the artificial (PBCO layers) anisotropy. The effect of this layered structure could be observed in the angular dependence of J_c vs H curves, even in the case the PBCO layers are few unit cells thick. When the applied magnetic field is increased an enhancement in J_c will appear at the applied field parallel to the PBCO layers, due to pinning coming from the artificial layers. These artificially induced pinning centers act as a new lock-in transition. They increase the J_c values when the magnetic field is applied perpendicular to the CuO_2 planes, just the opposite that happens in the usual lock-in transition (see Fig. 4).

Recently, Velez et al. (1996) have reported that the behavior of these PBCO layers could produce an angular dependence crossover which breaks the scaling laws for some critical angle and the superlattice behaves as a superconductor with a regular distribution of planar pinning centers. In the following, we are going to study in detail this behavior, and the possible

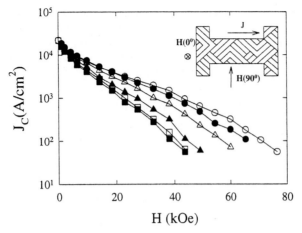

Fig. 4. Critical current vs magnetic field in a-axis oriented EBCO film at 0.75 T_c (θ = 0°, ○; 5°, ●; 15°, △; 35°, black △; 55°, □; 90°, black □). Inset shows the current and fiel d geometry respect to the 90° microdomains.

anisotropy scaling laws in this superlattice system. First of all, the tails of the resistivity transitions are a good probe to study the mixed state properties, see for instance Martin et al. (1995). Therefore, the experimental data and the scaling laws will be related to the $\rho(\theta)$ behavior. The scaling relation could be written, see Blatter et al. (1992) as:

$$\epsilon(\theta) = \frac{H_{c2}(90°)}{H_{c2}(\theta)} \tag{3}$$

$H_{c2}(90°)$ being the upper critical field with the applied magnetic field parallel to the c axis direction and $H_{c2}(\theta)$ the upper critical field, θ being the angle between the direction of the applied magnetic field and the normal to the film.

Scaling relations for the resistivity $\rho(\theta)$, in the frame of the Ginsburg-Landau theory, could be written as (see Blatter et al. (1992)):

$$\rho(H, \theta) = \rho(h) \tag{4}$$

and $h = H/H_{c2}$. Unfortunately, in HTS, one of the main experimental problems is the giant values of H_{c2}, even at temperatures close to T_c. But using (4) a suitable expression could be written to study the resistivity anisotropy

$$\rho(H, \theta) = \rho[H\epsilon(\theta)] \tag{5}$$

As an example, experimental data using this scaling could be seen in Fig. 5 (a) from a superlattice of 100 u. c. EBCO / 5 u. c. PBCO; for comparison the whole range scaling of a pure a-axis film is also shown, Fig. 5 (b). The data clearly show that in this superlattice the resistivity scaling breaks down above some critical angle, θ_c. That means, that the PBCO layers play a crucial role even when they are very thin in comparison with the EBCO layer thickness and the whole behavior of the superlattice is dominated by the artificially induced anisotropy, above a critical angle. It is striking that, although we can expect that the effect due to the PBCO layers appears when H is applied parallel or close to parallel to the insulating PBCO layers, the PBCO layers effect is playing a crucial role in a broad angular interval.

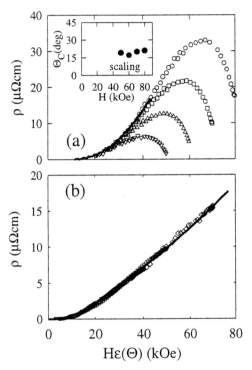

Fig. 5. (a) Resistivity vs reduced magnetic field for a-axis oriented 100 u. c. EBCO / 5 u. c. PBCO superlattice at 0.7 T_c (○, 80 kOe; □, 70 kOe; △, 60 kOe; ▽, 50 kOe; solid line $\rho(H,0°)$). Inset shows the critic al angle where the scaling breaks down. (b) Resistivity vs reduced magnetic field for a-axis oriented EBCO film. Magnetic fields between 10, and 70 kOe with steps of 10 kOe; solid line $\rho(H,90°)$.

The competition between the natural lock-in transition (CuO_2 planes) and the artificially induced lock-in transition (PBCO layers) could be easily observed in Fig. 6. This Fig. 6 shows the angular dependence of the resistivity for a-axis 40 u. c. EBCO / 5 u. c. PBCO superlattice. There are a minimum in the resistivity for $\theta = 0°$, where the field is parallel to the CuO_2 planes, due to the intrinsic anisotropy. The multilayer effect is shown when the applied magnetic field is parallel ($\theta = 90°$) to the PBCO layers.

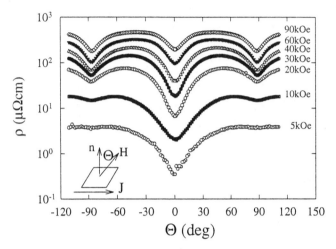

Fig. 6. Anisotropy of the resistivity in a-axis oriented 40 u. c. EBCO / 5 u. c. PBCO multilayer at 0.95 T_c.

In summary, the $\rho(\theta)$ curves show two dissipation minima, therefore two critical current, J_c, maxima (see Fig. 7), due to two different lock-in transition of the HTS vortex lattice.

The clear effect of the superlattice structure on the anisotropic behavior of the a-axis multilayers is remarkable, since this behavior is shown even in multilayers with very short thickness of the insulating layers; for instance, Fig. 6 shows the effect with only 5 unit cells thick (1.9 nm) of PBCO layers, which is comparable to the superconducting coherence length perpendicular to the EBCO superconducting layers $\xi_{ab} = 1.6$ nm and much smaller than the coupling length through a-axis oriented PBCO, which is estimated to be about 48 nm in a-axis oriented YBCO/PBCO superlattices, see Suzuki et al. (1994). The role of the PBCO layers can be described as planar pinning centers, characterized by a matching field H_Λ, very effective when the applied field is parallel to the substrate (see Velez et al. (1995a)). In a first approach, we can assume an Abrikosov vortex lattice which lattice parameter is a_0, and then, Λ being the superlattice modulation length, the matching field H_Λ is defined by

$$\Lambda = a_0 = (\frac{\Phi_0}{H_\Lambda})^{1/2} \qquad (6)$$

(Φ_0 being the fluxoid). The pinning force coming from the superlattice structure causes the smooth decreases of the critical current as the magnetic field

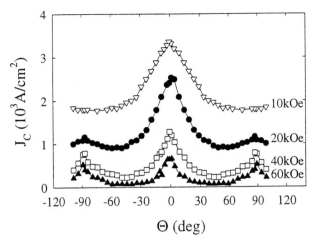

Fig. 7. Angular dependence of the critical current density of an a-axis oriented 40 u. c. EBCO / 5 u. c. PBCO superlattice at 0.4 T_c.

is increasing around H at $\theta = 90°$, since this pinning force is acting on a very soft HTS vortex lattice, see Velez et al. (1995a).

The influence of the PBCO layers on the behavior of the anisotropy is modified as the PBCO layer thickness is increased. The anisotropy of the a-axis multilayer could be completely reversed in comparison with the a-axis single film behavior, and the overall curve shape looks quite similar to the angular dependence of the resistivity of a c-axis single film, see Fig. 8.

Another interesting effect is the behavior of these strong coupled multi-layers when the superconducting layer thickness is decreasing. Fig. 9 (a) and (b) show the $H\epsilon(\theta)$ scaling of the resistivity for two superlattices with the same PBCO layer thicknesses, but decreasing the EBCO layer thicknesses (75 u. c. EBCO and 40 u. c. EBCO) in comparison with the superlattice of Fig. 5 (100 u. c. EBCO). The critical angle θ_c , where the scaling laws break down, is roughly the same, it does not change with the modulation length of the superlattice. These results seem to indicate that the effect is only related with the PBCO layers and these layers are effective in a broad angular range. Between $30°$ and $90°$ the artificially induced anisotropy dominates over the natural anisotropy. This could be due to the fact that we are dealing with a very soft vortex lattice in HTS and the vortices could accommodate to the PBCO pinning centers very easily,as we have said few lines above. This artificial lock-in transition is very strong and only when the magnetic field is applied very close to the CuO_2 planes direction the natural anisotropy is effective and the usual scaling laws appear.

Another remarkable effect is that, at the same temperature, there is a dimensional crossover in the scaling laws that fit the experimental data. The system follows a 3D type scaling law below a thickness of the EBCO layer of

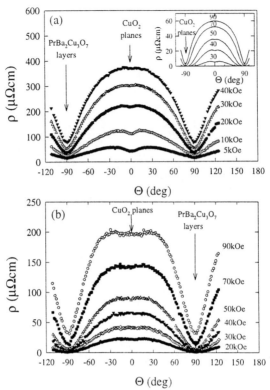

Fig. 8. Angular dependence of the resistivity for an a-axis oriented 40 u. c. EBCO / 5 u. c. PBCO superlattice, at (a) 0.9 T_c and (b) 0.7 T_c. The inset shows the angular dependence of a c-axis oriented EBCO film at 0.9 T_c.

75 u. c., while at the same temperature the sample with 100 u. c. of EBCO shows a 2D behavior. The latter (2D) could be understood in the frame of the usual decoupling of the CuO_2 planes below 0.85 T_c, the former (3D behavior) is not well understood at presen t. In the quasi-2D behavior the coherence length ξ_c is of the order of the distance between the CuO_2 planes and hence, these planes could be decoupled. The 75 u. c. EBCO / 5 u. c. PBCO multilayer has roughly the same ξ_c value than the 100 u. c. EBCO / 5 u. c. PBCO multilayer: In both films the critical temperature values T_c are similar, see Martin et al. (1996). Therefore, the comparison between ξ_c and the distance between the CuO_2 planes could be ruled out as the mechanism w hich governs the dimensional crossover 2D to 3D that we obtain decreasing the EBCO layer thickness in a-axis oriented EBCO / PBCO superlattices. If we remember that these multilayers are strong coupled and the vortex correlation length is of the order of the modulation length of the 100 u. c. EBCO / 5 u. c. PBCO superlattice, but much larger than the modulation length of the 75 u. c. EBCO / 5 u. c. PBCO multilayer, we can conclude

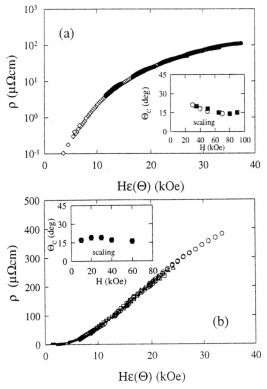

Fig. 9. (a) Logarithmic plot of the resistivity vs reduced magnetic field scale for a-axis oriented 75 u. c. EBCO / 5 u. c. PBCO superlattice at 0.7 T_c. The resistivities are measured in the range $\theta < \theta_c$. Open symbol $\rho(H, 0°)$. Filled symbols $\rho(\theta)$ measured at H = 35 kOe, 50 kOe, 65 kOe, 80 kOe and 90 kOe. Inset shows the critical angle where the scaling breaks down. Open symbol at 0.95 T_c, and filled symbol 0.7 T_c. (b) The same plot as (a) but in linear scale fo r an a-axis oriented 40 u. c. EBCO / 5 u. c. PBCO superlattice at 0.95 T_c.

that decreasing the thickness of the superconducting layers we can drive a crossover to a 3D behavior in the scaling laws of the dissipation mechanisms of the vortices.

Finally, we are going to deal with the onset of dissipation in this system. The onset of dissipation could be easily measured using the tails of the resisitivity transitions when a magnetic field is applied to the sample. In HTS a thermally activated motion of the vortices has been proposed (Palstra et al. (1988)), that follows a thermally activated flux creep model. In a-axis oriented 1:2:3 cuprate films the activation energy follows this behavior (Velez et al., 1995b) with a magnetic field dependence of the energy barriers given by

$$U(T, B) = \frac{U_0(T)}{B} \tag{7}$$

Fig. 10 shows the (ρ, B) behavior in the case of a 100 u. c. EBCO / 5 u. c. PBCO superlattice.

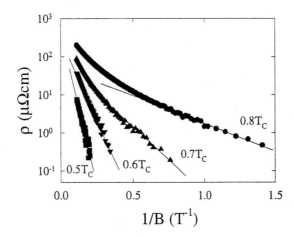

Fig. 10. Logarithmic plot of the resistivity vs 1/B for an a-axis oriented 100 u. c. EBCO / 5 u. c. PBCO multilayer.

The behavior is typical of an activated mechanisms the only difference with the case of single a-axis oriented film, is that in the multilayers the values of the activation energies are three times lower than in pure a-axis 1:2:3 films.

A peculiar crossover in the activation regimen occurs when the superconducting layer thickness decreases, see Fig. 11.

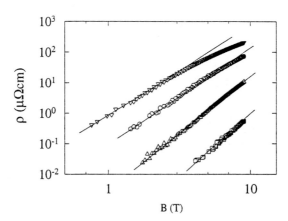

Fig. 11. Log ρ vs log B for an a-axis oriented 20 u.c. EBCO / 5 u. c. PBCO multilayer.

The onset of dissipation follows a logarithmic behavior below 60 u. c. EBCO. We have to remember that these multilayers are stongly coupled and therefore, an explanation of this effect could not be related to a crossover to 2D behavior as has been pr oposed to explain the log B dependence in c-axis oriented 1:2:3 superlattices (Brunner et al. (1991)), but with uncoupled superconducting layers.

In a-axis strong coupled superlattices the log B dependence of the activation energies could not be due to dimensionality problems, probably this behavior could be related to an uniform distribution of energy barriers of the pinning centers.

5   Conclusions

The 1:2:3 superconducting oxide cuprates show a rich and complex behavior in the mixed state. We have studied two aspects of these phenomena: 1) The roles that play the different vortex motion regimes and intrinsic parameters in the anomalous Hall effect behavior using c-axis oriented EBCO films. 2) The anisotropy and dissipation in a-axis oriented EBCO/PBCO superlattices. In these multilayers the natural anisotropy, coming from the CuO_2 planes, is decoupled from the artificial anisotropy induced by the PBCO insulating layers.

Concerning to the Hall effect in the mixed state, the following experimental facts have been found:

i) (I, V) characteristic data show that, in the ohmic and non-ohmic vortex motion regimes, the sign of the Hall effect is always negative and with the opposite sign to the positive normal state Hall effect.

ii) Intrinsic parameters as the mean free path, the coherence length and the carrier density seem to mediate in this effect. The maximum value of the sign reversal seems to be related with samples which are between the dirty and the clean limit. The decrease of the normal carrier density leads to a decrease of the Hall effect anomaly.

In closing, these results suggest that the vortex core size, the density of carriers and their mean free path could be relevant parameters that should be taken into account to understand the mechanisms that could explain this anomalous effect in single vortex dynamics.

a-axis oriented EBCO/PBCO superlattices have been used to study the role played by the insulating PBCO layers in the anisotropy, scaling laws and dissipation mechanisms of the EBCO/PBCO system. This sample texture allows us to separate the contributions coming from the CuO_2 planes and the PBCO layers. We have used multilayers with only 5 u. c. of PBCO; that is, we have long-range coupling across the PBCO layers. Therefore, when the magnetic field H is applied perpendicular to the substrate the superlattices show strong coupling of vortices along the a axis direction, that means a long vortex correlation length. In summary, the main results are the following:

i) The insulating PBCO layers induce an artificial lock-in transition that could overcome the effect of the intrinsic lock-in transition (CuO_2 planes). This new lock-in transition could reverse the anisotropy of the a-axis oriented EBCO / PBCO sup erlattices

ii) The angular dependence of the $\rho(H)$ in EBCO/PBCO only follows scaling laws in a very narrow interval between 0° - 30°, (0° being the direction with H perpendicular to the PBCO layers). This is due to that the PBCO layers act as very effective pinning centers and the vortex lattice in HTS is very soft. The PBCO layers dominate the anisotropy behavior in EBCO/PBCO multilayers, breaking the usual scaling laws, even in the case of being PBCO layer thickness very short and, therefore, with a strong vortex coupling along the a axis direction, that is, across the insulating PBCO layers.

iii) The activation energies and therefore the dissipation mechanisms show a crossover from the usual thermally activation regime to a logarithmic behavior when the thickness of the EBCO layers decrease.

In closing, in the a-axis oriented EBCO/PBCO system we have two clear situations. 1) For applied magnetic fields up to 30° off the normal to the substrate (close to the CuO_2 planes) the EBCO layer thickness controls the anisotropy behavior, the system follows scaling laws. 2) If the angle between the direction of the applied field and the normal to the film is higher than 30° , the PBCO layers rule the whole multilayer behavior, breaking the usual anisotropy scaling laws and a new lo ck-in transition appears, since the PBCO layers act as very effective pinning centers.

6   Acknowledgments

We thank Spanish CICYT (grant MAT 96 - 0904) for financial support.

6.1 References

Bardeen J., Stephen M. J. (1965) Phys. Rev. **140**, A1197.

Blatter G., Geshkenbein V. B. and Larkin A. I. (1992) Phys. Rev. Lett. **68**, 875.

Blatter G., Feigel'man M. V., Geshkenbein V. B., Larkin A. I. and Vinokur V. M. (1994) Rev. Mod. Phys. **66**, 1125.

Brunner O., Antognazza L., Triscone J. M., Mieville L. and Fisher O. (1991) Phys. Rev. Lett. **67**, 1354.

Civale L., Marwick A. D., Worthington T. K., Kirk M. A., Krusin-Elbaum L., Sun Y., Clem J. R. and Holtzberg F. (1991) Phys. Rev. Lett. **67**, 648.

Chaudari P. Koch R. H., Laibowitz R. B., McGuire T. R. and Gambino R. J. (1987) Phys. Rev. Lett. **58**, 2684.

Chun C. S. L., Zheng G. G., Vicent J. L. and Schuller I. K. (1984) Phys. Rev. B**29**, 4915.

Clem J. R. (1991) Phys. Rev. B**43**, 7837.

335

Clinton T. W., Smith A. W., Li Q., Peng J. L., Greene R. L., Lobb C. J., Eddy M. and Tsuei C. C. (1995) Phys. Rev. B52, 7046.

Coleman R. V., Eiserman G. K., Hillenius S. J., Mitchell A. T. and Vicent J. L. (1983) Phys. Rev. B27, 125.

Colino J., Sacedon J. L. and Vicent J. L. (1991) Appl. Phys. Lett. 59, 3327.

Colino, J., Gonzalez, M. A., Martin J. I., Velez M., Oyola D., Prieto P. and Vicent J. L. (1994) Phys. Rev. B49, 3496.

Eom C. B., Marshall A. F., Laderman S. S., Jacowitz R. D. and Geballe T. H. (1990) Science 248, 1549.

Fisher D. S., Fisher M. P. A. and Huse D. A. (1991) Phys. Rev. B43, 130.

Guimpel J., Hoghoj P., Schuller I. K., Vanacken J. and Bruynseraede Y. (1991) Physica C 175, 197.

Hagen S. J., Lobb C. J., Green R. L., Forrester M. G. and Kang J. H. (1990) Phys. Rev. B41, 11630.

Hagen S. J., Smith A. W., Rajeswari M., Peng J. L., Li Z. Y., Gree R. L., Mao S. N., Xi X. X., Bhattacharya S., Li Q. and Lobb C. J. (1993) Phys. Rev. B47, 1064.

Harris J. M., Ong N. P. and Yang Y. F. (1993) Phys. Rev. Lett. 71, 1455.

Iye Y., Nakamura S. and Tamegai T. (1989) Physica C159, 433.

Iye Y., Nakamura S., Tamegai T., Terashima T. and Bando Y. (1990) MRS Symposia Proceeding 169, 871.

Josephson B. D. (1965) Phys. Lett. 16, 242.

Khomskii D. I. and Freimuth A. (1995) Phys. Rev. Lett. 75, 1384.

Kwok W. K., Welp U., Vinokur V. M., Fleshler S., Downey J. and Crabtree G. W. (1991) Phys. Rev. Lett. 67, 390.

Lee M., Suzuki Y. and Geballe T. H. (1995) Phys. Rev. B51, 15619.

Li K., Zhang Y. and Adrian H. (1996) Phys. Rev. B53, 8608.

Lopez D., Nieva G. and de la Cruz F. (1994) Phys. Rev. B50, 7219.

Martin J. I., Wuyts B., Maenhoudt M., Osquiguil E., Vicent J. L., Moshchalkov V. V. and Bruynseraede Y. (1994) Physica C235-240, 1451.

Martin J. I., Velez M. and Vicent J. L. (1995) Phys. Rev. B52, 3872.

Martin J. I., Velez M. and Vicent J. L. (1996) Thin Solid Films 275, 119.

Miu L., Wagner P., Hadish A., Frey U. and Adrian H. (1995) J. Supercon. 8, 293.

Muller K. A., Takashige, M. and Bednorz, J. G. (1987) Phys. Rev. Lett. 58, 1143.

Nakamura O., Fullerton E. F., Guimpel J. and Schuller I. K. (1992) Appl. Phys. Lett. 60, 120.

Nelson D. R. and Vinokur V. M. (1992) Phys. Rev. Lett. 68, 2398.

Nozieres P. and Vinen W. F. (1966) Philos. Mag. 14, 667.

Palstra T. T. M., Batlogg B., Scheemeyer L. F. and Waszczack J. V. (1988) Phys. Rev. Lett. 61, 1662.

Qiu X. G., Jakob G., Moshchalkov V. V., Bruynseraede Y. and Adrian H. (1995) Phys. Rev. B52, 12994.

Reed W. A., Fawcett E. and Kim Y. B. (1965) Phys. Rev. Lett. 14, 790.

Safar H., Gammel P. L., Huse D. A., Bishop D. J., Rice J. P. and Ginsberg D. M., (1992) Phys. Rev. Lett. bf69, 824.

Suzuki Y., Triscone J. M., Eom C. B., Beasley M. R. and Geballe T. H. (1994) Phys. Rev. Lett. 73, 328.

Tachiki M. and Takahashi S. (1989) Solid State Commun. **70**, 291.

Triscone J. M., Fisher O., Brunner O., Antognazza L., Kent A. D. and Karkut (1990) Phys. Rev. Lett. **69**, 804.

Velez M., Martin J. I. and Vicent J. L. (1994) Appl. Phys. Lett. **65**, 2099.

Velez M., Martin J. I. and Vicent J. L. (1995a) Appl. Phys. Lett. **67**, 3186.

Velez M., Martin J. I. and Vicent J. L. (1995b) IEEE Trans. Appl. Superc. **5**, 1537.

Velez M., Gonzalez E. M., Martin J. I. and Vicent J. L. (1996) Phys. Rev. B**54**, 101.

Vicent J. L., Hillenius S. J. and Coleman R. V. (1980) Phys. Rev. Lett. **44**, 892.

Vinokur V. M., Geshkenbein V. B., Feigel'man M. V. and Blatter G. (1993) Phys. Rev. Lett. **71**, 1242.

Wu X. D., Foltyn S. R., Arendt P. N., Blumenthal W. R., Campbell I. H., Cotton J. D., Coulter J. Y., Hults W. L., Maley M. P., Safar H. F. and Smith J. L. (1995) Appl. Phys. Lett. **67**, 2397.

Wuyts B., Maenhoudt M., Gao Z. X., Libbrecht S., Osquiguil E. and Bruynseraede Y. (1993) Phys. Rev. B**47**, 5512.

Monte Carlo Study of a Three–Dimensional Vortex Glass Model with Screening

Carsten Wengel and A. Peter Young

Department of Physics, University of California, Santa Cruz, CA 95064, USA

1 Introduction

- ♠ Type–II Superconductors in magnetic fields $> H_{c_1}$ form a regular array of vortices (Abrikosov–Lattice)
- ♠ Vortices exposed to a current $J \perp H$ move collectively due to the Lorentz–Force, producing a voltage and thereby destroying superconductivity
- ♠ Point disorder pins vortices. This leads to
  - ⇒ destruction of Abrikosov–Lattice
  - ⇒ nonlinear response to J
  - ⇒ vanishing linear resistivity $\rho_{\mathrm{lin}} = \lim\limits_{J \to 0} E/J = 0$
  - ⇒ new superconducting phase: *Vortex Glass*
- ♠ Experiments and Simulations (Gauge Glass Model):
  - ⇒ in two dimensions $T_c \doteq 0$
  - ⇒ in three dimensions $T_c > 0$

2 Problem

- ♠ Vortices interact logarithmically up to length scale λ (penetration depth); in Type–II–SC, $\lambda \gg \xi$, where ξ is the correlation length
- ♠ Sufficiently close to T_c there is a crossover where the vortex interactions are *screened* due to critical fluctuations — this may alter the critical behavior (see fig. 1)

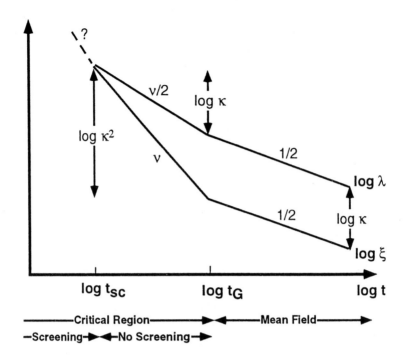

Fig. 1. A sketch of the logarithm of the correlation length, ξ, and penetration depth, λ, against $\log t$, where $t \equiv T - T_c$, for a type–II superconductor. The slopes of the curves are indicated: ν is the correlation length exponent in the part of the critical region where screening can be neglected. Two crossovers occur. The first, at the reduced Ginzburg temperature $t_G \equiv T_G - T_c$, is when critical fluctuations first start to be important. The second, at $t_{sc} \equiv T_{sc} - T_c$, is when λ and ξ become comparable. Screening effects are important for $t \leq t_{sc}$. For $T > T_G$, $\lambda/\xi = \kappa(> 1)$ is constant, while for $T_{sc} < T < T_G$, λ^2/ξ is constant. Between T_G and T_{sc}, ξ increases by a factor of κ^2 and λ by a factor of κ.

3 Aim & Questions

- ♠ Investigate a Vortex Glass Model with screening via finite temperature Monte Carlo in three dimensions
- ♠ Does screening destroy the finite temperature transition?
- ♠ Does the linear resistivity really vanish in Type–II–SC?

4 Model & Method

The *Gauge Glass* Model with Screening on a cubic lattice is defined as

$$\mathcal{H}_{GG} = -J \sum_{<i,j>} \cos(\phi_i - \phi_j - A_{ij} - \lambda_0^{-1} a_{ij}) + \sum_\square [\boldsymbol{\nabla} \times \mathbf{a}]^2$$

J coupling constant ($J = 1$)
ϕ_i phase of the order parameter on site i
A_{ij} random quenched vector potential $\in [0, 2\pi]$ uniformly
a_{ij} fluctuating vector potential
λ_0 bare screening length

For simulation purposes we go to the vortex representation of the above Hamiltonian:

♠ Perform Villain–Approximation
♠ Perform Dual Transformation
♠ Take the strong screening limit $\lambda_0 \to 0$

$$\mathcal{H}_V = -\frac{1}{2} \sum_i (\mathbf{n}_i - \mathbf{b}_i)^2$$

$\mathbf{n}_i$ vortices $\in \{0, \pm 1, ...\}$ defined on the links of the dual (simple cubic) lattice of size L^3
$\mathbf{b}_i$ quenched disorder given by the curl of A_{ij}
Constraints: $[\boldsymbol{\nabla} \cdot \mathbf{n}]_i = [\boldsymbol{\nabla} \cdot \mathbf{b}]_i = 0$, $\sum_i \mathbf{n}_i = \sum_i \mathbf{b}_i = 0$

5 Results

We measure the linear resistivity from the Kubo–Formula

$$\rho_{\text{lin}} = \frac{1}{2T} \int\limits_{-\infty}^{\infty} \langle V(t) V(0) \rangle \, dt \tag{1}$$

where $V(t)$ is the voltage drop (± 1) due to the movement of vortices at time t. At a continuous phase transition, ρ_{lin} has the scaling form

$$\rho_{\text{lin}} = L^{-(2-d+z)} \tilde{\rho}\left(L^{1/\nu}(T - T_c)\right) \tag{2}$$

with the scaling function $\tilde{\rho}$ and the dynamical exponent z.
At T_c, $\tilde{\rho}(0) = $ const. so that for different pairs (L, L')

$$\ln[\rho_{\mathrm{lin}}(L)/\rho_{\mathrm{lin}}(L')]/\ln[L/L'] = -(2-d+z) \quad \text{at } T_c. \tag{3}$$

i. e., all curves of the above quantity for different pairs (L, L') intersect at T_c. The value at the intersection point is $d - 2 - z(= 1 - z$ here).

5.1 Pure and disordered system

From plotting our data according to eq. (3) we obtain:

Pure System: $T_c = 0.331 \pm 0.002$, $z \simeq 3$.
Disordered System: $T_c \simeq 0 \Rightarrow z = \infty$.

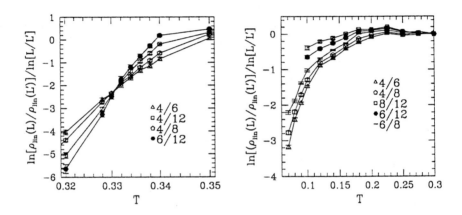

Fig. 2. Left: Pure System (all $b_i = 0$): this is the dual representation of the pure XY model with inverted temperature scale. We find the intersection at $T_c = 0.331 \pm 0.002$ and the y–value equals roughly -2 corresponding to $z \simeq 3$. **Right:** Disordered System: there appears to be no intersection over the entire temperature range, indicating the *absence* of a phase transition.

5.2 Finite–size scaling and dynamics

Finite–size scaling of the I–V–Characteristics according to

$$\frac{E}{J\rho_{\mathrm{lin}}} = \tilde{g}\left(\frac{J}{T^{1+2\nu}}, L^{1/\nu}T\right) \tag{4}$$

yields

$$T_c = 0, \quad \nu = 1.05 \pm 0.1$$

341

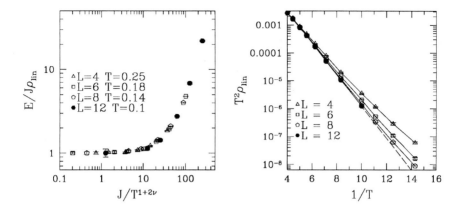

Fig. 3. Left: Scaling plot of the I-V–characteristics with $T_c = 0$ and $\nu = 1.05$, according to eq. (4), choosing sizes and temperatures such that $L^{1/\nu}T$ is roughly constant. **Right:** Plot of $T^2\rho_{\text{lin}}$ on a logarithmic scale versus $1/T$. The dashed line corresponds to an Arrhenius law, which, according to eq. (5), corresponds to a barrier exponent, ψ, equal to zero. The data for the largest size, $L = 12$, is well approximated by this.

At a zero temperature transition we expect the long time dynamics to be "activated"

$$T^{1+\nu}\rho_{\text{lin}} \propto \exp(-\Delta E/T) \quad \text{with } \Delta E \propto L^{\psi} \propto T^{-\psi\nu}, \tag{5}$$

where ψ is a barrier height exponent. From plotting $T^{1+\nu}\rho_{\text{lin}}$ with $\nu = 1$ versus $1/T$ we find approximately

$$\underline{\psi \simeq 0.}$$

Presumably, barriers increase logarithmically as $T \to 0$.

6 Conclusions

- ♠ Screening *destroys* the finite–temperature transition in the three– dimensional Gauge Glass
- ♠ This result is consistent with a domain wall renormalization study of the Gauge Glass with screening by Bokil & Young (PRL **74**, 3021 (1995)).
- ♠ Experimentally, the Vortex Glass transition would be rounded out — the linear resistance would not strictly vanish, only become extremely small as T approaches the nominal "T_c". So far experiments have lacked the sensitivity, but it would be interesting to look for this effect in the future.
- ♠ Critical Exponents: $\nu = 1.05 \pm 0.1$, $\psi \simeq 0$

Equilibrium Phase Transitions in Josephson Junction Arrays

S. Teitel

Department of Physics and Astronomy
University of Rochester, Rochester, NY 14627, USA

Abstract. I review several problems dealing with the equilibrium behavior of classical two dimensional Josephson junction arrays in applied magnetic fields. Specific attention is given to the cases of a uniform field with average flux density per unit cell of $f = 0$, $f = 1/2$, $f = 1/q$ and $f = 1/2 - 1/q$. Several models incorporating the effects of randomness on the Josephson array are also reviewed. These include the case of a random vortex pinning potential and its effects on vortex lattice order, and the spin glass, gauge glass, and positionally disordered array.

1 Introduction

Two dimensional arrays of coupled Josephson junctions have been a topic of much experimental and theoretical investigation for approximately the last fifteen years. As they are a system in which topological singularities, frustration, incommensurability, and randomness can all come into play, they serve as an excellent model for studying many diverse problems in statistical physics. More recently, Josephson arrays have received attention in connection with high temperature superconductors. Both are superconducting systems in which thermal fluctuations play a crucial role in determining the macroscopic behavior. The Josephson array, with its simpler phase space and well defined geometry, can therefore serve as a test case for understanding many issues of importance to high T_c superconductors, such as vortex pinning and the effects of randomness.

Yet despite nearly fifteen years of investigation, there remain fundamental unresolved issues in even some of the most simply posed problems. In this article I will review some of the theoretical work concerning the equilibrium phase transitions in classical Josephson arrays, placing emphasis on results from numerical simulations. I will try to point out what is understood, and what questions remain open.

1.1 The Josephson Junction Array Model

The system of the Josephson array can be conceptualized as follows. At each site of a grid of points there is a superconducting island. Nearest neighbor islands are coupled to each other by the tunneling of Cooper pairs (either though proximity effect barriers, or oxide layer barriers), thus producing a

Josephson junction on each bond of the grid. The relevant degrees of freedom of the system are then the phase angles θ_i of the superconducting wavefunction on sites i of the grid. If the array is placed in an external magnetic field given by the vector potential $\mathbf{A}$, one defines on nearest neighbor bonds $\langle ij \rangle$ the integrals $A_{ij} = (2\pi/\Phi_0) \int_i^j \mathbf{A} \cdot \mathbf{d\ell}$, where $\Phi_0 = 2e/hc$ is the flux quantum. The Hamiltonian of the array is then just the sum of Josephson energies for each nearest neighbor bond,

$$\mathcal{H}[\theta_i] = \sum_{\langle ij \rangle} V_{ij}(\theta_i - \theta_j - A_{ij}) \ , \tag{1}$$

where $V_{ij}(\phi)$ is the coupling energy of bond $\langle ij \rangle$, and its argument is the gauge invariant phase angle difference across the bond. $V_{ij}(\phi)$ is quadratic about its minimum at $\phi = 0$, has period 2π, and $dV_{ij}/d\phi$ is proportional to the supercurrent flowing through the bond. I consider here only the classical case in which charging energy, and hence quantum effects, can be ignored. The equilibrium behavior of the array is then obtained from the partition function, summing $e^{-\beta \mathcal{H}}$ ($\beta = 1/k_B T$) over all possible configurations of the phases $\{\theta_i\}$. If one ignores inductance effects, the A_{ij} remain fixed parameters determined from the applied magnetic field. Here I will consider only the case where the grid of sites i form a periodic two dimensional lattice, which unless stated otherwise, I take to be square. I will also assume that there is no randomness in the couplings, and hence the $V_{ij}(\phi) \equiv V_0(\phi)$ are all equal. Periodic boundary conditions will be imposed at the edges of the system.

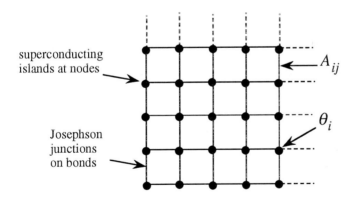

superconducting islands at nodes

Josephson junctions on bonds

Fig. 1. Schematic geometry of a Josephson junction array

The different junctions of the array are coupled to each other through a topological constraint on the θ_i. Since θ_i is the phase angle of a complex wavefunction, one must find that the sum of phase angle differences $[\theta_i - \theta_j] \in (-\pi, \pi]$, going around any closed path on the grid must sum to $2\pi n$, with n

integer. n is the net vorticity contained within the path. Such vortices are the key excitations of the Josephson array. In the absence of any applied magnetic field, $A_{ij} = 0$, and the ground state has all θ_i equal, so that $V_0(\phi)$ is minimized on each bond. When $A_{ij} \neq 0$, the above topological constraint will in general prevent the θ_i from adjusting so as to cancel out the effect of the A_{ij}. The result will be a ground state in which the θ_i have some spatially varying pattern, $V_0(\phi)$ is no longer minimized on all bonds, and so there is a flow of finite local supercurrents. Since each bond is now no longer able to achieve its minimum energy, such a model is said to be *frustrated*. For the case where the A_{ij} describe a uniform magnetic field perpendicular to the plane of the array, (1) is referred to as the *uniformly frustrated* 2D XY model.

1.2 Coulomb Gas Duality

For a Josephson junction one typically uses $V_0(\phi) = -J_0 \cos(\phi)$, where the coupling constant J_0 is related to the critical current of the junction I_0 by, $I_0 = (2e/\hbar)J_0$. However much theoretical simplification can be achieved by using instead the Villain function (Villain 1975), defined by

$$e^{-\beta V_0(\phi)} = \sum_{m=-\infty}^{\infty} e^{-\frac{1}{2}\beta J_0(\phi - 2\pi m)^2} \quad . \tag{2}$$

In this case, exact duality transformations (José et al. 1977, Fradkin et al. 1978, Vallat and Beck 1994) map (1) onto an equivalent problem of logarithmically interacting charges, given by the Hamiltonian,

$$\mathcal{H}[n_i] = \tfrac{1}{2}e^2 \sum_{i,j}(n_i - f_i)G(\mathbf{r}_i - \mathbf{r}_j)(n_j - f_j) \quad , \tag{3}$$

Here the *dual* sites i and j sit at the centers of the unit cells of the original grid; f_i is the sum of the $A_{ij}/2\pi$ going counterclockwise around the unit cell at i and gives the number of flux quanta of applied field penetrating cell i; $n_i = 0, \pm 1, \pm 2, \ldots$ are integers; the unit of charge is $e = \sqrt{2\pi J_0}$; and $G(\mathbf{r})$ is the solution to the lattice Laplacian with periodic boundary conditions, which can be explicitly written for a square lattice of length L as,

$$G(\mathbf{r}) = \frac{2\pi}{L^2} \sum_{k \neq 0} \frac{e^{i\mathbf{k} \cdot \mathbf{r}}}{4 - 2\cos k_x - 2\cos k_y} \quad . \tag{4}$$

Here the sum is over all wavevectors satisfying periodic boundary conditions, $k_\mu = \frac{2\pi}{L}m_\mu$, $m_\mu = 0, 1, \ldots, L-1$, and the grid spacing has been taken to be unity. For large separations, $1 \ll r \ll L$, $G(\mathbf{r}) \simeq -\ln r$.

The Hamiltonian (3) can be viewed as one of interacting integer charges n_i, superimposed on a quenched background charge distribution f_i, and is referred to as the 2D *Coulomb gas* (CG). The partition function is obtained summing over all charge configurations $\{n_i\}$ subject to the constraint of total charge neutrality $\sum_i (n_i - f_i) = 0$. In this duality mapping, the integer charges n_i are related to the vortices in the phase angles θ_i of the original Josephson array (Vallat and Beck 1994). Smooth spin wave fluctuations of the θ_i about a given vortex configuration give an additional contribution to the system energy, which is decoupled from the vortex part (3). This spin wave part is Gaussian, and so may be directly summed in the partition function, giving a non-singular additive contribution to the free energy which is henceforth ignored.[1] This Coulomb gas formulation of the Josephson array will form the basis for most of the following discussion. Henceforth, I will use the terms "charge" and "vortex" interchangeably.

The partition function derived from (3) is invariant under the transformation $f_i \rightarrow f_i + m_i$, where m_i is any integer, as such a change in f_i can always be canceled out by the addition of an appropriate integer charge n_i to site i. It is thus sufficient to consider only cases with $-\frac{1}{2} < f_i \leq \frac{1}{2}$.

1.3 The Kosterlitz-Thouless Instability Criterion

In terms of the Coulomb gas model (3) one can search for an insulator to conductor transition by considering the inverse dielectric function ϵ^{-1}. Defining the net charge on site i by $q_i \equiv n_i - f_i$, and its Fourier transform by $q_k = \sum_i e^{-i\mathbf{k}\cdot\mathbf{r}} q_i$, linear response theory gives (Minnhagen 1987),

$$\epsilon^{-1}(T) = \lim_{k \to 0} 1 - \frac{2\pi e^2}{L^2 T} \frac{\langle q_k q_{-k} \rangle}{k^2} . \tag{5}$$

The conducting state is one in which there are freely diffusing charges and $\epsilon^{-1} = 0$. In the insulating state, charges n_i are bound either to each other in neutral clusters, or to the background f_i, and $\epsilon^{-1} > 0$. Since a charge is identified with a vortex in the Josephson array, and each time a vortex crosses a path the total phase angle difference along that path changes by 2π, the free diffusion of charges (vortices) will correspond to large phase angle fluctuations that destroy superconducting coherence. Thus one can show that the conducting (insulating) phase of the Coulomb gas corresponds exactly to the normal (superconducting) phase of the Josephson array (Ohta and Jasnow 1979, Minnhagen 1987).

A criteria for predicting the instability of the insulating phase has been given in the pioneering work of Kosterlitz and Thouless (1973) (KT). Consider the energy U of a single *free* charge e, including the effect of the dielectric screening due to other *bound* charges. Since the dominant effects come from large length scales, one can make a continuum approximation, giving $U = \int d^2r |\mathbf{E}(\mathbf{r})|^2 / (4\pi\epsilon) = (e^2/2\epsilon) \ln(L/a)$, where $\mathbf{E} = e/|\mathbf{r}|$ is the electric field of

---

[1] Note however that a coupling between spin waves and vortices remains in the original XY model (1) if the cosine interaction is used in place of the Villain interaction (Ohta and Jasnow 1979).

the charge e, L is the radius of the system, and a is the hard core radius of the charge. The entropy of the free charge is just the logarithm of the number of non-overlapping positions in which the charge may be placed, $S = \ln(L/a)^2$. Thus the total free energy for a free charge to appear is,

$$F = U - TS = \left(\frac{e^2}{2\epsilon(T)} - 2T \right) \ln(L/a) \ . \tag{6}$$

As $L \to \infty$, $F \to \pm\infty$, depending of the sign of the prefactor of the logarithm. Thus when $T < e^2/4\epsilon$, it costs infinite free energy to create an isolated free charge, and the insulating phase is stable against such an excitation. When $T > e^2/4\epsilon$, however, the free energy for a free charge is infinite but negative; free charges proliferate and the insulating phase becomes unstable. If T_c is the true insulator to conductor transition temperature, one then has the following bound,

$$T_c \le e^2/4\epsilon(T_c) \ . \tag{7}$$

The above gives only an upper bound on T_c, as it is always possible that an excitation more complicated than the isolated free charge considered above, may drive the insulator to conductor transition.

The above bound on T_c also gives a very important result concerning the behavior of the dielectric function. One can rewrite (7) as,

$$\epsilon^{-1}(T_c) \ge 4T_c/e^2 \ . \tag{8}$$

Since one has $\epsilon^{-1} = 0$ for $T > T_c$ in the conducting state, (8) implies that for finite T_c, ϵ^{-1} must make a discontinuous jump to zero at the transition.

2 Uniform Frustration

In this section I focus on behavior for the case of a Josephson array in a uniform applied magnetic field (Teitel and Jayaprakash 1983a, 1983b). This corresponds to the CG with a uniform quenched background charge, i.e. all $f_i = f$, a constant. By the condition of charge neutrality, the ground state will consist of a finite density of integer charges (vortices) with $\sum_i n_i = fL^2$. For rational $f = p/q$ these ground state charges should be arranged in a periodic structure. However finding this ground state structure for a general $f = p/q$ remains an unsolved problem. The competition between the repulsive charge-charge interaction and the grid geometry of allowed charge sites, leads to complicated commensurability effects and discontinuous behavior as f is smoothly varied. A particularly clear experimental verification of such commmensurability effects has been seen in kinetic inductance measurements on a triangular Josephson array (Théron et al. 1994).

To find ground states in specific cases, one must resort to symmetry arguments combined with a numerical search through likely candidate states. The most extensive listing of ground states, considering all $f = p/q$ for $q \le 20$, has

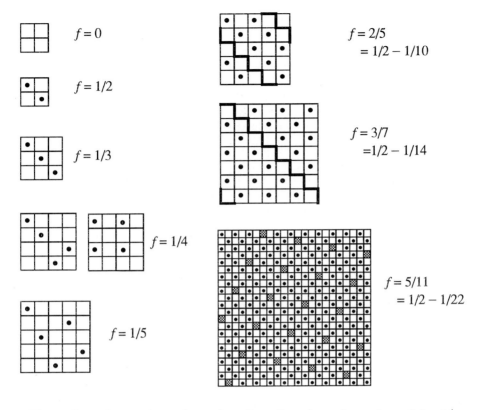

Fig. 2. Ground state charge (vortex) configurations for various values of $f = 1/q$ and $f = 1/2 - 1/q$. ($\bullet$) denotes a charge; for $f = 5/11$, a shaded box denotes a charge missing from the $1/2$–like background.

been given by Straley and Barnett (1993). In Fig. 2 are shown some selected ground states for two special cases: $f = 1/q$ and $f = 1/2 - 1/q$.

For $f = 0$, the ground state is a charge vacuum. For $f = 1/2$, the ground state is a checkerboard pattern of charges, with a double discrete degeneracy corresponding to the two possible sublattices which the integer charges $n_i = +1$ may occupy. For $f = 1/q$, the ground state is a periodic lattice that, as q gets larger, becomes an increasingly better approximation to the triangular lattice that would be found for vortices in a continuum. Note that for $f = 1/4$ there are two different degenerate configurations (Korshunov 1986, Straley and Barnett 1993). For $f = 1/2 - 1/q$, the ground state looks almost everywhere like the checkerboard pattern of $f = 1/2$ except for localized defect regions, which are required so as to give the correct charge density $f < 1/2$. For small values of q these defect regions take the form of domain walls between the two degenerate $f = 1/2$–like ground states; these are indicated by the heavy lines in Fig. 2 for $f = 2/5$ and $3/7$. For larger

q, the defects take the form of a supperlattice of missing charges in a single $f = 1/2$–like ground state; these are indicated by the shaded boxes in Fig. 2 for $f = 5/11$. Note that the ground state for $f = p/q$ is often described by a $q \times q$ unit cell. However this is not generally true. As shown in Fig. 2 for example, $f = 5/11$ is periodic with a $2q \times 2q$ unit cell.

How the ground state charge (vortex) structure melts upon increasing temperature is another question for which the general answer remains unknown. I will consider in greater detail below the cases $f = 0$, $f = 1/2$, $f = 1/q$ and $f = 1/2 - 1/q$ for *large* q. For $f = 1/5$ and $f = 2/5$ it is known that the charge lattice has a first order melting transition to a conducting liquid (Li and Teitel 1990, 1991). The simple fractions $f = 1/3$ and $1/4$ have received some study (Grest 1989, Lee and Lee 1995), however the critical behavior of these transitions remains poorly understood.

2.1   Ordinary XY Model: $f = 0$

The case $f = 0$ corresponds to the ordinary two dimensional XY model, originally studied by Kosterlitz and Thouless. The ground state is the vacuum and the low temperature excitations of the insulating phase consist of dipoles formed of bound $n_i = +1$, $n_j = -1$, charge pairs. As T increases, the average separation of the charges in these dipoles increases, until at a critical temperature T_c, the charge pairs unbind giving free charges and a conducting phase. The renormalization group (RG) recursion equations by Kosterlitz (1974) quantify this picture and yield the result that the KT instability criterion (7) is satisfied as an exact equality. Equivalently, the discontinuous jump to zero of $\epsilon^{-1}(T_c)$ has the *universal* value $4T_c/e^2$. Although ϵ^{-1} is discontinuous, the transition is second order with an infinite correlation length at T_c.

This KT transition has been well established both experimentally, as well as by high precision Monte Carlo simulations (Olsson 1995a). For a good theoretical and experimental review see Minnhagen (1987). However the Kosterlitz RG analysis is an expansion in charge fugacity that applies only at small average charge densities, $\rho \equiv L^{-2} \sum_i |n_i|$. At large ρ one can question if the KT mechanism continues to hold. Several authors, using continuum models, have attempted to extend the KT analysis to higher charge densities by including the screening effect of free charges on bound charges (Minnhagen and Wallin 1987, 1989, Thijssen and Knops 1988b, Levin et al. 1994, Friesen 1995). They have predicted that the KT transition becomes first order as the fugacity increases. In particular, Levin at al. predict that the KT transition becomes first order at the relatively low charge density of $\rho_c = 0.0039/a^2$, where a is the diameter of the hard core charge.

To investigate numerically the behavior of the $f = 0$ CG at large charge densities, one can add a chemical potential term to the Hamiltonian (3),

$$\mathcal{H}[n_i] = \tfrac{1}{2} \sum_{i,j} n_i G(\mathbf{r}_i - \mathbf{r}_j) n_j - u \sum_i n_i^2 + \sum_i (n_i^4 - n_i^2) . \qquad (9)$$

Here all the $f_i = 0$, and henceforth temperature is measured in units such that the magnitude of the integer charges is $e \equiv 1$. The last term in (9) effectively restricts the charges to the values $n_i = 0, \pm 1$, in which case the second term is equivalent to $-u\rho$; u is thus the chemical potential for a charge. Increasing u increases the average charge density in the system. Monte Carlo simulations for this CG on a square grid (Lee and Teitel 1992) then yield the phase diagram shown in Fig. 3 below.

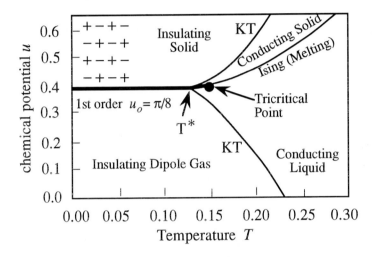

Fig. 3. Phase diagram of the dense $f = 0$ CG

There is indeed a first order transition (heavy solid line), however it separates the insulating gas of charge dipoles from an *insulating* charge solid of alternating $+1$ and -1 charges. As this charge solid is heated, there is first a KT transition to a conducting solid in which charge interstitials and vacancies can diffuse freely, followed by an Ising–like melting of the solid to a liquid. The two KT transition lines, in the solid phase and in the dilute gas phase, meet the first order line at the same temperature, $T^* \simeq 0.126$, somewhat below the tricritical point where the Ising melting line and the first order line meet.

The origin of the charge solid phase is easy to see by considering the Fourier transform of the first two terms of (9), $\mathcal{H} = L^{-2} \sum_k \left(\frac{1}{2} G_k - u \right) n_k n_{-k}$. As u increases, the ground state will change from the vacuum to an ordered charge solid when $2u$ first equals the smallest value of G_k. For the square grid Green's function (4) this occurs for $\mathbf{k} = \pi\hat{x} + \pi\hat{y}$ (thus giving the checkerboard pattern of the ground state) at $u_0 = \pi/8$.

As the first order line in Fig. 3 is a direct consequence of the formation of a charge solid, it is presumably unrelated to the first order transitions predicted by the above theoretical works, which are all built around charge pairs as the fundamental charge correlation.[2] It is thus worthwhile to give closer attention to the transition line from the gas of dipoles to the liquid, which is labeled as "KT" in Fig. 3. To test that this line does indeed remain a second order phase transition all the way up to the first order line at $u = u_0$, Gupta and Teitel (1996) have computed histograms of charge density, as temperature is varied for fixed $u = 0.39$ just below the first order line at $u_0 = \pi/8 \simeq 0.3937$. If the transition is first order, there should be a discontinuity in charge density as the first order line is crossed. One thus expects, in the vicinity of the transition, to see a charge density histogram with two distinct peaks corresponding to the two differing average charge densities of the two phases. If however the transition is second order, with no discontinuity in charge density, only a single peak should be present. The numerically computed histograms are shown in Fig. 4 below, for a system of size $L = 64$. As is seen, there is no hint at all of a bimodal distribution for any of the temperatures in the vicinity of the transition. While one can not rule out the possibility of a very weak first order transition, with finite correlation length $\xi(T_c) \gg L = 64$, the present evidence suggests that the transition remains second order, and presumably remains in the KT universality class. One can also see from Fig. 4 that the average charge density at this KT transition is $\rho \simeq 0.11$, well above the ρ_c estimated by Levin et al.

The phase diagram of Fig. 3, and in particular the presence of the charge solid phase, is very strongly influenced by the fact that the CG has been placed on a square grid of allowed charge sites. If a different geometry for the grid is used, the locations of the various phase boundaries shift, and charge solids with different symmetries can form (Lee and Teitel 1992). One can speculate whether for such another grid, or for charges in a continuum, it is possible that the melting transition line of the charge solid moves down to lower temperatures, so that it intersects the first order line at a temperature below T^*. One then could have a first order transition from an insulating gas of dipoles (or more complicated neutral clusters) to a very dense conducting liquid. Such a result has been reported by Caillol and Levesque (1986), who simulate a hard core CG in a continuum, placing the charges on the two dimensional surface of a sphere. Their KT line ends at a temperature and

---

[2] Nevertheless, it is interesting to note that the values of $u_0 = \pi/8$ and $T^* = 0.126$ in Fig. 3 do in fact lie fairly close to the values predicted by Minnhagen and Wallin. In order to compare the u_0 for the first order line of the CG on the square grid with Minnhagen and Wallin's results for a continuum CG, it is necessary to note (Kosterlitz and Thouless 1973) that if the interaction $G(\mathbf{r})$ on the grid is chosen so as to asymptotically match the continuum $-\ln r$ as $r \to \infty$, then the grid CG with $u = 0$ acts like a continuum model with a chemical potential $\mu_0 = -\frac{1}{2}(\gamma + \frac{3}{2}\ln 2) \simeq 0.8085$, where $\gamma \simeq 0.5772$ is Euler's constant. Thus a chemical potential $u_0 = \pi/8$ on the grid acts like a chemical potential $\mu = u_0 - \mu_0 = -0.416$ in the continuum. Minnhagen and Wallin predict that the KT line will end at $T^* = 0.144$, and chemical potential $\mu = T^* \ln z^* = -0.420$.

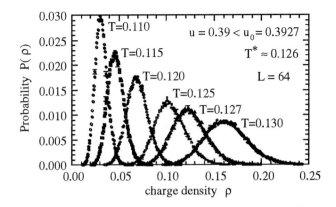

Fig. 4. Charge density histograms at various T, as the KT transition line is crossed for $u = 0.39$ just below $u_0 = \pi/8$

density comparable to the values found on the square grid, only they report no solid phase above the first order line. However, while the discrete square grid explicitly favors the formation of a charge solid, the surface of a sphere explicitly discourages it. In order to fit a periodic solid to the curvature of the spherical surface, it is necessary to introduce lattice defects, which might remain mobile even at low temperatures (Dodgson 1995).

Most recently, Lidmar and Wallin (1996) have carried out simulations of the hard core CG in a flat continuum with periodic boundary conditions. They find that the KT line remains second order down to quite low temperature and high density, until it finally hits a first order line at $T^* \simeq 0.032$. As in the case of the discrete grid, this first order line is associated with a transition in ground state from the vacuum to a charge solid. However unlike the discrete grid, their evidence suggests that the charge solid is melted at any finite temperature. Their first order line is thus a transition from an insulating gas to a dense conducting liquid.

2.2 Fully Frustrated Case: $f = 1/2$

Square Grid: On a square grid, the $f = 1/2$, or fully frustrated, CG consists of the checkerboard ground state shown in Fig. 2. This ground state breaks two distinct symmetries: (i) the continuous symmetry associated with uniform rotation of all phase angles θ_i of the original Josephson array model (1) - in the CG this is reflected in the insulating nature of the ground state; and (ii) the discrete translational symmetry broken by choosing one of the two equivalent sublattices on which to place the integer charges n_i. A natural question is whether these two symmetries are broken at one single, or two distinct, transition temperature(s)?

Since the background charge is $f = 1/2$, and the integer charges in the ground state are all $n_i = 0$ or $+1$, the net charge on a site is $q_i = n_i - f = \pm\frac{1}{2}$. One can therefore view the ground state as a checkerboard pattern of alternating positive and negative half integer charges. One can now identify two different types of excitations of the ground state, characteristic of the two types of broken symmetry. These are illustrated in Fig. 5 below. The first, referred to as a "KT–like" excitation, consists of the interchange of a given pair of $+\frac{1}{2}$ and $-\frac{1}{2}$ charges, with separation $\mathbf{r}$. The energy of such an excitation is proportional to $\ln r$. As long as these pair excitations remain bound with finite r, the inverse dielectric function ϵ^{-1} will stay finite. The unbinding of these pairs will lead to a vanishing of ϵ^{-1}, which might occur with the same KT mechanism as describes the unbinding of charge pairs in the $f = 0$ case. However, although the unbinding of such pairs restores symmetry (i) and yields a conducting phase, it does not necessarily destroy the translational order of the ground state. Such a conducting charge solid has already been seen in the dense region of the $f = 0$ model (see Fig. 3).

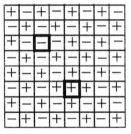
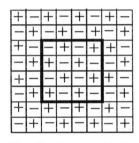

KT-like charge pair excitation   Ising-like domain excitation

Fig. 5. Two types of excitations of the $f = 1/2$ ground state

The second type of excitation, referred to as "Ising–like," consists of reversing the sign of all charges within a neutral domain. For domains which contain no net dipole moment, the energy of such an excitation is proportional to the perimeter of the domain. Such domain excitations, which can involve large numbers of charges, can not be accounted for within the small fugacity expansion of the KT analysis. It is these domain excitations that will melt the charge solid and restore the translational symmetry (ii). Since the ground state is doubly degenerate, one would naively expect this melting to be in the Ising universality class. Once the charge lattice has melted into a charge liquid, with freely diffusing charges, one expects $\epsilon^{-1} = 0$.

There are thus two likely scenarios: (i) Upon heating, the KT–like excitations cause a KT transition at T_{KT} with a universal jump in $\epsilon^{-1}(T_{\mathrm{KT}})$. This is followed at a higher T_{I} by the melting of the charge solid with Ising critical exponents. Or, (ii) the two types of excitations become sufficiently coupled that there is only a single transition T_c. In such a case the jump in $\epsilon^{-1}(T_c)$ might be larger than the universal value, and the melting could perhaps even be in a different universality class than Ising.

Early numerical works (Teitel and Jayaprakash 1983a, Miyashita and Shiba 1984, Lee et al. 1986, Thijssen and Knops 1988a, Grest 1989, Nicolaides 1991) focused, with inconclusive results, on determining whether there was one single, or two separate, transition(s). More recent works (Lee et al. 1991, Granato and Nightingale 1993, Ramirez-Santiago and José 1994, Lee 1994, Lee and Lee 1994, Knops et al. 1994) have focused on finite size scaling analyses of the transitions. They have reported critical exponents for melting distinctly different from Ising values, and jumps in ϵ^{-1} larger than the universal KT value, thus suggesting scenario (ii). Recently, Nightingale et al. (1995) have carried out Monte Carlo transfer matrix simulations on a related coupled XY-Ising model believed to be in the same universality class as the $f = 1/2$ CG. Considering systems of size $L \times \infty$, $L \leq 30$, they similarly find non-Ising melting and a larger than universal jump in ϵ^{-1}. However they find that the *central charge* (or conformal anomaly number) has not yet converged to its asymptotic large L limit, and the XY degrees of freedom similarly seem to show large corrections to scaling. They say that this " . . . calls into question the validity of the basic assumption of scaling theory, viz., that there is a single divergent length scale in this system as the critical point is approached. . ."

Most recently Olsson (1995b), using system sizes up to 128×128, has presented evidence supporting scenario (i). According to his arguments, the correlation length $\xi_{KT}(T)$ of the insulator to conductor transition, which diverges at T_{KT}, is still very large at the slightly higher melting transition T_I. This additional length scale at T_I complicates the finite size scaling analysis of the melting transition. Only for system sizes $L \gg \xi_{KT}(T_I)$ will one find a simple scaling characterized by Ising critical exponents. The non-Ising values found in previous works, according to his argument, merely reflect too small values of L. A conclusive resolution of the nature of the transition(s) in this model thus apparently awaits simulation of larger size systems.

Similar questions remain for other cases where the simple symmetry of the ground state charge lattice would naively suggest a melting transition in a well known universality class. One example is the $f = 1/3$ CG on a triangular grid, which might be expected to be in the 3–state Potts universality class (Korshunov 1986, Lee and Teitel 1992).

Triangular Grid: Another interesting model is the $f = 1/2$ CG on a triangular grid of sites (corresponding to a honeycomb Josephson array) (Korshunov 1986). For the triangular grid, the Green's function giving the charge interaction is given in terms of its Fourier transform by (Lee and Teitel 1992),

$$G_k = \frac{3\pi}{6 - 2\cos(\mathbf{k} \cdot \mathbf{a}_1) - 2\cos(\mathbf{k} \cdot \mathbf{a}_2) - 2\cos(\mathbf{k} \cdot \mathbf{a}_3)} , \tag{10}$$

where $\mathbf{a}_1$ and $\mathbf{a}_2$ are the two basis vectors of the triangular grid, and $\mathbf{a}_3 = \mathbf{a}_2 - \mathbf{a}_1$.

A natural candidate for the ground state would have alternating $\pm\frac{1}{2}$ charges along one of the grid directions $\mathbf{a}_\mu$ ($\mu = 1, 2, 3$) and be uniform in the remaining directions. Such a state is characterized by a wavevector with $\mathbf{k} \cdot \mathbf{a}_\mu = \pi$, but $\mathbf{k} \cdot \mathbf{a}_\nu = 0$, $\nu \neq \mu$. However, looking at the form (10) for G_k, one can easily see that all wavevectors such that $\mathbf{k} \cdot \mathbf{a}_\mu = \pi$ have degenerate values of G_k, regardless of their components along the other directions $\mathbf{a}_\nu$. The ground state is thus a configuration in which the charges oscillate in sign in one of the three directions $\mathbf{a}_\mu$, but are completely random in the other directions. The degeneracy[3] of the ground state is thus 3×2^L. An example of one such a ground state, which oscillates in the $\mathbf{a}_1$ direction, is shown in Fig. 6 below.

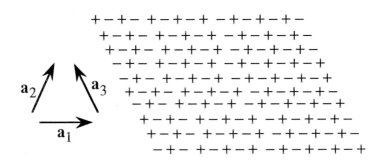

Fig. 6. $f = 1/2$ on a triangular lattice: one of the 3×2^L possible ground states

Such a large ground state degeneracy makes this system a good candidate for glassy behavior. However it remains to ask whether the system can have a true finite temperature equilibrium glass transition, or whether $T_c \to 0$ and so any glassiness would be a result of non-equilibrium effects. Lee and Teitel (1992) have carried out Monte Carlo simulations of this model, computing the dielectric function ϵ^{-1} and specific heat per site C. The results are shown in Fig. 7 below.

One sees in ϵ^{-1} an apparent transition at finite $T_c \simeq 0.035$. The non diverging peak and steep drop of C at T_c are characteristic of a structural glass

---

[3] Korshunov (1986) argues that the 2^L degeneracy may be lifted in the original XY model (1) with a cosine interaction, due to the coupling between spin waves and vortices that the cosine introduces.

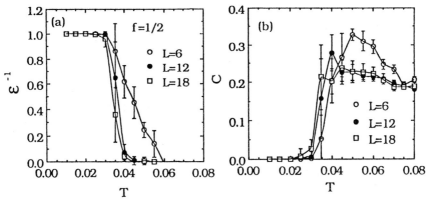

Fig. 7. $f = 1/2$ on a triangular lattice: $\epsilon^{-1}(T)$ and specific heat $C(T)$ for various system sizes L

transition. However to see whether this behavior reflects a true thermodynamic transition, or is rather a freezing out of equilibrium due to large energy barriers in phase space, consider the excitation of a ground state shown in Fig. 8 below. The excitation is formed by taking a $2 \times \ell$ domain and reversing the sign of all the charges. ℓ is chosen in the same direction as that in which the charges oscillate, and is chosen to be even, so that forming the domain creates no net dipole moment. The excitation energy $\Delta E(\ell)$ vs. ℓ is also plotted in Fig. 8. One sees that for $\ell > 10$, $\Delta E(\ell)$ saturates to a finite value $\Delta E(\infty)$. It is easy to see why this is so. If ℓ is parallel to direction $\mathbf{a_1}$ for example, then since all configurations with any sequence of charges in direction $\mathbf{a_2}$ are degenerate ground states, the domain wall segments parallel to $\mathbf{a_1}$ do not look *locally* like domain walls at all! Locally, they are consistent with the system being in one of the other degenerate ground states. It is only near the domain wall "end" segments parallel to the $\mathbf{a_3}$ direction that one notices one is not in a ground state. As the regions near these "end" segments are charge neutral and have no net dipole moment, any interaction between the two ends decays rapidly with increasing ℓ, and so contributions to $\Delta E(\ell)$ arise solely from the local distortion of the ground state near these ends.

Since $\Delta E(\infty)$ is finite, there is only a finite energy barrier to create infinitely long domains. As the growth of such infinitely long domains causes transitions between the 2^L different ground states which are all oscillating in the same direction, the system must in principle remain disordered among these 2^L states at any finite T. The time to hop between these ground states $\sim e^{\Delta E(\infty)/T}$, however, will get extremely long at low T.

Although the above domain excitations will disorder the system among the 2^L different ground states that oscillate in the same lattice direction, they will not cause transitions between ground states which oscillate in different directions. It thus remains an open question whether there can be a finite

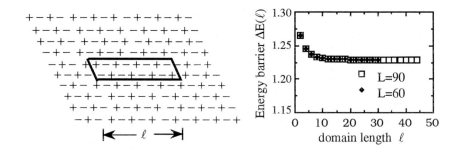

Fig. 8. $f = 1/2$ on a triangular lattice: domain excitation of ground state of length ℓ and excitation energy $\Delta E(\ell)$

temperature equilibrium transition which orders the system into one of the three possible classes of ground states (i.e. into the class of states in which the charges all oscillate in a particular direction $\mathbf{a}_\mu$), but leaves the system disordered among the 2^L degenerate ground states in that class.

2.3   Dilute Case: $f = 1/q$

Next consider the case of a dilute density of charges, $f = 1/q$. For large q, the ground state will be the closest approximation to a triangular charge lattice that is commensurate with the underlying square grid of allowed sites. The melting of this charge lattice, as $q \to \infty$, becomes a model for the melting of the vortex lattice in a thin superconducting film.

Consider as a example the specific case of $f = 1/51$. The ground state charge configuration is shown in Fig. 9a below. This case is chosen because, in contrast to other $f = 1/q$ with large q, where the ground state becomes very closely triangular, here the ground state remains very close to square, and so the influence of the discrete grid is strongest.

A convenient quantity for studying the melting of this charge lattice is the charge density structure function,

$$S(\mathbf{k}) = \frac{1}{fL^2} \sum_{i,j} e^{i\mathbf{k}\cdot(\mathbf{r}_i - \mathbf{r}_j)} \langle n_i n_j \rangle \ , \tag{11}$$

which gives the diffraction pattern that would be obtained from scattering off the charge positions. Together with the inverse dielectric function $\epsilon^{-1}(T)$, $S(\mathbf{k})$ will characterize the phase transitions of the system. Heating from the ground state, Franz and Teitel (1995) have computed $S(\mathbf{k})$ and ϵ^{-1} within Monte Carlo simulations. Shown in Figs. 9b-d are the resulting intensity plots of $S(\mathbf{k})$, for k_x, $k_y \in (-\pi, \pi]$, at three representative temperatures. $\epsilon^{-1}(T)$ is shown in Fig. 10. At low T, Fig. 9b, one sees a periodic lattice of *sharp* Bragg peaks, reflecting the *long range* translational order of the charges which

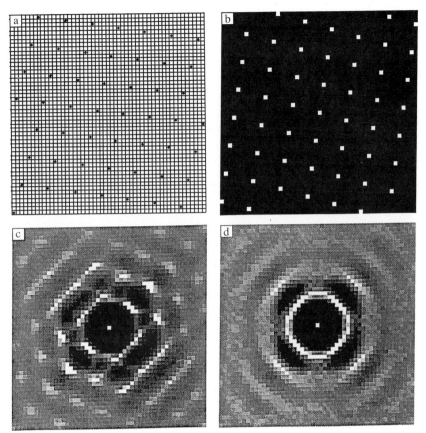

Fig. 9. (a) Ground state charge configuration for $f = 1/51$; (b-d) Structure function $S(\mathbf{k})$ for $T = 0.003, 0.0045$, and 0.006

remain in their ground state structure. The system is insulating, with $\epsilon^{-1} > 0$, reflecting the fact that the charges remain *pinned* to the discrete grid.

At an intermediate T, Fig. 9c, one sees a triangular lattice of peaks of *finite width*. This is characteristic of the *quasi-long range* translational order expected for a 2D solid in a uniform continuum (Kosterlitz and Thouless 1973). Note that the peaks in Fig.9c are in distinctly different locations compared to the peaks in Fig.9b. The inverse dielectric function ϵ^{-1} has now vanished indicating that the system is conducting; the charge lattice is free to diffuse as a correlated structure. Thus a transition has occurred from a pinned commensurate almost-square charge lattice, to a floating incommensurate triangular charge lattice. Although the translational coupling to the underlying grid has been destroyed, there remains a strong orientational coupling. The minimum energy corresponds to the case where one of the three basis directions of the triangular charge reciprocal lattice aligns with one of

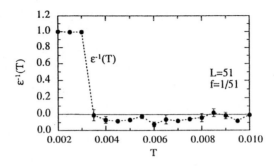

Fig. 10. Inverse dielectric function $\epsilon^{-1}(T)$ for $f = 1/51$

the two diagonal directions of the square grid. This results in two possible distinct orientations for the floating charge lattice (one of which is clearly seen in Fig. 9c) which break the cubic symmetry of the grid geometry.

Finally at high T, Fig. 9d, one has approximately circular intensity rings, indicating that the floating charge lattice has melted into a charge liquid. The 4–fold asymmetry in these rings is due to the square geometry of the underlying grid. Similar results have been found by Hattel and Wheatley (1995), who work directly in the XY representation (1) with a cosine interaction.

In terms of the Josephson array, the pinned vortex lattice corresponds to a state in which the system is truly superconducting with a vanishing linear resistivity. In the floating lattice phase, the vortex lattice is free to diffuse as a whole, thus giving finite linear "flux flow" resistance in the presence of any applied d.c. current. The floating lattice phase is no longer truly superconducting. However the breaking of cubic symmetry due to the orientational coupling of the vortex lattice to the grid will lead to an anisotropic mobility for the vortex lattice. The result should be an angular dependent resistivity, and in the case that the applied current is not aligned with a symmetry direction, a non zero Hall voltage. Once the floating vortex lattice has melted, the cubic symmetry of the grid is restored, and one expects to see an isotropic resistivity with a vanishing Hall voltage. The vanishing of the Hall voltage therefore should serve as a clear experimental signal for the melting of the floating vortex lattice.

Simulations for other values of $f = 1/q$ yield the phase diagram shown in Fig. 11. $T_c(f)$ denotes the transition between the pinned and floating lattices. As $f \to 0$, one sees that $T_c(f) \sim f$ vanishes, in agreement with results by Nelson and Halperin (1979), Korshunov (1986), and Hattel and Wheatley (1994). $T_m(f)$ denotes the melting temperature of the floating lattice. As $f \to 0, T_m \simeq 0.007$ approaches a finite constant. This value in agreement with estimates by Fisher (1980) for the melting of a vortex lattice in a continuous film. When the vortex density is too large, $f \gtrsim 1/30$, T_c and T_m merge, and there is only a single transition from pinned lattice to liquid.[4] Similar results have been obtained by Franz and Teitel (1995) for the case of the $f = 1/q$ CG on a triangular grid.

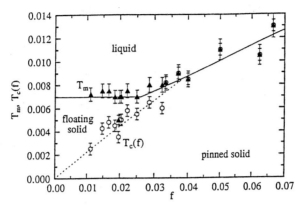

Fig. 11. Phase diagram for dilute densities $f = 1/q$

One can now ask about the nature of the depinning and melting transitions in these systems. The natural candidate for the melting transition is the theory of defect mediated melting in two dimensions, as developed by Kosterlitz and Thouless (1973), Nelson and Halperin (1979), and Young (1979) (KTNHY), and extended by Nelson and Halperin (1979) for a 2D solid on a periodic substrate. This KTNHY theory is essentially a modification of the $f = 0$ CG to *vector* charges, which are the Burger's vectors of lattice dislocations.

A two dimensional floating lattice is characterized by *quasi-long range*, or *algebraic*, translational order,

$$\langle e^{i\mathbf{G}\cdot(\mathbf{r}_i-\mathbf{r}_j)}\rangle \sim |\mathbf{r}_i - \mathbf{r}_j|^{-\eta_G(T)} \quad , \tag{12}$$

where $\mathbf{r}_i$ and $\mathbf{r}_j$ are two charge positions, and $\mathbf{G}$ is a reciprocal lattice vector of the charge lattice. The translational correlation exponent $\eta_\mathbf{G}$ is related to the shear modulus μ of the charge lattice by $\eta_\mathbf{G} = T|\mathbf{G}|^2/4\pi(\mu + \gamma)$, where γ describes the coupling to the periodic substrate and we have used the fact that for logarithmically interacting charges the compression modulus $\lambda \to \infty$. A main prediction of the KTNHY theory is that upon heating, $\eta_{\mathbf{G}_1}$ ($\mathbf{G}_1$ is the smallest reciprocal lattice vector) takes a discontinuous jump to infinity at T_m from the universal value $\eta_{\mathbf{G}_1}(T_m^-) = 1/3$. NH further show that the pinned commensurate to floating lattice transition is also described by the same type of vector CG as the KTNHY melting transition, but with

---

[4] The numerical result that the floating lattice exits only for $f \lesssim f^* = 1/30$ compares with the estimate of Nelson and Halperin (1979) that $f^* \simeq 1/12$

the temperature scale inverted. Cooling below the depinning temperature T_c, η_{G_1} must therefore take a discontinuous jump to zero. Assuming that there is only a small renormalization of the couplings of the vector charges describing the depinning transition, the results of NH give $\eta_{G_1}(T_c^+) \sim 4f$.

This KTNHY scenario has been investigated for the $f = 1/q$ CG on a triangular grid by Franz and Teitel (1995). Using (12) one can derive an equivalent scaling relation for the structure function (11),

$$S(\mathbf{G}) \sim L^{2-\eta_G(T)} \ . \tag{13}$$

One can then extract the exponent $\eta_{G_1}(T)$ by fitting to the form (13) for either (*i*) fixed size L and varying $\mathbf{G}$, or for (*ii*) fixed $\mathbf{G}_1$ and varying size L. Results for $\eta_{G_1(T)}$, using method (*i*) for the case $f = 1/100$, $L = 100$, are shown in Fig. 12 below. One clearly sees that upon increasing T, η_{G_1} blows up at $T_m \simeq 0.0066$, just after it has reached the value of $1/3$, in good agreement with the KTNHY prediction. At the depinning transition $T_c \simeq .0025$, η_{G_1} takes a sharp drop to zero from the value $\eta_{G_1} \simeq 0.1$. This is about twice the value of $4f = 0.04$ predicted by NH.[5]

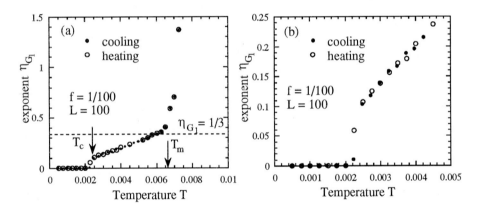

Fig. 12. Translational correlation exponent η_{G_1} for the $f = 1/100$ CG on a triangular grid: (*b*) is an expanded scale of (*a*) focusing on the depinning transition

A second key prediction of the KTNHY melting theory for a lattice in a continuum is that above T_m there exists a distinct hexatic liquid phase with algebraically decaying 6–fold orientational correlations. The hexatic to normal liquid transition T_i is brought about by the unbinding of disclinations,

---

[5] The assumption behind $\eta_{G_1}(T_c^+) \sim 4f$, that couplings experience only small renormalizations, may not be valid here. If it were, one would expect that at low temperatures above T_c, one would have $\eta_{G_1} \sim T$. Although the curve $\eta_{G_1}(T)$ in Fig. 12 does look roughly linear between T_c and T_m, note that it does not extrapolate through the origin.

and is of the ordinary KT type. For a lattice on a periodic square substrate, where the floating lattice has two possible angular orientations with respect the substrate (see discussion concerning Fig. 9c), NH predict that the transition at T_i becomes Ising–like. A triangular substrate however, which induces long range 6–fold orientational order at all T, eliminates in principle the transition at T_i. Nevertheless, if the substrate grid is a very fine compared to the spacing between particles (weak coupling), one might still expect to see continuum behavior over long intermediate length scales, with the orientational correlation saturating to a small finite constant only at large distances.

Because of the relatively large unit cells needed to describe the ground state of the $f = 1/q$ models on a square grid, it is not possible to do the finite size scaling needed to check for the Ising transition discussed above. Franz and Teitel have however searched for the hexatic liquid for the case of a triangular grid. For a density $f = 1/49$ they find no evidence of any hexatic liquid. Orientational correlations above T_m are found to decay exponentially over all lengths simulated. Although these correlations are expected to decay to a finite constant, this constant was found to be small and unmeasurable, indicating that at such dilute a density the grid is indeed very weakly coupled to the charge liquid.

A second possibility is that the melting transition is weakly first order. To investigate this possibility, Franz and Teitel have made histograms $P(E)$ of configuration total energy. Similar to the discussion in connection with Fig. 4, one expects to see a bimodal distribution if the transition is first order. In Fig. 13a below, are shown results for $F(E) \equiv -T \ln P(E)$, for $T = T_m$ and different sizes L. One sees that the energy barrier ΔF between the two minima continues to increase with increasing L, thus suggesting a first order transition. Similar results are shown in Fig. 13b regarding the depinning transition T_c, which also appears to be first order.

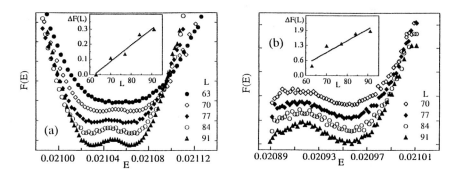

Fig. 13. Free energy histogram $F(E)$ at (a) melting transition and (b) depinning transition for various lattice sizes L, for the $f = 1/49$ CG on a triangular grid

The issue of whether melting in two dimensions is described by KTNHY or is first order has remained hotly contested for more ordinary systems with short range interacting particles (for a review, see Strandburg 1988). Recent simulations (Bagchi et al. 1996) have suggested that only at very large system sizes does the hexatic liquid appear. In the present case, it may also be that the discreteness of the grid has preempted the KTNHY transition and made it first order. Hattel and Wheatley (1994) have argued that the depinning transition must become KTNHY–like at small enough densities f. However simulations of the one component CG in a continuum (Caillol et al. 1982, Choquard and Clerouin 1983), as well as other formulations of the vortex lattice melting problem (Tešanović and Xing 1991, Hu and MacDonald 1993, Kato and Nagaosa 1993, Šášik and Stroud 1994), all suggest that the melting transition is weakly first order.

2.4   Near Full Frustration: $f = 1/2 - 1/q$

For a density $f = 1/2 - 1/q$, q large, the ground state is almost everywhere like the $f = 1/2$ checkerboard pattern except with a superlattice of missing charges (*defects*). Consider here the specific case $f = 5/11$. The correct ground state for this case, shown in Fig. 2, was first found by Kolahchi and Straley (1991). The finite temperature behavior was first studied by by Franz and Teitel (1995). In Figs. 14a-c are shown intensity plots of the structure function $S(\mathbf{k})$ at three representative temperatures. At low T, Fig. 14a, one finds a periodic structure of sharp Bragg peaks. Note that the peaks in the corners arise from the $f = 1/2$–like background; these are brighter than the other peaks, which arise from the defect superlattice.

At an intermediate T, Fig. 14b, one continues to see sharp Bragg peaks in the corners, however the other peaks have been replaced by circular rings. The defect superlattice has melted into a defect liquid, but the $f = 1/2$–like background remains ordered. Finally, at high T, Fig. 14c, the peaks in the corners broaden, the $f = 1/2$–like background has melted, and one finds an isotropic liquid.

Looking more closely at Fig. 14b for the defect liquid, one sees that $S(\mathbf{k})$ is symmetric with respect to the Bragg planes that bisect the diagonals from the origin to the corners. This indicates that the defects, while freely diffusing, are still constrained to sit on only one sublattice of the original square grid of sites; equivalently, one never has two charges on two nearest neighbor sites. Since this sublattice has half the number of sites as the original grid, and since the defects interact with the same logarithmic interaction as do charges, the problem of $f = 1/2 - 1/q$ at low temperatures becomes equivalent to that considered in the previous section: the $f = 1/2$–like background remains ordered and can be ignored; the dilute density of mobile defects behaves in the same way as a dilute density of charges with $f' = 2/n$. This leads to the expectation that, for q sufficiently large, one will find an additional phase not observed for $f = 5/11$. Upon heating, one will first have a transition T_c from a pinned to a floating defect superlattice, followed by a melting transition

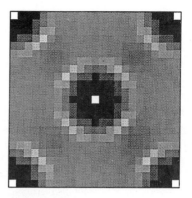

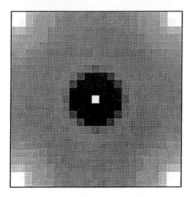

Fig. 14. (*a*) Structure function $S(\mathbf{k})$ for the case $f = 5/11$ at (*a*) $T = 0.010$, (*b*) $T = 0.018$, and (c)$T = 0.055$

T_{m} to a defect liquid, followed finally by the melting $T_{\mathrm{m'}}$ of the $f = 1/2$–like background into an isotropic liquid. The phase boundaries for T_{c} and T_{m} may be inferred from Fig. 11. The boundary for $T_{\mathrm{m'}}$ remains in general unknown. The depinning transition T_{c} will mark the vanishing of ϵ^{-1} and hence the transition between the superconducting and the normal states in the Josephson array.

2.5　General f

For general rational $f = p/q$, Teitel and Jayaprakash (1983b) (TJ) presented arguments that the loss of superconductivity in a Josephson array would occur at a temperature $T_{\mathrm{c}}(f) \sim 1/q$. This argument was based on considering the effect of applying a uniform twist gradient δ to the phases θ_i of the model (1), and studying the periodicity of resulting supercurrents as δ is varied. However, for large systems $L \to \infty$, the maximum twist gradient that can be applied is $\delta = 2\pi/L \to 0$. The TJ argument that assumes it is meaningful to consider a finite δ ensemble in the thermodynamic limit thus becomes ill posed. TJ also assumed that the transition temperature $T_c(f)$ should be proportional to the zero temperature critical current $I_c(f)$. This also is incorrect, as is discussed below.

Nevertheless, if one identifies the depinning transition with the loss of superconductivity in the Josephson array, the preceding sections show that the TJ conjecture remains valid for the special cases of $f = 1/q$ and $f = 1/2 - 1/q$. It is thus interesting to speculate what might happen near other simple rational f. For example, the ground state for $f = 1/3 - 1/q$, for q sufficiently large, is likely to consist of an $f = 1/3$–like background with a superlattice of defects. A schematic[6] of such a case is shown in Fig. 15. At low T the defect superlattice is pinned. Increasing T, one expects that the defect superlattice will unpin at a $T_c \sim 1/q$. For large enough q however, the floating defect lattice is most likely to be free to move only in the diagonal direction of the $f = 1/3$–like background; i.e. a charge in the $f = 1/3$–like background may hop to a vacancy site within a given diagonal stripe, but will not hop to a site between adjacent stripes. The result will be a Josephson array which remains superconducting in one direction, but has flux flow resistance in the orthogonal direction. At higher T, the floating defect lattice will melt into a defect liquid with anisotropic mobility, and only at a still higher temperature will one find an isotropic liquid.

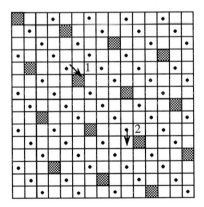

Fig. 15. Schematic of a ground state for $f = 1/3 - 1/q$. ($\bullet$) denotes a charge; a shaded box denotes a charge missing from the 1/3–like background. At low T, moves of type "1" are more likely than moves of type "2".

One can easily imagine similar scenarios near other simple rational frac-

---

[6] Fig. 15 is a schematic only. Straley and Barnett (1993) note that, for $f = 1/3 - 1/q$, one expects the defect superlattice to be the correct ground state only at *larger* values of q than depicted in Fig. 15.

tions. If $f = p/q$ is near some simpler $f_0 = p_0/q_0$, then the ground state of f will look almost everywhere like that of f_0, with a superlattice of defect regions. When $f - f_0$ is sufficiently small, these defects should unpin at a $T_c(f) \ll T_c(f_0)$, thus leading to a very discontinuous phase boundary $T_c(f)$. However the detailed dependence of T_c on f remains in question. In particular, for the dilute density $f = 1/q$ it has been shown that $T_c \sim 1/q$. It would be interesting to see if the depinning transition for a dilute density $f = p/q$, $p \ll q$, is proportional to the charge density p/q, or if there are additional commensurability effects that cause $T_c(p/q) \sim 1/q$ as in the TJ conjecture.

A related question is the behavior at irrational values of f. Approximating an irrational f by a very high order rational p/q, $q \gg 1$, TJ argued that $T_c(f) \rightarrow 0$. The specific case of $f = (3 - \sqrt{5})/2$ has been studied numerically by Halsey (1985,1988) who argued in favor of a finite T_c. Halsey's evidence consists of the observation that (i) the Josephson array appears to have a finite zero temperature critical current, and (ii) that an Edwards-Anderson-like order parameter appears to be finite at low temperature. However it has been shown (Lobb et al. 1983, Straley 1988, Rzchowski et al. 1990, Vallat and Beck 1992) that a finite $T = 0$ critical current persists even for the case of a single vortex, i.e. $f = 1/L^2$. This is just a single body effect of the vortex moving in an effective two dimensional periodic pinning potential created by the grid of the array. At any finite T however, thermal activation over the energy barriers of this single body pinning potential will lead to finite linear resistivity, in agreement with the observation that for $f = 1/q$, $T_c \sim 1/q \rightarrow 0$ as $q = L^2 \rightarrow \infty$. Thus (in contrast to TJ's assumption) a finite $T = 0$ critical current does not imply a finite T_c. A finite Edwards-Anderson order parameter at low T would be a more convincing demonstration of a finite T_c. However Halsey leaves open the question of whether his results reflect a true equilibrium transition, or rather a freezing out of equilibrium due to a finite cooling rate in the presence of large energy barriers between many metastable nearly degenerate states. Thus behavior at irrational f remains an open question.

3 Arrays with Randomness

The introduction of randomness into the Josephson array, while posing many new challenges for theory, is also a topic of great practical importance. As vortex diffusion is a source of "flux flow" resistivity in superconductors, the introduction of random vortex pinning impurities has been viewed as a way of increasing critical currents and enhancing superconductivity. On the other hand, introducing randomness into statistical models often has the effect of reducing the transition temperature, or even of driving $T_c \rightarrow 0$. In this section I will review several simple models in which the effects of randomness have been included in the Josephson array.

3.1 Random Point Pinning

A topic that has received renewed interest in connection with behavior in high T_c superconductors, has been the effect of random point pinning sites on the structure of a vortex lattice. This question can be easily addressed within the CG model for a Josephson array, by adding a random potential to the Hamiltonian (3),

$$\mathcal{H}[n_i] = \frac{1}{2} \sum_{i,j} (n_i - f) G(\mathbf{r}_i - \mathbf{r}_j)(n_j - f) + \sum_i V_i n_i \ . \tag{14}$$

Here V_i is a random pinning potential on each site of the grid, with averages,

$$\overline{V_i} = 0, \qquad \overline{V_i V_j} = w^2 \delta_{ij} \ , \tag{15}$$

where the overbar denotes an average over different realizations of the random potential V_i. Note that V_i can be viewed as arising from random magnetic fluxes δf_j, $V_i = \sum_j G(\mathbf{r}_i - \mathbf{r}_j)\delta f_j$. The δ_{ij} correlations of (15) then imply long range correlations between these random fluxes, $\overline{\delta f_i \delta f_j} = \sum_m G^{-1}(\mathbf{r}_i - \mathbf{r}_m)G^{-1}(\mathbf{r}_m - \mathbf{r}_j)$, were $G^{-1}(\mathbf{r})$ is the inverse of the Green's function, proportional to the two dimensional lattice Laplacian, $G^{-1}(\mathbf{r}) = \frac{1}{2\pi} \sum_\mu [\delta_{\mathbf{r},\hat{\mu}} - \delta_{\mathbf{r},0}]$, where $\mu = \pm x, \pm y$. In contrast, localized randomness in a bond parameter of (1), for example A_{ij} or a bond coupling J_{ij}, would in general lead to a longer range correlation in the pinning potential V_i (Cohn et al. 1991). Equation (14) is thus an idealized model of random pinning, rather than a true model of the effects of randomness in a physical Josephson array. For a more realistic model treating the effects of bond dilution on a Josephson array, see Li and Teitel (1991).

The effects a of random point pinning potential as in (14) have been treated in a classic paper by Larkin and Ovchinnikov (1979). They argue that any amount of random pinning will lead to exponentially decaying translational correlations on long enough length scales (see also Chudnovsky 1991). The length scale L_p on which the random pins disorder the vortex lattice is estimated as follows. In a domain of size L, there will be $M = (L/a_v)^d$ vortices, where a_v is the average distance between vortices. If the vortices in the domain are still ordered, then the total pinning energy felt by these M vortices will be the sum of M uncorrelated values of V_i. The root mean square average of this pinning energy is,

$$U_{\text{pin}} = \sqrt{M w^2} = \left(\frac{L}{a_v}\right)^{d/2} w \ , \tag{16}$$

where d is the spatial dimension. The competing elastic energy to distort the vortices in the domain by one lattice constant a_v over the length L is approximately,

$$U_{\text{el}} = \frac{1}{2}\mu \left(\frac{a_v}{L}\right)^2 L^d \ , \tag{17}$$

where μ is the shear modulus of the vortex lattice. The domain will remain ordered provided $U_{pin} < U_{el}$. The criteria $U_{pin} = U_{el}$ thus determines the length L_p beyond which elastic distortions destroy the long range order of the vortex lattice,

$$\frac{L_p}{a_v} = \left(\frac{\mu a_v^d}{2w}\right)^{2/(4-d)} = \frac{\mu a_v^2}{2w} \quad \text{for } d = 2 \ . \tag{18}$$

Using the KTNHY criteria for the vortex lattice melting transition, $\eta_{G_1} = 1/3$ (which by Fig. 12 appears to be well obeyed, even if the transition is weakly first order), one gets $\mu a_v^2 = 4\pi T_m$, yielding $L_p/a_v = 2\pi T_m/w$.

More recently, Giamarchi and Le Doussal (1994), and Korshunov (1993), have challenged the Larkin-Ovchinnikov result. Their variational renormalization group analysis finds that at low temperatures disorder does not lead to exponential decay, but rather to algebraically decaying correlations, as in the case of the pure (non-random) vortex lattice in two dimensions. The translational correlation exponent η_{G_1} of this algebraic decay, instead of vanishing linearly in T as in the pure case, now saturates at low T to a finite constant, independent of the strength of the disorder w.

Both the Larkin-Ovchinnikov and variational RG analyses treat only elastic distortions, ignoring the possibility of disorder induced free dislocations. Such dislocations are expected to appear, and lead to exponentially decaying translational correlations, on some length scale L_d. One can then ask how the length L_d compares to the length L_p, and in the case that $L_p \ll L_d$, how do the correlations decay in the intermediate region $L_p < L < L_d$.

In order to investigate this question, Franz and Teitel (1996) have carried out Monte Carlo simulations of the Hamiltonian (14) for the $f = 1/100$ CG on a triangular grid of length $L = 100$. In Fig. 16 below are shown the results for η_{G_1} vs. T, for several values of pinning strength w, as determined by the same procedure as used in Fig. 12. In agreement with the variational RG calculations, for temperatures smaller than the pure case T_m, one has a good fit to an algebraic decay. Upon cooling, $\eta_{G_1}(T)$ tracks the pure case value, until saturating to a constant at low temperature. However, in opposition to the variational RG results, this low temperature value continues to increase with increasing w. For the largest value of w shown in Fig. 16, η_{G_1} stays constant at the critical value of $1/3$ for all $T < T_m$.

In Fig. 17 are shown results for η_{G_1} vs. w, for fixed $T = 0.004$, midway between the pure case melting and depinning transitions. Comparing increasing with decreasing w for $L = 100$, one sees a strong hysteresis as η_{G_1} crosses the critical KTNHY value of $1/3$. $\eta_{G_1} = 1/3$ signals the appearance of disorder induced free dislocations. For comparison, the same results are shown for the smaller system length $L = 50$. While again one finds a good fit to algebraic decay in the dislocation free region $\eta_{G_1} < 1/3$, the exponent η_{G_1} is now found to be significantly smaller than the value found for $L = 100$. If correlations truly decayed algebraically, one would expect the results from

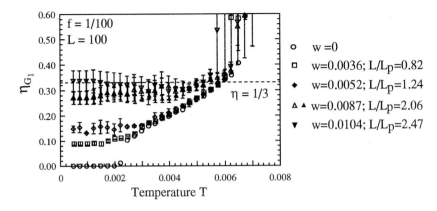

Fig. 16. Translational correlation exponent η_{G_1} vs. T for $f = 1/100$, $L = 100$, and various strengths of disorder w

$L = 50$ to agree with those from $L = 100$, except for relatively small finite size effects. In contrast, Fig. 17 shows that the effective algebraic decay exponent is increasing as L increases. Correlations in the dislocation free region are thus decaying faster than algebraically. Whether this indicates a failure of the variational RG approach, or rather that one is in the so-called "random manifold" cross-over region (the appearance of disorder induced free dislocations preventing one from reaching the larger length scales on which algebraic decay is expected) remains for further investigation.

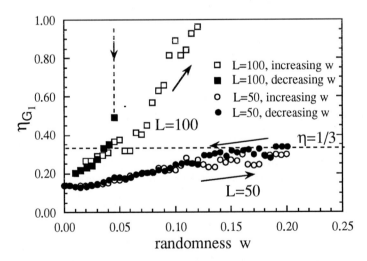

Fig. 17. Translational correlation exponent η_{G_1} vs. disorder strength w for $f = 1/100$ and $L = 100, 50$

Finally, Franz and Teitel find that for all temperatures $T_c^{\text{pure}} < T < T_m^{\text{pure}}$, $\epsilon^{-1} = 0$. Thus the random pinning potential, in addition to disordering the vortex lattice, fails to pin vortices to the substrate.

3.2 Random Gauge Models

Finally, I will mention some other random models that have received attention recently. These are all models in which the variables A_{ij} of the Hamiltonian (1) are taken as random variables, hence the term *random gauge* model.

XY Spin Glass: If one takes the A_{ij} to be independent random variables with equally likely values 0 and π, this corresponds to the ordinary XY model with equally likely ferromagnetic and antiferromagnetic bond couplings. This is the two dimensional XY *spin glass* for which it has been established that $T_c = 0$. Nevertheless, an interesting complexity remains. Just as in the $f = 1/2$ case there existed the possibility of separate transitions with respect to the continuous and discrete symmetries, here it appears that the two correlation lengths, that describe the phase and vortex (chiral) orderings as $T \rightarrow 0$, diverge with distinctly different exponents (Kawamura and Tanemura 1987, Ray and Moore 1992, Bokil and Young 1996).

Gauge Glass: If one takes the A_{ij} to be independent random variables uniformly distributed on the interval $(-\pi, \pi]$, this corresponds to independently correlated random magnetic fluxes through each unit cell of the array, and is referred to as the *gauge glass*. Here again it appears that $T_c = 0$ (Fisher et al. 1991, Cieplak et al. 1992, Gingras 1992).

Gaussian Phase Shifts: If one takes the A_{ij} to be independent Gaussian distributed random variables, with average zero and standard deviation σ, this is known as the *random Gaussian phase shift* model, or also the *positionally disordered* Josephson array. The latter name refers to one way such a model can be physically realized. If the position of either a node or a bond of the Josephson array (see Fig. 1) is randomly displaced, then in the presence of a uniform applied magnetic field, there will be a randomly shifted value for A_{ij} on the effected bonds. This type of disorder corresponds to the case of random quenched dipole fluxes, $\delta f_i = +\epsilon$ and $\delta f_j = -\epsilon$, on nearest neighbor sites i and j. To see this, note that changing A_{ij} will increase the magnetic flux through the unit cell on one side of the bond $\langle ij \rangle$, while decreasing the flux by the same amount through the unit cell on the opposite side of the bond. For a given positionally disordered array, one can increase the strength of the randomness by increasing the uniform applied magnetic field by an amount equal to an integer number of flux quanta per unit cell. This has no effect on the average fluxes f_i (which can always be shifted back to the interval $(-\frac{1}{2}, \frac{1}{2}]$ leaving all physical quantities invariant), but increases the strength ϵ of the quenched random fluxes, δf_i.

For very large σ, this model reduces to the gauge glass, and $T_c = 0$. For $\sigma < \sigma_c$ however, a finite temperature transition has been predicted. This model was first analyzed by Rubinstein et al. (1983), and applied to the Josephson array problem by Granato and Kosterlitz (1986). For the case where the average f_i is integer (or equivalently zero), they find that as T is decreased for $\sigma < \sigma_c$, there is first a KT transition to an ordered state, followed at lower temperature by a reentrant transition back to the disordered state. Numerical simulations by Forrester et al. (1988, 1990) and by Chakrabarti and Dasgupta (1988), as well as experiments (Forrester et al. 1988, Benz et al. 1988), failed to find any evidence for a reentrant phase. More recently, Ozeki and Nishimori (1993) have presented results which claim to rigorously rule out such a reentrant phase, but leave open the question of whether there remains a finite ordering transition T_c, or if $T_c \to 0$. Most recently, Nattermann et al. (1995), Scheidl (1996), and Tang (1996), all report arguments that a finite KT type transition should exist. This transition in the $\sigma - T$ plane should lie near the KT transition line as found by Rubinstein et al. The reentrant transition of Rubinstein et al. disappears, but is replaced by a cross-over region below which behavior becomes glassy. Numerical simulations that confirm these new results remain to be done.

The case where random Gaussian phase shifts are superimposed upon a fractional average f_i (Choi et al. 1987), remains a largely unexplored problem.

4   Conclusions

To conclude, I have attempted to review some of the rich equilibrium critical phenomena exhibited by classical two dimensional arrays of Josephson junctions. Many interesting old and new questions remain to be resolved. In particular, it would be interesting to have a better understanding of how frustration induced by quenched randomness, as in Sect. 3, differs from frustration induced by geometrical effects, such as in the case of uniform irrational f, or the $f = 1/2$ CG on a triangular grid. The former are more akin to spin glass problems, while the latter appear more akin to the structural glass problem.

I have omitted mention of many other interesting areas of ongoing research on Josephson junction arrays. In the area of equilibrium behavior of classical arrays these include such topics as fractal, incommensurate, anisotropic, and three dimensional geometries. In the area of dynamics these include, understanding the critical behavior of I-V characteristics, giant Shapiro steps, mode locking, chaos, the contrasting behavior of underdamped as opposed to overdamped junctions, and plastic flow of vortex lattices. Quantum effects become important when charging energy must be considered. A good review of recent work on many of these topics may be found in Giovannella and Tinkham (1995).

Clearly the study of Josephson junction arrays will continue to yield much new interesting physics well into the future.

Acknowledgements

I am indebted to C. Jayaprakash, D. Stroud, and J. Garland for my initial introduction to the topic of Josephson arrays. Since then I have benefited greatly from working with my students and colleagues, Y.-H. Li, J.-R. Lee, M. Franz, and P. Gupta. This work has been supported by U. S. Department of Energy grant DE-FG02-89ER14017.

References

Bagchi K., Anderson H. C., Swope W. (1996): Computer simulation study of the melting transition in two dimensions. Phys. Rev. Lett. **76**, 255

Benz S. P., Forrester M. G., Tinkham M., Lobb C. J. (1988): Positional disorder in superconducting wire networks and Josephson junction arrays. Phys. Rev. B **38**, 2869

Bokil H. S., Young A. P. (1996): Study of chirality in the two-dimensional XY spin glass. Preprint, cond-mat/9512042

Caillol J. M., Levesque D., Weis J. J., Hansen J. P. (1982): A Monte Carlo study of the classical two-dimensional one-component plasma. J. Stat. Phys. **28**, 325

Caillol J. M., Levesque D. (1986): Low-density phase diagram of the two-dimensional Coulomb gas. Phys. Rev. B **33**, 499

Chakrabarti A., Dasgupta C. (1988): Phase transition in positionally disordered Josephson-junction arrays in a transverse magnetic field. Phys. Rev. B **37**, 7557

Choi M. Y., Chung J. S., Stroud D. (1987): Positional disorder in a Josephson-junction array. Phys. Rev. B **35** 1669

Choquard Ph., Clerouin J. (1983): Cooperative phenomena below melting of the one-component two-dimensional plasma. Phys. Rev. Lett. **50**, 2086

Chudnovsky E. M. (1991): Orientational and positional order in flux lattices of type-II superconductors. Phys. Rev. B **43**, 7831

Cieplak M., Banavar J. R., Li M. S., Khurana A. (1992): Frustration, scaling, and local gauge invariance. Phys. Rev. B **45**, 786

Cohn M. B., Rzchoswki M. S., Benz S. P., Lobb C. J. (1991): Vortex-defect interactions in Josephson-junction arrays. Phys. Rev. B **43**, 12823

Dodgson M. J. W. (1995): Investigation on the ground states of a model thin film superconductor on a sphere. Preprint, cond-mat/9512124

Fisher D. S. (1980): Flux-lattice melting in a thin-film superconductor. Phys. Rev. B **22**, 1190

Fisher M. P. A., Tokuyasu T. A., Young A. P. (1991): Vortex variable-range-hopping resistivity in superconducting films. Phys. Rev. Lett. **66**, 2931

Forrester M. B., Lee H. J., Tinkham M., Lobb C. J. (1988): Positional disorder in Josephson-junction arrays: Experiments and simulations. Phys. Rev. B **37**, 5966

Forrester M. G., Benz S. P., Lobb C. J. (1990): Monte Carlo simulations of Josephson-junction arrays with positional disorder. Phys. Rev. B **41**, 8749

Fradkin E., Huberman B. A., Shenker S. H. (1978): Gauge symmetries in random magnetic systems. Phys. Rev. B **18**, 4789

Franz M., Teitel S. (1995): Vortex-lattice melting in two-dimensional superconducting networks and films. Phys. Rev. B **51** 6551

Franz M., Teitel S. (1996): Effect of random pinning on 2D vortex lattice correlations. In preparation

Friesen M. (1995): Critical and non-critical behavior of the Kosterlitz-Thouless-Berezinskii transition. Phys. Rev. B **53**, R514

Giamarchi T., Le Doussal P. (1994): Elastic theory of pinned flux lattices. Phys. Rev. Lett. **72**, 1530

Gingras M. J. P. (1992): Numerical study of vortex-glass order in random-superconductor and related spin-glass models. Phys. Rev. B **45**, 7547

Giovannella C., Tinkham M., eds. (1995): Macroscopic Quantum Phenomena and Coherence in Superconducting Networks. (World Scientific, Singapore)

Granato E., Kosterlitz J. M. (1986): Quenched disorder in Josephson-junction arrays in a transverse magnetic field. Phys. Rev. B **33**, 6533

Granato E., Nightingale M. P. (1993): Chiral exponents of the square-lattice frustrated XY model: a Monte Carlo transfer-matrix calculation. Phys. Rev. B **48**, 7438

Grest G. S. (1989): Critical behavior of the two-dimensional uniformly frustrated charged Coulomb gas. Phys. Rev. B **39**, 9267

Gupta P., Teitel S. (1996): Phase diagram of the 2D dense lattice Coulomb gas. In preparation

Halsey T. C. (1985): Josephson-junction array in an irrational magnetic field: A superconducting glass? Phys. Rev. Lett. **55**, 1018

Halsey T. C. (1988): On the critical current of a Josephson junction array in a magnetic field. Physica B **152**, 22

Hattel S. A., Wheatley J. M. (1994): Depinning phase transitions in two-dimensional lattice Coulomb solids. Phys. Rev. B **50** 16590

Hattel S. A., Wheatley J. M. (1995): Flux lattice melting and depinning in the weakly frustrated two-dimensional XY model. Phys. Rev. B **51**, 11951

Hu J., MacDonald A. H. (1993): Two-dimensional vortex lattice melting. Phys. Rev. Lett. **71**, 432

José J. V., Kadanoff L. P., Kirkpatrick S., Nelson D. R. (1977): Renormalization, vortices, and symmetry-breaking perturbations in the two-dimensional planar model. Phys. Rev. B **16**, 1217

Kato Y., Nagaosa N. (1993): Monte Carlo simulation of two-dimensional flux-line-lattice melting. Phys. Rev. B **48**, 7383

Kawamura H., Tanemura M. (1987): Chiral order in a two-dimensional XY spin glass. Phys. Rev. B **36**, 7177

Knops Y. M. M., Nienhuis B., Knops H. J. F., Blöte J. W. J. (1994): 19-vertex version of the fully frustrated XY model. Phys. Rev. B **50**, 1061

Kolahchi M. R., Straley, J. P. (1991): Ground state of the uniformly frustrated two-dimensional XY model near $f = 1/2$. Phys. Rev. B **43**, 7651

Korshunov S. E. (1986): Phase transitions in two-dimensional uniformly frustrated XY models. II. General scheme. J. Stat. Phys. **43**, 17

Korshunov S. E. (1993): Replica symmetry breaking in vortex glasses. Phys. Rev. B **48**, 3969

Kosterlitz J. M., Thouless D. (1973): Ordering, metastability and phase transitions in two-dimensional systems. J. Phys. C **6**, 1181

Kosterlitz J. M. (1974): The critical properties of the two-dimensional xy model. J. Phys. C. **7**, 1046

Larkin A. I., Ovchinnikov Yu. N. (1979) Pinning in type II superconductors. J. Low Temp. Phys. **34**, 409

Lee D. H., Joannopoulos J. D., Negele J. W., Landau D. P. (1986): Symmetry analysis and Monte Carlo study of a frustrated antiferromagnetic planar (XY) model in two dimensions. Phys. Rev. B **33**, 450

Lee J., Kosterlitz J. M., Granato E. (1991): Monte Carlo study of frustrated XY models on a triangular and square lattice. Phys. Rev. B **43**, 11531

Lee J.-R. (1994): Phase transitions in the two-dimensional classical lattice Coulomb gas of half-integer charges. Phys. Rev. B **49**, 3317

Lee J.-R., Teitel S. (1992): Phase transitions in classical two-dimensional lattice Coulomb gases. Phys. Rev. B **46**, 3247

Lee S., Lee K.-C. (1994): Phase transitions in the fully frustrated XY model studied by the microcanonical Monte Carlo technique. Phys. Rev. B **49**, 15184

Lee S., Lee K.-C. (1995): Phase transitions in the uniformly frustrated XY model with frustration parameter $f = 1/3$ studied with use of the microcanonical Monte Carlo technique. Phys. Rev. B **52**, 6706

Levin Y., Li X., Fisher M. E. (1994): Coulombic criticality in general dimensions. Phys. Rev. Lett. **73**, 2716

Li Y.-H., Teitel S. (1990): Flux flow resistance in frustrated Josephson junction arrays. Phys. Rev. Lett. **65**, 2595

Li Y.-H., Teitel S. (1991): The effect of random pinning sites on behavior in Josephson junction arrays. Phys. Rev. Lett. **67**, 2894

Lidmar J., Wallin M. (1996): Monte Carlo simulation of a two-dimensional continuum Coulomb gas. Preprint, cond-mat/9607025

Lobb C. J., Abraham D. W., Tinkham M. (1983): Theoretical interpretation of resistive transition data from arrays of superconducting weak links. Phys. Rev. B **27**, 150 (1983)

Minnhagen P. (1987): The two-dimensional Coulomb gas, vortex unbinding, and superfluid-superconducting films. Rev. Mod. Phys. **59**, 1001

Minnhagen P., Wallin M. (1987): New phase diagram for the two-dimensional Coulomb gas. Phys. Rev. B **36**, 5620

Minnhagen P., Wallin M. (1989): Results for the phase diagram of the two-dimensional Coulomb gas. Phys. Rev. B **40**, 5109

Miyshita S., and Shiba H. (1984): Nature of the phase transition of the two-dimensional antiferromagnetic plane rotor model on the triangular lattice. J. Phys. Soc. Jpn. **53**, 1145

Nattermann T., Scheidl S., Korshunov S. E., Li M. S. (1995): Absence of reentrance in the two-dimensional XY-model with random phase shifts. J. Phys. I France **5**, 565

Nelson D. R., Halperin B. I. (1979): Dislocation-mediated melting in two dimensions. Phys. Rev. B **19**, 2457

Nicolaides D. B. (1991): Monte Carlo simulation of the fully frustrated XY model. J. Phys. A **24**, L231

Nightingale M. P., Granato E., Kosterlitz J. M. (1995): Conformal anomaly and critical exponents of the XY Ising model. Phys. Rev. B **52**, 7402

Ohta T., Jasnow D. (1979): XY model and the superfluid density in two dimensions. Phys. Rev. B **20**, 139

Olsson P. (1995a): Monte Carlo analysis of the two-dimensional XY model. II. Comparison with the Kosterlitz renormalization group equations. Phys. Rev. B **52**, 4511

Olsson P. (1995b): Two phase transitions in the fully frustrated XY model. Phys. Rev. Lett. **75**, 2758

Ozeki Y., Nishimori H. (1993): Phase diagram of gauge glasses. J. Phys. A **26**, 3399

Ramirez-Santiago G., José J. V. (1994): Critical exponents of the fully frustrated two-dimensional XY model. Phys. Rev. B **49**, 9567

Ray P., Moore M. A. (1992): Chirality-glass and spin-glass correlations in the two-dimensional random-bond XY model. Phys. Rev. B **45**, 5361

Rubinstein M., Shraiman B., Nelson D. R. (1983): Two-dimensional XY magnets with random Dzyaloshinskii-Moriya interactions. Phys. Rev. B **27**, 1800

Rzchowski M. S., Benz S. P., Tinkham M., Lobb C. J. (1990): Vortex pinning in Josephson-junction arrays. Phys. Rev. B **42**, 2041

Šášik R., Stroud D. (1994): Calculation of the shear modulus of a two-dimensional vortex lattice. Phys. Rev. B **49**, 16074

Scheidl S. (1996): Glassy vortex state in a two-dimensional XY-model. Preprint, cond-mat/9601131

Straley J. P. (1988): Magnetic field effects in Josephson networks. Phys. Rev. B **38**, 11225

Straley J. P., Barnett G. M. (1993): Phase diagram for a Josephson network in a magnetic field. Phys. Rev. B **48**, 3309

Standburg K. J. (1988): Two-dimensional melting. Rev. Mod. Phys. **60**, 69

Tang L.-H. (1996): Vortex statistics in a disordered two-dimensional XY model. Preprint, cond-mat/9602162

Teitel S., Jayaprakash C. (1983a): Phase transitions in frustrated two dimensional XY models. Phys. Rev. B **27**, 598

Teitel S., Jayaprakash C. (1983b): Josephson junction arrays in transverse magnetic fields. Phys. Rev. Lett. **51**, 1999

Tešanović Z., Xing L. (1991): Critical fluctuations in strongly type-II quasi-two-dimensional superconductors. Phys. Rev. Lett. **67**, 2729

Théron R., Korshunov S. E., Simond J. B., Leemann Ch., Martinoli P. (1994): Observation of domain-wall superlattice states in a frustrated triangular array of Josephson junctions. Phys. Rev. Lett. **72**, 562

Thijssen J. M., Knops H. J. F. (1988a): Monte Carlo study of the Coulomb-gas representation of frustrated XY models. Phys. Rev. B **37**, 7738

Thijssen J. M., Knops H. J. F. (1988b): Analysis of a new set of renormalization equations for the Kosterlitz-Thouless transition. Phys. Rev. B **38**, 9080

Thijssen J. M., Knops H. J. F. (1990): Monte Carlo transfer-matrix study of the frustrated XY model. Phys. Rev. B **42**, 2438

Vallat A., Beck H. (1992): Classical frustrated XY model: Continuity of the ground-state energy as a function of the frustration. Phys. Rev. Lett. **68**, 3096

Vallat A., Beck H. (1994): Coulomb-gas representation of the two-dimensional XY model on a torus. Phys. Rev. B **50**, 4015

Villain J. (1975): Theory of one- and two-dimensional magnets with an easy magnetization plane. II. The planar, classical, two-dimensional magnet. J. Phys. (Paris) **36**, 581

Young A. P. (1979): Melting of the vector Coulomb gas in two dimensions. Phys. Rev. B **19**, 1855

An Experimentally Realizable Weiss Model for Disorder-Free Glassiness

P. Chandra,[1] M.V. Feigelman,[2] M.E. Gershenson[3] and L.B. Ioffe[2,3]

[1] NEC Research Institute, 4 Independence Way, Princeton NJ 08540i
[2] Landau Institute for Theoretical Physics, Moscow, RUSSIAi
[3] Department of Physics, Rutgers University, Piscataway, NJ 08855

Abstract. We summarize recent work on a frustrated periodic long-range Josephson array in a parameter regime where its dynamical behavior is *identical* to that of the $p = 4$ disordered spherical model. We also discuss the physical requirements imposed by the theory on the experimental realization of this superconducting network.

The identification of the key features underlying the physics of glass formation remains an outstanding problem of statistical mechanics. A glassy system has a "memory" of its past behavior and does not explore all of phase space; it breaks ergodicity without thermodynamic selection of a unique state thereby defying description by the standard Gibbs methods. Though glass formation in the absence of intrinsic disorder is a widespread phenemenon, it remains poorly understood even at the mean-field level. Because vitrification is a dynamical transition that is not necessarily accompanied by a static one, it remains outside the framework of conventional Landau theory. The crucial links between glasses with intrinsic and self-generated disorder remain an area of active discussion.[1] Recently several microscropic non-random models for glassiness have been proposed; most were studied via mapping to disordered systems.[2] What common/distinct features are characteristic of glasses with and without randomness? How similar/different are glasses with short- vs. long-range interactions? What aspects of existing theoretical models are relevant for experiment? In this Proceeding we summarize our current efforts to address these questions in a periodic model that is both analytically accessible[3], [4] and experimentally realizable.[5]

The physical system under study is a stack of two mutually perpendicular sets of N parallel thin wires with Josephson junctions at each node (Figure 1) that is placed in an external tranverse field H. The classical thermodynamic variables of this array are the $2N$ superconducting phases associated with each wire. In the absence of an external field the phase differences would be zero at each junction, but this is not possible for finite H and the phases are thus frustrated. Here we assume that the Josephson couplings are sufficiently small so that the induced fields are negligible in comparison with H; we shall return to this topic when discussing the experimental realization of this network. We can therefore describe the array by the Hamiltonian

$$\mathcal{H} = -\sum_{m,n}^{2N} s_m^* \mathcal{J}_{mn} s_n \tag{1}$$

where $\mathcal{J}_{mn}$ is the coupling matrix

$$\hat{\mathcal{J}} = \begin{pmatrix} 0 & \hat{J} \\ \hat{J}^\dagger & 0 \end{pmatrix} \tag{2}$$

with $J_{jk} = \frac{J_0}{\sqrt{N}} \exp(2\pi i \alpha jk/N)$ and $1 \le (j,k) \le N$ where $j(k)$ is the index of the horizontal (vertical) wires; $s_m = e^{i\phi_m}$ where the ϕ_m are the superconducting phases of the $2N$ wires. Here we have introduced the flux per unit strip, $\alpha = NHl^2/\Phi_0$, where l is the inter-node spacing and Φ_0 is the flux quantum; the normalization has been chosen so that T_G does not scale with N.

Because every horizontal (vertical) wire is linked to every vertical (horizontal) wire, the number of nearest neighbors in this model is N and it is accessible to a mean-field treatment. For the same reason, the free energy barriers separating its low-temperature solutions scale with N. This situation is in marked contrast to that in conventional 2D arrays where the coordination number and hence the barriers are low.[6], [7] A similar long-range network with positional disorder was studied previously.[8] For $\alpha \gg 1/N$ the latter system displays a spin glass transition which was mapped onto the Sherrington-Kirkpatrick model[9] for $\alpha \gg 1$; in this field regime there is no residual "ferromagnetic" phase coherence between wires. Physically this glassy behavior occurs because the phase differences associated with the couplings J_{jk} acquire random values and fill the interval $(0, 2\pi)$ uniformly for $\alpha \gg 1/N$. More specifically, there will be no commensurability if the sum

$$\sum_{k=1}^{N} J_{jk} J_{kl}^\dagger = \sum_{k=1}^{N} \exp 2\pi i \left\{ \frac{\alpha(j-l)}{N} \right\} k \tag{3}$$

is a smooth function of $(j-l)$ where $\{j, l\}$ (k) are indices labelling horizontal (vertical) wires; this will occur only if the expression in curly brackets on the r.h.s. of (3) is *not* an integer for all k, a condition always satisfied for the disordered array. For the periodic case, this situation is realized in the "incommensurate window" $1/N \ll \alpha \le 1$; here the phase-ordering unit cell is larger than the system size so that the "crystalline" phase is inaccessible.[3] There are thus no special field values where the number of low-temperature solutions are finite, in contrast to the situation for $\alpha > 1$.

In the thermodynamic limit of $N \to \infty$ (with fixed array area), the high-temperature approach to the glass transition in the periodic array has been studied[3], [4] using a modified Thouless-Anderson-Palmer (TAP) method.[10] Here we discuss the qualitative picture that emerges from these results (cf.

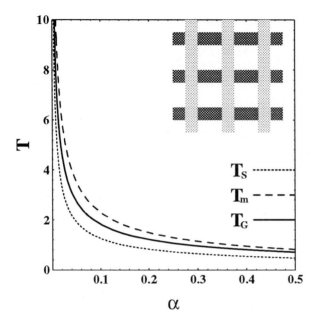

Fig. 1. The phase diagram of the array (insert) where T_G indicates the temperature associated with the dynamical instability discussed in the text, T_S is the speculated equilibrium transition temperature and T_m is the "superheating" temperature where the low-temperature metastable states cease to exist.

Fig. 1), referring the interested reader elsewhere for a more detailed quantitative treatment. [3], [4] As T approaches T_m^+, where $T_m^+ \sim T_0 \approx \frac{J\sqrt{N}}{2\sqrt{\alpha}}$, there appear a number of metastable states in addition to the paramagnetic free-energy minimum. There is no dynamical transition at this temperature, indicating that the system is not "trapped". We speculate that this is because these states are energetically unfavorable and/or they are separated by low barriers. At lower temperatures, as $T \to T_G^+$, the paramagnetic minimum is "subdivided" into an extensive number of degenerate metastable states separated by effectively infinite barriers, and the system is dynamically localized into one of them. Qualitatively, in the interval $T_m > T > T_G$ there appear many local minima in the vicinity of the paramagnetic state separated by *finite* barriers; these barriers increase continuously and become infinite at $T = T_G$. Each of these minima is characterized by a finite "site magnetization" $m_i = \langle s_i \rangle_T$ where "site" refers to a wire. When $T > T_G$ thermal fluctuations average over many states so that $\langle m_i \rangle \equiv 0$. At $T = T_G$ the system is localized in one metastable state and there is an associated jump in the Edwards-Anderson order parameter, $\left(q = \frac{1}{N} \sum_i \langle m_i \rangle^2 \right)$. The low-temperature phase is characterized by a finite q and by the presence of a memory, $\lim_{t' \to \infty} \Delta(t, t') \neq 0$ where $\Delta(t, t')$ is the anomalous response. We expect that at $T = T_G$, the metastable states are degenerate and thus there can be no thermodynamic selection. However at lower temperatures interactions will probably break this degeneracy and select a few states from this manifold; then we expect an $(t \to \infty)$ equilibrium first-order transition (T_S) which should be accompanied by a jump in the local magnetization. In order to observe this transition at T_S the array must be equilibrated on a time-scale (t_W) longer than that (t_A) necessary to overcome the barriers separating its metastable states; t_A scales exponentially with the number of wires in the array. Thus the equilibrium transition at T_S is observable *only* if $t_W \to \infty$ *before* the thermodynamic limit $(N \to \infty)$ is taken; in the opposite order of limits only the dynamical transition occurs.

The periodic array thus exhibits a first-order thermodynamic transition *preceeded* by a dynamical instability; the glass transition at T_G is characterized by a diverging relaxation time and an accompanying jump in the Edwards-Anderson order parameter. Furthermore the dynamical coupled equations for the response and correlation functions of this *non-random* network in the regime $(1/N \ll \alpha < 1)$ are *identical* to those obtained for the $p = 4$ (disordered) spherical Potts model.[11] By contrast the behavior of the disordered array is similar to that of the $p = 2$ case with coinciding static and dynamic transitions. The possible connection between non-random glasses and $p \geq 3$ (disordered) spherical models has been previously suggested in the literature,[12], [13], [14] and the periodic array results give support to those conjectures.[4] Furthermore these long-range Josephson arrays can be built in the laboratory, allowing for parallel theoretical and experimental studies of the "simplest spin glasses" that have previously been an abstraction.

Since the experimental realization of these arrays is important, we would like to identify a number of simplifying assumptions inherent in the theoretical treatment discussed here; in order to test its predictions and probe beyond its realm (finite-size effects etc.) the fabricated system must satisfy certain physical requirements. In particular the model has been only considered in the classical limit where quantum mechanical fluctuations of the phase variables are negligible. Furthermore the fields induced by the Josephson currents must always be small compared to H so that it can frustrate the phases effectively. These constraints put some strong restrictions on the choice of array parameters, which we elaborate below.

We begin with the first condition, that of maintaining weak quantum fluctuations. The strength of the latter is controlled by two parameters: the dimensionless normal resistance $e^2 R/\hbar$ and the ratio E_C/E_J where E_C and E_J are the Coulomb and the Josephson energy scales respectively associated with each individual phase variables. The state of any array (or junction) is conveniently described by its position on the plane defined by these dimensionless parameters, E_C/E_J and $e^2 R/\hbar$; the resulting plot is called a Schmid diagram.[15] For a conventional short-range array $R \sim R_0$ and $E_C \sim e^2/C_0$ where R_0 (C_0) is a resistance (capacitance) of an individual junction. We require that the capacitance of the individual wires and the dissipation in the system should be sufficiently large to suppress the quantum fluctuations of phase and charge effects.

According to the Ambegaokar-Baratoff relation,[16] the normal-state resistance of a single junction in the fabricated long-range arrays should be large ($R_0 \gg \hbar/e^2$) so that the associated screening current is small; for the sake of example we shall assume $R_0 \approx 100 \ kOhm$ and an individual capacitance of $C \approx 1. \ 10^{-15} \ F$, parameters associated with preliminary $Al - Al_2O_3 - Al$ tunnel-junction arrays already fabricated.[5] We note that the Josephson effect for a single tunnel-junction of such a large resistance (or for a short-range array of such junctions) would be completely overwhelmed by quantum fluctuations and Coulomb blockade effects. [17], [18], [19], [20] Indeed, it is known that conventional nearest-neighbor arrays exhibit insulating behavior if their individual junction resistance is more than $15 kOhm$.[19] However in the long-range Josephson arrays all the junctions associated with a particular wire are connected in parallel, so that the effective resistance and capacitance determining their position on the Schmid diagram[15] are $R = R_0/N$ and $C = NC_0$ respectively. Even for a small array with $N = 100$ the effective resistance and capacitance are $R \sim 1 \ kOhm$ and $C \sim 1. \ 10^{-13} \ F$ (corresponding to the charging energy $E_C \sim 10 \ mK$) respectively so that quantum fluctuations and charging effects are negligible for these networks. We note, however, that if the number of wires (and hence the number of neighbors) were decreased significantly the parameters of the individual junctions would have to be altered to avoid complications of charging effects.

We would also like to minimize the effects of everpresent screening currents to ensure that the external field frustrates the superconducting phases effectively. More specifically we demand that the induced flux in the array is less than a flux quantum

$$Li_{eff} < \Phi_O \tag{4}$$

so that all fields and phase gradients produced by the diamagnetic currents are negligible. In order to determine the maximum induced flux in the array we need only look at its largest loop; our condition then becomes

$$(Nl)i_{eff} < \Phi_0 \tag{5}$$

where l is the distance between nodes in the array. For the frustrated case, the maximum field occurs when the currents add incoherently in one direction and coherently in the other[8] so that

$$i_{eff} = \left(\sqrt{N}N\right)i_c. \tag{6}$$

where i_c is the critical current associated with an individual junction. Using

$$i_c \sim \frac{2e}{\hbar}E_J = \frac{2e}{\hbar}\left(\frac{k_BT_c}{\sqrt{N}}\right) \tag{7}$$

we find that the necessary condition for weak induced fields becomes

$$E_J \lesssim \frac{\Phi_0^2}{lN^{5/2}}. \tag{8}$$

Thus there is a delicate balance to be obtained between the need for many neighbors (and high free-energy barriers) and the requirement (8) that the effective London penetration length be substantial larger than the system size; unfortunately early experimental attempts to realize these arrays were not able to satisfy these restrictions due to resolution limitations associated with the optical lithography methods used.[21], [22] Furthermore the predicted glassy behavior is only observable in the field range $\Phi_0/(Nl)^2 < H < \Phi_0/l^2$; where the minimum and maximum correspond to a single flux quantum in the whole array and in a single plaquette. Even with the use of the e-beam lithography ($l = 0.5~\mu m$) the requirement (8) can only be satisfied for an array of $Al - Al_2O_3 - Al$ junctions ($E_J = 1.~10^{-17}~Erg$) with the number of wires N less than 300. For $N = 100$ and these parameters, $\Phi_0/(Nl)^2$ is of the order of $0.01~Oe$, and a careful screening of the earth's magnetic field is necessary.

Once the arrays are fabricated to specification what measurements should be performed? The field-dependence of the critical temperature of both disordered and periodic arrays is crucial in establishing the effectiveness of of the external field in frustrating the phases. It is also a good test of "phase freezing" into a glassy phase, and the functional dependence $T_c(H) \propto H^{-1/2}$ is expected over a wide field and temperature range for both types of arrays. In order to firmly establish the presence of large barriers imperative for glassiness, the history-dependence of the critical current j_c should be studied (i.e. its dependence on the *path* in the $T-H$ plane) in both arrays. We note that below $T_c(\alpha)$ we expect a significantly faster increase in $j_c(T)$ for the periodic network than for its disordered counterpart;[8] specifics for the former case have yet to be determined. We expect that the diverging relaxation time at T_c can be accessed experimentally via the a.c. response to a time-varying magnetic field $H(t)$; the associated ac susceptibility is

$$\chi_\omega = \frac{\partial M_\omega}{\partial H_\omega} = \frac{C(\alpha, N)\omega}{\omega + i/t_R(T)} \tag{9}$$

where t_R is the longest response of the system that diverges at $T = T_G$ and $C(\alpha, N)$ is to be found elsewhere.[4] The $\omega \to 0$ limit of the a.c. susceptibility jumps to a finite value at $T = T_G$, indicating the development of a finite superconducting stiffness at the transition. Therefore measurement of this a.c. response in a fabricated array would probe its predicted glassiness.

Work is currently in progress to study the physical properties of the periodic model in its non-ergodic regime $(T < T_G)$.[23] We note that the glass transition temperature T_G described above corresponds to the system going out of equilibrium as it is cooled infinitesmally slowly from high temperatures. In practice there will always be a finite-rate of cooling and the effective glass transition will occur when the system drops out of equilibrium. If we define the reduced temperature $\Theta \equiv \left(\frac{(T-T_G)}{T_G} \right)$ then the effective glass transition will occur when the time associated with cooling, $t_c \approx \frac{\Theta}{d\Theta/dt}$ is equal to the relaxation time at that temperature, $t_R \approx t_0 \theta^{-\nu}$ where t_0 and ν are determined from the high-temperature analysis elsewhere;[4] the effective glass transition temperature and the associated time are indicated as $\tilde{T}_G$ and $\tilde{t}_G$ in Figure 2. There is evidence for history-dependence; the system's response is very different if an additional field is turned on at a time, t_H, during or after the cooling process and a subsequent measurement is taken at time t_M (cf. Figure 2), reminiscent of the zero-field cooled vs. field-cooled susceptibility observed in spin glasses. Furthermore an "ageing" time-scale t^* seems to exist in this system (cf. Fig. 2); if the field is turned on during the cooling process at time t_H and a subsequent measurement is taken at time t_M such that $t_H - t_M < t^*$ the system "remembers" the cooling process; otherwise $(t_H - t_M > t^*)$ it "forgets" it completely. Preliminary results suggest that t^* has a decreasing temperature-dependence and probably vanishes at low temperatures. This presence of an additional "ageing" time-scale does not exist in the Sherrington-Kirkpatrick theory, though it is indeed observed in experimental spin glasses; it is thus amusing that it emerges from the study of a non-random model.

In conclusion we have presented a summary of recent work on a periodic model that displays "freezing" into one of an extensive number of metastable states without thermodynamic selection. This glass transition is characterized by a diverging relaxation time and an accompanying jump in the Edwards-Anderson order parameter. At low field strengths the dynamical behavior of this system is identical to that of the $p = 4$ (disordered) spherical model. A key advantage of this system is that it can be built in the laboratory, thereby allowing for the possibility of detailed interplay between theory and experiment. The behavior of a similar array with a reduced number of neighbors is a completely open question that could be important in understanding the relevance of long-range models to real experimental glasses. Finally, since any uncertainty in the position of the wires introduces randomness in the array, the continued study of this network could also offer an opportunity to study the crossover between glasses with spontaneously generated and quenched disorder.

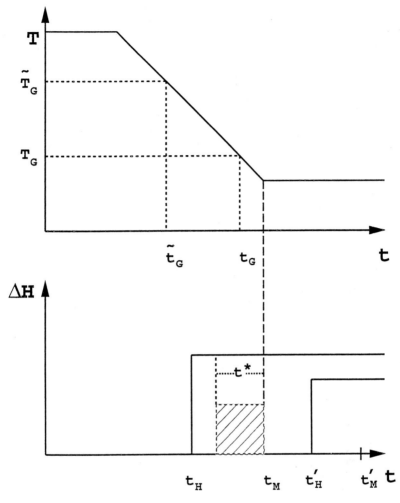

Fig. 2. Schematics of (a) Temperature (T) (b) Application of an Additional Field (ΔH)vs time (t) indicating the time-scales involved in a finite-cooling experiment as discussed in the text; $\tilde{T}_G$ and $\tilde{t}_G$ refer to the "effective glass transition" when the system goes out of equilibrium, and t_H, t_M and t^* are the time-scales associated with the onset of an additional field, a subseqent measurement and ageing.

References

[1] e.g. T.R. Kirkpatrick and D. Thirumalai, PRB **36**, 5388 (1987); J.P. Bouchaud and M. Mezard, J. Phys. I **4**, 1109 (1994); S. Franz and J. Herz, PRL **74**, 2115 (1995); M. Potters and G. Parisi, unpublished (cond-mat preprint 9503009) and references therein.

[2] e.g. E. Marinari, G. Parisi and F. Ritort, *J. Phys. A*, **27**, 7647 (1994); M. Mezard and G. Parisi, unpublished.

[3] P. Chandra, L. B. Ioffe and D. Sherrington, Phys. Rev. Lett. **75**, 713 (1995).

[4] P.Chandra, M.V. Feigelman and L.B. Ioffe, Phys. Rev. Lett. **76**, 4805 (1996).

[5] M.E. Gershenson et al., To Be Published.

[6] e.g. J.E. Mooij and G.B.J. Schön ed., "Coherence in Superconducting Networks," Physica **152B**, 1 (1988).

[7] J. P. Carini, Phys. Rev B **38**, 63 (1988).

[8] V. M. Vinokur, L. B. Ioffe, A. I. Larkin, M. V. Feigelman, Sov. Phys. JETP **66**, 198 (1987).

[9] D. Sherrington and S. Kirkpatrick, Phys. Rev. B **35**, 1792 (1975).

[10] D.J. Thouless, P.W. Anderson and R.G. Palmer, *Phil. Mag.*, **35**, 593 (1977).

[11] A. Crisanti, H. Horner, H.-J. Sommers, Z. Phys. B, **92**, 257 (1993).

[12] T.R. Kirkpatrick and D. Thirumalai, *Phys. Rev. B* **36**, 5388 (1987); T.R. Kirkpatrick, D. Thirumalai and P.G. Wolynes, *Phys. Rev. B* **40**, 1045 (1989).

[13] G. Parisi in J.J. Brey, J. Marro, J.M. Rubi and M. San Miguel, eds. *Twenty Five Years of Non-Equilibrium Statistical Mechanics; Proc. of the Thirteenth Sitges Conference*, (Springer-Verlag, Berlin 1995), pp. 135-42.

[14] S. Franz and J. Herz, *Phys. Rev. Lett.* **74**, 2115 (1995).

[15] A. Schmid, Phys. Rev. Lett. **51**, 1506 (1983); R. Fazio and G. Shon, Phys. Rev. B **43**, 5307 (1991).

[16] V. Ambegaoker and A.Baratoff, Phys. Rev. Lett. **10**, 486 (1963).

[17] P. Delsing, C. D. Chen, D. B. Haviland, Y. Harada and T. Cleson, Phys. Rev. B **50**, 3959 (1994); J. E. Mooij and G. Shon in "Single Charge Tunneling: Coulomb Blocade Phenomena in Nanostructures" ed. by H. Grabert and M. Devoret, Plenum NY (1992).

[18] M.Iansiti, A.T.Johnson, C.J.Lobb, and M.Tinkham. Phys. Rev. Lett. **60**, 2414 (1988).

[19] J. E. Mooij, B. J. van Wess, L. J. Geerligs, M. Peters, R. Fazio and G. Shon, Phys. Rev. Lett. **65**, 645 (1990).

[20] M. Tinkham in in "Single Charge Tunnelling: Coulomb BlocKade Phenomena in Nanostructures" ed. by H. Grabert and M. Devoret, Plenum NY (1992).

[21] L. L. Sohn, M. T. Tuominen, M. S. Rzchowski, J. U. Free, and M. Tinkham. Phys.Rev.B **47**, 975 (1993).

[22] L. L. Sohn, M. S. Rzchowski, J.U.Free, and M.Tinkham. Phys. Rev. B **47**, 967 (1993).

[23] P. Chandra, M.V. Feigelman, L.B. Ioffe and D.M. Kagan, To be published.

Randomly Charged Polymers

Mehran Kardar[1] and Yacov Kantor[2]

[1]Department of Physics, Massachusetts Institute of Technology, Cambridge, MA 02139, USA
[2]School of Physics and Astronomy, Tel Aviv University, Tel Aviv 69 978, Israel

Abstract. Polyampholytes (PAs) are polymers with a random sequence of positive and negative charges along their backbone. We have studied systematically the dependence of internal energy and shape of the PA on its excess charge by combining analytic arguments, Monte Carlo simulations, and exact enumeration of all configurations of short chains. The results indicate that the overall excess charge, Q, is the main determinant of the size of the PA. A polymer composed of a mixture of N positive and negative charges $\pm q_0$, is compact for $Q < Q_c \approx q_0 \sqrt{N}$, and expanded otherwise. The transition between the two states at low temperatures is reminiscent of the Rayleigh shape instability of a charged drop. A uniform excess charge causes the breakup of a fluid drop; we show that a uniformly charged polymer stretches out to a *necklace shape*. The inhomogeneities in charge distort the shape away from an ordered necklace. The freezing transition of a PA, and its relevance to proteins is also discussed.

1 Introduction

Randomly charged polymers, or *polyampholytes* (PAs), are of interest to statistical mechanics for several reasons: (1) While the physics of homogeneous polymers is fairly well understood[1], considerably less is known about heteropolymers. The seemingly random sequence of monomers in the latter, provides an interesting challenge to the physics of inhomogeneous and glassy systems[2]. Simple heteropolymers have been studied in connection with the folding of proteins.[3] Since real proteins incorporate charged amino acids, they are examples of PAs[5]. (3) The long–range nature of the Coulomb interactions profoundly effects the behavior of polymers. Experiments on charged heteropolymers[6] and gels[7] reveal interesting and puzzling aspects in need of explanation.

Our model PAs are polymers of length N, where each monomer (pictured as a sphere of size a) has a charge $q_i = \pm q_0$ $(i = 1, \cdots, N)$. Configurations of a polymer are completely specified by listing the position vectors $\{\mathbf{r}_i\}$ of its monomers. The energy of each configuration is then given by

$$\mathcal{H} = \sum_{i=1}^{N-1} \mathcal{V}\left(\mid r_{i+1} - r_i \mid\right) + \sum_{i<j}^{N} \left[\mathcal{U}\left(\mid r_i - r_j \mid\right) + \frac{q_i q_j}{\mid r_i - r_j \mid^{d-2}} \right]. \quad (1)$$

The first term describes the covalent bonding of neighboring monomers. The potential V diverges at short and long distances, to forbid monomer overlap, and bond breaking, respectively. The excluded volume between non-neighboring pairs of monomers is described by the potential U. Finally, the last term in Eq.(1) describes the (unscreened) Coulomb interaction between the charges q_i and q_j. Note that, when generalized to d–dimensional space, the Coulomb interaction falls off with separation as $1/r^{d-2}$.

We have performed several numerical and analytical studies of the above Hamiltonian[8], [9], [10], [11]. Our analytical studies included a high temperature expansion on a continuum version of the problem, *without excluded volume effects*, i.e. starting with

$$\frac{\mathcal{H}}{T} = \frac{K}{2} \int_0^N dx \left(\frac{d\mathbf{r}}{dx}\right)^2 + \frac{1}{2T} \int dx dx' \frac{q(x)q(x')}{|\mathbf{r}(x) - \mathbf{r}(x')|^{d-2}}. \tag{2}$$

We also employed an analogy to charged droplets which will be described in later sections.

For purposes of *Monte Carlo* (MC) simulations[10], we employed a "tether potential" V, which allows the bonded monomers to explore configurations in which their separation is within an allowed range (without any additional energy cost). In these simulations the relaxation time at high temperatures scales as N^4, and we typically looked at chains with $N \leq 128$. However, as the PA falls out of equilibrium at low temperatures, we only examined temperatures $T \geq 0.01 q_0^2/a$ (the basic energy scale of the problem). To perform quench averages, we typically examined 10 different realizations of random sequences.

For exact enumeration studies[11], we looked at *all* self-avoiding random walk configurations on cubic lattices. We also examined all sequences. Since the numbers of configurations (and sequences) grows exponentially with N, we were limited to lengths $N \leq 13$. The exact enumeration allows us to look at all temperatures, and to do exact quench averages, at the expense of severe finite size limitations.

2 Expansion versus Collapse

The spatial extent of a polymer is characterized by the critical exponent ν, which relates its radius of gyration (r.m.s. size) R_g to the number of monomers N, through the power law $R_g \propto N^\nu$. A polymer is "compact" if $\nu = 1/d$, where d is the dimension of the embedding space, and fully "stretched" if $\nu = 1$. For self-avoiding polymers $\nu \approx 0.6$ in $d = 3$[1]. Perhaps the simplest question regarding the PA is whether the introduction of Coulomb interactions causes the polymer to be more or less expanded. There was initially some controversy regarding this question, which is now resolved. If the charges within a PA behave as in a plasma, they will rearrange themselves so as to neutralize excess charge in any region of space. The energy gain from such a screening will provide an inward pressure which compact-ifies the PA[12]. On the other hand, for a sequence in which each monomer takes a charge $q_i = \pm q_0$ independent of the others; there is a "typical" excess charge of $Q_c \equiv q_0 N^{1/2}$. Independent of its sign, the excess charge produces a repulsive energy, which should swell the PA[8].

The apparent contradiction between the above results was resolved in Ref.[9]: If the monomers comprising the chain are assembled together in an organic solvent, the strong Coulomb forces ensure the neutrality of the re-sulting polymer. The neutral polymer in any environment then behaves as a plasma and compactifies to screen the charges. There is no such bias towards neutrality if the polymer is prepared in a solution where the Coulomb forces are screened. The typical polyampholyte will then have a net charge due to the random imbalance of positive and negative components. Such net charge is actually sufficient to stretch the chain in an environment with no Coulomb screening. This is nicely confirmed in MC simulations: the size of a neutral PA monotonically decreases with temperature, while typical unconstrained sequences expand at low temperatures due to their excess charge.

In a high temperature expansion, the Coulomb interaction is treated as a perturbation, and various quantities are obtained as a series in inverse temperature. For example, the radius of gyration is given by

$$\overline{\langle R^2 \rangle} = \langle R^2 \rangle_0 \left[1 + A\left(\frac{T_0}{T}\right) + B\left(\frac{T_0}{T}\right)^2 + \cdots \right], \tag{3}$$

where $\overline{\cdots}$ and $\langle \ldots \rangle$ denote averaging over sequences and conformations re-spectively. The characteristic temperature is obtained by dimensional analy-sis, and given by

$$T_0 \equiv \frac{q_0^2 N}{\langle R^2 \rangle_0^{\frac{d-2}{2}}} \propto N^{1-(d-2)\nu}. \tag{4}$$

In $d \leq 4$, T_0 grows with N signalling the breakdown of perturbation theory, and the relevance of interactions. In Ref.[9], the coefficients A and B were calculated, starting from an ideal (non self–avoiding) chain at infinite tem-perature: for a neutral PA, $A = -\sqrt{192/4050\pi}$, while for a random sequence $A = 0$ and B is negative.

Quite generally, the coefficient A is proportional to the correlator of charges, i.e. $A \propto \overline{q_i q_j}$. While $\overline{q_i q_j} = 0$ for independently selected charges, con-straining the net charge to Q leads to $\overline{q_i q_j} = (Q^2 - q_0^2 N)/N^2$. The first term in the high temperature expansion thus changes sign at $Q = Q_c = q_0 \sqrt{N}$. We confirmed this behavior by MC simulations of PAs with fixed charge. For $N = 64$, chains with $Q > 8q_0$ expand upon lowering temperature, while those with $Q \leq 8q_0$ collapse. Rather surprising, the radii of gyration appear to vary monotonically with temperature, and the above trends continue to low temperatures. Histograms constructed at $T = 0.05q_0^2/a$ suggested that Q is a good determinant of the size of the PA, with polymers of different Q falling into distinct bands[10].

3   Excess Charge and the Necklace Model

The high temperature expansion *does not* explain why PAs are compact for $Q \approx Q_c$ at low temperatures. Instead, we use the observation that PAs with $Q = 0$ form dense globules, as a starting point for the investigation of the dependence of low–T shapes on Q in $d = 3$. The energy (or rather the quench–averaged free energy) of the PA is then phenomenologically related to its shape by

$$E = -\epsilon_c N + \gamma S + \frac{bQ^2}{R}. \tag{5}$$

The first term is a condensation energy proportional to the volume (assumed compact) with $\epsilon_c \sim q_0^2/a$; the second term is proportional to the surface area S (with a surface tension $\gamma \sim q_0^2/a^3$), while the third term represents the long–range part of the electrostatic energy due to an excess charge Q. (For a uniformly charged spherical drop $b = 0.6$, while the migration of charges to the surface reduces b to 0.5 in a conducting drop.)

The optimal shape is obtained by minimizing the overall energy in Eq.(5). The first term is the same for all compact shapes, while the competition between the surface and electrostatic energies is controlled by the dimensionless parameter

$$\alpha \equiv \frac{Q^2}{16\pi R^3 \gamma} = \frac{Q^2}{12V\gamma} \equiv \frac{Q^2}{Q_R^2}. \tag{6}$$

Here, R and V are the radius and volume of a spherical drop of N particles, and we have defined the *Rayleigh charge* Q_R. We note that in the case of PAs[10], [11] $Q_R \approx q_0 N^{1/2} = Q_c$. The behavior of a system described by Eq.(5) was first analyzed more than a century ago in the context of charged liquid drops by Lord Rayleigh [13], [14]. He showed that a conducting drop is locally unstable to an infinitesimal sphroidal distortion for $\alpha > 1$. In fact, if the possible shapes of a globule are restricted to spheroids, this instability is preempted by a discontinuous transition[10] to a prolate shape for $\alpha \gtrsim 0.899$. Both MC simulations[10] and experimental results[7] indicate that for $Q < Q_R$, the R_g of PAs at low temperature is almost independent of Q, while it increases extremely fast as a function of Q for larger charges. The question of the stable shape of the PA for $Q > Q_R$ is taken up next.

Although the ellipsoidal globule may have a lower energy than the spherical one, the former is *not* the minimum energy configuration of a charged drop. This was first noted in the context of liquid drop models of atomic nuclei[15] A uniformly charged nucleus minimizes its energy by undergoing *fission*, and splitting into two equal parts [16], [17], [18]. If the resulting parts (droplets) can be infinitely separated, this instability occurs for $\alpha > 0.3$. Obviously, the PA must maintain its connectivity and cannot undergo fission. We proposed in Ref.[10] that as a compromise a charged polymer splits into two droplets that are connected by a narrow tube of length L and diameter a (a dumbbell shape). As long as $La^2 \ll R^3$, most of the charge remains in the spherical globules. The total electrostatic energy is proportional to Q^2/L, while the energy cost of creating the tube (from surface or condensation energy) grows as $\gamma a L$; equating the two gives $L \propto Q$. For larger Q, a liquid drop will split into a larger number of droplets. The corresponding picture for the PA is a *necklace* of globules connected by narrow strands. This picture has been independently confirmed by MC simulations of uniformly charged polymers[19].

The astute reader will note that the picture developed in the previous paragraph in fact applies to a uniformly charged polymer, such as a polyelectrolyte in a poor solvent[19]. Inhomogeneities in charge are expected to modify this scenario. If a random PA with $\alpha > 0.3$ splits into globules of charges Q_1 and Q_2, it can be shown that for each globule $\overline{\alpha_{\text{subchain}}} = (1+\alpha)/2$, while the mean product of the charges is $\overline{Q_1 Q_2} = q_0^2 N(\alpha - 1)/4$. For $\alpha = 1$, each globule has, on average, values of α close to unity, while the average value of the product of charges vanishes. We thus have the paradoxical situation in which most spherical shapes are unstable, while there is on average no energetic gain in splitting the sphere into two equal parts.

While charge inhomogeneities drastically modify the necklace picture, the resulting PA is probably still composed of rather compact globules connected by a (not necessarily linear) network of tubes. The globules are selected preferentially from segments of the chain that are approximately neutral (or at least below the instability threshold), while the tubes are from subsequences with larger than average excess charge. A few globules can indeed be identified in recent MC simulations of PAs[20]. It can be shown[21] that the probability of finding a very large neutral segment in a random sequence of charges is large, i.e. it is possible to build a configuration consisting of one very large neutral globule with highly charged "tails" sticking out of it.

4   Freezing Transition

Compact heteropolymers with short range interactions are known to undergo a freezing transition in which the number of thermodynamically relevant states goes from an exponentially large value ($\mathcal{O}(e^N)$) in the random globule state, to only a few ($\mathcal{O}(1)$) conformations in the frozen state[3]. This transition is believed to be well described by the Random Energy Model (REM)[22]. The principal assumption of REM is that, for a given sequence, the energies of different configurations are unrelated, i.e. $\overline{E_\alpha E_\beta} \approx 0$. This is certainly not true, as configurations related by simple moves will have correlated energies. The approximate validity of REM resides in such configurations making a negligible contribution to the overall probability distribution for overlaps $\mathcal{P}(E_\alpha E_\beta)$, which is expected to approach a delta function as N becomes large.

In REM the single energy density of states is assumed to be a Gaussian of width $u\sqrt{N}$. Typically the number of conformations of a polymer scales as $\mathcal{M} \sim e^{\omega N}$, with ω of the order of unity, resulting in an entropy

$$S_{\text{REM}}(E) = \ln\left[\exp\left(\omega N - \frac{E^2}{2Nu^2}\right)\right].\tag{7}$$

On lowering temperature, the number of available states decreases, leading to an entropy crisis, and hence a freezing transition at $T_f = u/\sqrt{2\omega}$.

The exact enumeration studies of short chains[11] also provided some preliminary evidence of a freezing transition in PAs. However, it is unclear how Coulomb interactions modify freezing. Indeed the basic assumption of REM is not valid as the energies in a PA are highly correlated. In fact dimensional analysis, supported by numerical studies[23], indicates that

$$\overline{E_\alpha E_\beta} \approx \frac{N^2}{R^{2(d-2)}} \propto u^2 N^{4/d}.\tag{8}$$

Enumeration of energy levels for chains of up to $N = 36$ indicates that while the mean energy fluctuates strongly with sequence, each sequence has an approximately Gaussian density of states. If we naively apply the REM freezing criterion, we thus obtain a $T_f \propto N^{(4-d)/2d}$, i.e. the freezing temperatures goes up with N, and all chains are frozen in the thermodynamic limit for $d \leq 4$.

It is tempting to compare this conclusion with an earlier result[9] that PA interactions also remove the collapse (θ) transition of chains in $d < 4$, replacing it by a smooth crossover. This is probably just a coincidence as the result for freezing is obtained from energy *correlations* ($N^2/[N^{1/d}]^{d-2}$) for *compact* chains of the size $\sim N^{1/d}$, while the earlier result is obtained by considering the interaction of two ideal chains, with charge imbalance $N^{1/2}$, over the Gaussian separation $N^{1/2}$, i.e. $[N^{1/2}]^2/[N^{1/2}]^{d-2}$.

An important element left out of the previous analysis is screening. It is likely that at temperatures lower than T_0 in Eq.(4), the charges are screened over a Debye length $r_D \sim (TR^d/Ne^2)^{1/2}$. Assuming that in this case the entropy is proportional to the number of independent Debye volumes $\mathcal{N} \approx (R/r_D)^d$, we obtain

$$S_{\text{PA}}(T) = \ln\left[\exp\left(\omega N - c\left(\frac{e^2 N}{TR^{d-2}}\right)^{d/2}\right)\right].\tag{9}$$

In this case, freezing (signalled by $S(T_f) = 0$) occurs for an N independent value of $T_f = e^2/a^{d-2}\cdot(c/\omega)^{2/d}$. However, the results of lattice simulations[23] appear to indicate that chains of up to $N = 36$ are still in a short chain regime where screening effects are not important.

5   Conclusions

We have demonstrated that while neutral PAs form compact globules, a small net charge of $Q_c \propto \sqrt{N}$ is sufficient to cause instabilities in the globular shape. Beyond this threshold, a uniformly charged polymer breaks up into a regular sequence of globules arranged like pearls on a stretched necklace. The inhomogeneities in charge of a random PA are expected to destroy the regularity of the necklace. However, MC simulations indicate that the picture of globules connected by tubes continues to remain applicable.

The experimental signature of the necklace shape is the appearance of multiple length scales in scattering probes of PAs. There is some indication of this happening in the experiments of Ref.[6]. Another intriguing observation is that gels composed of PAs exhibit multiple phases with different volumes[7], [25]. A possible explanation of this observation is that the different phases are composed of globules of different sizes at the mesoscopic level, the transitions between them induced by surface tension instabilities similar to those of individual PAs. Again a possible experimental test is to search for inhomogeneities in density at short scales.

Finally, we note that *proteins* are weakly charged PAs: Of the 20 natural amino-acids, three are positively charged (Lys, Arg, His), two are negatively charged (Asp, Glu), and the rest are neutral [5]. It is often assumed that Coulomb interactions are not essential to proteins, as the screening length in biological solvents is often quite small. It is less clear that screening is also effective in compact globular configurations with little or no solvent in their interiors. Secondary structural elements such as α-helices effectively reduce the conformational flexibility of proteins. Thus, while the total charge on a given protein may be small, in solvents with few counter ions, this may be sufficient to lead to a REM-violating correlated energy landscape, making the results of this study relevant.

We can also (indirectly) examine the relevance of Coulomb interactions in protein structure, by searching for correlations of charged amino acids along the tabulated sequences. Firstly, the typical net charge of proteins is less than that expected on the basis of statistical fluctuations. Secondly, protein sequences are anti-correlated with respect to their charge [24]. This indicates that perhaps protein evolution was not dictated solely by the degree of hydrophobicity of monomers (which depends on the degree of charge, not its sign), and that Coulomb interactions also played an important role.

acknowledgments This work was supported by the US–Israel BSF grant No. 92–00026, by the NSF through grant No. DMR–94–00334. Discussions with V.J. Pande and A.Yu. Grosberg are gratefully acknowledged. Many thanks are due to Prof. Miguel Rubi for organizing an interesting and productive conference in the beautiful setting of Sitges.

References

[1] P.G. de Gennes, *Scaling Concepts in Polymer Physics*, Cornell Univ. Press, Ithaca (1979).

[2] M. Mezard, G. Parisi, and M. A. Virasoro, *Spin Glass Theory and Beyond*, World Scientific, Singapore (1987).

[3] D.L. Stein, Proc. Natl. Acad. Sci. USA **82**, 3670 (1985); J.D. Bryngelson and P.G. Wolynes, Proc. Natl. Acad. Sci. USA **84**, 7524 (1987); H.S. Chan and K.A. Dill, Physics Today **46**(2), 24 (1993); T. Garel and H. Orland, Europhys. Lett. **6**, 307 (1988); E.I. Shakhnovich and A.M. Gutin, Europhys. Lett. **8**, 327 (1989); M. Karplus and E.I. Shakhnovich, in *Protein Folding*, ed. by T.E. Creighton, ch.4, p. 127, (Freeman & Co., New York, 1992).

[5] D. Dressler and H. Potter, *Discovering Enzymes*, (Scientific American Library, NY, 1990).

[6] J. Copart and F. Candau, *Macromolecules* **26**, 1333 (1993); M. Scouri, J.P. Munch, S.F. Candau, S. Neyret, and F. Candau, Macromol. **27**, 69 (1994).

[7] X.-H. Yu, A. Tanaka, K. Tanaka, and T. Tanaka, J. Chem. Phys. **97**, 7805 (1992); Yu X.-H., Ph. D. thesis, MIT (1993); A. E. English, S. Mafe, J. A. Manzanares, X.-H. Yu, A. Yu. Grosberg, and T. Tanaka, *J. Chem. Phys* **104**, 8713 (1996).

[8] Y. Kantor and M. Kardar, Europhys. Lett. **14**, 421 (1991).

[9] Y. Kantor, H. Li, and M. Kardar, Phys. Rev. Lett. **69**, 61 (1992); Y. Kantor, M. Kardar, and H. Li, Phys. Rev. **E49**, 1383 (1994).

[10] Y. Kantor and M. Kardar, Europhys. Lett. **27**, 643 (1994); Y. Kantor and M. Kardar, Phys. Rev. **E51**, 1299 (1995).

[11] Y. Kantor and M. Kardar, Phys. Rev. **E52**, 835 (1995).

[12] P.G. Higgs and J.-F. Joanny, J. Chem. Phys. **94**, 1543 (1991); J. Wittmer, A. Johner and J.F. Joanny, Europhys. Lett. **24**, 263 (1993).

[13] Lord Rayleigh, Phil. Mag. **14**, 184 (1882).

[14] G. Taylor, Proc. R. Soc. London **A280** 383 (1964).

[15] J.M. Blatt and V.F. Weisskopf, *Theoretical Nuclear Physics*, ch. 7, p. 303, Willey, New York (1952); R.D. Evans, *The Atomic Nucleus*, ch. 11, p. 387, McGraw–Hill, New York (1955).

[16] N. Bohr and J. A. Wheeler, Phys. Rev. **56**, 426 (1939).

[17] E. Feenberg, Phys. Rev. **55**, 504 (1939).

[18] F. Weizsacker, Naturwiss. **27**, 133 (1939).

[19] A.V. Dobrynin, S.P. Obukhov, and M. Rubinstein, preprint (1995).

[20] N. Lee and S.P. Obukhov, work in progress (1996); see http://www.phys.ufl.edu/~nlee/research.html.

[21] Y. Kantor and D. Ertaş, J. Phys. **A27**, L907 (1994).

[22] B. Derrida, Phys. Rev. Lett. **45**, 79 (1980).

[23] V.S. Pande, A.Yu. Grosberg, C. Joerg, M. Kardar, and T. Tanaka, preprint (1996).

[24] V.S. Pande, A.Yu. Grosberg, and T. Tanaka, Proc. Nat. Acad. Sci. USA **91**, 12976 (1994).

[25] M. Annaka and T. Tanaka, Nature **355**, 430 (1992).

Copolymer melts in disordered media

A. V. Dobrynin[1], S. Stepanow[2], T. A. Vilgis[3]

[1]Department of Chemistry, University of North Carolina, Chapel Hill, NC, USA
[2]Martin-Luther-Universität Halle-Wittenberg, Fachbereich Physik,
D-06099 Halle/Saale, Germany
[3]Max-Planck-Institut für Polymerforschung, Postfach 3148,
D-55021 Mainz, Germany

Abstract. The symmetric AB block copolymer melt in a gel matrix with preferential adsorption of A monomers on the gel gives an example of a random-field system, which is described near the point of the microphase separation transition by the random field Landau-Brazovskii Hamiltonian. By using the technique of the 2-nd Legendre transform, the phase diagram of the system is calculated. We found that the preferential adsorption of the copolymer on the gel results in two effects: a) It decreases the temperature of the first order phase transition between disordered and ordered phase. b) There exists a region on the phase diagram at some small but finite value of the adsorption energy in which the replica symmetric solution for two replica correlation function is unstable with respect to replica symmetry breaking. We interpret this state as a glassy state and calculate a spinodal line of this transition. We also consider the stability of the lamellar phase in the weak segregation limit by mapping the copolymer Hamiltonian onto the Hamiltonian of the random field XY model. We suggest that the long range order is always destroyed by weak randomness.

1 Introduction

Diblock copolymer melts consist of macromolecules composed of the two blocks of chemically distinct repeat units of A and B monomers covalently bonded with each other. The copolymers can be characterized by the overall degree of polymerization N, the composition $f = N_A/N$, and the interaction energies $V_{\alpha\beta}(r)=V_{\alpha\beta}\delta(r)$ between the monomers. In the case when $V_{AB} > V_{AA}$, V_{BB} the contacts between A and B monomers are unfavorable, which drives the system to segregate. Quantitatively the tendency to segregation can be expressed by the Flory-Huggins parameter $\chi = 2V_{AB}-V_{AA}-V_{BB}$. Since the entropic and enthalpic contributions to the free energy scales as N^{-1} and χ, respectively, block copolymer phase state is dictated by the product χN. For many materials the Flory-Huggins parameter has the temperature dependence $\chi \approx a/T+b$, where $a > 0$ and b is a constant, so the tendency for segregation will be increased with decreasing the temperature T. Due to the covalent junction between blocks of A and B monomers the macrophase separation cannot occur. Instead of this the system separates locally on A-rich and B-rich regions. Such local segregation is called microphase separation.

The domain structure appears as a result of competition between short-range monomer-monomer interaction that want to decrease the number of unfavorable contacts between A and B monomers and long-range correlation due to chemical bonds between those part of the chains that tends to segregate into domains. The first stage of the domain structure formation, when the amplitude of the local composition fluctuations is small in comparison with its average value, can be described in terms of Landau-Brazovskii effective Hamiltonian [1]-[6]. This Hamiltonian describes the phase transitions in the systems such as weakly anisotropic antiferromagnets [4], the isotropic-cholesteric and nematic-smectic C transition in liquid crystals [6], pion condensation in neutron stars [7]. The homogeneous state of these systems is unstable with respect to fluctuations of the finite wave number q_0 that below phase transition point results in formation of the modulated structure with period $L = 2\pi/q_0$ (for review see [2], [3]).

The description of the copolymer melt is based on the introduction instead of the individual monomeric coordinates $\mathbf{r}_i(s_i)$ (the index $i = 1, ...n_c$, counts the copolymer chains, s_i gives the position of the segment along the chain) the densities of polymer segments (collective coordinates) $\rho_A(\mathbf{r}) = \sum_{i=1}^{n_c} \int_0^{fN} ds\delta(\mathbf{r} - \mathbf{r}_i(s))$ and $\rho_B(\mathbf{r}) = \sum_{i=1}^{n_c} \int_{fN}^N ds\delta(\mathbf{r} - \mathbf{r}_2(s))$. For the incompressible copolymer melt $(\rho_A(\mathbf{r}) + \rho_B(\mathbf{r}) = \rho_m$, ρ_m is the average density of the melt) the order parameter $\Phi(r)$ is the deviation of the local density of A monomers $\rho_A(r)$ from its average concentration $f\rho_m$. There is one characteristic length scale of the system under consideration, the gyration radius of the copolymer chain $R_g = lN^{1/2}/6^{1/2}$, where l is the bond size. So, it is useful to normalize the length scales on the gyration radius and introduce the following dimensionless variables

$$q = qR_g, \Phi^2(r) = \Phi^2(r)/(\rho_m N R_g^3), \ \lambda = u_0 N/(\rho_m R_g^3), \ \tau = 2(\chi_s - \chi)N. \quad (1)$$

χ_s is the value of Flory-Huggins parameter on the spinodal computed by Leibler [1] ($\chi_s N = 10.495$ at $f = 1/2$), τ is the effective temperature, and the parameter u_0 depends on the copolymer composition [1]. Near the phase transition point the free energy of the system can be expanded in the power series of the order parameter $\Phi(r)$ [4], [8]

$$H(\Phi) = \frac{1}{2} \int_q ((|\mathbf{q}| - q_0)^2 + \tau)\Phi(\mathbf{q})\Phi(-\mathbf{q}) + \quad (2)$$

$$\frac{\lambda}{4!} \prod_{i=1}^{4} \int_{q_i} \Phi(\mathbf{q}_i)\delta(\sum_{i=1}^{4} \mathbf{q}_i) + \int_q \frac{(N\rho_m)^{1/2}}{R_g^{3/2}} h(-\mathbf{q})\Phi(\mathbf{q}),$$

where q_0 (= 1.94 at $f = 1/2$) is the peak position of the scattering factor and $\Phi(\mathbf{q})$ is the Fourier transform of the order parameter. The last term in (2) describes the interaction of the copolymers with an external field; $h(r) = U_A(r) - U_B(r)$ with $U_\alpha(r)$ being the interaction energy of the αth monomer. For AB copolymer melt that is immersed in a gel matrix the random field $U_\alpha(\mathbf{r})$ describes interaction of monomers of αth type with the gel monomers. In this case the random field $U_\alpha(\mathbf{r})$ is

$$U_\alpha(\mathbf{r}) = -v\epsilon_\alpha \Phi_g(r), \tag{3}$$

where ϵ_α is the adsorption energy of αth monomer in the units of the thermal energy kT, v is the excluded volume of the gel-copolymer interaction, which is assumed to be the same for all types of interactions, $\Phi_g(r)$ is the local gel concentration. The gel structure can be characterized by the two first correlation functions [9]

$$< \Phi_g(r) > = \overline{\rho_g} = N_g/R_m^3, < \Phi_g(r)\Phi_g(r') > = G(r - r'), \tag{4}$$

where N_g is the number of monomers between cross-links and R_m is mesh size distance being for rigid network proportional to N_g and for Gaussian one $\sim N_g^{1/2}$. Taking into account the relations (4) we can write two first moments of the random fields $U_\alpha(r)$ as follows

$$< U_\alpha(r) > = v\epsilon_\alpha \overline{\rho_g}, < U_\alpha(r)U_\beta(r) > = v^2 \epsilon_\alpha \epsilon_\beta N_g^2 R_m^{-3}\delta(r - r'). \tag{5}$$

One should note that the last equality in the r.h.s. of Eq.(5) is valid as long as the characteristic length scale of the copolymer melt L is larger than R_m. The average over the random field $h(q)$ can be carried out by using the replica trick. As a result the multireplica effective Hamiltonian is obtained as

$$H_n(\{\Phi\}) = \frac{1}{2}\int_q \sum_{a,b=1}^{n} \Phi_a(\mathbf{q})(\delta_{ab}G_0^{-1}(\mathbf{q}) - \varDelta)\Phi_b(-\mathbf{q})$$

$$+\frac{\lambda}{4!}\sum_{a=1}^{n}\prod_{i=1}^{4}\int_{q_i} \Phi_a(\mathbf{q}_i)\delta(\sum_{i=1}^{4}\mathbf{q}_i), \tag{6}$$

where n is number of replicas, and $\varDelta = N\rho_m v^2(\epsilon_1 - \epsilon_2)^2 N_g^2/R_m^3$. To calculate the free energy F_n we use a variational principle based on the second Legendre transformation [10]-[13]. In the framework of this approach the n-replica free energy of the system under consideration is

$$F_n = \min W_n(\{\Phi_a(\mathbf{q})\}, \{G_{ab}(\mathbf{q})\}) \tag{7}$$

with

$$W_n = -\frac{1}{2}Sp\int_q \ln G_{ab}(\mathbf{q}) + \frac{1}{2}\sum_{a,b=1}^{n}\int_q (\delta_{ab}G_0^{-1}(q) - \varDelta)G_{ab}(-\mathbf{q}) +$$

$$\frac{1}{2}\sum_{a,b=1}^{n}\int_q < \Phi_a(-\mathbf{q}) > (\delta_{ab}G_0^{-1}(q) - \varDelta) < \Phi_b(\mathbf{q}) > +$$

$$\sum_{a=1}^{n}\int_q < \Phi_a(-\mathbf{q}) > (\frac{\lambda}{4}\int_{q'} G_{aa}(\mathbf{q}')) < \Phi_a(\mathbf{q}) > +\frac{\lambda}{4!}\sum_{a=1}^{n}\int d^3r < \Phi_a(\mathbf{r}) >^4$$

$$+\frac{\lambda}{8}\sum_{a=1}^{n}(\int_q G_{aa}(\mathbf{q}))^2 - \frac{\lambda^2}{48}\sum_{a,b=1}^{n}\int d^3r G_{ab}^4(\mathbf{r}), \tag{8}$$

where the minimum of the functional W_n is to be sought with respect to functions $<\Phi_a(\mathbf{q})>$ and the renormalized Green's function $G_{ab}(\mathbf{q})$ that are considered as independent variables at the fixed parameters λ, τ, Δ. The functional W_n contains only 2-irreducible diagrams that cannot be cut into two independent parts by removing any two lines between vertices λ.

2   Replica symmetric solution. The phase diagram

In this section we will calculate the phase diagram of the copolymer melt in the gel taking into account only one-loop diagrams in the functional W_n. In this case there is only a replica symmetric solution. In the replica symmetric case the effective propagator is expected to have the following form

$$G_{ab}(q) = g_B(q)\delta_{ab} + \Delta g_B^2(q), \tag{9}$$

with $g_B(q) = 1/((|\mathbf{q}| - q_0)^2 + r)$, where r is the renormalized temperature. In the ordered phase there is a nonzero average value of the order parameter $<\Phi(q)>$, which describes the appearance of the domain structure. The Fourier transform of the order parameter has the form

$$<\Phi_a(q)> = A(\delta(q_z - q_0) + \delta(q_z + q_0)). \tag{10}$$

Here we will consider only the ordered phase with the lamellar type of symmetry of the microphase structure that has the lowest free energy for the effective Hamiltonian (6). Substituting the trial functions (9-10) into expression of the free energy we can write the free energy as a functional of A and $g_B(q)$. In order to find the values A and $g_B(q)$ corresponding to the extremum of the functional W_n we have to take variational derivatives of W_n with respect to these functions. The analysis of the extremal equations can be simplified if one introduces the new reduced variables

$$t = \tau/(\lambda q_0^2/2\pi)^{2/3}, \quad \delta = \Delta/(\lambda q_0^2/2\pi)^{2/3}, \quad z = r/(\lambda q_0^2/2\pi)^{2/3}. \tag{11}$$

In the homogeneous state the extremal equation for the renormalized temperature r reads

$$z = t + \frac{1}{2}z^{-1/2}(1 + \frac{\delta}{2z}). \tag{12}$$

Repeating the same calculations for the ordered phase for which $A^2 = 2r/\lambda$, one can obtain

$$-t = z + \frac{1}{2}z^{-1/2}(1 + \frac{\delta}{2z}). \tag{13}$$

To calculate the phase diagram of the system under consideration the energies of the homogeneous and ordered phases have to be compared. Substituting trial functions given by Eqs.(9) into expression for n-replica free energy F_n

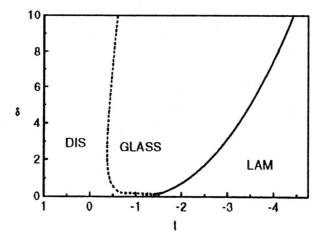

Fig. 1. The phase diagram of the copolymer melt in a random field environment obtained within the one-loop replica symmetric solution. The continuous curve is the coexistence curve between the ordered (lamellar) and disordered states. It is computed by setting equal Eqs.(14-15). The dashed line is the stability line of the replica symmetric solution. It is derived by solving Eqs.(21,12).

and taking limit $< F >_{av} = \lim_{n \to 0} F_n/n$ one can write the expressions for the free energy of homogeneous phase

$$< F >_{av} = (\frac{q_0^2}{2\pi})^{4/3} \lambda^{1/3} (\frac{1}{2}\sqrt{z} - \frac{\delta}{\sqrt{z}} + \frac{1}{2}z^2 - \frac{1}{2}t^2) \qquad (14)$$

and for the ordered one

$$< F >_{av} = (\frac{q_0^2}{2\pi})^{4/3} \lambda^{1/3} (\frac{1}{2}\sqrt{z} - \frac{\delta}{\sqrt{z}} - \frac{1}{2}z^2 - \frac{1}{2}t^2). \qquad (15)$$

The phase diagram of the copolymer system in random media is sketched in Fig.1 in the plane $(\delta = \Delta/((\lambda q_0^2/2\pi)^{2/3}), t = \tau/((\lambda q_0^2/2\pi)^{2/3}))$. In the limit of weak random field the first-order phase transition occurs at $\tau_{tr} \simeq -1.3(\lambda q_0^2/2\pi)^{2/3}$. At the strong random field the temperature of the phase transition is proportional to $\Delta^{2/5}$.

3   Stability of the Replica symmetric solution

In this section we consider the stability of the replica symmetric solution (9) with respect to replica symmetry breaking. With this purpose let us represent the renormalized two-replica correlation function $G_{ab}(q)$ in the following form

$$G_{ab}(\mathbf{q}) = g_B(q)\delta_{ab} + \Delta g_B^2(q)e_a^T e_b + \delta Q_{ab}(q), \qquad (16)$$

where $e = (1, ..., 1)$ and $\delta Q_{ab}(q)$ is a function that equals to zero for $a = b$. Substituting the function $G_{ab}(q)$ into the r.h.s. of the n-replica free energy (8) and expanding $Sp\ln(G_{ab})$ in the power of the function $\delta Q_{ab}(q)$ one finds

$$\Delta F_n = \frac{1}{4}[\{\int_q [\frac{\delta_{bc}}{g_B^2(q)} - \frac{2\Delta}{g_B(q)}e_b^T e_c]\delta Q_{ab}(q)\delta Q_{ca}(q)\} - \tag{17}$$
$$\frac{\lambda^2}{2}\int_q \int_k \int_p \delta Q_{ab}(k)\delta Q_{ab}(p)f(q)f(q-k-p)],$$

where in the r.h.s. of Eq.(17) the summation over all repeat indices is assumed. The analysis of the Dyson equation for off-diagonal part of the two-replica correlation function including the second-order diagram in the power of vertex λ shows that the fluctuations with $|\mathbf{q}| = q_0$ give the main contribution to the renormalization of the bare characteristic of the system, so that we can choose $\delta Q_{ab}(q)$ in the form $\delta Q_{ab}(q) = Q_{ab}g_B^2(q)$, where Q_{ab} is a $n \times n$ matrix. Introducing the Parisi's function $q(x)$ [14] defined in the interval $[0, 1]$ and connected to Q_{ab} by

$$\int_0^1 q^k(x)dx = \lim_{n\to 0} \frac{1}{n(n-1)} \sum_{a,b} Q_{ab}^k \quad \forall, k \tag{18}$$

the quadratic part of the free energy in power of Q_{ab} is obtained as

$$\Delta F = -\frac{1}{4}\{\int_0^1 \int_0^1 ((I_1 - I_3)\delta(x - y) + I_2)q(x)q(y)dxdy\}, \tag{19}$$

where

$$I_1 = \frac{q_0^2}{4\pi}r^{-3/2}, I_3 = \frac{1}{512\pi^2}\lambda^2\Delta^2 q_0^5 r^{-6}, I_2 = \frac{3}{8\pi}q_0^2\Delta r^{-5/2}. \tag{20}$$

The replica symmetric solution is unstable in the region of parameters where the matrix $(I_1 - I_3)\delta(x-y) + I_2$ has a negative eigenvalue $\lambda_- = I_1 + I_2 - I_3 < 0$ (or in the reduced variables (11))

$$\lambda_- \sim (1 + \frac{3}{2}\frac{\delta}{z} - c\frac{\delta^2}{z^{9/2}}) \leq 0 \tag{21}$$

where $c = ((2\pi)^{2/3}/64)(\lambda/q_0)^{1/3} < 1$. Eq.(21) determines the spinodal line of the replica symmetric solution. The boundary of the RS solution, which is computed by using Eqs.(21,12) for the value $c = 0.5$, is also plotted in Fig.1. It follows that RSB occurs already in the disordered phase.

4 Disorder vs. ordering in the lamellar phase

In the weak segregation limit for which the local composition fluctuations is small in comparison with its average value we can use the Hamiltonian (2) to describe the system below microphase separation transition. Keeping only gradient and random field terms the effective Hamiltonian is

$$H\left(\{\Psi\left(r\right)\}\right) = \frac{1}{8q_0^2} \int_r \left(\left(\nabla^2 + q_0^2\right)\Phi\left(r\right)\right)^2 - \frac{(N\rho_m)^{1/2}}{R_g^{3/2}} \int_r h(r)\Phi\left(r\right). \quad (22)$$

For periodic structure in z direction the average value of the order parameter $\langle \Phi\left(r\right)\rangle$ is

$$\langle \Phi\left(r\right)\rangle = 2A\cos\left(q_0\left(z + u(r)\right)\right), \quad (23)$$

where scalar function $u(r)$ describes the deformation of the layers in z-direction. Substituting Eq.(23) into the r.h.s. of Eq. (22), one obtains after averaging over all possible configurations of the random field $h(r)$ the effective Hamiltonian for function $u(r)$ as

$$H_n\left(\{u_a\left(r\right)\}\right) = \sum_{a=1}^{n} \frac{A_a^2}{4} \int_r \left(\left(\Delta_{x,y}u_a(r)\right)^2 + 4q_0^2\left(\nabla_z u_\alpha(r)\right)^2\right)$$

$$- \sum_{a,b=1}^{n} \Delta A_a A_b \int_r \cos\left(q_0\left(u_a(r) - u_b(r)\right)\right), \quad (24)$$

where the first term in the r.h.s. of the Eq.(24) describes the deformation of the lamellar layers and the second one couples this deformation in different replicas.

It is interesting to note that the cosine-like coupling term between fluctuations of the order parameter in two different replicas appears in disordered physical systems such as an array of flux-line in type II superconducting film in magnetic film [15], a crystalline surface with a disordered substrate [16], random field XY model [17]. In all these systems the cosine term results in the breaking of the long-range order and spontaneous replica symmetry breaking [18]- [19].

In the framework of the gaussian variational principle [20], [19] the contribution to the free energy of the system due to fluctuations of the displacement $u_a(r)$ in different replicas is

$$F_{var} = -\frac{1}{2}Sp\int_q \ln(G_{ab}(q)) + \langle H\left(\{u_a(q)\}\right) - H_0\rangle_0, \quad (25)$$

where we introduced

$$H_0 = \frac{1}{2}\sum_{a,b=1}^{n}\int_q G_{ab}^{-1}(q)u_a(q)u_b(-q) \quad (26)$$

and $G_{ab}^{-1}(q)$ is the two replica trial function which form has to be defined self-consistently and the brackets $\langle ... \rangle_0$ denote the thermal averaging with the weight $\exp(-H_0)$. Analyzing the last equation one can conclude that function $G_{ab}^{-1}(q)$ does not depend on the momentum q for $a \neq b$. So, we can define $G_{ab}^{-1}(q) = -\sigma_{ab}$ $(a \neq b)$. In the case of the one-step replica symmetry breaking for which the elements of the matrix σ_{ab} are assumed to have two different values σ_0 and σ_1 depending on whether the two indices a and b belong to the same blocks of the length m or not one can rewrite the extremal equations (see for details [19])

$$\sigma_1 = Y \exp\left(\eta \ln t\right), \sigma_0 = Y \exp\left(\eta \ln t - \frac{1}{m}\left(2\eta \ln L q_0 + \eta \ln t\right)\right), \quad (27)$$

where L is the linear size of the system and we introduced the following parameters assuming A_a to be the same in the all replicas and equal A

$$\eta = \frac{q_0}{16\pi A^2}; t = 2\frac{m(\sigma_1 - \sigma_0)}{A^2 q_0^4}; Y = 2\Delta q_0^2 A^2. \quad (28)$$

In the thermodynamic limit $L \to \infty$ Eq.(27) gives $\sigma_0 = 0$. Substitution of the solution for the trial function $G_{ab}(q)$ into variational free energy yields (see [19], [8])

$$f_{var} = \lim_{n \to 0} \frac{1}{n}\left(F_{var}(t) - F_{var}(0)\right) = \frac{VA^2 q_0^2}{4}\left((1 - \frac{1}{m})\eta t + Y'(1 - m)t^\eta\right), \quad (29)$$

where $Y' = 2Y/(A^2 q_0^4) = 4\Delta/q_0^2$ is introduced. The equilibrium values of the parameters m and t can be found from the system of the equations

$$\frac{1}{m^2}\eta t - Y't^\eta = 0, \quad (1 - \frac{1}{m}) + Y'(1 - m)t^{\eta - 1} = 0. \quad (30)$$

For $\eta > 1$ this system has only trivial solution $m = 1$ and $t = 0$ that corresponds to the replica symmetric solution with all off diagonal elements of the matrix $G_{ab}(q)$ equal to zero. The nontrivial solution appears for $\eta \leq 1$, that reads

$$m = \eta, \quad t = (Y'\eta)^{\frac{1}{1-\eta}}. \quad (31)$$

In other words at $\eta = 1$ the system undergoes a phase transition for which the correlation function $G_{ab}(q)$ changes the form. We can rewrite the condition $\eta = 1$ in terms of the parameters of the system. In the mean field approximation $A^2 = 2|\tau|/\lambda$ that gives the effective temperature of the phase transition $|\tau| = q_0\lambda/32\pi$. Comparing this temperature with the temperature of the first order phase transition $|\tau| \approx (q_0^2\lambda\Delta)^{2/5}$ one can see that for $\Delta \geq q_0^{1/2}\lambda^{3/2}$ there is the first order phase transition between the disordered state $(A = 0)$ and the ordered phase $(A \neq 0)$ with one step replica symmetry breaking for the correlation function $\langle u_a(q)u_b(-q)\rangle$. The form of the correlation function, which was computed in [8], shows that there are two different regions. Inside the domains of size $\left| x_\perp - x'_\perp \right| < \xi_x$ and $\left| z - z' \right| < \xi_z$, where $\xi_z q_0 = t^{-1/2}$ and $\xi_x q_0 = t^{-1/4}$ are the correlations lengths in z and x, y directions, the system behaves like smectic A [21]. On the larger length scales the fluctuations of the layer displacement $u(r)$ wash out the long-range modulated order and result in formation of the highly anisotropic translational incoherent domains with $\xi_z/\xi_x \sim \Delta^{-\frac{1}{4(1-\eta)}}$

5 Conclusions

The symmetric AB copolymers in the gel matrix with a weak preferential adsorption of A monomers gives an example of a random-field system. By using the method of the 2nd Legendre transform we have shown that two replica correlation function is unstable with respect to replica symmetry breaking at a finite values of the random field. We interpret this phase as glassy state. The stability of the lamellar phase in the weak segregation limit was considered by mapping the copolymer Hamiltonian onto the Hamiltonian of the random field XY model. In the ordered phase $(A \neq 0)$ very weak random field $\Delta \sim L^{-4(1-\eta)}$ destroys the long-range modulated order, resulting in formation of the highly anisotropic translational incoherent domains.

Acknowledgments. The work presented here was done in delightful collaboration with Kurt Binder, whose ideas and insights have been invaluable.

References

[1] L. Leibler, Macromolecules **13**, 1602 (1980).

[2] F. S. Bates, G. H. Fredrickson, Annu. Rev. Phys. Chem. **41**, 525 (1990).

[3] K. Binder, Adv. Pol. Sci., **112**, 181 (1994).

[4] S. A. Brazovskii, Soviet Phys JETP **41**, 85 (1975).

[5] S. A. Brazovskii, I. R. Dzayloshinskii and A. R. Muratov, Sov. Phys. JETP **66** 625 (1987).

[6] S. A. Brazovskii, I. R. Dmitriev, Sov. Phys. JETP **42** 497 (1976); D. L. Jonhson, J. H. Flack and P. P. Crooker, Phys. Rev. Lett. **45** 641 (1980); L. J. Martinez-Miranda, A. R. Kortan and R. J. Birgeneu, Phys. Rev. Lett **56** 2264 (1986); E. I. Kac, V. V. Lebedev and A. R. Muratov, Sov. Phys. Solid. State **30** 775 (1988).

[7] R. F. Sawyer, Phys. Rev. Lett. **29** 382 (1972); A. N. Dugaev, Sov. Phys. Lett. JETP **22** 83 (1975).

[8] S. Stepanow, A. D. Dobrynin, T. A. Vilgis, K. Binder, J. Phys. I France **6**, 837 (1996).

[9] P. G. de Gennes, J. Phys. Chem. **88**, 6469 (1984).

[10 J. M. Cornwall, R. Jackiw, E. Tomboulis, Phys. Rev. D 10, 2428 (1974).

[11] C. De Dominicicis, H. Orland, T. Temesvari, J. Phys. I France **5**, 987 (1995).

[12] A. V. Dobrynin, I. Ya. Erukhimovich, J. Phys. II France **1**, 1387 (1991); Sov. Phys. JETP **77**, 307 (1993).

[13 A. V. Dobrynin, J. Phys. France **5**, 657 (1995).

[14] G. Parisi, J. Phys. A **13**, 1887 (1980).

[15] J. L. Cardy and S. Ostlund, Phys. Rev. B **25**, 6899 (1983); L. Golubovich, M. Kulic, Phys. Rev. B **37**, 7582 (1988).

[16] Y.-C. Tsai and Y. Shapir, Phys. Rev. Lett. **69**, 1773 (1992).

[17] M. P. A. Fisher, Phys. Rev. Lett. **62**, 1415 (1989).

[18] P. Le Doussal, T. Giamarchi, Phys. Rev. Lett. **74**, 606 (1995).

[19] S. E. Korshunov, Phys. Rev. B **48**, 3969 (1993).

[20] M. Mézard, G. Parisi, J. Phys. I (France) **1**, 809 (1991).

[21] P. G. de Gennes, J. Prost, The Physics of Liquid Crystals, (Clarendon Press, Oxford, 1993).

Cross-Linked Polymer Chains: Scaling and Exact Results

Thomas A. Vilgis and Michael P. Solf

Max-Planck-Institut für Polymerforschung, Postfach 3148, D-55021 Mainz, F.R.G.

Abstract. The paper discusses the size of a randomly cross-linked polymer chain. The calculations are based on an *exact* theorem for the characteristic function of a polydisperse phantom network that allows for treating the cross-links between pairs of randomly selected monomers as quenched variables without resorting to replica methods. By variation of the cross-linking potential from infinity (hard constraints) to zero (free chain), we have studied the cross-over of the radius of gyration from the branched polymer regime where $R_g \simeq \mathcal{O}(1)$ to the extended regime $R_g \simeq \mathcal{O}(\sqrt{N})$. In the cross-over regime the network size R_g is found to be proportional to $(N/M)^{1/4}$, where M is the total number of cross-links and N the number of monomers in the system. Our exact results can also be understood in terms of simple scaling arguments.

1 Introduction

The physics of polymer networks poses one of the fundamental open questions in polymer theory. It was already noted that polymer networks belong to a special class of glassy, disordered systems (Edwards 1982). The challenge that arises in any rigorous mathematical approach is that the permanent junction-points between macromolecules are frozen variables and cannot be treated within the classical framework of Gibbsian statistical mechanics. This was realized by Edwards, 1971 and further developed by Deam and Edwards, 1976. Their *replica* formalism has by now become the standard approach in the field of polymer networks. Recently various variational studies analyzed the vulcanization process (Goldbart and Goldenfeld, 1987, 1989) mainly in terms of an anisotropic field theory (Vilgis and Solf, 1995) at various levels of complication (Panyukov and Rabin, 1996). Earlier studies have focused at the deformation behavior (Higgs and Ball, 1988), derived from the scattering function (Warner and Edwards, 1978). Most of the "classical" theories have been reviewed in the reference by Edwards and Vilgis, 1988.

Our present study avoids the replica formalism completely. It has recently been shown (Solf and Vilgis, 1995) that for randomly cross-linked Gaussian structures substantial progress can be made by invoking quite different mathematical tools from linear algebra and matrix calculus. Our working model is an ideal polymer chain of N monomers with M randomly selected pairs of monomers that are constrained to be in close neighborhood. These distance constraints are modeled by harmonic potentials. By varying the strength of this cross-linking potential we are able to continuously switch from a network situation with "strong" cross-links to the case of an ideal free chain.

Although the above model is highly idealized, the system is most interesting in its own right because

1. it poses a fundamental problem in polymer networks,
2. the model contains quenched degrees of freedom,
3. the system is "glassy" because of the randomness of the cross-link positions,
4. without excluded volume it can be solved exactly.

The statistics of a huge macromolecule cross-linked to itself (at random or not) has also attracted a lot of attention recently because of its possible implications to protein structure reconstruction from NMR data (Gutin and Shakhnovich, 1993). Methods, however, resorted either to crude variational estimates (Bryngelson and Thirumalai, 1996) or MC simulations (Kantor and Kardar, 1996). To the best of our knowledge this study gives the first exact results for ideal networks that have been reported.

2 The standard Edwards model

Let us first review the classical network model introduced by Edwards, 1971 and Deam and Edwards, 1976. The partition function of a chain with one particular cross-link configuration reads

$$Z(\mathrm{C}) = \int_V \mathcal{D}\mathbf{R}\, e^{-\beta(H_W + H_I)} \prod_{e=1}^{M} \delta\left(\mathbf{R}_{i_e} - \mathbf{R}_{j_e}\right) , \tag{1}$$

where the monomer positions are given by a set of vectors $\mathbf{R}_i$ and the index i runs over all monomers along the contour, i.e., $i = 0, ..., N$. The exponent contains two different contributions. The first defines the connectivity of the chain and is expressed by the discrete Wiener measure,

$$\beta H_W = \frac{d}{2a^2} \sum_{i=1}^{N} (\mathbf{R}_i - \mathbf{R}_{i-1})^2 . \tag{2}$$

The second part stems from the excluded volume interaction between different monomers

$$\beta H_I = v \sum_{0 \le i < j}^{N} \delta\left(\mathbf{R}_i - \mathbf{R}_j\right) . \tag{3}$$

In remainder of the paper we will ignore the excluded volume term deliberately, because of the following reasons: Firstly, we want to concentrate on an exact solution of the ideal network problem. Secondly, we aim for the exact bare "propagator" of the cross-linked chain. As for free chains without cross-links the knowledge of the bare propagator is the basis for further advancements, such as renormalization group theory in the theory of polymer solutions.

The main difficulty that arises in the Edwards formulation of the problem is the handling of the cross-link constraints which are expressed in the multiple Delta function in Equation (1). This constraint imposes quenched disorder on the system that a monomer, say at position $\mathbf{R}_{i_e}$, is always linked to a monomer at $\mathbf{R}_{j_e}$. Thus the free energy depends on the actual cross-link configuration, $F(C) = -k_B T \log Z(C)$ which has to be averaged over the distribution of the frozen variables $C = (i_1, j_1), ..., (i_M, j_M)$:

$$F = \langle F(C) \rangle_C = -k_B T \langle \log Z(C) \rangle_C \tag{4}$$

A standard way to treat the cross-link term is to employ the replica trick, which allows to average the quenched system using the mathematical identity $\log Z = \lim_{n \to 0} \frac{1}{n}(Z^n - 1)$. Instead of averaging the logarithm of Z, the nth power of the partition function has to be evaluated

$$\langle Z^n(C) \rangle_C = \int_V \prod_{\alpha=1}^{n} \mathcal{D} \mathbf{R}^\alpha \, e^{-H_{\text{eff}}} \, , \tag{5}$$

with an effective (annealed) Hamiltonian

$$H_{\text{eff}} = \sum_{\alpha=1}^{n} \left(\frac{d}{2a^2} \sum_{i=1}^{N} (\mathbf{R}_i^\alpha - \mathbf{R}_{i-1}^\alpha)^2 + v \sum_{0 \leq i < j}^{N} \delta \left(\mathbf{R}_i^\alpha - \mathbf{R}_j^\alpha \right) \right) \tag{6}$$

$$\tag{7}$$

$$-\mu \sum_{0 \leq i < j}^{N} \prod_{\alpha=1}^{n} \delta \left(\mathbf{R}_i^\alpha - \mathbf{R}_j^\alpha \right) \, ,$$

where μ is an average cross-link density. The resulting problem is that the replica trick has coupled *all* $\alpha = 1, \ldots, n$ replicas in a complicated way, and there is at the present no elegant way to treat this problem theoretically.

We had already remarked that the problem without excluded volume is exactly solvable. To do so, we aim for an alternative treatment of the cross-link term. The basic idea is to use another representation of the Delta function in Equation (1). One such possibility is to employ

$$\delta(\mathbf{x}) \propto \lim_{\varepsilon \to 0} \sqrt{\frac{d}{2\pi\varepsilon^2}} \exp\left(-\frac{d\mathbf{x}^2}{2\varepsilon^2} \right)$$

which is the basis for our following treatment.

3   The ideal network - a novel representation

We picture a huge Gaussian chain that is M times cross-linked to itself at random as proposed by Edwards *et al.* In the Hamiltonian only terms that model chain connectedness and contributions due to cross-linking are retained. Since we are searching for the ideal propagator we ignore complications, such as entanglements and excluded volume. An appropriate Hamiltonian to begin with is now given by

$$\beta H_0 = \frac{d}{2a^2} \sum_{i=1}^{N} (\mathbf{R}_i - \mathbf{R}_{i-1})^2 + \frac{d}{2\varepsilon^2} \sum_{e=1}^{M} (\mathbf{R}_{i_e} - \mathbf{R}_{j_e})^2 . \tag{8}$$

We have assumed $N + 1$ monomers whose locations in space are given by d-dimensional vectors $\mathbf{R}_i$ ($i = 0, 1, ..., N$). The distance constraints exist between pairs of monomers labeled by i_e and j_e. For further use we introduce the inverse strength of the cross-linking potential

$$z = \left(\frac{\varepsilon}{a}\right)^2 \tag{9}$$

as the mean squared distance between monomers that form the cross-links measured in units of the persistence length a of the chain. Limiting cases are given by $z = 0$ (hard δ-constraints) and $z \to \infty$ (free chain). The whole cross-linking topology is specified by a set of $2M$ integers C$=\{i_e, j_e\}_{e=1}^{M}$. It has been shown earlier (Solf and Vilgis, 1995) that the model in (8) is equivalent to the classical model by Deam and Edwards, 1976 without excluded-volume interaction if averages are understood in the following sense

$$\langle .... \rangle_0 = \lim_{z \to 0} \frac{\int \prod_{i=0}^{N} d\mathbf{R}_i \, e^{-\beta H_0} ....}{\int \prod_{i=0}^{N} d\mathbf{R}_i \, e^{-\beta H_0}} . \tag{10}$$

To model M *uncorrelated* cross-links the distribution of frozen variables C is in the following assumed to be uniform

$$\prod_{e=1}^{M} \left\{ \frac{2}{N^2} \sum_{0 \le i_e < j_e \le N} \right\} . \tag{11}$$

Other types of distributions can also be used but are not considered here. Our further strategy is not to start with the quenched average over the frozen variables from the beginning by employing for instance the replica trick, but to keep explicitly all cross-link coordinates C during the calculation. Only at the very end the physical observable of interest is evaluated for a particular realization of C which will be generated by the distribution in (11). Clearly both approaches are equivalent if only self-averaging quantities are considered.

4   Scaling estimates of the free energy

Before we discuss exact results let us estimate the free energy of (8) by classical Flory type arguments, to see what can be expected. Here we claim that the Hamiltonian (8) lead to three different scaling regimes.

1. The free polymer regime $\varepsilon \to \infty$. This case has a trivial solution and corresponds to very weak constraints. Thus the free chain result for the radius of gyration

$$R_g^2 = \frac{a^2}{6} N$$

   must be recovered.

2. Branched polymer regime $\varepsilon \sim \mathcal{O}(N^{1/2})$. The connectivity term in (17) represents the standard entropic elasticity of a Gaussian chain $R^2/(Na^2) + Na^2/R^2$, where R is a measure of the size of the system. The first term accounts for stretching, whereas the second term describes the response due to compression (de Gennes, 1976).

   For relatively soft cross-links ($\varepsilon \gg a$) the cross-link term requires more attention. In this regime the second term of the Hamiltonian is estimated by $M(R/\varepsilon)^2$, because the mean squared distance between a pair of constrained monomers is of order ε^2. The relevant part of the total Flory free energy is then given by

$$F \approx \frac{a^2 N}{R^2} + M \frac{R^2}{\varepsilon^2} \, .$$

   Minimization yields a surprising novel result

$$R_g \cong a \sqrt{\frac{\varepsilon}{a}} \left(\frac{N}{M} \right)^{1/4} \, . \tag{12}$$

   Here the typical scaling exponent for branched polymers occurs. This is indeed physically reasonable since the soft cross-links change the connectivity of the linear chain to a randomly branched object.

3. Network regime $\varepsilon \to 0$. The case of hard cross-links $\varepsilon \simeq \mathcal{O}(a)$ is the most difficult to obtain. To find a reasonable free energy estimate we picture the system as a coarse-grained random walk over the M cross-links with an effective step length proportional to N/M, i.e., the mean number of monomers between cross-links. Hence the cross-link term is estimated to be of the order $M[R^2/(a^2 N/M)]$. The latter expression has the effect that it tries to shrink the chain upon cost of confinement entropy. A suitable Flory free energy is

$$F \approx \frac{a^2 N}{R^2} + M \frac{R^2}{a^2 (N/M)} \, ,$$

   whence

$$R_g \cong a \left(\frac{N}{M} \right)^{1/2} . \tag{13}$$

Below we show that these simple estimates are indeed confirmed by rigorous calculations.

5 Characteristic function

Here we give a brief review of the central mathematical theorem for the characteristic function of a Gaussian structure with internal δ-constraints. We also discuss an extension to arbitrary cross-linking potential z. The characteristic function for the problem is formally introduced by

$$Z_0(\mathbf{E}; C) = \left\langle e^{i \mathbf{E} \cdot \mathbf{r}} \right\rangle_0 , \tag{14}$$

from which in principle all expectation values can be obtained via differentiation. Simplifying notation has been adopted, where $\mathbf{r}_j \equiv \mathbf{R}_j - \mathbf{R}_{j-1}$ $(j = 1, ..., N)$ denote bond vectors along the backbone of the chain, and $\mathbf{E} = (\mathbf{E}_1, ..., \mathbf{E}_N)$, $\mathbf{r} = (\mathbf{r}_1, ..., \mathbf{r}_N)$ are N-dimensional super-vectors with d-dimensional vector components. Thus $Z_0(\mathbf{E}; C)$ is also the partition function of an ideal network in the presence of external fields $\mathbf{E}_j$. Note that $Z_0(\mathbf{E}; C)$ depends explicitly on all external field vector $\mathbf{E}_j$ which are contained in one super-vector $\mathbf{E}$, as well as on all cross-link positions C. Without going into all the mathematical details, it is now possible to proof the following analytically *exact* projection theorem (Solf and Vilgis, 1995)

$$Z_0(\mathbf{E}; C) = \exp \left(-\frac{a^2}{2d} \mathbf{E}_\perp^2 \right) , \tag{15}$$

where $\mathbf{E}_\perp$ is the length of the external field vector $\mathbf{E}$ projected perpendicular to the vector space spanned by "crosslink vectors"

$$\mathbf{p}_e = (0, \ldots, 0, \underbrace{1, 1, \ldots, 1, 1}_{i_e + 1 \text{ to } j_e}, 0, \ldots, 0) . \tag{16}$$

In the above definition each cross-link is uniquely specified by a pair of integers $1 \le i_e < j_e \le N$ from which a corresponding N-dimensional vector $\mathbf{p}_e$ is constructed. The 1's in (16) run from the $(i_e + 1)$th to the j_eth position. The rest of the N components are filled up with 0's. The whole set of M vectors $\mathbf{p}_1, ..., \mathbf{p}_M$ defines a characteristic vector space for the problem, say U. There is an unique decomposition of any field vector $\mathbf{E}$ parallel and perpendicular to U, viz., $\mathbf{E} = \mathbf{E}_\parallel + \mathbf{E}_\perp$. The operator that projects $\mathbf{E}$ on U can be constructed from $\mathbf{p}_e$ for any realization of cross-links C. It is given by $\mathcal{P}\mathcal{P}^+$, where $\mathcal{P}$ is the $N \times M$ rectangular matrix associated with the cross-link vectors $\mathbf{p}_e$ $(e = 1, ..., M)$,

$$\mathcal{P} \equiv (\mathbf{p}_1, ..., \mathbf{p}_M) \ , \tag{17}$$

while $\mathcal{P}^+$ is a generalized inverse of $\mathcal{P}$ (see, for example, the book by Lancaster and Tismenetsky, 1958).

The projection theorem (15) is only valid for hard cross-linking constraints when $z = 0$. A generalization to arbitrary cross-linking potential z is, however, possible using the methods discussed in (Solf and Vilgis, 1995). Only the final result for the characteristic function is quoted here which reads

$$Z_0(z, \mathbf{E}; C) = \exp\left[-\frac{a^2}{2d}\left(\mathbf{E}^2 - \sum_{e=1}^{M} \frac{(\mathbf{x}_e \mathbf{E})^2}{1 + (z/w_e^2)}\right)\right] \ , \tag{18}$$

The unit vectors $\mathbf{x}_e$ denote *any* orthonormal basis associated with $\mathbf{p}_e$ ($e = 1, ..., M$), and w_e are the corresponding singular values (Lancaster and Tismenetsky, 1958). For tetrafunctional cross-links it was shown that w_e is always positive. From (18) it is straightforward to derive expressions for the radius of gyration and structure factor.

By the scattering function (form factor) we denote density fluctuations (Doi and Edwards, 1986) normalized to one

$$S_0(\mathbf{k}, z; C) = \langle |\rho_{\mathbf{k}}|^2 \rangle_0 \tag{19}$$

$$\equiv \frac{1}{N^2} \sum_{i,j=0}^{N} \left\langle \exp\left(i\mathbf{k}(\mathbf{R}_i - \mathbf{R}_j)\right)\right\rangle_0 \ ,$$

where $\mathbf{k}$ is the scattering wave vector, and the average is over the measure in (10). In the following no further distinction between N and $N + 1$ will be made since N is always assumed to be large. From Equation (18) an exact expression for S_0 can be obtained by introducing the external field vector $\mathbf{E} = k\,\mathbf{c}_{ij}$, where

$$\mathbf{c}_{ij} = (0, \ldots, 0, \underbrace{1, 1, \ldots, 1, 1}_{i+1 \text{ to } j}, 0, \ldots, 0) \ , \tag{20}$$

and $k = |\mathbf{k}|$. From (18) and (19) we find

$$S_0(\mathbf{k}, z; C) = \frac{1}{N} \tag{21}$$

$$+ \frac{2}{N^2} \sum_{i<j}^{N} \exp\left[-\frac{a^2 k^2}{2d}\left(j - i - \sum_{e=1}^{M} \frac{(\mathbf{x}_e \mathbf{c}_{ij})^2}{1 + (z/w_e^2)}\right)\right] \ .$$

A similar result is obtained for the radius of gyration (Doi and Edwards, 1986)

$$\left(\frac{R_g(z; C)}{a}\right)^2 = \frac{N}{6} - \frac{1}{N^2} \sum_{i<j}^{N} \sum_{e=1}^{M} \frac{(\mathbf{x}_e \mathbf{c}_{ij})^2}{1 + (z/w_e^2)} \ . \tag{22}$$

It is worthwhile to note that the applicability of these formulas is not limited to random networks since all cross-linking coordinates are still implicit in the formulas through $\mathbf{x}_e$ and w_e. By selecting different ensembles for C=$\{i_e, j_e\}_{e=1}^M$ (random or not) any generalized Gaussian structure with internal cross-linking constrained can be treated by the same method.

6   Exact results and discussion

In applying Equations (21) and (22) to random networks, one has to deal with a sufficient number of monomers N and cross-links M. Moreover, the positions of cross-links C=$\{i_e, j_e\}_{e=1}^M$ have to be chosen at random. The simplest scenario is to pick $2M$ integers i_e, j_e ($e = 1, ..., M$) from the uniform distribution, Equation (11), defined on the interval $[0, N]$. For a given realization C we compute an orthonormal basis $\mathbf{x}_e$ and singular values w_e of $\mathbf{p}_e$ ($e = 1, ..., M$). Any standard technique like singular value decomposition (Press et al., 1992) will suffice, for the orthonormalization process presents only a minor numerical task. We find that for number of monomers $N > 10000$ and cross-links $M > 200$ fluctuations between different realizations of C differ by less than 1 percent.

6.1   Scaling behavior of scattering function

Within the framework of the Deam-Edwards model ($z = 0$) it was demonstrated (Solf and Vilgis, 1995) that the scattering function S_0 of an ideal network is an universal function of wave vector $\mathbf{k}$ and mean cross-link density M/N as long as N and M are sufficiently large to ensure self-averaging. A scaling form for S_0 is motivated by the following argument. For a linear polymer without cross-links the scattering intensity $S_0(x)$ depends only on the product $x = k^2 R_g^2$, where $R_g^2 = a^2 N/6$ is the radius of gyration of the chain and $S_0(x) = 2(e^{-x} - 1 + x)/x^2$ the Debye function (Doi and Edwards, 1986). In close analogy it was shown that for hard constraints ($z = 0$) the radius of gyration of an ideal network is given by $R_g^2 \simeq 0.26 \, a^2 N/M$ which suggests a scaling behavior, similar to that of linear chains without cross-links. This scaling hypothesis for ideal networks was confirmed by our calculation based on the expressions in Equations (21) and (22).

The numerically *exact* result for the scattering function is presented in figure 6.1. In the Kratky plot of figure 3, $xS_0(x)$ has been evaluated as a function of $\sqrt{x} \equiv kR_g$ for $z = 0$. Independent of details of cross-linking topology C, all networks investigated fall on the same master curve (solid line). Statistical fluctuations between different networks were too small in the self-averaging regime to be seen on the scales used in figure 6.1. No indication of a power-law decay for intermediate wave vectors k or other simplifying feature was detected, besides the pronounced maximum in the Kratky plot

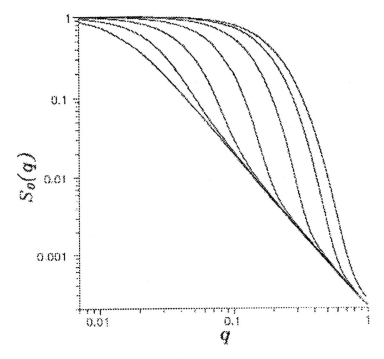

Fig. 1. Structure function S_0 for different cross-linking potentials z. From right to left $z = 0, 10, 10^2, 10^3, 10^4, 10^5, \infty$ for a network size of $N = 10000$ and $M = 200$ plotted in dimensionless units $q = ka/\sqrt{2d}$. The left curve ($z \to \infty$) is the Debye function of a linear chain.

at $\sqrt{x} \simeq 2$ which reflects the strong correlations of the monomers due to cross-linking. For comparison, the case of a linear polymer with $z \to \infty$ (Debye function, dashed line) was also computed from (21).

The results of S_0 for different values of cross-linking potential z are illustrated in figure 6.1 for a network with $N = 10000$ and $M = 200$. Upon increasing the strength of the constraint from left to right, large deviations of S_0 from ideal chain behavior (Debye function, left curve) arise at smaller and smaller length scales. For $z = 0$ the network character persists down to even the shortest length scale ($k \approx 1$) as a consequence of the high degree of cross-linking in the system ($M/N = 0.02$). For sufficiently large wave vectors all curves decay as k^{-2} as expected when the scanning wavelength becomes small compared to the mesh size of the network. Again, a master curve for S_0 can be obtained by plotting the x-axis in units of R_g.

6.2 The size of the cross-linked chain

In figure 9 we have calculated the radius of gyration of a network of N monomers and M cross-links as a function of $\sqrt{z} = \varepsilon/a$ by use of Equa-

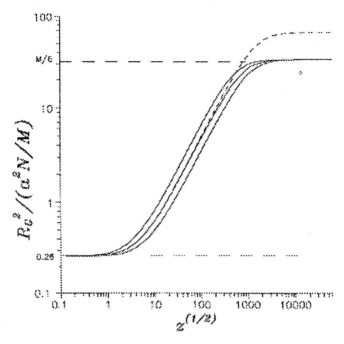

Fig. 2. Radius of gyration of a polymer chain as a function of $\sqrt{z} = \varepsilon/a$. For the three solid curves the number of monomers was varied from top to bottom $N = 5000, 10000, 20000$; M was kept constant at 200. The short dashed line shows a network with $N = 20000$ and $M = 400$.

tion (22). From this investigation we can clearly distinguish between three different scaling regimes

$$\left(\frac{R_g}{a}\right)^2 \simeq \begin{cases} 0.26\,N/M \,, & \text{if } \varepsilon \ll \varepsilon_1 \\ 0.34\,(\varepsilon/a)\,(N/M)^{1/2} \,, & \text{if } \varepsilon_1 \ll \varepsilon \ll \varepsilon_2 \\ N/6 \,, & \text{if } \varepsilon \gg \varepsilon_2 \,, \end{cases} \qquad (23)$$

with cross-overs at $\varepsilon_1 \simeq a\sqrt{N/M}$ and $\varepsilon_2 \simeq a\sqrt{MN}$. The plateau values in figure 6.2 correspond to the two extremes $R_g^2/a^2 = 0.26\,N/M$ ($z \to 0$ at the left) and $R_g^2/a^2 = N/6$ ($z \to \infty$ at the right).

In particular our investigation showed that the cases $z = 0$ (hard constraints) and $z = 1$ (constraints of the order of the persistence length a) only differ by a numerical prefactor which varies from 0.26 for $z = 0$ to about 0.27 for $z = 1$. From this we conclude that an ideal network subject to *uncorrelated* cross-linking constraints is collapsed in a sense that its size is proportional to the square root of N/M. Therefore $R_g/a \simeq \mathcal{O}(1)$, since M/N is the mean cross-link density and of order unity in the thermodynamic limit $N, M \to \infty$. This finding seems to be at variance with current speculations regarding the collapse transition of macromolecules (Bryngelson and

Thirumalai, 1996), where it was argued that a critical number of cross-links $M \geq M_c \simeq N/\log N$ will force the system to collapse. On the other hand our result for $z = 0$ is in agreement with recent Monte Carlo simulations by (Kantor and Kardar, 1996) who found for the mean squared end-to-end distance $R^2/a^2 \simeq 1.5\,N/M$. This suggests the same one to six ratio for $(R_g/R)^2$ in ideal networks as for linear polymers without cross-linking constraints and excluded-volume interaction (Doi and Edwards, 1986).

Conversely, a free chain ($z \to \infty$) is an extended object with $R_g/a \simeq \mathcal{O}(\sqrt{N})$. Between the collapsed and the extended regime we find a smooth cross-over with $(R_g/a)^2$ being proportional to $(\varepsilon/a)\sqrt{N/M}$. Remarkably this is the same scaling as for randomly branched polymers without excluded volume interaction as was suggested by our simple scaling argument before.

Acknowledgments TAV would like to thank the organizers of the meeting for the pleasant and stimulating atmosphere during the conference. The authors gratefully acknowledge financial support by the DFG, Sonderforschungsbereich 262.

References

Edwards S.F. (1982): Ann. N.Y. Soc. **371**, 210.

S. F. Edwards in *Polymer Networks*, eds. A. J. Chompff and S. Newman (Plenum Press, New York, 1971).

R. T. Deam and S. F. Edwards, Proc. Trans. R. Soc. London A **280** (1976) 317.

Goldbart P. and Goldenfeld N., (1989): Phys. Rev. Lett. **58** (1987) 2676; Phys. Rev. A **39** 1402; *ibid* 1412.

Vilgis T. A. and Solf M. P. (1995): J. Phys. I France **5** 1241.

Panyukow S. V. and Rabin Y. (1996): Phys. Rep. **269** no. 1 & 2.

Higgs P. G. and Ball R. C. (1988): J. Phys. France **49** 1785.

Warner M. and Edwards S.F. (1978): J. Phys. A **11** 1649.

Edwards S.F. and Vilgis T.A. (1988): Rep. Prog. Phys. **51** 243.

Solf M.P. and Vilgis T. A. (1995): J. Phys. A: Math. Gen. **28** 6655.

Gutin A. M. and Shakhnovich E. I. (1993): J. Chem. Phys. **100** 5290.

Bryngelson J. D. and Thirumalai D. (1996): Phys. Rev. Lett. **76** 542.

Kantor Y. and Kardar M. (1996): preprint.

Lancaster P. and Tismenetsky M. (1985): The Theory of Matrices (Academic Press, New York, 2nd Ed.).

Doi M. and Edwards S. F. (1986): The Theory of Polymer Dynamics (Clarendon Press, Oxford).

Press W. H., Teukolsky S. A., Vetterling W. T., Flannery B. P. (1992): Numerical Recipes (University Press, Cambridge).

de Gennes P. G. (1979): Scaling Concepts in Polymer Physics (Cornell University Press, Ithaca).

Magnetic Properties of Geometrically Frustrated Systems

[1]B. Martínez, [1]X. Obradors, [1]F. Sandiumenge, and [2]A. Labarta

[1]Instituto de Ciencia de Materiales de Barcelona-CSIC. Campus UAB, Bellaterra 08193, [2] Dept. de Física Fundamental, Facultat de Física, Univ. de Barcelona, Diagonal 647, Barcelona 08028. SPAIN

1 Introduction

Frustration is defined as the inability of a system to simultaneously minimise the competing interaction energies between its components which in
the case of magnetic materials is reflected in the lack of long range magnetic
ordering. Usually, magnetic frustration is found together with site disorder
giving place to the well known spin glass syndrome. Nevertheless, these two
features may appear independently and its interplay gives place to different
magnetic behaviours. The most common types of magnetic ordering, namely,
ferromagnetism (FM), antiferrromagnetism (AF) [1] and ferrimagnetism [2]
corresponds to both low disorder and weak frustration. As disorder increases,
keeping the frustration weak, the phenomena of random fields and percolation
effects dominate the magnetic behaviour. When both, disorder and frustration, are high the spin glass behaviour appears [3]. Finally, systems with
low or not disorder at all and high frustration exhibit a magnetic behaviour
distinct from the rest and not well understood yet [4]. One of the most outstanding features of frustrated magnets is the high degeneracy of the ground
state integrated by a quasi-continuos succession of equivalent states. This
high degeneracy of ground states is the origin of the very special behaviour
of the frustrated systems in the limit of very low temperatures ($T \ll J$),
since as $T \to 0$ the system should evolve towards a particular state. This
particularity has generated a considerable theoretical interest suggesting the
appearance of spin-nematic ordering [5], [6], the absence of any magnetic ordering at all [7], [8] or de formation of an "spin liquid" in which as $T \to 0$
strongly interacting spins fluctuate rather than being in a frozen static configuration.

Systems are highly frustrated when competing interactions of similar magnitude coexist. In most of the cases competition arises due to the existence of
FM and AF interactions, together with some degree of site disorder, giving
place to the moderate degree of frustration that is observed in spin glasses [3].
On the contrary, geometrical frustration arises due to the competition origi-
nated in the spatial arrangement of the spins with very little or not disorder

at all [4]. In spin glasses a random combination of FM and AF interactions is generated by disorder and the competition between these interactions leads to a degenerated ground state in which different states are separated by energy barriers. On the other hand, in an ideal geometrically frustrated magnet without disorder ground states are not necessarily separated by energy barriers leading to a continuum of equivalent ground states. Is in this sense that geometrically frustrated magnets are qualitatively different from traditional spin glasses and what makes its ground state properties (low temperature behaviour) so fascinating from both the theoretical and experimental point of view.

The simplest realisation of a geometrically frustrated unit is a two-dimensional triangular lattice of spins with antiferromagnetic interactions between them. In fact, all of the materials that have been experimentally found to be highly frustrated are based in combinations of this basic unit (corner-sharing or side-sharing triangles or tetrahedra) preserving the intrinsic frustration of the triangular lattice with short range AF interactions.

As pointed out by Schiffer et al. [4], geometrical frustration requires a high degree of symmetry in the interactions between the spins to avoid the reduction of the number of degenerate ground states. Therefore, geometrically frustrated magnets are systems having isotropic nearest-neighbor exchange interactions between Heisenberg rather than Ising-like spins. Apart from exchange anisotropy, any other anisotropic microscopic interactions, such as further-neighbors exchange or dipolar interactions, or what is more important, chemical and structural disorder further reduce the symmetry of the lattice, thus reducing the degree of frustration. On the other hand, since no geometrical arrangement can generated magnetic frustration with only FM interactions, the existence of AF interactions is inherent to geometriclally frustrated magnets. Since in real materials chemical and structural disorder are always present the problem is how frustrated magnets can be experimentally identified. As we have comented above the first condiction is the existence of AF interactions identified through the negative sign of the extrapolated Weiss temperature, θ, of the high temperature Curie-Weiss law.

In an AF material as T approaches $|\theta|$ the system will undergo a phase transition to the long-range AF ordered state indicated by a peak in the magnetic susceptibility. On the other hand, an ideal geometrically frustrated AF material, in which the ground state is highly degenerated, should not display any ordering transition down to $T = 0$. Nevertheless, real frustrated materials have some degree of disorder and therefore do show a magnetic transition, reflected in a peak in the magnetic susceptibility, but at a temperature T_f well below $|\theta|$. The ratio between these two temperatures, $f = |\theta|/T_f$, has been adopted as a mesure to quantify the degree of frustration of the material.

It has also been observed experimetally that in the case of highly frustrated materials the inverse of the magnetic susceptibility does show a linear dependence in T down to temperatures well below $|\theta|$, which is not the expected behavior in an ordinary magnet since the linearity of $\chi^{-1}(T)$, predicted by mean field theory, shoudl be observed in the high temperature limit only ($T \gg |\theta|$).

Taking into account all these considerations it has been propossed [4] that a geometrically frustrated magnet should acomplish the following three rules: i) The materials must have AF interactions, that is $\theta < 0$. ii) The material should not show any long-range magnetic ordering down to $T \ll |\theta|$. iii) $\chi^{-1}(T)$ should be linear in T down to $T \ll |\theta|$. Among the compounds that exhibit geometrical frustration the most widely studied by far are pyrochlores [9] (in which metal sublattices form an infinite three-dimensional network of corner-sharing tetrahedra) and the $SrCr_8Ga_4O_{19}$ compound (where antiferromagnetically coupled Cr atoms are arranged in a network of corner-sharing triangles, the so-called Kagome lattice).

In this work we will report a detailed experimental study of the magnetic behaviour of one of the systems with the highest degree of magnetic frustration ever reported ($f \approx 140$), the $SrCr_8Ga_4O_{19}$ compound, first discovered by Obradors et al. [10], showing that the low temperature magnetic behaviour is clearly different from that of a traditional spin glass ordering.

2 Experimental and Results

Samples of $SrGa_{12-x}Cr_xO_{19}(2 \leq x \leq 9)$ where prepared by the ceramic method, further details may be found in [11]. Neutron-diffraction patterns were collected in the high resolution D2B diffractometer (ILL, Grenoble). DC magnetic measurements were performed by using a SQUID magnetometer in the temperature range from 1.6 to 340 K. AC magnetic susceptibility data were measured in the temperature range from 2.5 to 300 K in a LakeShore susceptometer, where temperatures lower than 4.2 K were reached by pumping over the He bath.

Regarding to the structure of the $SrCr_8Ga_4O_{19}$ compound the most important feature is that Cr^{3+} ions do only enter in the three different octahedral sites, namely 2a, 12k and $4f_{VI}$, with an almost random distribution between them (2a sites are occupied with slight preference) [11] (see Fig. 1). The magnetic sublattice is made up of a sequence of c-stacked two-dimensional Kagom layers (12k sites). 12k layers are connected between them through the more diluted 2a and $4f_{VI}$ layers, the latter acting as bottleneck since there are two $4f_{VI}$ layers between two consecutive 12k layers ($4f_{VI}$-12k-2a-12k-$4f_{VI}$).

In order to provide a quantitative measure of the variation of the magnetic dimensionality with dilution, we have calculated the probability of formation of superexchange paths along the c axis, either through the $4f_{VI}$ sites (X_R) or the 2a sites (X_S). In this way, the composition dependence of X_R and X_S gives a measure of the influence of dilution on the probability to find a path for magnetic interactions between two consecutive 12k layers. For these

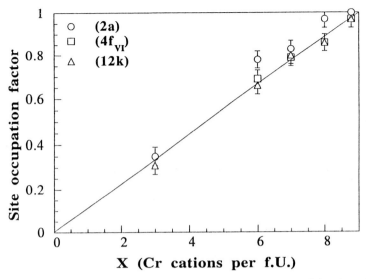

Fig. 1. Cr_{3+} site-occupation factors vs. the Cr content per formula unit.

calculations, Eqs. (1) and (2) have been used:

$$X_R = P_{4f_{VI}}^2 [\sum_{\xi=1}^{6} (\frac{6}{6-\xi})(1 - P_{12k})^{6-\xi} P_{12k}^{\xi}]^2 \qquad (1)$$

$$X_s = P_{2a} [\sum_{\zeta=1}^{3} (\frac{3}{3-\zeta})(1 - P_{12k})^{3-\zeta} P_{12k}^{\zeta}]^2 \qquad (2)$$

where the p's represent the Cr^{3+} site occupation factors in the corresponding sites; ξ is the number of Cr^{3+} cations located in 12k sites which are first neighbors of a $4f_{VI}$ site, and similarly, ζ is the number of Cr^{3+} cations located in 12k sites which are first neighbors of a 2a site. In summary, both expressions represent the probability of finding uninterrupted "chromium superexchange paths" between 12k layers (see Fig.2). The result clearly indicates that $X_S > X_R$ even for the $x = 8$ compound, i.e., the magnetic interactions are preferentially destroyed at $4f_{VI}$ sites. Actually this effect is slightly enhanced by the small preference of Cr^{3+} for 2a sites. It is straightforward to note, however, that the difference $X_R - X_S$ remains rather constant all over the compositon range, and thus indicating that the possible variation of the magnetic dimensionality does not play any role in the magnetic behavior of the system. As we will see later on, this is indeed what is actually observed through the compositions dedendence of the frustration parameter f that turns out to be independent of the degree of dilution, thus making evident the intrinsic structural origin of the magnetic frustration. This idea has been recently confirmed by neutron diffraction studies that clearly show tha Cr^{3+} ions in $4f_{VI}$ sites form spin pairs very weakly coupled to the 12k planes [12]. Therefore, the magnetic structure may be considered as a c-stacked series of slabs, with very littel interaction between them, integrated by kagome planes (12k sites) coupled through a triangular lattice (2a sites).

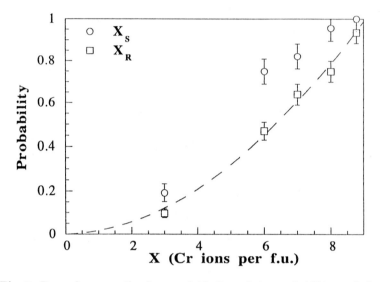

Fig. 2. Dependence on the degree of dilution of the probability to find a path for magnetic interactions between to consecutive 12k planes of Cr_{3+} ions.

Regarding to the magnetic behavior, all the samples are paramagnetic at high temperature following a Curie-Weiss law for an effective manetic moment $\mu_{eff} \approx 4\mu_B$ typical of the spin-only 3/2 Cr^{3+} ions. The Weiss temperature, θ, varies between -90 K and -537 K as a function of x indicating the existence of strong antiferromagnetic interactions for all the members of the series, but there are no signals of any kind of magnetic ordering down to 4.2 K. On the other hand, it is worth mentioning that $\chi^{-1}(T)$ does show a linear dependence on T down to temperatures well below $|\theta|$ (see Fig. 3). Therefore, we observe that the three points proposed as the hallmarks of the geometrically frustrated magnets are fully accomplish in this series of samples.

On lowering temperature below 4.2 K the appearance of a blocking process of the spins, similar in some aspects to that found in canonical spin glasses, is observed. Following the well known zero-field-coling (ZFC) field cooling (FC) process the ZFC susceptibility shows a peak at a temperature T_f that decreases with dilution. The cusp of the peak also indicates the onset of irreversibility between the ZFC and FC branches of the susceptibility. Nevertheless, it is worth mentioning that magnetic frustration remains almost constant for those samples of the series in which T_f is accesible with our experimental facilities (see inset of Fig. 3) $f = |\theta|/T_f \approx 140 \pm 5$, in contrast with what is observed in spin glasses in which frustration decreases as dilution increases. This facts strongly points to the intrinsic geometrical nature of the magnetic frustration in this material.

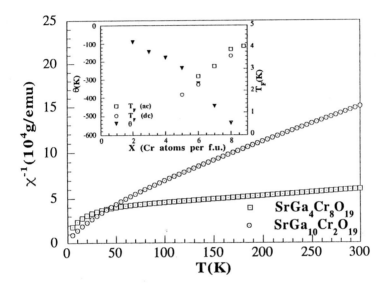

Fig. 3. Reciprocal susceptibility of some of the samples as a function of temperature. Inset: Dependence of the Weiss temperature, θ, and the freezing temperature, T_f, as a function of the Cr content.

Once we have made a general picture of the magnetic behavior of our samples let us focalize our attention in the main aim of this work, that is, to study the nature of the freezing phenomenon that takes place at T_f. To determine the nature of the freezing phenomenon at T_f we have studied the dependence of field cooled (FC) magnetic susceptibility χ, as a function of both temperature and applied magnetic field from which we derive the static critical behavior. In Fig. 4 we show dc susceptibility curves from 3K to 15K with the applied field varying from 20 Oe to 50 kOe; it can be observed from the figure that as a result of the non-linear effects, the peak associated with the freezing phenomena broadens and becomes rounded as the field increases.

To study the nature of the transition that takes place at T_f, we have analysed the non-linear part of the magnetic susceptibility by developing the magnetization above T_f in terms of the odd powers of the field.

$$M = \chi_0 H - b_3(\chi_0 H)^3 + b_5(\chi_0 H)^5 - ... \tag{3}$$

being χ_0 the linear susceptibility, while the rest of the coefficients stand for the

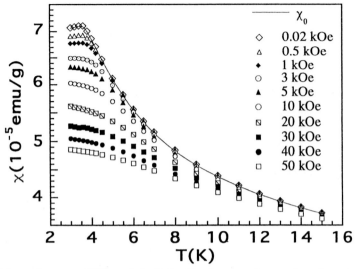

Fig. 4. Magnetic susceptibility of the $SrCr_8Ga_4O_{19}$ sample as a function of the applied field obtained after a FC process. Solid line corresponds to linear susceptibility obtained by fitting the experimental data by using Eq. (3).

non-linear contribution. The least-squares fit of the isothermal magnetization curves obtained from the field cooled data to Eq. (3) allows the determination of the temperature dependence of χ_0, b_3 and b_5. An slight increase of the b_3 coefficient as T_f is approached from above is observed [13]. Nevertheless, this increase is very small when compared with that observed for a typical spin glass (three orders of magnitude [14]) and that is usually taken as a proof of the existence of a true spin glass transition at T_f [14]. The most relevant test of the critical behavior of a spin glass system is obtained by measuring and analyzing the non-linear part of the magnetic susceptibility defined by the equation:

$$\chi_{nl}(H,T) = \chi_0(T) - M(H,T)/H \qquad (4)$$

being χ_0 the linear susceptibility obtained trough the fitting of FC experimental data to Eq. (3). In the case of a spin glass transition it is expected that should scales following the single-parameter relation [15]:

$$\chi_{nl}(H,T) \propto H^{2/\delta} f(\epsilon/H^{2/\phi}) \qquad (5)$$

where $\epsilon = (T-T_c)/T_c$, δ and ϕ are the critical exponents that characterize the spin glass transition, T_c is the critical temperature at which the phase transitions occurs and $f(x)$ is an arbitrary scaling function with the following asymptotic behavior:

$$f(x) = constant, x \rightarrow 0 f(x) = x^{-\gamma}, x \rightarrow \infty \qquad (6)$$

The critical exponent δ may be determined making use of the above asymptotic relations by the following equation:

$$\chi_{nl}(H, T_c) \propto H^{2/\delta} \tag{7}$$

that represent the asymptotic behavior of the critical isotherm as x approaches 0. In the inset of Fig. 5 we show the log-log plot of χ_{nl} as a function of H, from which, by using Eq. (7), $\delta \approx 5 \pm 0.3$ have been obtained from the slope of the linear regime. The data points for this plot have been taken at $T = 3.4$ K that corresponds to the nearest measured temperature to that at which the b_3 coefficient shows its maximum [13]. Then we have used Eq. (5) to scale our FC data, in the temperature range $1.1Tc < T < 2Tc$, varying the values of T_c and ϕ in order to get the best data collapsing. The result is depicted in Fig. 5 and correspond to the following set of critical exponents $\delta \approx 5 \pm 0.4, \phi \approx 4.4 \pm 0.5$ and a critical temperature $T_c \approx 3.45 \pm 0.1$.

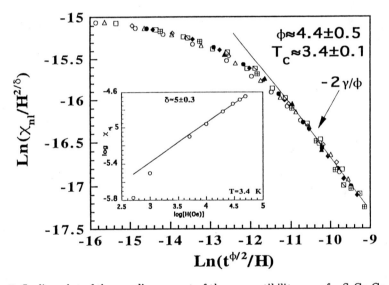

Fig. 5. Scaling plot of the non-linear part of the susceptibility, χ_{nl}, for $SrCr_8Ga_4O_{19}$ compound at several fields ranging from 2 kOe to 50 KOe. Inset: Determination of the d value from the slope of the log-log plot of cnl as a function of H. Data at T=3.4 K have been used.

The reliability of the scaling behavior and the set of critical exponents can be checked by studying the asymptotic behavior of the scaling functions given by eq. (6). In the limit of x large with constant magnetic field (at small fields or for large values of the reduced temperature) an asymptotic behavior of the form $x^{-\gamma}$ has to be observed, where γ is the susceptibility exponent that is related to δ and ϕ exponents through the following hyperscaling relation [16]:

$$\gamma = \phi(1 - 1/\delta) \tag{8}$$

In our case, the asymptotic slope that we obtain from the scaling plot (see Fig. 5) is $-2\gamma/\phi \approx 0.68 \pm 0.1$ given a value of $\gamma \approx 1.5 \pm 0.3$ that is two times smaller that the corresponding value obtained through the hyperscaling relation expressed in Eq. (8). Eventhough an asymptotic constant value is observed in the limit $x \to 0$, the fact that the scaling relation given by Eq. (8) is not fulfilled with the value obtained for γ from the asymptotic slope indicates that the scaling plot we have obtained is just a mathematical relation but it lacks of physical meaning [16]. This fact together with the non-divergence of the b_3 and b_5 coefficients of Eq. (3) as T_f is approached from above indicates that the transition observed T_f does not correspond to the transition at a conventional spin glass state, i.e. that the low temperature state of the $SrCr_8Ga_4O_{19}$ is not a conventional spin glass, as its low temperature specific heat of the T^2 form indicates[4].

In order to get a more complete comprehension of the nature of the low temperature magnetic state of the $SrCr_8Ga_4O_{19}$ compound we have also studied its dynamical critical properties by measuring the real , χ' , and imaginary, χ'' , parts of the ac magnetic susceptibility as a function of the frequency of the ac field. In-phase magnetic susceptibility shows a peak that defines the freezing temperature,$T_f(\omega)$, that has a smooth dependence on frequency of measurement (see Fig. 6). Taking the temperature corresponding to the cusp of χ',$T_f(\omega)$, as the onset of strong irreversibility of each measuring time $\tau = 1/\omega$ (being ω the frequency of the ac magnetic field) and studying its dependence on the ac field frequency the dynamical properties of the transition may be checked. There are basically two different interpretations of the freezing phenomenon: one is the cluster model, in which the system is considered as a set of superparamagnetic clusters with each cluster having a probability to overcome the anisotropy energy barrier E, following an Arrhenius or Vogel-Fulcher law [17]. The other model assumes the existence of a true equilibrium phase transition, in which case the divergence of the relaxation time τ as the critical temperature is approached from above is given by [18]:

$$\tau \propto [\frac{T_f(\omega)}{T_c} - 1]^{z\nu} \tag{9}$$

where ν is the critical exponent for the correlation length ξ,z is the dynamic exponent relating ξ and τ, and T_c is the phase transition temperature. In the framework of this second model we have studied the dynamic scaling of the quantity

$$\Delta\chi' = [\chi_0(T) - \chi'(\omega, T)]/\chi_0(T) \tag{10}$$

where $\chi_0(T) = \chi'(0, T)$ is the equilibrium susceptibility that cam be assumed to be equal to the linear susceptibility deduced from Eq. (3). In Fig. 6 the real part of the ac magnetic susceptibility $\chi'(T, \omega)$ is depicted for different frequencies together with the linear susceptibility $\chi_0(T)$. The above assumption that $\chi_0(T) = \chi'(0,T)$ is fully confirmed in this picture since, as temperature increases, the asymptotic behavior of $\chi'(\omega,T)$ perfectly matches the $\chi_0(T)$ curve (solid line). It follows from Eq. (9) and (10) that $\Delta\chi'(\omega,T)$ data should assume, as $T \to T_c$ and $\omega \to 0$, the following scaling form:

$$\Delta\chi'(\omega,T) \approx (T/T_c - 1)^\beta F[\omega(T/T_c - 1)^{-z\nu}] \qquad (11)$$

where β is the exponent of the order parameter and $F(x)$ a general scaling function. In Fig. 7 we depicted the results obtained by using the power law scaling given by (11) in the temperature range $T_c < T < 1.5T_c$. The best collapsing of the data points in a single curve is obtained with the following set of critical exponents: $z\nu \approx 9 \pm 0.3$, $\beta \approx 0.6 \pm 0.1$ and $T_c \approx 3.5 \pm 0.1$. Nevertheless, the poor quality of the fitting is evident as well as the fact that the asymptotic behavior is not observed. It is also worth mentioning that the value of the exponent $\beta \approx 0.6$, obtained from the scaling plot does not obey at all the hyperscaling relation $\beta = \phi/\delta \approx 0.9$, being ϕ and δ the critical exponents deduced from the analysis of the non-linear part of the dc susceptibility. This facts lead us to conclude that, as in the case of the non-linear part of the dc susceptibility, the scaling fit is merely mathematics [16]. Then, the existence of a true phase transition to an spin glass state at T_c seems to be also precluded from the point of view of the dynamical behavior of the system.

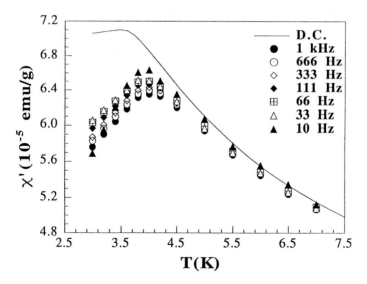

Fig. 6. Thermal dependence of the real part of the ac susceptibility, χ', as function of the frequency of the ac filed for $SrCr_8Ga_4O_{19}$ sample. The solid line represents the equilibrium susceptibility obtained by fitting the dc FC data to Eq. (3).

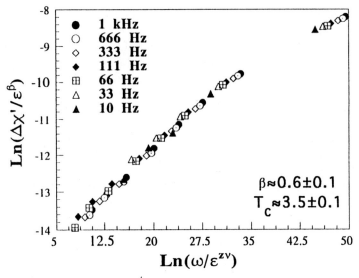

Fig. 7. Power law scaling of $\Delta\chi'(\omega, T)$ data according to Eq. (11) for $SrCr_8Ga_4O_{19}$ sample. Data from T=3.6 K to 6 K have been used.

3 CONCLUSIONS

We have carefully analyzed both dc and ac magnetic susceptibility in the highly frustrated $SrCr_8Ga_4O_{19}$ compound to study the properties of its low temperature magnetic state and to determine whether or not a conventional spin glass state exists below T_f. The high magnetic frustration in this compound arises from the Cr_{3+} ions, interacting antiferromagnetically, arranged in a corner-sharing triangular lattice (Kagome lattice) in the 12k planes..

Eventhough the scaling of the non-linear part of the dc susceptibility χ_{nl}, appears to be of good quality and the critical exponents are similar to other reported for spin glasses, it turns out to be a mathematical fit lacking of its physical meaning, since the value of the γ exponent obtained from the asymptotic slope in the limit of $x \to \infty$ does not fulfill the hyperscaling relation given by Eq. (8) [15]. Furthermore, b_3 and b_5 coefficients of Eq. (3) show only an small increase approaching T_f from above while the divergence of both coefficients should occur in the case of a true spin glass transition. These facts seem to indicate that the magnetic state below T_f of the $SrCr_8Ga_4O_{19}$ compound does not correspond to a conventional spin glass state. Similarly, the analysis of the dynamical critical behavior also points to a non-conventional spin glass state below T_f.

References

[1] J.S. Smart. *Effective Field Theories of Magnetism.* Ed. Saunders. Philadelphia (1966)

[2] W. Wolf. Rep. Prog. Phys. 24, 212 (1961)

[3] K. Binder, A.P. Young. Rev. Mod. Phys. 58, 801 (1986)

[4] A.P. Ramirez. Annu. Rev. Mater. Sci. 24, 453 (1994); P. Schiffer and A.P. Ramirez. Comm. Condens. Matt. Phys. (in press)

[5] P. Chandra and P. Coleman. Phys. Rev. Lett. 66, 100 (1991); P. Chandra, P. Coleman and A.I. Larkin. J. Phys.:Condens. Matter 2, 7933 (1990); P. Chandra, P. Coleman and I. Ritchey. J. Appl. Phys. 69, 4974 (1991)

[6] J.T. Chalker, P.C.W. Holdsworth and E.F. Shender. Phys. Rev. Lett. 68, 855 (1992); A.B. Harris, C. Kallin and A.J. Berlinsky. Phys. Rev. B-45, 2899 (1992)

[7] J.T. Chalker and J.F.G. Eastmond. Phys. Rev. B-46, 14201 (1992)

[8] P.W. Leung and Veit Elser. Phys. Rev. B-47, 5459 (1993)

[9] B.D. Gaulin, J.N. Reimers, T.E. Mason, J.E. Greedan and Z. Tun. Phys. Rev. Lett 69, 3244 (1992); J.N. Reimers, J.E. Greedan, C.V. Stager and M. Bjorgvinnsen. Phis. Rev. B-43, 5692 (1991); J.E. Greedan, J.N. Reimers, C.V. Stager and S.L. Penny. Phys. Rev. B-43, 5682 (1991); J.N. Reimers, J.E. Greedan, R.K. Kremer, E. Gmelin and M.A. Subramanian. Phys. Rev. B-43, 3387 (1991)

[10] X. Obradors, A. Labarta, A. Isalgué, J. Tejada, J. Rodríguez and M. Pernet. Solid State Commun. 65, 189 (1988)

[11] B. Martínez, F. Sandiumenge, A. Rouco, A. Labarta, J. Rodríguez, M. Tovar, M.T. Causa, S. Galí and X. Obradors. Phys. Rev. B-46, 10786 (1992). A. Rouco, F. Sandiumenge, B. Martínez, S. Galí, M. Tovar and X. Obradors. J. Magn. Magn. Mat. 104-107, 1645 (1992)

[12] S.H. Lee, C. Broholm, G. Aeppli, T.G. Perring, B. Hessen and A. Taylor. Phys. Rev. Lett. 76, 4424 (1996)

[13] B. Martínez, A. Labarta, R. Rodríguez-Sola and X. Obradors. Phys. Rev. B-50, 15779 (1994)

[14] R. Omari, J.J. Préjean and J. Souletie. J. Phys. (Paris) 44, 1069 (1983)

[15] B. Barbara, A.P. Malozemoff and Y. Imry. Phys. Rev. Lett. 47, 1852 (1981)

[16] A. Gavrin, J.R. Childress, C.L. Chien, B. Martinez and M.B. Salamon. Phys. Rev. Lett. 64, 2438 (1990). E. Carré, E. Puech, J.J. Préjean. Heidelberg Colloquium on Glassy Dynamics. Springer-Verlag Lecture notes. Heidelberg (1986). A. Mauger, J. Villain, Y. Zhou. Phys. Rev. B-41. 4587 (1990)

[17] L. Néel. Ann. Geophys. 5, 99 (1949). J.L. Tholence. Solid State Commun. 35, 113 (1980)

[18] P.C. Hohenberg and B.I. Halperin. Rev. Mod. Phys. 49, 435 (1977)

Fractal Growth with Quenched Disorder

L. PIETRONERO[1], R. CAFIERO[1] and A. GABRIELLI[2]

[1] Dipartimento di Fisica, Universitá di Roma "La Sapienza",
   P.le Aldo Moro 2, I-00185 Roma, Italy; Istituto Nazionale di Fisica della Materia,
   unitá di Roma I
[2] Dipartimento di Fisica, Universitá di Roma "Tor Vergata",
   Via della Ricerca Scientifica 1, I-00133 Roma, Italy

Abstract. In this lecture we present an overview of the physics of irreversible fractal growth process, with particular emphasis on a class of models characterized by *quenched disorder*. These models exhibit self-organization, with critical properties developing spontaneously, without the fine tuning of external parameters. This situation is different from the usual critical phenomena, and requires the introduction of new theoretical methods. Our approach to these problems is based on two concepts, the Fixed Scale Transformation, and the quenched-stochastic transformation, or Run Time Statistics (RTS), which maps a dynamics with quenched disorder into a stochastic process. These methods, combined together, allow us to understand the self-organized nature of models with quenched disorder and to compute analytically their critical exponents. In addition, it is also possible characterize mathematically the origin of the dynamics by *avalanches* and compare it with the *continuous growth* of other fractal models. A specific application to Invasion Percolation will be discussed. Some possible relations to glasses will also be mentioned.

1   Introduction

The introduction of the fractal geometry (Mandelbrot (1982)) has changed the way physicists look at a vast class of natural phenomena which produce irregular structures. Many models have been introduced since the early eighties trying to relate these structures to well defined physical phenomena. These are the Diffusion Limited Aggregation (DLA) (Witten and Sander (1981)), the Dielectric Breakdown Model (DBM) (Niemeyer and Pietronero (1984)), the Invasion Percolation (IP) (Wilkinson and Willemsen (1983)), the Sandpile (Bak, Tang and Wiesenfeld (1987)), the Bak and Sneppen model (BS) (Bak and Sneppen (1993)), just to give some examples. All these models lead spontaneously (for a broad range of parameters) to the development of critical properties and fractal structures.

In the last years these has been a great interest on fractal models characterized by *quenched disorder*. These models are generally characterized by an intermittent dynamics, with bursts of activity of any size concentrated in a region of the system (*avalanches*), and by memory effects induced by the presence of quenched disorder and by the dynamical rules (Paczuski, Bak and

Maslov (1996); Cafiero, Gabrielli, Marsili, and Pietronero (1996); Vendruscolo and Marsili (1996); Marsili, Caldarelli and Vendruscolo (1996)). These memory effects in fractal growth processes with quenched disorder, may resemble an element characteristic of the spin glasses and glass dynamics. In fact, also in spin glasses, memory effects (aging) are due to the presence of quenched disorder, and are relied to the typical Kolrausch stretched exponential relaxation dynamics. However, at the moment it is hard to develop such an analogy in a more concrete way.

The study of physical phenomena leading to fractal structures can be classified by three different levels:

- *Mathematical Level: Fractal Geometry.* This is a descriptive level, at which one simply recognizes the fractal nature of the phenomena and extimates the fractal dimension D.
- *Physical Models.* One develops a model of fractal growth based on the physical process. This level is the analogue of the Ising model in equilibrium statistical mechanics.
- *Physical Theories.* This level corresponds to a fully understanding of the origin of fractals in nature, their self- organization etc. The corresponding level for phase transitions is the Renormalization Group.

The analogy we make with phase transitions is quite natural, because, like ordinary critical phenomena, fractal growth models are scale invariant. However, some profound differences, like irreversible, non-equilibrium dynamics and SOC, make unavoidable the development of new theoretical concepts.

Fractal physical models can be classified into two main groups:

1. Irreversible stochastic models
2. Irreversible quenched models

In the next section we discuss these models in relation with standard critical phenomena.

2 Physical Models for Self-similar Growth

We briefly mention below some examples of these two classes of models, with a particular emphasis on models with quenched disorder.

1) Irreversible stochastic models

- *Diffusion Limited Aggregation (DLA)* (Witten and Sander (1981)). This is the first physical model of fractal growth. Particles performing a Brownian motion aggregate and form complex fractal structures.
- *Dielectric Breakdown Model (DBM)* (Niemeyer and Pietronero (1984)). Is a generalization of DLA via the relation between potential theory and random walk.

– *Sandpile Models* (Bak, Tang and Wiesenfeld (1987)). These models are inspired by the marginal stability of sandpiles. The random addition of sand grains drives the system into a stationary state with a scale invariant distribution of avalanches.

2) Irreversible Quenched Models

– *Invasion Percolation (IP)* (Wilkinson and Willemsen (1983)). This model was developed to simulate the capillary displacement of a fluid in a porous medium. The porous medium is represented by a lattice where to each bond i is assigned a quenched value x_i of its conductance. At each time step the dynamics of the fluid evolves by occupying the bond with the smallest conductance between all its perimeter bonds. We call this kind of dynamics *extremal dynamics*. IP is known to reproduce asymptotically the Percolation cluster of standard critical Percolation.
The main characteristics of this model are:
1. *Deterministic dynamics.* Once a realization of the quenched disorder is chosen, the dynamical rule selects in a deterministic way the bond to be invaded.
2. *Self-organization.* The process spontaneously develops scale-invariant structures and critical properties. In the limit $t \to \infty$ both long range space and time correlations appear.
3. *Avalanches.* The asymptotic dynamical evolution consists of local, scale-invariant macro-events, composed by elementary growth steps spatially and causally connected, called *avalanches*. When an avalanche stops, the activity is transferred to another region of the perimeter.

– *Bak and Sneppen Model (BS)* (Bak and Sneppen (1993)). This model is similar to IP and has the same properties exposed in points 1-3. In fact, points 1-3 are characteristic of the whole class of SOC models with quenched disorder and extremal dynamics (for a review see Paczuski, Bak and Maslov (1996)). The BS model has been introduced to describe scale free events in biological evolution.

– *Quenched models with an external modulating field.* This is a particular class of models, the prototype of which is the *Quenched Dielectric Breakdown Model (QDBM* (Family and Zhang (1986); De Arcangelis, Hansen, Herrmann and Roux (1989)). This model is a sort of combination between IP (quenched) and DBM (stochastic). To each bond of a lattice is assigned a quenched random number x_i, representing the local resistivity. In addition, an external electric field E is introduced. The ratio $y_i = x_i/E_i$ between quenched disorder and local electric field is the inverse of the current flowing in the bonds. At each time step the dynamics breaks the bond with the smallest y_i, that is to say with the biggest current. The QDBM modelizes the dielectric breakdown of a disordered solid. A similar model can be formulated to modelize the propagation of fractures in a inhomogeneous solid (Hansen, Hinrichsen, Roux, Herrmann and De Arcangelis (1990)).

All the models in these classes share some important properties, that differentiates them from ordinary critical phenomena. First of all they are characterized by an *irreversible dynamics*, so that they cannot be described in hamiltonian terms, and the statistical weight of a configuration depends on the complete growth history (see. fig. 1).

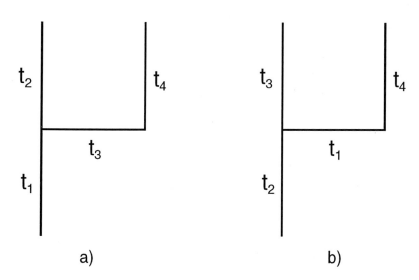

Fig. 1. For systems with irreversible dynamics, the statistical weight of a configuration depends on the whole growth history. The two identical configuration shown here have, in general, different statistical weights because their histories, indicated by the times t_i $i = 1, 2, 3, 4$, are different.

Quenched models have an additional problem, in that they have a deterministic dynamics, and the stochasticity enters only in the choice of the realization of the disorder. So, it is difficult to define transition probabilities for the dynamics.

Another fundamental difference, is that these models are self- organized. Their dynamics evolves spontaneously in the phase space towards an *attractive* fixed point. No fine tuning of any parameter is needed. In addition, quenched models with extremal dynamics have an *avalanche dynamics*, that is to say the system tends to concentrate its activity in a well localized region of the perimeter, during an avalanche. When an avalanche stops, the activity transfers to another region of the perimeter and a new avalanche starts. On the contrary, in stochastic models, like DLA, there is a continuous growth process, that is to say for large systems the probability to have two nearby subsequent growth events tends to zero. The dynamical activity is diffused at each time on the whole growth interface.

In table. 1 we propose a scheme of comparison between the properties of ordinary critical phenomena and the most popular stochastic and quenched fractal growth models.

3 New Theoretical Concepts

The application of the standard theoretical methods of statistical physics (field theory and renormalization group) is, in general, not possible for the main fractal growth problems, like DLA, DBM, Invasion Percolation, the sandpile model, which are characterized by an intrinsically irreversible dynamics.

Here we discuss two theoretical methods we have developed in the past few years, as a step towards the construction of a physical theory for self-organized fractal growth processes. The exposition will be colloquial, and mainly devoted to the theoretical analysis of quenched models. For more details readers should refer to the bibliography.

3.1 Fixed Scale Tranformation (FST)

This approach combines a technique of lattice path integral, to take into account the irreversible dynamics, with the study of the scale invariant dynamics inspired by the RG theory. It permits a description of the scale invariant properties of fractal growth
models.

The method focuses on the dynamics at a given scale and analyzes the nearest neighbours correlations at this scale using *lattice path integral* approach by which one can calculate the elements of a probability matrix, the FST matrix. The fixed point of the FST matrix gives the nearest neighbour correlations, at that scale. If one uses the scale invariant dynamics of the system, that can be obtained by Real Space Renormalization Group approaches (Cafiero, Vespignani and Pietronero (1993)), one can generalize these correlations to all scales and compute the fractal dimension (Erzan, Pietronero and Vespignani (1995)).

The basic point of the FST is the separation of the long time limit $(t \to \infty)$ for the dynamical process at a given scale, from the large scale limit $(r \to \infty)$, that defines the scale invariant dynamics. The interesting feature of FST is that it works at a fixed scale, so it is possible to include the fluctuation of the boundary conditions, that in systems with long range interactions, like *Diffusion Limited Aggregation* (DLA), have a great influence on the fractal properties. For this reason, the FST approach allows to reach a remarkable level of accuracy in the calculation of the fractal dimension.

At the moment, the FST framework, eventually combined with the RTS method that we discuss below, seems to be the only general approach to

Table 1. Comparison between the properties of ordinary critical phenomena represented by the Ising model and the most popular stochastic and quenched models generating fractal or scale invariant structures in a self-organized way.

SELF-SIMILARITY: PHYSICAL MODELS		
Ising-Type (70's)	**DLA-DBM (81)**	**Invasion Percolation (82-83)**
Equilibrium Statistical Mechanics	NON LINEAR, IRREVERSIBLE DYNAMICAL EVOLUTION. Statistical weight of configurations depend on the whole growth history. In extremal models they are difficult to compute because of deterministic dynamics	
Ergodicity		
Boltzmann Weight		
Standard Critical behaviour Fine Tuning: $T=T_c$	CRITICAL BEHAVIOUR IS SELF-ORGANIZED ATTRACTIVE FIXED POINT	
Repulsive Fixed Point	Asymptotically frozen fractal structure	Fractal structure
$\xi = (T - T_c)^{-v}$	continuous growth	Avalanche dynamics driven by extremal statistics, with avalanches of all sizes
Approach to the critical point	Long range interactions (Laplacian)	Cognitive time memory
$\Gamma(r) = \dfrac{1}{r^{(d-2+\eta)}}$		
Anomalous dimension exactly at $T = T_c$	Problem: understand the SOC origin of fractal properties and computation of the fractal dimension D	Problems: origin of SOC and avalanche dynamics. Computation of D distribution of avalanche sizes $P(s) = s^{-\tau}$
Theory: Renormalization Group	THEORY: NEW CONCEPTS ARE NEEDED, LIKE **FST** AND **RTS**	

understand the self-organized critical nature of a broad class of models going from DLA, to Percolation, to sandpile, to Invasion Percolation (Erzan, Pietronero and Vespignani (1995); Cafiero, Gabrielli, Marsili, and Pietronero (1996)). This situation therefore supports the idea that these models pose new questions for which one would like to develop a common theoretical scheme.

3.2 Quenched-Stochastic Transformation (RST)

As we mentioned above, quenched models with extremal dynamics, like Invasion Percolation (IP), have a deterministic dynamics. This makes it impossible to address directly these class of model with the FST method or any other microscopic theory. Here we describe a general theoretical approach which addresses the basic problems of extremal models: (i) the understanding of the scale-invariance and self-organization;(ii) the origin of the avalanche dynamics and (iii) the computation of the independent critical exponents. We will discuss in particular a specific application to Invasion Percolation (IP) (Cafiero, Gabrielli, Marsili, and Pietronero (1996)), but these ideas can be easily extended to other models of this type like the Bak and Sneppen model (Marsili (1994b)).

In order to overcome the problem represented by quenched disorder, we introduced a mapping of the quenched extremal dynamics into a stochastic one with cognitive memory, the *quenched-stochastic transformation*, also called Run Time Statistics (RTS) (Marsili (1994a)). This approach was improved in various steps (Pietronero, Schneider and Stella (1990); Pietronero and Schneider (1990); Marsili (1994a); Marsili (1994b)) and now we can develop it into a general theoretical scheme, that we call RTS-FST method (Cafiero, Gabrielli, Marsili, and Pietronero (1996)). Its essential points are:
- Quenched-stochastic transformation.
- Identification of the microscopic fixed point dynamics. This point clarifies the *SOC nature* of the problem.
- Identification of the scale invariant dynamics for block variables. This elucidates the origin of fractal structures.
- Definition of *local* growth rules for the extremal model. This clarifies the origin of *avalanche dynamics*.
- Use of the above elements in a real space scheme, like the FST, to compute analytically the relevant exponents of the model.

A general stochastic process is based on the following elements: a) a set of time dependent dynamical variables $\{\eta_{i,t}\}$; b) a Growth Probability Distribution (GPD) for the single growth step $\{\mu_{i,t}\}$, obtained from the $\{\eta_{i,t}\}$; c) a rule for the evolution of the dynamical variables $\eta_{i,t} \rightarrow \eta_{i,t+1}$.

Therefore, in order to map IP onto a stochastic process we have to: a) find the correct dynamical variables (the $\{\eta_{i,t}\}$'s); b) determine the GPD $\{\mu_{i,t}\}$ in terms of these variables; c) find the evolution rule of the $\{\eta_{i,t}\}$

A simple example can be useful to get an insight into the essence of the problem. Consider two independent random variables X_1, X_2, with uniform distribution $p_0(x_1) = p_0(x_2) = 1$ in $[0, 1]$ and let us eliminate the smallest, for example X_2. Clearly the probability that $X_2 < X_1$ is $1/2$. At the second "time step", we compare the surviving variable X_1 with a third, uniform, random variable X_3 just added to the game and, again, we eliminate the smallest one. At first sight one might think that, since both variables are independent, the probability that X_1 survives again is $1/2$, but this is actually incorrect. In this case we indeed need to calculate the probability μ_3 that $X_3 < X_1$ given that $X_2 < X_1$. This, using the rules of conditional probability, reads:

$$\mu_3 = \tilde{P}(X_3 < X_1) = P(X_3 < X_1 | X_2 < X_1) =$$
$$= \frac{P(X_3 < X_1 \bigcap X_2 < X_1)}{P(X_2 < X_1)} = \frac{2}{3}, \tag{1}$$

where $P(A|B)$ is the probability of the event A, given that B occurred, and $P(A \cap B)$ is the probability of occurrence of both A and B. The point is that the distribution of the variable X_1 is no longer uniform when it is compared with X_3, even though they are independent. The information that $X_2 < X_1$ *changes in a conditional way* the *effective* probability density $p_1(x)$ of X_1. Indeed the probability that $x \leq X_1 < x + dx$ must now account for the fact that $X_2 < x$. By imposing this condition, we get: $p_1(x) = 2x$. An analogous calculation for the distribution of X_2 gives $p_2(x) = 2(1 - x)$. Qualitatively, the event $X_2 < X_1$ decreases the probability that X_1 has small values. On the contrary, the probability that X_2 is small is enhanced.

The above example contains the essential idea of the *quenched-stochastic transformation*. Extremal dynamics establishes, at each time step t, an order relation between quenched variables ($X_2 < X_1$ in the example). This *information* on the statistical properties of the variables involved in the process (*active* variables) can be conditionally stored in the form of their *effective densities*. Variables which have experienced the same dynamical history, will have the same effective density, irrespectively of their spatial position. This *memory* is represented by the *age* $k = t - t_0$, where t is the actual time and t_0 is the time at which the variable became active. The effective densities $p_{k,t}(x)$ ($p_{0,t}(x) = p_{0,0}(x) = 1$) of variables of age k at time t are *the dynamical variables of the stochastic process we are looking for*.

A generalization of our simple example (Eq. 1) leads to the following equation for the growth probability $\mu_{k,t}$ of a variable of age k at time t (GPD):

$$\mu_{k,t} = \int_0^1 dx \, p_{k,t}(x) \prod_\theta (1 - P_{\theta,t}(x))^{n_{\theta,t} - \delta_{\theta,k}}, \qquad (2)$$

where $P_{\theta,t}(x) = \int_0^x dy p_{\theta,t}(y)$, the product is intended over all the ages of the active variables and $n_{\theta,t}$ is the number of active variables of *age* θ at time t. The meaning of this expression is that the product inside the integral takes into account of the competition of the selected variable with each of the other active variables. The density $m_{k,t}(x)$ of this (smallest) variable after its growth is conditioned by the information that it has grown, and can be computed from Eq. 2. The temporal evolution of the densities of the still active variables is then given by:

$$p_{\theta+1,t+1}(x) = p_{\theta,t}(x) \int_0^x \frac{m_{k,t}(y)}{1 - P_{\theta,t}(y)} dy. \qquad (3)$$

Equations 2, 3 *accomplish our goal to describe a quenched extremal process as a stochastic process with time memory*. The presence of memory is enlighted by the dependence of the GPD on the parameter k. A mean field like expansion of Eq.(2) in the limit $t \to \infty$ gives (Marsili (1994a)): $\mu_{k,\infty} \sim \frac{1}{(k+1)^\alpha}$. This result is also confirmed by simulations (Marsili, Caldarelli and Vendruscolo (1996)). Memory is at the origin of *screening effects* in the GPD $\{\mu_{k,t}\}$.

The power law behaviour of $\mu_{k,t}$ guarantees that screening is preserved at all scales, which is the condition to generate holes of all sizes in a growing pattern, leading to fractal structures (Cafiero, Vespignani and Pietronero (1993)).

This mapping, applied to models like IP, allows us to characterize mathematically the self-organization. In fact, the following *histogram equation* can be derived by the RTS equations 2, 3:

$$\partial_x \Phi_t(x) = \beta \Omega_t \Phi_t^2(x) \left[1 - \frac{\omega_t}{\omega_t + 1} \Phi_t(x) \right] \tag{4}$$

where $\omega_t = \langle N_{t+1} - N_t \rangle$, $\Omega_t = \langle N_t \rangle$, N_t is the number of interface variables at time t, and β is the solution of: $\beta = 1 - e^{-\beta \Omega_t}$. This equation describes the time evolution of the distribution $\Phi_t(x)$ of quenched disorder on the growth interface. The solution of eq. 4 becomes asymptotically (fig.2) (Marsili (1994a)):

$$\lim_{t \to \infty} \Phi_t(x) = \frac{1}{1 - p_c} \theta(x - p_c) \tag{5}$$

where p_c is a critical threshold of the original extremal dynamics ($p_c = 1/2$ for $2d$ bond IP), in agreement with numerical simulations (Wilkinson and Willemsen (1983)). Note that, in order to obtain the asymptotic behaviour 5, no fine tuning of any parameter is needed. This clarifies the SOC nature of the problem.

The scale invariant dynamics can be shown to coincide with the microscopic one (Cafiero, Gabrielli, Marsili, and Pietronero (1996)). The RTS approach permits also to characterize the origin of avalanche dynamics and to write down a set of equations describing the evolution of a single avalanche, by a straightforward modification of equations 2, 3 (Cafiero, Gabrielli, Marsili, and Pietronero (1996)). From simulations (Maslov (1995)), and from the histogram equation, one deduces that, asymptotically, each avalanche starts with a variable (initiator) equal to the threshold p_c ($p_c = 1/2$ for $2d$ bond IP). All other variables in the avalanche have values smaller than p_c. In view of the above arguments, the RTS equations for the *local avalanche dynamics* are obtained from Eqs. (2, 3) by taking into account only the variables which become active after the initiator's growth and by integrating in Eq.(2) only in $[0, p_c]$.

By using the equations for the local avalanche dynamics together with the FST method we have been able to compute with a very good accuracy (tipically $1 - 2\%$, depending on approximations), the relevant critical exponents of Invasion Percolation, that is to say the fractal dimension and the avalanche exponent (Cafiero, Gabrielli, Marsili, and Pietronero (1996)). The method can be applied successfully also to the Bak and Sneppen model (Marsili (1994b)). In table 2 we show the theoretical values of the exponents of IP, that we have computed with the RTS-FST method, compared with numerical simulations

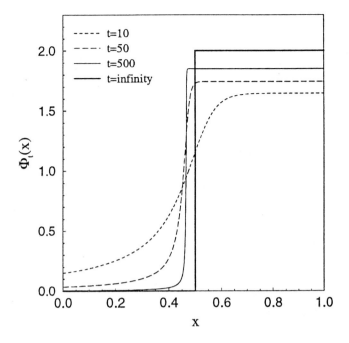

Fig. 2. Time evolution of the solution of the equation for $\Phi_t(x)$ (for $2-d$ bond IP). $\Phi_t(x)$ tends asymptotically to a theta function with discontinuity at $p_c = 1/2$.

Table 2. Theoretical values of the fractal dimension of IP, with (D_f^{trap}) and without trapping (D_f) and of directed IP (D_f^{DIP}), and of the avalanche exponent τ. These values are compared with numerical simulations.

	D_f	D_f^{trap}	D_f^{DIP}	τ^{IP}	τ^{trap}
$RTS-FST$	1.8879	1.8544	1.7444	1.5832	1.5463
$simul.$	~ 1.89	~ 1.86	~ 1.75	~ 1.60	~ 1.53

Recently, we have extended the RTS-FST scheme to QDBM, and we have obtained interesting, although preliminar, results, which allow us to elucidate some important characteristics of the class of models to which QDBM belongs (Cafiero, Gabrielli, Marsili, Torosantucci and Pietronero (1996)).

4   Further Developments

In this lecture we have discussed two recently introduced theoretical methods, the FST and the RTS. These approaches have been applied sucesfully to many models for fractal growth, and allow to make a significative step towards the formulation of a common theoretical scheme for the physics of self-organized fractal growth.

At the moment, we are studying the application of the RTS mapping to interface dynamics in quenched disorder (Sneppen (1992)), and to glassy type dynamics. An interesting work, for what concerns the last point, is and RTS-type analysis of the statistical and dynamical properties of the random walk in quenched disorder (RRW) (Vendruscolo and Marsili (1996)), which has been studied by many authors as a toy model for localization (Tosatti, Zannetti and Pietronero (1988)), depinning transitions (Bouchaud, Comtet, Georges and Le Doussal (1990)), and aging effects (Marinari and Parisi (1993)). In this work, the authors map, by using the RTS method, the RRW dynamics into a stochastic dynamics with cognitive memory and recover all the characteristics of the original model. This suggests a link between stochastic dynamics with memory and the realizations of a dynamics with quenched disorder.

References

Mandelbrot B. B. (1982): The fractal Geometry of Nature. W. H. Freeman, New York.

Witten T. A. , Sander L. M. (1981): Phys. Rev. Lett. **47**, 1400.

Niemeyer L. , Pietronero L. , Wiesmann H. J. (1984): Phys. Rev. Lett. **52**, 1033.

Bak P. , Sneppen K. (1993):Phys. Rev. Lett. **71**, 4083.

Wilkinson D. , Willemsen J. F. (1983): J. Phys. **A 16**, 3365.

Bak P. , Tang C. , Wiesenfeld K. (1987): Phys. Rev. Lett. **59**, 381.

Sneppen K. (1992): Phys. Rev. Lett. **69**, 3539.

Family F. , Zhang Y. C. , Vicsek T. (1986): J. Phys. **A 19**, L733.

De Arcangelis L., Hansen A. , Herrmann H. J. , Roux S. (1989): Phys. Rev. **B 40**, 877.

Hansen A. , Hinrichsen E. L. , Roux S. , Herrmann H. J. , De Arcangelis L. (1990): Europhys. Lett. **13**, 341.

Cafiero R., Pietronero L., Vespignani A. (1993): Phys. Rev. Lett. **70**, 3939.

Erzan A., Pietronero L., Vespignani A. (1995): Rev. Mod. Phys. **67**, no. 3.

Pietronero L., Schneider W. R., Stella A. (1990): Phys. Rev.**A 42**, R7496.

Pietronero L., Schneider W. R. (1990): Physica **A 119**, 249-267.

Marsili M. (1994): J. Stat. Phys. **77**, 733.

Gabrielli A. , Marsili M., Cafiero R., Pietronero L. (1996): J. Stat. Phys. **84**, 889-893.

Marsili M. (1994): Europhys. Lett. **28**, 385.

Paczuski M., Bak P., Maslov S. (1996): Phys. Rev. **E** 53, 414.

Cafiero R., Gabrielli A., Marsili M., Pietronero L. (1996): Phys. Rev. **E 54**, 1406.

Vendruscolo M., Marsili M. (1996): Phys. Rev. E **54**, R1021.

Maslov S. (1995): Phys.Rev.Lett. **74**, 562.

Marsili M., Caldarelli G., Vendruscolo M. (1996): Phys. Rev. **E 53**, 13.

Cafiero R., Gabrielli A., Marsili M., Torosantucci L., Pietronero L. (1996): in preparation.

Tosatti E., Zannetti M., Pietronero L. (1988): Z. Phys. **B 73**, 161.

Bouchaud J. P., Comtet A., Georges A., Le Doussal P. (1990): Ann. Phys. **201**, 285.

Marinari E., Parisi G. (1993): J. Phys. **A** Math. Gen. **26**, L1149.

Data Clustering and the Glassy Structures of Randomness

E. Lootens and C. Van den Broeck

LUC, B-3590 Diepenbeek, Belgium

Abstract. Using techniques, borrowed from statistical mechanics of spin glasses, we investigate the properties of cluster algorithms applied to random and non-random data points.

1 Introduction

We will review a new approach, one based on the statistical mechanics of systems with disorder, of an old problem, namely the clustering of data points. The general purpose is to discover whether or not a given set of data points is purely random, and, if not, to extract information about the underlying structure. The main message is that one can, by applying the replica formalism developed in the context of spin glasses, obtain analytic results illuminating some surprising properties of currently used clustering algorithms and allowing for the exploration of new algorithms.

As our starting point, we have at our disposition a set of p data points $\{\boldsymbol{\xi}^\mu, \mu = 1, ..., p\}$. For simplicity, we will assume that every data point can be represented by an N-dimensional vector $\boldsymbol{\xi} = \{\xi_1, ..., \xi_i, ..., \xi_N\}$ with binary components $\xi_i = +1$ or $= -1$. The resulting vector thus lies on the surface of a sphere with radius $\sqrt{N}$. To discuss the structure of these date points, we will focus our attention on the overlaps of their vectors with specific other vectors that lie on the sphere. For example, consider first a randomly chosen vector $\mathbf{J}$ with $\mathbf{J}^2 = N$. Clearly the overlap λ:

$$\lambda = \mathbf{J}\boldsymbol{\xi} \tag{1}$$

with any vector of the data set is, by virtue of the central limit theorem, a normally distributed random variable (cf. Fig. 1a). Suppose now that one would like to investigate whether the data points are clustered along a preferential direction. A good and simple guess of this direction is the one pointing along the center of mass:

$$\mathbf{J}^* = C \sum_{\mu=1}^{p} \boldsymbol{\xi}^\mu. \tag{2}$$

where the constant C is chosen such that $|\mathbf{J}^*|^2 = N$. In the theory of neural networks, where $\mathbf{J}$ represents the vector of the synaptic strengths, this choice corresponds to Hebbian learning [1].

To illustrate how this construction can be misleading, we consider data points that are chosen at random and independently of each other. The overlap of $\mathbf{J}^*$ with any vector $\boldsymbol{\xi}$ of the data set can, in the limit $N \to \infty$, $p \to \infty$ with $\alpha = p/N$ fixed, easily be obtained by separating out the contribution of $\boldsymbol{\xi}$ to $\mathbf{J}^*$. One finds that the resulting overlap λ is again Gaussian random variable with unit dispersion but with non-zero mean equal to $1/\sqrt{\alpha}$, cf. Fig. 1b. This result creates the false impression that the data points are generated from a non-uniform probability distribution with a preferential orientation along the center of mass. The observed structure however is just the product of chance, even though it will appear, in the above described limit, with probability one and be perfectly reproducible!

The above considerations point to the need for a more systematic study of how data points are distributed when viewed from one or more special directions, that are correlated to these very points. This can be achieved by reformulating the above questions in the framework of statistical mechanics. To make this transition, we note that the center of mass vector $\mathbf{J}^*$ can alternatively be obtained as follows:

$$\mathbf{J}^* = \arg\mathbf{J}^2 = NE(\mathbf{J}) \tag{3}$$

with

$$E(\mathbf{J}) = -\mathbf{J} \cdot \sum_{\mu=1}^{p} \frac{\boldsymbol{\xi}^\mu}{\sqrt{N}}. \tag{4}$$

In other words, $\mathbf{J}^*$ is the "ground state" of the energy function E. The formulation and generalization of the problem within the context of statistical mechanics is now straightforward. We will search for the minimum of an energy function whose general form is as follows:

$$E(\mathbf{J}) = \sum_{\mu=1}^{p} V(\lambda^\mu). \tag{5}$$

where $\lambda^\mu = \frac{\mathbf{J}.\boldsymbol{\xi}^\mu}{\sqrt{N}}$. The properties of vectors minimizing the energy function Eq. (5) can be investigated using the techniques developed in the statistical mechanics of spin glasses and neural networks, as we proceed to show in the next section.

2   Statistical Mechanics

We start by defining a partition function, associated to the energy function E :

$$Z = \int d\mathbf{J} \; e^{-\beta E(\mathbf{J})} \; \delta(\mathbf{J}^2 - N). \tag{6}$$

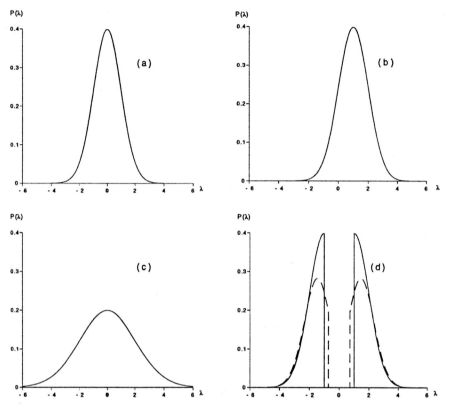

Fig. 1. Overlap distributions for random and independent patterns with the **J** vector that minimizes $E(\mathbf{J})$, cf. Eq. (5), for (a) $V = 0$, (b) $V = -\lambda$, (c) $V = -\lambda^2$ and (d) $V = -|\lambda|$ (with the dashed line corresponding to the 1RSB result).

This quantity is a random variable through its dependence on the randomly chosen data points, but one expects that the corresponding free energy F is self-averaging in the limit $N \to \infty$, $p \to \infty$ with $\alpha = p/N$ finite:

$$F = -\frac{1}{\beta} \ln Z = -\frac{1}{\beta} < \ln Z > . \tag{7}$$

F can be calculated using the techniques of the replica method. We will focus here on the case of data points that are sampled independently from a uniform distribution on the sphere. Since we are interested in the minima E_0 of E, we only mention here the results in the 'zero temperature limit' $\beta \to \infty$. Assuming replica symmetry (RS) one finds [2] :

$$\lim_{N \to \infty} \frac{F_0}{N} = \lim_{N \to \infty} \frac{E_0}{N} = -x \left[\frac{1}{2x} - \alpha t \lambda \left\{ V(\lambda) + \frac{(\lambda - t)^2)}{2x} \right\} \right], \tag{8}$$

where $\mathcal{D}t = \exp(-t^2/2)/\sqrt{2\pi}$ stands for the Gaussian measure. Of particular interest is the probability distribution $P(\lambda)$ of overlaps between the vector $\mathbf{J}^*$ that minimizes E, and the data points $\boldsymbol{\xi}^\mu$ that enter in the construction of E. One finds :

$$P(\lambda) = t\,\delta(\lambda - \lambda_0(t, x)), \tag{9}$$

where $\lambda_0(t, x)$ is the function of t and x that minimizes the expression between curly brackets in Eq. (8), and x is the function of α that extremizes the free energy, cf. Eq. (8).

It is easy to verify that one recovers the result, discussed in the previous section, for the particular choice $V(\lambda) = -\lambda$. This specific form of V reflects the intention to find a direction along which the overlap λ is as large as possible. We mention two other choices that incorporate a quite different aim. Maximal variance learning, which can be implemented by Oja's rule [3], is characterized by the following potential:

$$V(\lambda) = -\lambda^2. \tag{10}$$

The corresponding overlap distribution reflects the aim built into the form of V, namely that of maximizing the variance of the overlap: $P(\lambda)$ is a broadened Gaussian with zero mean and dispersion $<\lambda^2> = (1 + \sqrt{\alpha})^2/\alpha$ [4], cf. Fig. 1c.

As another variant, one that does not correspond to any specific algorithm, but will be important in the further discussion, we consider a potential that penalizes the overlap according to the absolute value of the overlap:

$$V(\lambda) = -|\lambda|. \tag{11}$$

The corresponding overlap distribution reads, cf. Fig. 1d:

$$P(\lambda) = \begin{cases} \dfrac{e^{-\frac{1}{2}(\lambda+x)^2}}{\sqrt{2\pi}} & \lambda \le -x \\ 0 & -x < \lambda < x \\ \dfrac{e^{-\frac{1}{2}(\lambda-x)^2}}{\sqrt{2\pi}} & \lambda \ge x \end{cases} \tag{12}$$

with $x = 1/\sqrt{\alpha}$. This result contains two surprises. Firstly, the overlap distribution possesses a genuine gap of width $2/\sqrt{\alpha}$: there are no data points whatsoever in the band $-x < \lambda < x$, which occupies a finite fraction of the total surface of the sphere. One concludes that the data points, even though they are sampled from a uniform distribution on the sphere, are highly structured when viewed from the special direction $\mathbf{J}^*$. Secondly, it turns out that so-called Almeida-Thouless (AT) condition [5] for the stability of the replica-symmetric point is violated. This condition can be written under the following general form for the problem under consideration [6]:

$$\alpha t \left[\frac{\partial \lambda_0(t, x)}{\partial t} - 1 \right]^2 < 1, \tag{13}$$

where x is again taken in the extremum of Eq. (8). As can be seen by comparing Eqs. (9) and (13), the existence of a gap in the overlap distribution, corresponding to a discontinuity of $\lambda_0(t, x)$ as a function of t, implies replica symmetry breaking. The results of a one-step replica symmetry breaking calculation (1RSB) are included in Fig. 1d. One observes that the width of the gap is slightly reduced, but we do not expect that the gap will disappear, even with further steps of replica symmetry breaking. The fact that replica symmetry breaking occurs indicates that the energy function corresponding to Eq. (11) has a large number of minima, revealing the underlying glassy structure appearing at finite α in the realizations of a uniform distribution.

3 Clustering of Random Data

As the word implies, clustering is about grouping data points into different sets. Here we will consider the simplest situation corresponding to clustering into 2 different sets, each one characterized by a cluster axis $\mathbf{J}_i, i = 1, 2$. The problem is now to find these two directions such that the data points are separated in some optimal way in two classes. Building on our experience with the single $\mathbf{J}$-case, it is natural to define the optimal directions $\mathbf{J}_i^*, i = 1, 2$ as the minima of an energy function $E(\mathbf{J}_1, \mathbf{J}_2)$ of the following form :

$$E(\mathbf{J}_1, \mathbf{J}_2) = \sum_{\mu=1}^{p} V(\lambda_1^\mu, \lambda_2^\mu), \tag{14}$$

with :

$$\lambda_1^\mu = \mathbf{J}_1 \boldsymbol{\xi}^\mu, \qquad\qquad \lambda_2^\mu = \mathbf{J}_2 \boldsymbol{\xi}^\mu. \tag{15}$$

Note that the numbering of the clusters is interchangeable, hence we will assume that the potential reflects this symmetry, $V(\lambda_1, \lambda_2) = V(\lambda_2, \lambda_1)$. Furthermore, to make the connection with the previous sections, we will also assume that the potential has the following specific form : $V(\lambda_1, \lambda_2) = V_+(\frac{\lambda_1+\lambda_2}{\sqrt{2}}) + V_-(\frac{\lambda_1-\lambda_2}{\sqrt{2}})$. As before, we introduce a partition function :

$$Z = \int d\mathbf{J}_1 \int d\mathbf{J}_2 e^{-\beta E(\mathbf{J}_1, \mathbf{J}_2)} \delta(\mathbf{J}_1^2 - N)\delta(\mathbf{J}_2^2 - N), \tag{16}$$

and calculate the corresponding free energy F, which is expected to be self-averaging in the limit $N \to \infty$, $p \to \infty$ with $\alpha = p/N$ finite. Adopting a RS ansatz one obtains the following result for the zero-temperature (free) energy [7]:

$$\lim_{N\to\infty} \frac{F_0}{N} = \lim_{N\to\infty} \frac{E_0}{N} = -Q_{,x_+,x_-} \left[\frac{1}{2x_+} - \alpha t_+ \lambda_+ \left\{ V_+(\lambda_+) + \right. \right. \tag{17}$$

$$\left. \frac{(\frac{\lambda_+}{\sqrt{1+Q}} - t_+)^2}{2x_+} \right\} + \frac{1}{2x_-} - \alpha t_- \lambda_- \left\{ V_-(\lambda_-) + \frac{(\frac{\lambda_-}{\sqrt{1-Q}} - t_-)^2}{2x_-} \right\} \right],$$

where $Q = (\mathbf{J}_1.\mathbf{J}_2)/N$ has the meaning of the overlap between the cluster axes. While comparing this result with Eq. (8), we note that finding the minima for E can be performed in two steps. First one fixes the value of Q and finds the vectors $\mathbf{J}_+ = (\mathbf{J}_1 + \mathbf{J}_2)/\sqrt{2(1 + Q)}$ and $\mathbf{J}_- = (\mathbf{J}_1 - \mathbf{J}_2)/\sqrt{2(1 - Q)}$ that minimize $V_+(\lambda_+ \sqrt{1 + Q})$ and $V_-(\lambda_- \sqrt{1 - Q})$ respectively. These 2 minimization procedures are of exactly the same type as the ones encountered in section 2. In a second step one minimizes the resulting total cost with respect to Q, and determines in this way the optimal angle between $\mathbf{J}_1$ and $\mathbf{J}_2$. The AT condition for stability of the replica symmetric solution reads :

$$
\begin{cases}
\alpha t_+ \left[\dfrac{1}{\sqrt{1+Q}} \dfrac{\partial \lambda_+^0(x_+, t_+, Q)}{\partial t_+} - 1 \right]^2 < 1 \\[4mm]
\alpha t_- \left[\dfrac{1}{\sqrt{1-Q}} \dfrac{\partial \lambda_-^0(x_-, t_-, Q)}{\partial t_-} - 1 \right]^2 < 1
\end{cases}
, \tag{18}
$$

where $\lambda_+^0 = (\lambda_1^0 + \lambda_2^0)/\sqrt{2}$ and $\lambda_-^0 == (\lambda_1^0 - \lambda_2^0)/\sqrt{2}$ are the functions of Q, x_+, x_-, t_+ and t_- that minimize the expressions between curly brackets in Eq. (18). Finally, the overlap distribution $P(\lambda_1, \lambda_2)$ reads:

$$
P(\lambda_1, \lambda_2) = t_+ t_- \delta(\lambda_1 - \lambda_1^0)\delta(\lambda_2 - \lambda_2^0). \tag{19}
$$

4 Competitive learning

One of the standard algorithms to cluster data points is the so-called K-means algorithm [8], also know as competitive learning in the theory of neural networks [1]. The important observation for us is that this algorithm (or at least its batch mode version) reduces to the minimization of an energy function with the following potential:

$$
V(\lambda_1, \lambda_2) = -\lambda_1 \theta(\lambda_1 - \lambda_2) - \lambda_2 \theta(\lambda_2 - \lambda_1) \tag{20}
$$

$$
= -\frac{\lambda_1 + \lambda_2}{2} - \frac{|\lambda_1 - \lambda_2|}{2}. \tag{21}
$$

To make the connection with the discussion of the single $\mathbf{J}$-case, it is revealing to view this potential in two ways. In its first form (20) one recognizes in each term a Hebbian type of contribution $-\lambda_i$, i=1,2. This Hebbian contribution is multiplied by a theta function which "classifies" the pattern as belonging either to cluster 1 or 2, depending on which overlap $\lambda_i^\mu = \mathbf{J}_i \cdot \boldsymbol{\xi}^\mu/\sqrt{N}$ being largest. Alternatively, one can write the potential as the sum of a pure Hebbian term (along the direction of $\mathbf{J}_1 + \mathbf{J}_2$, no theta function) and a maximal absolute value term (following the direction of $\mathbf{J}_1 - \mathbf{J}_2$), so one expects to find back the characteristics of the corresponding single $\mathbf{J}$-cases discussed previously. This is indeed exactly what happens. Consider for example the overlap distribution. For random orientations of $\mathbf{J}_i, i = 1, 2$, this distribution

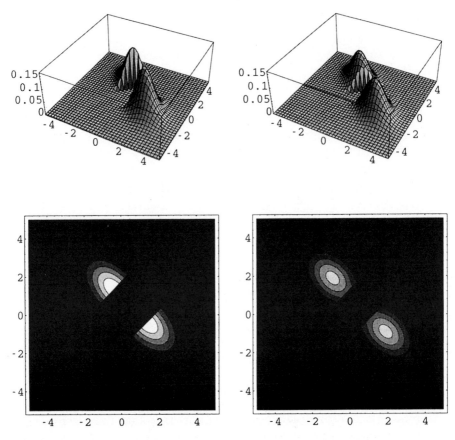

Fig. 2. The overlap distribution $P(\lambda_1, \lambda_2)$, for competitive learning applied to random data, as predicted by RS (left hand side) and 1RSB (right hand side) for $\alpha = 1$.

is a bi-normal. Due to the potential contribution of the Hebb type along $\lambda = \lambda_1 + \lambda_2$, one observes that the distribution is shifted along this line over a distance equal to $1/\sqrt{\alpha}$. Furthermore, the maximal absolute value potential acting along $\lambda = \lambda_1 - \lambda_2$, results in the overlap distribution being cut in half and shifted apart along this direction, cf Fig. (2). The width of this gap decreases as $2/\sqrt{\alpha}$. The overlap Q between $\mathbf{J}_1^*$ and $\mathbf{J}_2^*$ is found to be (cf. Fig. (3)):

$$Q(\alpha) = -\frac{2\sqrt{\frac{\alpha}{2\pi}}(1 + \sqrt{\frac{\alpha}{2\pi}})}{1 + 2\sqrt{\frac{\alpha}{2\pi}}(1 + \sqrt{\frac{\alpha}{2\pi}})}, \tag{22}$$

We conclude that the angle between $\mathbf{J}_1^*$ and $\mathbf{J}_2^*$ decreases from $\pi/2$ to π for $\alpha \to \infty$.

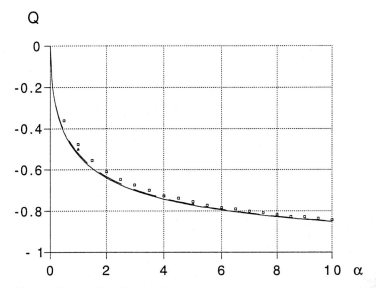

Fig. 3. The overlap Q as a function of α, as obtained from RS (full line, cf. Eq. (22)), 1RSB (dashed line) and simulations (circles: system size $N = 50$ starting from 10^2 initial configurations, triangles: system size $N = 100$ starting form 2.10^4 initial configurations, both using the second algorithm).

As in the single **J**-case, the maximal absolute value contribution to the potential leads to a discontinuity in overlap structure and, as a result, replica symmetry is violated. We therefore performed a 1RSB calculation, the results of which are included in Figs. (2) and (3). It turns out that the corrections due to RSB are minor for the free energy and Q, but the overlap distribution and the gap between the two clusters are significantly modified.

5   Maximal variance competitive learning

The fact that RS is broken in competitive learning is due to the contribution of V_- in the potential. In section 2, we mentioned the maximal variance potential that also penalizes zero overlaps but does not lead to RSB. This prompts us to introduce a new clustering algorithm, defined by the following potential:

$$V(\lambda_1, \lambda_2) = -\lambda_1(1 + \gamma(\lambda_1 - \lambda_2)) - \lambda_2(1 + \gamma(\lambda_2 - da_1))$$

$$= -(\lambda_1 + \lambda_2) - \gamma(\lambda_1 - \lambda_2)^2. \tag{23}$$

Performing the calculations for this particular case leads to the following results:

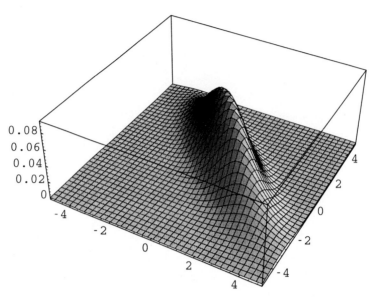

Fig. 4. The overlap distribution $P(\lambda_1, \lambda_2)$, for maximal variance competitive learning applied to random data, for $\alpha = 1$ and $\gamma = 1/8$.

$$Q = \begin{cases} -1 + \frac{\alpha}{8\gamma^2(1+\sqrt{\alpha})^4} & \alpha \leq \alpha_- \text{ or } \alpha > \alpha_+ \\ 1 & \alpha_- \leq \alpha \leq \alpha_+ \end{cases} \tag{24}$$

where α_+ and α_- are defined as the positive real roots of the equation $4\gamma(1 + \sqrt{\alpha})^2 - \sqrt{\alpha} = 0$. These roots do not exist for $\gamma > 1/16$. In this case the first line in Eq. (24) holds for all values of α, and $Q(\alpha)$ attains its maximal value (< 1) at $\alpha = 1$, see Fig. (5). Note that in contrast to what happens for competitive learning, Q is now non-monotonic in α. The overlap distribution reads:

$$P(\lambda_1, \lambda_2) = \frac{1 + \sqrt{\alpha}}{2\pi\sqrt{\alpha}} e^{-\frac{1}{2}(\frac{\lambda_1+\lambda_2}{\sqrt{2(1+Q)}} - \frac{1}{\sqrt{\alpha}})^2} e^{-\frac{1}{2}(\frac{1+\sqrt{\alpha}}{\sqrt{\alpha}} \frac{\lambda_1-\lambda_2}{\sqrt{2(1-Q)}})^2} \tag{25}$$

As expected, one finds a Gaussian, shifted in the direction of $\mathbf{J}_+$ and widened in the direction of $\mathbf{J}_-$, see Fig. (4). Finally, it is found that AT condition involving V_+ is always fulfilled, while the part involving V_- predicts only marginal stability. Also, the agreement with simulations is very good . We conclude that the RS results are probably exact.

Q

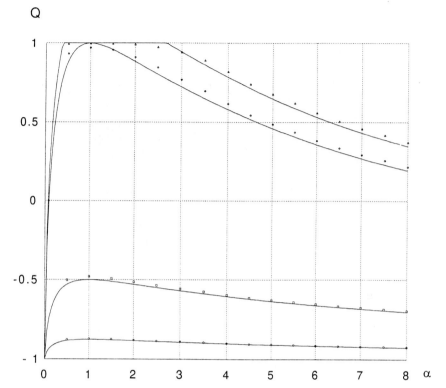

Fig. 5. The overlap Q as a function of α. RS results (full line, cf. Eq. (24)) simulations for N=400 (squares), for $\gamma = 1/4$ (lower curve), 1/8, 1/16 and 1/17 (upper curve)

6 Clustering of Non-Random Data Points

We will now investigate how the clustering algorithms introduced in the two previous sections perform on non-random data. For simplicity we focus on the following non-uniform probability density :

$$P(\boldsymbol{\xi}) = \frac{1}{2} \prod_{i=1}^{N} \frac{e^{-\frac{1}{2}(\xi_i - \frac{\rho B_i}{\sqrt{N}})^2}}{\sqrt{2\pi}} + \frac{1}{2} \prod_{i=1}^{N} \frac{e^{-\frac{1}{2}(\xi_i + \frac{\rho B_i}{\sqrt{N}})^2}}{\sqrt{2\pi}}, \tag{26}$$

where $\mathbf{B}$ is a vector whose length is normalized to $\sqrt{N}$. Since the input patterns have a preferential overlap equal to $+\rho$ or $-\rho$ with structure generating vector $\mathbf{B}$, we expect that the cluster axes will reflect this structure. Since the calculations and analytic expressions are relatively tedious, we only review the final results.

We first turn to competitive learning. It is found that there exists a critical α-value α_c below which $R_- = (\mathbf{J}_1 - \mathbf{J}_2)/N$ is identically zero and all the statis-

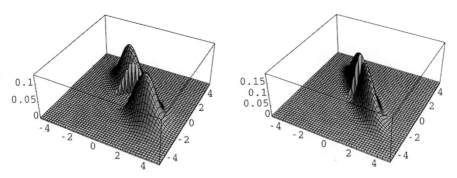

Fig. 6. The overlap distribution $P(\lambda_1, \lambda_2)$, for competitive learning applied to data sampled from Eq. (26), for $\alpha = 1$ (left) and $\alpha = 5$ (right) at $\rho = 1$ as predicted by 1RSB.

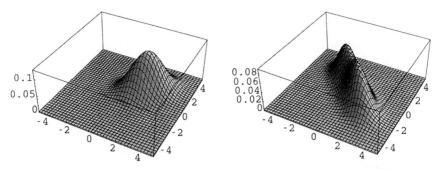

Fig. 7. The overlap distribution $P(\lambda_1, \lambda_2)$, , for maximal variance competitive learning applied to data sampled from Eq. (26), for $\alpha = 0.5$ (left) and $\alpha = 7$ (right) at $\rho = 1.3$ for $\gamma = 1/16$.

tical properties are identical to those for random patterns. In other words, the underlying non-random structure of the data points is not discovered at all! Within RS one finds $\alpha_c = \frac{\pi}{2\rho^4}$, whereas 1RSB predicts a somewhat smaller value $\alpha_c \approx 1.13/\rho^4$. For $\alpha > \alpha_c$, the overlap R_- ceases to be zero and in fact converges to 1 for $\alpha \to \infty$. An example of the overlap distribution below and above α_c is given in Fig. (6).

Turning to the alternative algorithm, maximal variance competitive learning, we find a similar situation. Below a critical α-value, which is however smaller, namely $\alpha_c = 1/\rho^4$, all the properties are identical to those for random points. Above α_c, the overlap R_- is different form zero and converges to 1 for $\alpha \to \infty$. An example of the overlap distribution below and above α_c is given in Fig. (7).

7 Discussion

We investigated, using techniques from statistical mechanics, how various clustering techniques reveal the properties of random and non-random data in high dimensional systems. One basic lesson is that one risks finding what one is searching for. For example, competitive learning, which is a often used algorithm, will reveal a reproducable and sharp cluster like distribution of data points, even when these points are generated independently from a uniform distribution. This structure inherent to typical realizations of random data can mask a genuine underlying structure. For the case we studied here, namely bi-gaussian pattern distributions, this masking is so efficient that up to a critical size of the data set no information about the underlying structure can be extracted. This phenomenon is not an exception, but has been proven to hold for a large class of pattern distributions [9], [10].

Bearing in mind this limitation, one can set out to find the "best" clustering algorithm, building on the detailed information provided by a statistical mechanical analysis. The algorithm that we introduced, namely maximal variance competitive learning, is an interesting option because : the distribution found for the case of random data is not misleading ; the cost function has a single minimum (no RSB + allows for easy and fast minimization) ; there is perfect detection for large α, and finally one can show that the algorithm achieves the smallest possible α_c.

References

[1] J. Hertz A. Krogh and R. G. Palmer. *Introduction to the Theory of Neural Computation*, volume 1 of *Lecture Notes Volume in the Santa Fe Institute Studies in the sciences of complexity*. Addison-Wesley, Santa Fe, 1 edition, 1991.

[2] M. Griniasty and H. Gutfreund. *J. Phys. A*, 24:715, 1990.

[3] E. Oja. *J. Math. Biology*, 15:267, 1992.

[4] M. Biehl and A. Mietzner. Statistical mechanics of unsupervised learning. *Europhys. Lett.*, 24:421, 1993.

[5] J.R.L. Almeida and D.J. Thouless. Stability of the sherrington-kirpatrick solution of a spin glass model. *J. Phys. A*, 11(5):983, 1978.

[6] M. Bouten. *J. Phys. A*, 27(17):6021, 1994.

[7] E. Lootens and C. Van den Broeck. *Europhys. Lett.*, 30(7):381, 1995.

[8] P. R. Krishnaiah and L. N. Kanal. *Classification Pattern Recognition and Reduction of Dimensionality, Handbook of Statistics*, volume 2. North Holland, Amsterdam, 1982.

[9] T. Watkin and J.P. Nadal. *J. Phys. A*, 27(6):1899, 1994.

[10] P. Reimann and C. Van den Broeck. *Phys. Rev. E*, 53:3989, 1996.

A Kinetic Description of Disorder

P.L. Garrido[1], J. Marro[1] and M.A. Muñoz[1,2]

1 Instituto Carlos I de Física Teórica y Computacional. Facultad de Ciencias. Universidad de Granada. 18071-Granada, España.

2 T. J. Watson IBM Research Center, P. O. Box 218, Yorktown Heights, New York 10598, U.S.A.

Abstract. Some strategies to study the kinetics of disorder in Ising-like systems are reviewed.

1. Introduction

It has been known for a long time that the behavior of a physical system may drastically be affected by the presence of impurities. In order to get some insight into the role played by impurities, defects or, in general, *disorder*, many different models have been proposed in the context of equilibrium statistical mechanics. One of the most straightforward ways to implement disorder into well-known pure equilibrium models, consist on modifying its interaction Hamiltonian, in such a way that its translation invariance symmetry is broken. This can be achieved by introducing local variables in the Hamiltonian that take random values from site to site. Being more specific, let us consider, for example, a situation in which the translation invariant pure system is a d-dimensional lattice with sites occupied by spin variables. Each spin configuration, $\underline{s} \equiv \{s_{\underline{x}} = \pm 1; \underline{x} \in Z^d\}$, has an associated energy given by the Hamiltonian:

$$H(\underline{s}; \underline{J}) = - \sum_{A \subset Z^d} J_A s_A \quad , \quad s_A \equiv \prod_{\underline{x} \in A} s_{\underline{x}} \quad , \qquad (hamil)$$

where $\underline{J} = \{J_A | A \subset Z^d\}$. The translation invariance property reflects in the fact that the Hamiltonian has the symmetry: $T_{\underline{z}} H(\underline{s}; \underline{J}) = H(T_{\underline{z}}\underline{s}; \underline{J}) = H(\underline{s}; \underline{J}) \quad \forall \underline{z} \in Z^d$, where the operator $T_{\underline{z}}$ displaces any configuration by $\underline{z}$: $\underline{s}' \equiv T_{\underline{z}}\underline{s} = \{s'_{\underline{x}} = s_{\underline{x}-\underline{z}}; \underline{x} \in Z^d\}$. This property, if applied to the Hamiltonian (hamil) implies that $J_{T_{\underline{x}}A} = J_A \quad \forall \underline{x}, A$.

Let us assume first that the set A in (hamil) consists of all possible different pairs of nearest-neighbor sites in a hypercubic lattice. The translation invariance property implies in this case that $J_A = J$ $\forall A$, and equation (hamil) reduces to the Ising Hamiltonian. On the other hand, when the translation invariance symmetry is broken, several models of disorder are recovered. For example:

(i) Assuming that, for every set A, J_A is a random variable distributed around zero, we get the *Edwards-Anderson model* (Edwards and Anderson 1975). This exhibits *frustration* (the spatial competition between positive and negative couplings, J_A, prevents that all the exchange interactions in eq.(hamil) are simultaneously minimized) that induces *rare* macroscopic behavior observed in a class of materials known as *spin glasses*; see, for instance, the review book by Fischer and Hertz (1991).

(ii) If J_A is a random variable taking only two different discrete values, J or 0 with probabilities p and $1 - p$, respectively, we get the *bond dilute Ising model* which describes the physics of impure magnetic systems; see, for instance, the review by Stinchcombe (1983). These two models assume that impurities are quenched in the lattice, *i.e.* , their kinetics is neglected.

A simple way to induce some kinetics of disorder in Ising-like models, is to consider the couplings J as thermal variables. In this case, there is time variation of the spatial distribution of J's that is determined by the need to reach equilibrium with the spin degrees of freedom. This is the *anneal Ising model* (Thorpe and Beeman 1976). However, impurities tend in this case to be strongly correlated (for instance, located at the interfaces below the critical point) which is not observed in general.

The above (equilibrium) models and some variations of them exhibit a rich macroscopic behavior, and their study has helped the understanding of many phenomena first observed in real materials. Nevertheless, in order to study the broad variety of natural phenomena which appears to be related with disorder, a less restrictive scenario needs to be considered. In particular, many real systems under consideration are open to the environment, namely, driven by some external non-Hamiltonian force and/or interacting with external subsystems. Such general conditions may induce changes with time of the disorder variables. In practice, however, one needs to pay both analitically and conceptually a high price for the consideration of kinetic disorder: steady *non-equilibrium* situations ensue due to added competition and extra randomness.

2. Definition of the basic model

We focus our attention here on the study of some non-equilibrium lattice models as defined via a master equation. Consider a system consisting of a lattice and spin, $\underline{s}$, and *disorder, $\underline{J}$*, degrees of freedom, which evolve stochastically according to a Markovian process (see, for instance, Garrido and Marro 1994). That is, the probability distribution, $\mu_t(\underline{s}, \underline{J})$, that the system has at time t the configuration $(\underline{s}, \underline{J})$ satisfies the master equation:

$$\partial_t \mu_t(\underline{s}, \underline{J}) = \left(L_{\underline{s}} + \Gamma L_{\underline{J}} \right) \mu_t(\underline{s}, \underline{J}) \quad , \tag{ME}$$

where

$$L_{\underline{s}} g(\underline{s}, \underline{J}) = \sum_x [c(\underline{s}^x \to \underline{s}|\underline{J}) g(\underline{s}^x, \underline{J}) - c(\underline{s} \to \underline{s}^x|\underline{J}) g(\underline{s}, \underline{J})] \tag{Ls}$$

$$L_{\underline{J}} g(\underline{s}, \underline{J}) = \sum_{\underline{J}'} [w(\underline{J}' \to \underline{J}) g(\underline{s}, \underline{J}') - w(\underline{J} \to \underline{J}') g(\underline{s}, \underline{J})] \quad . \tag{Lj}$$

Here, $g(\underline{s}, \underline{J})$ stands for an arbitrary function, and $\underline{s}^x$ is the configuration $\underline{s}$ with the spin at site x flipped, *i.e.* $s_x \to -s_x$. The Glauber operator, $L_{\underline{s}}$, describes stochastic spin flips with an associated transition probability per unit time (*rate*) $c(\underline{s} \to \underline{s}^x|\underline{J})$ for given $\underline{J}$. $L_{\underline{J}}$ induces stochastic changes on $\underline{J}$ with rate $w(\underline{J} \to \underline{J}')$, and it is assumed to be independent of the spin configuration (further possibilities have been studied in Garrido and Marro 1994).

Both kinetic and stationary properties of the model depend in general on the rate. However, if dynamics of $\underline{J}$ is suppressed, which corresponds to the *quenched* case, $\Gamma = 0$, and the rates $c(\underline{s} \to \underline{s}^x|\underline{J})$ in eq.(Ls) satisfy the *detailed balance property*, namely,

$$c(\underline{s} \to \underline{s}^x|\underline{J}) = c(\underline{s}^x \to \underline{s}|\underline{J}) \exp\left[-\Delta H_{\underline{x}}\right], \ \Delta H_{\underline{x}} \equiv H(\underline{s}^x, \underline{J}) - H(\underline{s}, \underline{J}), \tag{DB}$$

the stationary state corresponds to an equilibrium Gibbsian distribution, characterized by the Hamiltonian (hamil), *i.e.* , $\mu_{st}(\underline{s}, \underline{J}) \propto \exp[-H(\underline{s}, \underline{J})]$. Let us notice that, in this case, the detailed balance property (DB) implies that the stationary state does not depend on the particular choice of $c(\underline{s} \rightarrow \underline{s}^{\pm}|\underline{J})$. A simple, commonly considered choice, satisfying Eq. (DB), is

$$c(\underline{s} \rightarrow \underline{s}^{\pm}|\underline{J}) = \Phi(\Delta H_{\underline{r}}) \qquad\qquad (rate)$$

where Φ is an arbitrary function with the property $\Phi(\lambda) = \Phi(-\lambda)\exp(-\lambda) \geq 0$. In particular, some specific examples are: $\Phi(\lambda) = 1 - \tanh(\lambda/2)$, $\min(1, \exp(-\lambda))$ and $\exp(-\lambda/2)$.

For $\Gamma > 0$, (ME) induces kinetics of the disorder degrees of freedom. In this case, the simultaneous action of the two kinetic mechanism in (ME), makes that, *a priori*, the steady state is not Gibbsian, and, contrary to what happens in the equilibrium, there is strong dependence on the analytical form of the dynamical mechanisms (rates).

There is not a general theory, analogous to the equilibrium ensembles theory, which relates in a simple and systematic way the microscopic structure of non-equilibrium systems and their macroscopic properties. The lack of a general theory makes the theoretical analysis of such systems to be quite complex, and usually approximate schemes are required to get some insight into their macroscopic behavior.

Under this circumstances, there are two different alternative approaches to deal with this kind of systems. On one hand, sometimes it is possible to find the exact solution of the probabilistic model by using some kind of statistical analysis (specially in dimension $d = 1$) or mapping the system somehow into an equilibrium one with effective parameters.

On the other hand, approximate schemes can be used. In particular, mean-field type approximations can be constructed. The problem with that sort of approach is that it is mainly based on the uncontrolled truncations of local system correlations. It is common to use numerical simulations together with mean field approaches to test one each other in order to get a comprehensive understanding of the system-macroscopic-behavior . An alternative type of approach is based in the representation of the original discrete Master equation, in terms of some continuous stochastic or partial differential equation. For the latter equation further analytical methods are available.

In the next two section we discuss some recent results in the two previously mentioned directions: namely, the derivation of exact results, and the construction of continuous descriptions of microscopic non-equilibrium models.

3. Some exact results: models with effective Hamiltonian

Let us consider the limiting case: $\Gamma \to \infty$. This limit represent physical situations in which the time scale for the evolution of the impurities is much faster that the characteristic time scale for the spin evolution. In other words, during the interval elapsed between two consecutive spin-flip processes, the disorder degrees of freedom undergo enough changes to assure that their associated probability distribution, $p(\underline{J})$, reaches its steady state,

$$L_{\underline{J}} p_{st}(\underline{J}) = 0. \qquad (ste)$$

One can easily argue (see, for instance Garrido and Marro 1994), that in this case, the spin probability distribution, $\rho_t(\underline{s}) = \sum_{\underline{J}} \mu_t(\underline{s}, \underline{J})$, is solution of the effective Master equation

$$\partial_t \rho_t(\underline{s}) = L_{\underline{s}}^{eff} \rho_t(\underline{s}) \qquad (MEef)$$

where

$$L_{\underline{s}}^{eff} = \sum_{\underline{J}} p_{st}(\underline{J}) L_{\underline{s}} \qquad (Leff)$$

is an effective Glauber spin flip operator similar to $L_{\underline{s}}$ in eq. (Ls).

Assuming that $c(\underline{s} \to \underline{s}^x | \underline{J})$ has the equilibrium analytical form given by equation (rate), the effective rate associated with the Glauber operator $L_{\underline{s}}^{eff}$ can be written as

$$c_{eff}(\underline{s} \to \underline{s}^x) = \sum_{\underline{J}} p_{st}(\underline{J}) \Phi(\Delta H_{\underline{x}}(\underline{J})). \qquad (ceff)$$

That is, an effective dynamics has been defined which is an stochastic superposition of different mechanisms, $\Phi(\Delta H_{\underline{x}}(\underline{J}))$, weighted with $p_{st}(\underline{J})$. Each of the mechanisms, acting by itself, would drive the system to a different equilibrium state, and the competition of many of them introduces a kind of *dynamical frustration*, which drives the system in general to a non-equilibrium stationary state.

The exact stationary solution of the master equation (MEef) can be found in many cases by working out whether the effective spin flip rate (ceff) holds the detailed balance property (DB) with respect some *effective Hamiltonian* (Garrido and Marro 1989). The disorder distribution, the spatial dimension and the structure of the competing Hamiltonian determine the existence or non-existence of such an effective Hamiltonian [3].

To explicitly show that influence, let us consider as an specific example, the reference Hamiltonian (that is, the Hamiltonian used to define the competing rates (ceff)) to be the Ising one,

$$H(\underline{s}, J) = -J \sum_{|\underline{x}-\underline{y}|=1} s_{\underline{x}} s_{\underline{y}} \qquad (IS)$$

where the coupling J is a random variable with a given stationary distribution $p_{st}(J)$.

¿From the detailed balance property (DB) one can show in the one dimensional case that the stationary distribution associated to the effective dynamics, defined by (MEef) and (Leff), is a Gibbsian distribution with an Ising Hamiltonian, that is

$$\rho_{st}(\underline{s}) \propto \exp\left[-H(\underline{s}, J_{eff})\right] \qquad (roef)$$

where

$$J_{eff} = \frac{1}{4} \ln \left[\frac{\ll \Phi(-4J) \gg}{\ll \Phi(4J) \gg} \right] \qquad (keff)$$

and $\ll g(J) \gg \equiv \sum_J p_{st}(J) g(J)$ stands for the average of a given function $g(J)$ over the J stationary distribution.

In this way, the competition of different equilibrium rates [4] drives the system to a Gibbsian state with the same reference Ising Hamiltonian but with an effective coupling parameter. Such effective coupling depends on the analytical form of the competing rates and on the disorder stationary state as can be concluded from (keff).

This simple picture is specific of the one-dimensional model and cannot be extended to higher dimensions (there are, however, some Monte Carlo

---

[3] Notice that even in cases in which a the detailed balance property does not hold, it is possible to have an stationary Gibbsian measure. A well known example is given by the rate $c(\underline{s} \to \underline{s}^{\underline{x}}) = \exp[-s_{\underline{x}}(s_{\underline{x}+\underline{i}}+s_{\underline{x}+\underline{j}})]$ defined on a two dimensional square lattice with periodic boundary conditions. $\underline{i}$ and $\underline{j}$ are the unit lattice vectors in the X and Y axis directions respectively. This rate has not the detailed balance property but the master equation stationary solution is a Gibbsian measure with Ising Hamiltonian $H = -\sum_{NN} s_{\underline{x}} s_{\underline{y}}$.

[4] In the sense that each one acting alone drive the system to a Gibbsian state characterized by the one dimensional Ising Hamiltonian with *different* coupling.

computer simulations performed in the two-dimensional version of the previous model which show that the stationary solution is an equilibrium one with effective parameters, except at low temperatures (González-Miranda et al. 1994)).

To show the lack of robustness of the effective detailed balance property, even in one dimension, we point out that, there is no effective-Hamiltonian description when a fixed magnetic field is included in the Ising Hamiltonian (IS).

Nevertheless, there exists one particular case in which we are able to find an effective Hamiltonian for any dimension. It corresponds to: *1)* the particular function $\Phi(\lambda) = \exp(-\lambda/2)$ is considered in (ceff) *2)* the couplings J_A of the Hamiltonian (hamil) are stochastically independent variables, which evolve uncorrelately. In other words, $p_{st}(\underline{J}) = \prod_{A \subset Z^d} p_A(J_A)$. The effective rate (ceff) is then written

$$c_{eff}(\underline{s} \rightarrow \underline{s}^x) = \prod_{B \cap \{\underline{x}\}} \left[\sum_{J_B} p_B(J_B) \exp\left(-J_B s_B\right) \right] \qquad (dd1)$$

In this situation one may show that the stationary state in any dimension is a Gibbsian one $\rho_{st}(\underline{s}) \propto \exp(-H(\underline{s}; \underline{J}^{eff}))$ (see for instance reference Garrido and Muñoz 1993), where

$$J_A^{eff} = \frac{1}{2} \ln \left[\frac{\ll \exp(J_A) \gg}{\ll \exp(-J_A) \gg} \right] \qquad (JAeff)$$

Let us emphasize that we are not able to find an effective Hamiltonian when another analytical form of Φ is considered in (ceff), or if it is assumed that J_A are correlated variables.

As a particular realization of equation (JAeff) let us choose the reference Hamiltonian to be the general d-dimensional Ising Hamiltonian,

$$H(\underline{s}, \underline{J}) = - \sum_{|\underline{x} - \underline{y}| = 1} J_{\underline{x}\underline{y}} s_{\underline{x}} s_{\underline{y}} \qquad (GIM)$$

where now the set B in equation (dd1) denotes pairs of nearest-neighbor sites in the lattice. For any disorder distribution of the form

$$p_{\underline{x}\underline{y}}(J) = f(J - J_{\underline{x}\underline{y}}^0) \qquad (dist)$$

with $f(J)$ being any probability distribution symmetric around zero, and $\{J^0_{\underline{xy}}\}$ being a given set of couplings, it is straightforward to show that an effective Ising Hamiltonian exits with effective couplings given by: $J^{eff}_{\underline{xy}} = J^0_{\underline{xy}}$ *independently of any parameter of the distribution* f (for instance, if f is a Gaussian distribution with variance σ, the result will be σ-independent).

Other similar realizations of (JAeff) can easily be worked out (Garrido and Muñoz 1993). For instance, the *non-equilibrium impure Ising model* corresponding to $p_{\underline{xy}}(J) = p\delta(J - J^0) + (1-p)\delta(J)$, or the *non-equilibrium spin glass Ising model* defined by $p_{\underline{xy}}(J) = p\delta(J - J^0) + (1-p)\delta(J + J^0)$.

Let us finally mention a case in which we relax the aforementioned property, *2)*, in which we assumed that the disorder was totally uncorrelated, and we still have an effective Hamiltonian. This occurs in the so-called *kinetic ANNNI* model (López-Lacomba and Marro 1994). In this case, the Hamiltonian that defines the effective rate (dd1) is

$$H(\underline{s}; J, J') = -J \sum_{|\underline{x}-\underline{y}|=1} s_{\underline{x}} s_{\underline{y}} - J' \sum_{\underline{x}} s_{\underline{x}} s_{\underline{x}+2\underline{z}} \qquad (ANNI)$$

where $\underline{z}$ is the unit vector pointing to one of the lattice directions. Namely, the model is defined on a d-dimensional Ising model in which a next-nearest-neighbor interaction with strength J' is added in one of the lattice directions. It is assumed that the coupling J is fixed and J' is a random variable, that is, p_{st} depends only on J'. The effective Hamiltonian is found to be $H(\underline{s}; J, J'_{eff})$ with

$$J'_{eff} = \frac{1}{2} \ln \left[\frac{\ll \exp(J') \gg}{\ll \exp(-J') \gg} \right] \qquad (Jpe)$$

All these models have interesting macroscopic behaviors which have been studied in the above commented references where we refer the reader for more details.

4. The continuum description: looking for a suitable starting point

This is a more broadly applicable method. It is suitable for the study of phase transitions and critical phenomena appearing in non-equilibrium Master equations. It is based on the observation that the strategies used in the study of the critical dynamic properties of systems evolving towards an equilibrium state can be extended to the analysis of non-equilibrium dynamical models. The idea is to construct a continuum version of the lattice model whose equation of motion is a stochastic differential equation (Langevin equation) [5].

The Langevin equation is a simplified representation of the microscopic system that contains the most relevant features to describe properties that depend on large scales in space and time. As the nature of critical phenomena is usually determined only by large-scale properties and not by specific microscopic details of the models, this approach is a natural way to study phase transitions in both equilibrium and non-equilibrium situations.

In principle, such continuum description should be derived from the microscopic Master Equation by means of a coarse-graining procedure. That is, changing in the original Master equation the spin variables by some new local variables defined as averages of the spins over a large region. One expects that, in an adequate limit, the microscopic details are averaged away and in the resulting Langevin description only the large scale properties remain. However, the coarse-graining procedure cannot be, in general, applied successfully without introducing some extra assumptions [6]. Nevertheless, for systems evolving towards an equilibrium distribution, it is used a very simple method to construct a Langevin equation. First, one knows that the system stationary distribution is given by a Gibbsian measure, $\rho \propto \exp[-H]$ for a

---

[5] We indistinctly will use the Langevin or the Fokker-Planck equations. Both are stochastically equivalent. The first one is a stochastic differential equation where the unknowns are the local fields. The Fokker-Planck is a second order functional differential equation in which the unknown is the probability to find the systems at a given time with a given configuration.

[6] A. de Masi, P. Ferrari and J.L. Lebowitz managed to make in a rigorous manner a coarse-graining procedure in a reaction diffusion model (de Masi et al. 1985) in which the diffusion process is infinitely faster than the reaction one.

given interaction Hamiltonian H. And second, one expects that the critical phenomena will be independent on the particular analytical form of the dynamics. That is, one should write down a Langevin equation such that it is guaranteed that the stationary state is the one given by the Gibbsian measure. That is the strategy followed by Hohenberg and Halperin (1977) in their, so called, model A, in order to describe the critical dynamics of systems with a non-conserved order parameter. The equation defining the model A is:

$$\partial_t \varphi_t(\underline{r}) = -\lambda \frac{\delta H}{\delta \varphi_t(\underline{r})} + \eta_t(\underline{r}) \qquad (HH)$$

where $\varphi_t(\underline{r})$ is the coarse-graining density field at a given time t on the point $\underline{r}$ of the continuum space, and η is a stochastic white noise that reflects the fluctuations on the density field due to the action of the microscopic dynamics. Equation (HH) is the starting point for the study of large scale dynamic properties in equilibrium systems.

Coming back to our non-equilibrium problem, we have obviously the same aforementioned technical problems in the coarse-graining procedure. Moreover, any extra assumption may now change dramatically the system macroscopic behavior. Finally, we cannot apply the Hohenberg-Halperin strategy because *we don't know what is the measure corresponding to the stationary state*, which is coherent with the fact that such state depends on the microscopic dynamics. Nevertheless, it is possible to apply the idea of constructing a Langevin equation such that it is guaranteed, without solving the equations, that its stationary distribution is equal to the exact one, solution of the non-equilibrium Master equation, at least in some basic aspects. To achieve that, it is necessary to introduce a suitable continuum version of our Master Equation (MEef).

Let us introduce now the continuum version of the non-equilibrium models we are going to deal with. The system consists of a d-dimensional lattice where at each site there is a spin-like variable. We define at each point, $\underline{r} \in R^d$, a field variable, $\varphi(\underline{r}) \in R$, which is the averaged value of the spins on a region of volume Ω around $\underline{r}$. When Ω is large enough, φ is assumed to be a continuous function on $\underline{r}$. The probability to find a field configuration, $\underline{\varphi} = \{\varphi(\underline{r}); \underline{r} \in R^d\}$, at time t, say $P_t^\Omega(\underline{\varphi})$, evolves according to a Markovian Master equation:

$$\partial_t P_t^\Omega(\underline{\varphi}) = \int_{R^d} d\underline{r} \int_R d\eta f(\eta) \left[w^\Omega(\underline{\varphi}^{\eta,\underline{r}} \to \underline{\varphi}) P_t^\Omega(\underline{\varphi}^{\eta,\underline{r}}) \right.$$
$$\left. - w^\Omega(\underline{\varphi} \to \underline{\varphi}^{\eta,\underline{r}}) P_t^\Omega(\underline{\varphi}) \right], \qquad (MME2)$$

where $f(\eta)$, an even and analytical real function around the origin, stands for the field increments distribution, $\underline{\varphi}^{\eta,\underline{r}} = \{\varphi(\underline{r}') + \frac{\eta}{\Omega}\delta_{\underline{r},\underline{r}'}, \underline{r}' \in R^d\}$, and $w^\Omega(\underline{\varphi} \to \underline{\varphi}')$ represents the transition probabilities per unit time. These are defined as in eq.(ceff):

$$w^\Omega(\underline{\varphi} \to \underline{\varphi}') = \int_{R^d} d\underline{K} \ p_{st}(\underline{K}) w^\Omega(\underline{\varphi} \to \underline{\varphi}'; \underline{K}), \qquad (rate2)$$

where $w^\Omega(\underline{\varphi} \to \underline{\varphi}'; \underline{K}) = \Phi[H^\Omega(\underline{\varphi}'; \underline{K}) - H^\Omega(\underline{\varphi}; \underline{K})]$, and now $H^\Omega(\underline{\varphi}; \underline{K}) = \Omega \int_{R^d} d\underline{r} \ h(\underline{\varphi}(\underline{r}); \underline{K})$ is a continuum interaction Hamiltonian which depends on the parameters $\underline{K} = \{K_i, i = 1, .., n\}$.

When we rescale the time variable $\tau = \Omega^{-1}t$, and we do the limit $\Omega \to \infty$, the solution of the Master Equation (MME2) is $P_\tau(\underline{\varphi}) = \delta(\underline{\varphi} - \underline{v}_\tau)$ where $\underline{v}_\tau$ is the solution of the so called *deterministic equation*:

$$\partial_\tau v_\tau(\underline{r}) = -\Xi_0^{exact}(\underline{v}_t(\underline{r})) \equiv \int_R d\eta f(\eta)\eta \ll \Phi(U_\eta(\underline{v}_t(\underline{r}); \underline{K})) \gg, \qquad (det2)$$

being $U_\eta(\underline{\varphi}(\underline{r}); \underline{K}) = \eta \frac{\delta \hat{H}(\underline{\varphi}; \underline{K})}{\delta\varphi(\underline{r})}$, and $\ll A \gg \equiv \int_{R^n} d\underline{K} \ p_{st}(\underline{K}) A(\underline{K})$. In general, for large enough Ω, the stationary probability distribution solution of the Master Equation (MME2) can be written as: $P_{st}^\Omega(\underline{\varphi}) \propto \exp[-V_{st}^\Omega(\underline{\varphi})]$ where $V_{st}^\Omega(\underline{\varphi}) = \Omega V_{0,st}(\underline{\varphi}) + V_{1,st}(\underline{\varphi}) + O(\Omega^{-1})$. The non-equilibrium potential $V_{st}^\Omega(\underline{\varphi})$ is expected to be continuous but not differentiable in some small regions in the phase space (see for instance Graham and Tel, 1984, 1985, 1986) and it can be shown that it is a Lyapunov function for the underlying deterministic dynamical system (Jauslin 1987).

In this conditions, the following theorem can be proven (Garrido and Muñoz 1995; Muñoz and Garrido 1994):

Theorem: Let any Fokker-Planck equation

$$\partial_t P_t^\Omega(\underline{\varphi}) = \frac{1}{\Omega} \int_{R^d} d\underline{r} \frac{\delta}{\delta\varphi(\underline{r})} \left[\Xi_0(\underline{\varphi}(\underline{r})) + \frac{1}{\Omega}\Xi_1(\underline{\varphi}(\underline{r})) \frac{\delta}{\delta\varphi(\underline{r})} \right] P_t^\Omega(\underline{\varphi}), \qquad (FP)$$

such that its coefficients have the form

$$\Xi_0(\underline{\varphi}(\underline{r})) = \frac{1}{2}\left[D_-(\underline{\varphi}(\underline{r})) - D_+(\underline{\varphi}(\underline{r}))\right]$$

$$\Xi_1(\underline{\varphi}(\underline{r})) = \frac{D_-(\underline{\varphi}(\underline{r})) - D_+(\underline{\varphi}(\underline{r}))}{2\ln\left[\frac{D_-(\underline{\varphi}(\underline{r}))}{D_+(\underline{\varphi}(\underline{r}))}\right]} \tag{Th3}$$

where $D_\eta(\underline{\varphi}(\underline{r})) = \ll \Phi(U_\eta(\underline{\varphi}(\underline{r}); \underline{K})) \gg$. When $f(\eta) = \frac{1}{2}\Big[\delta(\eta - 1) + \delta(\eta + 1)\Big]$, then

i) It reproduces the exact deterministic dynamics given by eq.(det2).

ii) The $V_{0,st}$ and $V_{1,st}$ parts of its stationary solution almost coincide with the exact one in a suitable neighborhood of all spatially homogeneous deterministic solutions $\underline{v}^*$.

Using this theorem, it is possible to find a Fokker-Planck type of description for the competing-dynamics models which represent exactly the original stationary distribution in some regions of phase space. These regions are very relevant because they determine the stationary critical behavior of the system. With this description as starting point it is feasible to perform suitable analysis of the critical properties by using well known methods developed for the study of Fokker-Planck and Langevin equations (Jansen et al 1976).

Summarizing, we have defined a general model that includes a kinetic mechanism to describe the disorder (impurities) time evolution. That induces the associated stationary probability distribution to be a non-equilibrium one. The theoretical tools available to study such non-equilibrium systems are scarce and not too powerful. Nevertheless we have shown that one can get interesting results in the particular limit in which the disorder evolves in a time scale much shorter that the spins. In such situation, and for a particular analytical form of the microscopic dynamics, we have shown that one can find the exact stationary distribution which is in fact a Gibbsian one characterized by an effective Hamiltonian. However, the latter particular results cannot be generalize for other rates and one should go to a much simpler mesoscopic continuous descriptions to get some valid information about the system behavior. In that context, we have shown that it is possible to explicitly write down a Fokker Planck equation whose stationary state coincides with the exact one in some relevant parts of the phase space.

References

de Masi, A., Ferrari, P. and Lebowitz, J.L. (1985): Phys. Rev. Lett. **55**, 1947.

Edwards, S.F. and Anderson, P.W. (1975): J. of Phys. F, **5** 965; see also Sherrington, D. and Kirpatrick, S. (1975): Phys. Rev. Lett. **35** 1972.

Fischer, K.H. and Hertz, J.A. (1991): *Spin Glasses* (Cambridge Lectures in Magnetism 1, Cambridge University Press, Cambridge).

Garrido, P.L. and Marro, J. (1989): Phys. Rev. Lett. **62** 1929.

Garrido, P.L. and Marro, J. (1994): J. Stat. Phys. **74** 663.

Garrido, P.L. and Muñoz, M.A. (1993): Phys. Rev. E **48** R4153.

Garrido, P.L. and Muñoz, M.A. (1995): Phys. Rev. Lett. **75** 1875.

González-Miranda, J.M., Labarta, A., Puma, M., Fernández, J.F., Garrido, P.L. and Marro, J. (1994): Phys. Rev. E **49** 2041.

Graham, R. and Tél, T (1984): J. Stat. Phys. **35** 729; (1985) Phys. Rev. A **31**, 1109; (1986) **33**, 1322.

Hohenberg, P.C. and Halperin, B.J. (1977): Rev. Mod. Phys. **49**, 435.

Jauslin, H.R. (1987): Physica **144A**, 179.

López-Lacomba, A.I. and Marro, J. (1994): J. of Phys. A **27** 1111.

Muñoz, M.A. and Garrido, P.L. (1994): Phys. Rev. E **50** 2458.

Stinchcombe, R.B. (1983): *Phase Transitions and Critical Phenomena* (C. Domb and J.L. Lebowitz, vol. 7 p. 151, Academic Press, London).

Thorpe, M.F. and Beeman, D. (1976): Phys. Rev. B **14** 188.

Janssen, H.K. (1976): Z. Phys. **B23**, 377; Bausch, R., Janssen, H.K. and Wagner, H. (1976): Z. Phys. **B24**, 113; Zinn-Justin, J. (1989): *Quantum Field Theory and Critical Phenomena*, (Oxford Science).

LIST OF POSTERS

M.A. Montagna and M. Ferrari, *Size determination of nanoclusters in glasses by low frequency Raman spectroscopy of their acoustic vibrations.*

P. Molinàs, M.A. Muñoz, D.O Martínez and A. L. Barabasi, *Ballistic random walker.*

R.J.P. Keijsers, O.I. Shklyarevskii and H. van Kampen, *Point-contact studies of metallic glasses: fast and slow two-level fluctuators.*

N. Israeloff, *What is a glass?: The 1/f fluctuation connection.*

I. Echevarría, S.L. Simon, D.J. Plazek, P.C. Su and E. McGregor, *Study of time dependent properties of a polyetherimide.*

K. Wiesenfeld, P. Colet and S.H. Strogatz, *Synchronization transitions in a disordered Joshepson series array.*

J.V. Alvarez and F. Ritort, *Critical and dynamical behavior of the infinite-ranged Ising spin glass.*

T. Aste, *Amorphous packing of atoms.*

J. C. Ciria and C. Giovanella, *Order-disorder crossover in Joshepson junction arrays with screening.*

A.V. Smirnov and A.M. Bratkovsky, *Topological disorder and magnetism in Al-Mn system.*

J.J. Mazo, F. Falo and L.M. Floría, *Slow relaxation in a frustrated ordered Joshepson junction ladder.*

J.J. Ruiz-Lorenzo, *Simulation of 3-d Ising sping glass model using three replicas: study of Binder cumulants.*

S. Bravo Yuste, M. López de Haro and A. Santos, *Structure of hard-sphere metastable fluids.*

J.J. Arenzon, M. Nicodemi and M. Sellito, *Equilibrium properties of the ising frustrated lattice gas.*

D.A. Stariolo, *Heisenberg spin glass on a hypercubic cell.*

B. Tadic, *Self-organization and nonequilibrium phase transitions in complex systems.*

U. C. Täuber and D.R. Nelson, *Superfluid bosons and flux liquids: disorder and finite-temperature/finite-size corrections.*

C. Wengel and A.P. Young, *Monte Carlo study of the three-dimensional vortex glass mode with screening.*

M.A. Martín Delgado, *Snakes and ladders.*

J.P. Rodriguez, *Random frustration in the 2D s=1/2 Heisenberg antiferromagnet.*

L.L. Bonilla, F.G. Padilla, G. Parisi and F. Ritort, *Closure of the Monte carlo dynamical equations in the spherical Sherrington-Kirkpatrick model.*

C.J. Pérez Vicente, A. Corral, A. Díaz-Guilera and A. Arenas, *Quenched disorder induces self-organized criticality.*

K. Christensen, A. Corral, V. Frette, J. Feder and T. Jossang, *Transit time distribution in a critical granular system.*

T. Wasiutyinsky, *Calorimetric study of glassy transitions in cyclooctanol.*

LIST OF PARTICIPANTS

V. ALVAREZ.   UNIVERSIDAD CARLOS III (SPAIN)
PROF. C.A. ANGELL.   ARIZONA STATE UNIVERSITY (USA)
DR. A. ARENAS   URV (SPAIN)
DR. J.J ARENZON.   UNIVERSITY OF NAPLES (ITALY)
DR. T. ASTE.   UNIV. LOUIS PASTEUR (FRANCE)
DR. V.AZCOITI CERN (SWITZERLAND)
DR. J. BAFALUY UAB (SPAIN)
PROF. J. BERMEJO.   CSIC (SPAIN)
PROF. K. BINDER.   UNIVERSITY OF MAINZ (GERMANY)
DR. F. BADIA. UNIVERSIDAD DE BARCELONA (SPAIN)
DR. J. BONET. UNIVERSIDAD DE BARCELONA (SPAIN)
DR. S. BRAVO.   UNIVERSIDAD DE EXTREMADURA (SPAIN)
DR. J. CAMACHO UAB (SPAIN)
PROF. J. CARDY.   UNIVERSITY OF OXFORD (UNITED KINGDOM)
DR. P. CHANDRA.   NEC RESEARCH INSTITUTE (USA)
PROF. E.M. CHUDNOWSKY.   LEHMAN COLLEGE, CUNY (USA)
DR. J. CIRIA.   UNIVERSITY OF ROME I (ITALY)
DR. P. COLET.   CSIC (SPAIN)
A. CORRAL UNIVERSITY OF BARCELONA (SPAIN)
PROF. C. DASGUPTA.   INDIAN INSTITUTE OF PHYSICS (INDIA)
R. DOMINGUEZ UAB (SPAIN)
I. ECHEVERRIA.   UNED (SPAIN)
DR. E. ENCISO.   UNIV. COMPLUTENSE MADRID (SPAIN)
DR. J. FERNANDEZ. CSIC (SPAIN)
PROF. E.W. FISCHER.   MAX PLANCK INSTITUTE (GERMANY)
J. FARAUDO UAB (SPAIN)
E. FOLLANA.   UNIVERSITY OF ZARAGOZA (SPAIN)
PROF. A. FONTANA.   UNIVERSITY OF TRENTO (ITALY)
PROF. J. FONTCUBERTA. ICM-CSIC (SPAIN)
DR. S. FRANZ.   ICTP TRIESTE (ITALY)
DR. A. GARCIA. UNIVERSITY OF BARCELONA (SPAIN)
DR. M.A. GARCIA BACH   UNIVERSITY OF BARCELONA (SPAIN)
DR. P. GARCIA FERNANDEZ. CSIC (SPAIN)
DR. P.L. GARRIDO.   UNIVERSITY OF GRANADA (SPAIN)
PROF. L. GARRIDO.   UNIVERSITY OF BARCELONA (SPAIN)
PROF. M. GIURA.   UNIVERSITY OF ROME I (ITALY)
G. GOMILA UNIVERSITY OF BARCELONA (SPAIN)
DR. J.M. GONZALEZ. UNIVERSITY OF BARCELONA (SPAIN)
DR. N. ISRAELOFF.   NORTHEASTERN UNIV. BOSTON (USA)
DR. S. JAIN.   UNIVERSITY OF DERBY (UNITED KINGDOM)
DR. B. JEROME.   FOM AMSTERDAM (THE NETHERLANDS)

PROF. D. JOU UAB (SPAIN)
PROF. M. KARDAR. MIT (USA)
R.J.P. KEIJSERS. KUN (THE NETHERLANDS)
PROF. T. KIRKPATRICK. UNIVERSITY OF MARYLAND (USA)
DR. W. KOB. UNIVERSITY OF MAINZ (GERMANY)
DR. R. KUHN. UNIVERSITY OF HEIDELBERG (GERMANY)
DR. A. LABARTA UNIVERSITY OF BARCELONA (SPAIN)
M. LEHRER. HARVARD UNIVERSITY (USA)
PROF. E. MARINARI. UNIVERSITY OF CAGLIARI (ITALY)
DR. M.A. MARTIN. UNIV. COMPLUTENSE MADRID (SPAIN)
DR. B. MARTINEZ. ICM- CSIC (SPAIN)
J.J. MAZO. UNIVERSITY OF ZARAGOZA (SPAIN)
PROF. M. MEZARD. ECOLE NORMALE SUPERIEURE (FRANCE)
DR. A.A MIDDLETON. SYRACUSE UNIVERSITY (USA)
DR. C. MIGUEL MIT (USA)
DR. P. MOLINAS. UNIVERSITY PARIS-SUD (FRANCE)
PROF. M. MONTAGNA. UNIVERSITY OF TRENTO (ITALY)
DR. C. MORKEL. TU MUNICH (GERMANY)
PROF. D. NELSON. HARVARD UNIVERSITY (USA)
DR. M. NEY-NIFLE. ECOLE NORMALE SUPERIEURE (FRANCE)
M. NICODEMI. UNIVERSITY OF NAPLES (ITALY)
DR. T.NIEUWENHUIZEN.   UNIVERSITY OF AMSTERDAM (THE
NETHERLANDS)
PROF. X. OBRADORS ICM-CSIC (SPAIN)
PROF. R. OPPERMANN. UNIVERSITY OF WURZBURG (GERMANY)
F. J. PADILLA. UNIVERSIDAD CARLOS III (SPAIN)
PROF. J.A. PADRO. UNIVERSITY OF BARCELONA (SPAIN)
DR. I. PAGONABARRAGA. UNIVERSITY OF AMSTERDAM (SPAIN)
PROF. G. PARISI. UNIVERSITY OF ROME I (ITALY)
DR. A. PEREZ-MADRID UNIVERSITY OF BARCELONA (SPAIN)
DR. C.J. PEREZ-VICENTE UNIVERSITY OF BARCELONA (SPAIN)
DR. J.J. PEREZ. UPC-ETSII (SPAIN)
DR. R. PASTOR UNIVERSITY OF BARCELONA (SPAIN)
DR. M. PICCO. LPTHE, UNIV. PARIS 6 (FRANCE)
PROF. A. PLANES UNIVERSITY OF BARCELONA (SPAIN)
PROF. L. PIETRONERO. UNIVERSITY OF ROME I (ITALY)
DR. H. RIEGER. HLRZ JULICH (GERMANY)
DR. F. RITORT. UNIVERSITY OF AMSTERDAM (THE NETHERLANDS)
A. ROBINSON UNIVERSITY OF BARCELONA (SPAIN)
DR. J. RODRIGUEZ. UNIVERSIDAD COMPLUTENSE MADRID (SPAIN)
DR. V. ROMERO UNAM (MEJICO)

PROF. J.M. RUBI   UNIVERSITY OF BARCELONA (SPAIN)
DR. J.J. RUIZ.   UNIVERSITY OF ROME I (ITALY)
DR. M. J. RUIZ.   UNIVERSITY OF SEVILLA (SPAIN)
DR. C. SALUEÑA. UNIVERSITY OF BARCELONA (SPAIN)
PROF. J.M. SANCHO   UNIVERSITY OF BARCELONA (SPAIN)
DR. S. SARTI.   UNIVERSITY OF ROME I (ITALY)
DR. H. SCHOBER.   ILL GRENOBLE (FRANCE)
DR. F. SCIORTINO.   UNIVERSITY OF ROME I (ITALY)
DR. J.S. SELLAR.   MONASH UNIVERSITY (AUSTRALIA)
M. SELLITO.   UNIVERSITY OF NAPLES (ITALY)
DR. A. SMIRNOV.   TH DARMSTADT (GERMANY)
DR. D.A. STARIOLO.   UNIVERSITY OF ROME I (ITALY)
DR. S. STEPANOW.   UNIVERSITY OF HALLE (GERMANY)
DR. A. SUAREZ.   ULB (BELGIUM)
DR. B. TADIC.   J. STEFAN INSTITUTE (SLOVENIA)
PROF. P. TARTAGLIA.   UNIVERSITY OF ROME I (ITALY)
DR. U. TAUBER.   UNIVERSITY OF OXFORD (UNITED KINGDOM)
PROF. S. TEITEL.   ROCHESTER UNIVERSITY (USA)
PROF. J. TEJADA   UNIVERSITY OF BARCELONA (SPAIN)
DR. O. TERZIDIS.   UNIVERSITY OF HEIDELBERG (GERMANY)
PROF. C. VAN DEN BROECK.   LIMBURGS UNIVERSITY (BELGIUM)
PROF. J.L. VICENT.   UNIV. COMPLUTENSE MADRID (SPAIN)
DR. T. VILGIS.   MAX PLANCK INSTITUTE (GERMANY)
J.M. VILAR UNIVERSITY OF BARCELONA (SPAIN)
PROF. E. VINCENT.   CEA SACLAY (FRANCE)
DR. E. VIVES. UNIVERSITY OF BARCELONA (SPAIN)
DR. U. YARON.   AT&T BELL LABS (USA)
DR. T. WASIUTYNSK.   INSTITUTE OF NUCLEAR PHYSICS (POLAND)
DR. C. WENGEL.   UNIVERSITY OF CALIFORNIA (USA)
PROF. P. WOLYNES.   UNIVERSITY OF ILLINOIS (USA)
PROF. M. ZANNETTI.   UNIVERSITY OF SALERNO (ITALY)
DR. W. ZIPPOLD.   UNIVERSITY OF HEIDELBERG (GERMANY)

Printing: Druckhaus Beltz, Hemsbach
Binding: Buchbinderei Schäffer, Grünstadt

Lecture Notes in Physics

For information about Vols. 1–461
please contact your bookseller or Springer-Verlag

New Series m: Monographs